Artificial Intelligence-Driven Methods for Remote Sensing Target and Object Detection II

Artificial Intelligence-Driven Methods for Remote Sensing Target and Object Detection II

Guest Editors

Yanni Dong
Xiaochen Yang
Qian Du

MDPI

Basel • Beijing • Wuhan • Barcelona • Belgrade • Novi Sad • Cluj • Manchester

Guest Editors
Yanni Dong
School of Resource and Environment Sciences
Wuhan University
Wuhan
China

Xiaochen Yang
School of Mathematics and Statistics
University of Glasgow
Glasgow
UK

Qian Du
Department of Electrical and Computer Engineering
Mississippi State University
Starkville
USA

Editorial Office
MDPI AG
Grosspeteranlage 5
4052 Basel, Switzerland

This is a reprint of the Special Issue, published open access by the journal *Remote Sensing* (ISSN 2072-4292), freely accessible at: https://www.mdpi.com/journal/remotesensing/special_issues/E936A0S61N.

For citation purposes, cite each article independently as indicated on the article page online and as indicated below:

Lastname, A.A.; Lastname, B.B. Article Title. *Journal Name* **Year**, *Volume Number*, Page Range.

ISBN 978-3-7258-3605-5 (Hbk)
ISBN 978-3-7258-3606-2 (PDF)
https://doi.org/10.3390/books978-3-7258-3606-2

Contents

About the Editors

Yanni Dong

Yanni Dong is a Professor and PhD supervisor at the School of Resources and Environmental Science, Wuhan University, as well as being a recipient of the National Excellent Young Scientist Fund. Her research primarily focuses on the intelligent interpretation of remote sensing imagery and geospatial intelligence applications. She has published more than 60 SCI-indexed papers in prestigious journals such as *IEEE Transactions on Image Processing (TIP)*, *IEEE Transactions on Cybernetics (TCybern)*, and *IEEE Transactions on Geoscience and Remote Sensing (TGRS)*. She serves as an Associate Editor for the SCI journals *IEEE TGRS* and *IEEE GRSL*, and is a senior program committee member for international image analysis conferences such as ICCV, AAAI, and IJCAI, as well as a senior member of IEEE. She is also the Vice Chair of the IEEE GRSS Wuhan Chapter and the Vice President of the Excellent Doctoral Club of the China Society of Image and Graphics.

Xiaochen Yang

Xiaochen Yang, PhD, is a Lecturer in the School of Mathematics and Statistics at the University of Glasgow. She received her PhD in Statistical Science from University College London in 2020. Her research focuses on statistical machine learning and image analysis, with a strong interest in medical applications. Her research has been published in top conferences and journals such as Neural Information Processing Systems (NeurIPS), the International Conference on Learning Representations (ICLR), and *Transactions on Pattern Analysis and Machine Intelligence* (*TPAMI*); she is also on the Editorial Boards of *IEEE Transactions on Circuits and Systems for Video Technology* (*TCSVT*) and *Neural Computing*.

Qian Du

Qian Du received her PhD in electrical engineering from the University of Maryland, Baltimore, MD, USA, in 2000. She is currently the Bobby Shackouls Professor with the Department of Electrical and Computer Engineering, Mississippi State University, Starkville, MS, USA. Her research interests include hyperspectral image analysis and applications, pattern classification, data compression, and neural networks. She is a Fellow of the SPIE-International Society for Optics and Photonics. From 2016 to 2020, she was the Editor-in-Chief for the *IEEE Journal of Selected Topics in Applied Earth Observations and Remote Sensing (JSTARS)*. She is currently the Editor-in-Chief of the *Journal of Applied Remote Sensing*.

remote sensing

MDPI

Article

Exploiting Remote Sensing Imagery for Vehicle Detection and Classification Using an Artificial Intelligence Technique

Masoud Alajmi [1], Hayam Alamro [2,*], Fuad Al-Mutiri [3], Mohammed Aljebreen [4], Kamal M. Othman [5] and Ahmed Sayed [6]

1 Department of Computer Engineering, College of Computers and Information Technology, Taif University, P.O. Box 11099, Taif 21944, Saudi Arabia
2 Department of Information Systems, College of Computer and Information Sciences, Princess Nourah bint Abdulrahman University, P.O. Box 84428, Riyadh 11671, Saudi Arabia
3 Department of Mathematics, Faculty of Sciences and Arts, King Khalid University, Abha 63311, Saudi Arabia
4 Department of Computer Science, Community College, King Saud University, P.O. Box 28095, Riyadh 11437, Saudi Arabia
5 Department of Electrical Engineering, Umm Al-Qura University, Makkah 21955, Saudi Arabia
6 Research Center, Future University in Egypt, New Cairo 11835, Egypt
* Correspondence: hmamro@pnu.edu.sa

Abstract: Remote sensing imagery involves capturing and examining details about the Earth's surface from a distance, often using satellites, drones, or other aerial platforms. It offers useful data with which to monitor and understand different phenomena on Earth. Vehicle detection and classification play a crucial role in various applications, including traffic monitoring, urban planning, and environmental analysis. Deep learning, specifically convolutional neural networks (CNNs), has revolutionized vehicle detection in remote sensing. This study designs an improved Chimp optimization algorithm with a DL-based vehicle detection and classification (ICOA-DLVDC) technique on RSI. The presented ICOA-DLVDC technique involves two phases: object detection and classification. For vehicle detection, the ICOA-DLVDC technique applies the EfficientDet model. Next, the detected objects can be classified by using the sparse autoencoder (SAE) model. To optimize the SAE's hyperparameters effectively, we introduce an ICOA which streamlines the parameter tuning process, accelerating convergence and enhancing the overall performance of the SAE classifier. An extensive set of experiments has been conducted to highlight the improved vehicle classification outcomes of the ICOA-DLVDC technique. The simulation values demonstrated the remarkable performance of the ICOA-DLVDC approach compared to other recent techniques, with a maximum accuracy of 99.70% and 99.50% on the VEDAI dataset and ISPRS Postdam dataset, respectively.

Keywords: artificial intelligence; object detector; computer vision; remote sensing; target detection; deep learning

Citation: Alajmi, M.; Alamro, H.; Al-Mutiri, F.; Aljebreen, M.; Othman, K.M.; Sayed, A. Exploiting Remote Sensing Imagery for Vehicle Detection and Classification Using an Artificial Intelligence Technique. *Remote Sens.* **2023**, *15*, 4600. https://doi.org/10.3390/rs15184600

Academic Editors: Yanni Dong, Xiaochen Yang and Qian Du

Received: 30 July 2023
Revised: 2 September 2023
Accepted: 7 September 2023
Published: 19 September 2023

1. Introduction

Remote sensing target detection is used to mark the objects of interest in remote sensing imagery (RSIs) and to predict the location and type of these targets [1]. Based on the perspective of the Earth vision platform, the object strength in the aviation images always appears in a random direction and the target is only concentrated in the conventional detection dataset [2]. The object detection (OD) technique is used to detect samples of semantic objects of specific classes (for example, humans, birds, or airplanes) in digital videos and images. Small target detection has often become a hot and challenging field in target detection tasks. Transport planning, environmental management, military, and disaster control are crucial applications of RSIs [3]. Moreover, vehicles in RSIs, as a special class (whether transportation, civilian, or military), are of particular significance and increasingly difficult. First, vehicle targets in RSIs are fewer than twenty pixels or even ten pixels in the target detection task; the smaller target is generally a target that has fewer than thirty pixels in an image [4]. Next, weather and environment images, including shadow, building, and atmospheric occlusions, and other factors, including similar colors amongst vehicles, dissimilar sizes of vehicle targets in similar images, different overhead views, and their environments, can all lead to the poor detection accuracy of car targets [5].

Vehicle detection in RSI aims to identify each instance of a vehicle [6]. In previous approaches, researchers often developed and extracted vehicle features manually and then classified them to attain vehicle detection [7]. The fundamental objective is to extract vehicle features and utilize traditional machine learning (ML) techniques for classification. Generally, the integration channel features, the scale-invariant feature transform (SIFT), and the histogram of the oriented gradient (HOG) are the features utilized in the detection process. [8]. The approaches utilized for classification are intersection kernel support vectors (IKSVM), AdaBoost, SVM, and so on. However, conventional target detection techniques pay greater consideration to completing the tasks of RSI vehicle detection, and it is challenging to balance speed and accuracy. In contrast to the tremendous growth of deep learning (DL) techniques, there is a big difference in the efficiency and accuracy of detection [9]. Network models based on DL approaches can map complex nonlinear relationships and extract richer features. Two categories of target detection network models are continually formed and optimized due to the development of hardware technology and enormous data: single-stage networks (i.e., SSD and YOLOv3) and two-stage networks (i.e., cascade RCNN and fast RCNN) [10].

This study designs an improved chimp optimization algorithm with a DL-based vehicle detection and classification (ICOA-DLVDC) technique on RSIs. The presented ICOA-DLVDC technique focuses on the utilization of the DL model for the detection of vehicles on the RSI with a hyperparameter tuning strategy. First, the ICOA-DLVDC method exploits the EfficientDet model for OD purposes. Next, the detected objects are classified using the sparse autoencoder (SAE) model. Finally, the hyperparameter tuning of the SAE method can be chosen by ICOA. An extensive set of experiments has been conducted to highlight the improved vehicle classification outcomes of the ICOA-DLVDC technique. In short, the key contributions of the paper are listed as follows.

- An intelligent ICOA-DLVDC technique comprising an EfficientDet object detector, SAE classification, and ICOA-based hyperparameter tuning for RSI has been presented, and to the best of our knowledge, the proposed model will not be found in the literature;
- SAE is able to learn informative and discriminative features with the reduction of the data dimensionality, which is helpful in handling large and complex remote sensing datasets;
- The integration of the EfficientNet object detector with SAE classification can significantly accomplish enhanced generalization and adaptability over various RSI datasets;
- Hyperparameter optimization of the SAE model using the ICOA algorithm using cross-validation helps to boost the predictive outcome of the ICOA-DLVDC model for unseen data.

The rest of the paper is organized as follows: Section 2 provides the related works and Section 3 offers the proposed model. Then, Section 4 gives the result analysis and Section 5 concludes the paper.

2. Related Works

Ahmed et al. [11] designed an IoT-assisted smart surveillance solution for multi-OD using segmentation. In particular, the study proposes the utilization of DL, IoT, and collaborative drones to enhance surveillance applications in smart cities. The study proposed an AI-based technique using a DL-based pyramid scene parsing network (PSPNet) for multiple-object segmentation and applied an aerial drone dataset. The authors in [12] developed a new one-phase OD technique termed MDCT based on a transformer block and multi-kernel dilated convolution (MDC) blocks. Initially, in the single-phase OD technique, a feature enhancement model, the MDC block, was introduced. Next, a transformer block was incorporated into the neck network of the single-phase OD technique. Finally, a depth-wise convolutional layer was incorporated into the MDC block for reducing the computation cost. Qiu, Bai, and Chen [13] designed a new technique called YOLO-GNS for vehicle detection. First, the SSH (single-stage headless) model was devised to facilitate the detection of smaller objects and optimize the feature extraction.

The authors in [14] developed an OD technique based on YOLOv5 for aerial RSI, named KCFS-YOLOv5. The K-means++ algorithm was used for optimizing the initial cluster point to attain the suitable anchor box. Coordinate attention (CA) was embedded with the backbone network of YOLO_v5 to develop the Bi-directional FPN (BiFPN) architecture. Ye et al. [15] designed a convolution network using an adaptive attention fusion module (AAFM). Initially, the stitcher was used for developing one image with objects of different scales according to the features of object distribution in the dataset. Moreover, a spatial attention module was developed, and the semantic data of the feature map was attained. Xiaolin et al. [16] presented an S^2ANET-SR model based on the S^2A-NET network. The original and reduced images were fed to the detection model; later, a super-resolution enhancement model for the reduced images was developed for enhancing the feature extraction of smaller objects, and the texture matching loss and perceptual loss were introduced as supervision.

Javadi et al. [17] investigated the ability of 3D feature maps for enhancing the accuracy of DNN for the recognition of vehicles. First, they introduced a DNN by using YOLOv3 with the base network, involving DenseNet201, DarkNet53, SqueezeNet, and MobileNetv2. Next, 3D depth maps were produced. Later, FCNN was trained on 3D feature mapping. Wu et al. [18] introduced a GCWNet (global context-weaving network) for object recognition in RSIs. Then, two novel modules were introduced for refinement and feature extraction.

Several automated vehicle detection and classification models have been presented in the literature. Despite the benefits of the earlier studies, it is still required to boost the vehicle classification performance. Because of the continual deepening of the model, the number of parameters of DL models also increases quickly, which results in model overfitting. At the same time, different hyperparameters have a significant impact on the efficiency of the CNN model. Particularly, hyperparameters such as epoch count, batch size, and learning rate selection are essential to attaining an effective outcome. Since the trial-and-error method for hyperparameter tuning is a tedious and erroneous process, metaheuristic algorithms can be applied. Therefore, in this work, we employ the ICOA algorithm for the parameter selection of the SAE model.

3. The Proposed Model

In this work, the ICOA-DLVDC technique is established for automated vehicle detection and classification on RSI. In the proposed ICOA-DLVDC technique, a DL-based object detector and classifier are applied. Figure 1 shows the working flow of the ICOA-DLVDC algorithm. The presented ICOA-DLVDC technique involves two phases: EfficientDet-based object detector and ICOA with SAE-based classification. Initially, the input images are passed into the EfficientDet model for the detection of vehicles. Next, the detected objects are classified by the use of SAE model. Finally, the ICOA is applied for the hyperparameter tuning of the SAE model.

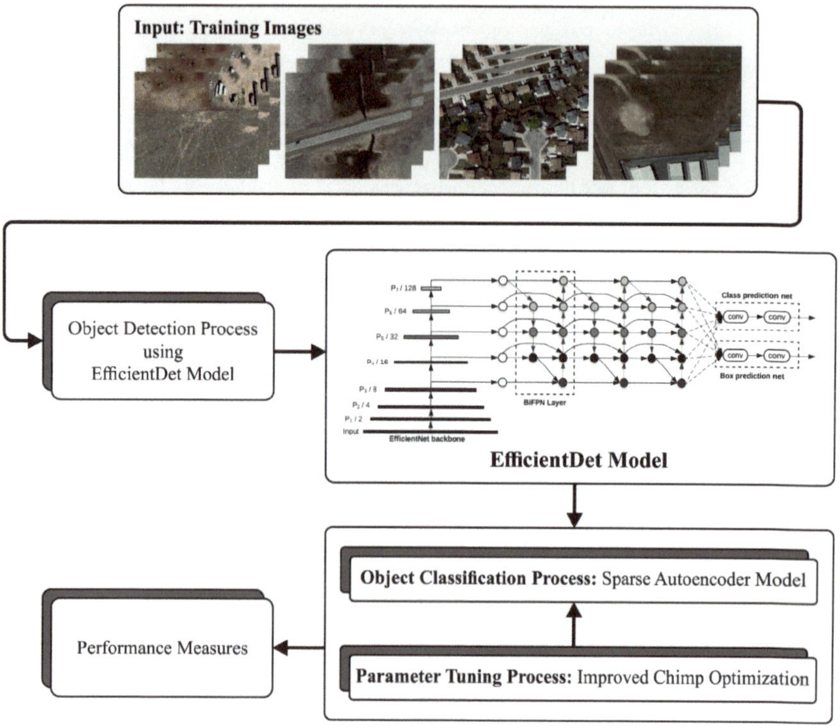

Figure 1. Working flow of the ICOA-DLVDC approach.

3.1. Stage I: Object Detector

The EfficientDet model is used to detect the objects (i.e., vehicles) in the RSI. For combining features with a top-down direction, a conventional approach, named Feature pyramid network (FPN), was used [19]. The PANet (path aggregation network) allows for the forward and reverse flows of feature fusion from low to high resolution. Lastly, the Efficient-Det architecture stacks this BiFPN block. Scaling issues were addressed for resizing the weighted BiFPN, backbone, input quality of the image, and class/box. The EfficientDet model was validated on 100,000 photographs. The network automatically scales from EfficientNetB0 to EfficientNetB6; therefore, the quantity of BiFPN stacks might affect the depth and width of the networks. In most instances, EfficientDet outperforms other OD techniques.

$$\underset{m}{\text{maximize}} ACC(m) \cdot \left[\frac{FLOPS(m)}{T}\right]^{w}, \tag{1}$$

where T refers to the target of FLOPS $ACC(m)$ and is defined as the accuracy of the algorithm m; $FLOPS(m)$ denotes the FLOPS (floating point operations per second) of the algorithm m; and $w = -0.07$ denotes the hyperparameter that controls the exchange amongst FLOPS and accuracy. The EfficientNet seems to be a solid foundation.

As a feature network, the BiFPN function accepts levels 3–7 components $(P3,\ P4,\ P5,\ P6,\ P7)$ from the EfficientNet (backbone network).

$$W_{BiFPN} = 64 \cdot (1.35^{\varphi}), D_{BiFPN} = 3 + \varphi \quad (2)$$

The width of BiFPN was exponentially expanded because the levels of BiFPN should be transformed into small integers, but it gradually enhances the depth. The depth was continuously increased; however, the width was retained at the accurate levels of BiFPN and formulated as follows:

$$D_{box} = D_{class} = 3 + \left[\frac{\varphi}{3}\right]. \quad (3)$$

Considering that BiFPN exploits feature levels 3–7, the input resolution should be dividable by $2^7 = 128$, which implies that it linearly improves the resolution by using the following equation:

$$R_{input} = 512 + (\varphi)(128). \quad (4)$$

Generally, a compound scaling method for OD was introduced, which exploits the φ compound coefficient to enhance each feature of the input image resolution and the backbone, featured, and class/box networks.

The Efficient-Det structure is based on the backbone network EfficientNet42. The class/box net layers and feature network BiFPN are repeated to constitute resource constraints of different magnitudes.

3.2. Stage II: Classification Model

Once the objects are detected, the SAE model is utilized for classification purposes. AE has the potential to duplicate (without learning the hidden representation) the input dataset in the output layer because of the hidden representation, $L1(x)$, and to maximize the mutual information of the input dataset, x [20]. Therefore, the application of sparsity was used to constrain AE in order to learn the hidden representation for the input dataset. Figure 2 demonstrates the infrastructure of SAE.

The hidden unit was constrained to have a small pre-determined activation value, z. The calculated sparsity parameter, $\check{z}$, for j hidden units, was attained via Equation (5):

$$\check{z}_j = \frac{1}{N}\sum_{n=1}^{N} o_j x_n. \quad (5)$$

In Equation (5), N indicates the number of training samples, O_j denotes the activation (or output) of the hidden module, and x_n shows the training sample with index n. The sparsity is used to limit the j hidden units so that $\check{z} = z$.s. The KL (Kullback–Leibler) divergence is used for measuring the distribution deviation z from $\check{z}$ and thus enhances the algorithm.

$$KL(z\|\check{z}_j) = z\log\frac{z}{\check{z}_j} + (1-z)\log\frac{(1-z)}{(1-\check{z}_j)} \quad (6)$$

Note that $(z\|\check{z}) = 0$ for $\check{z} = z$. KL divergence is added to the MSE for the minimization of cost. Thus, the cost function $C(x, y; \theta)$ is formulated in (7):

$$C(x, y, \theta) = \arg\min\sum_{=n1}^{N}\{\sum_{i=1}^{u}(x_i - y_i)^2 + \gamma(\sum_{j=1}^{h} KL(z\|\check{z}_j))\}. \quad (7)$$

The SAE with the convolution operation can be represented as a sparse CAE (SCAE).

The ICOA is used to finetune the hyperparameter value of the SAE technique. COA is derived from the predatory behaviors of the chimp population [21]. Attacker, driver, barrier, and chaser are four different groups based on their behaviors during hunting. Chasing and attacking prey are the two different hunting methods of chimps, which corresponds to the exploration and development phases. Each chimpanzee participating in predation randomly changes its location to move closer to the prey as follows:

$$D = \left| c \cdot x_{prey}(t) - m \cdot x_{chimp}(t) \right| \quad (8)$$

$$x_{chimp}(t+1) = x_{prey}(t) - a \cdot d, \quad (9)$$

where x_{chimp} shows the chimp's location vector, D denotes the distance between the prey and the chimps, x_{prey} indicates the prey's location vector t signifies the existing amount of iteration, and a, m, and c represent coefficient vectors.

$$a = 2 \cdot f \cdot r_1 - f \quad (10)$$

$$c = 2 \cdot r_2 \quad (11)$$

$$m = Chaotic\_value \quad (12)$$

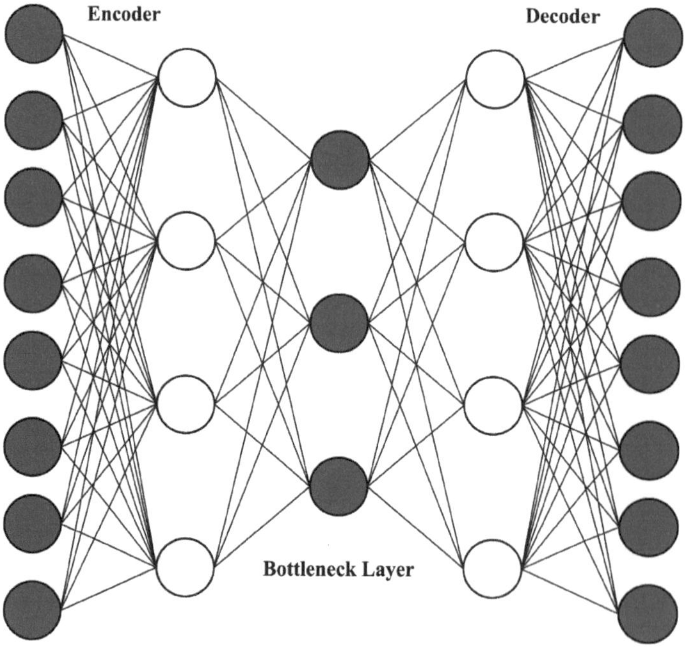

Figure 2. SAE structure.

During the iteration, the value of f reduces from 2.5 to 0, r_1 and r_2 denotes the random vector within $[0, 1]$, and m refers to the chaotic vector computed based on the chaotic map.

The present optimum solution (the first attacker), barrier, chaser, and driver are informed about the target position, and other members are forced to update the locations based on the optimum location of chimps.

$$\begin{cases} d_{Attacker} = |c_1 x_{Attacker} - m_1 x| \\ d_{Barrier} = |c_2 x_{Barrier} - m_2 x| \\ d_{Chaser} = |c_3 x_{Chaser} - m_3 x| \\ d_{Driver} = |c_4 x_{Driver} - m_4 x| \end{cases} \tag{13}$$

$$\begin{cases} V_1 = x_{Attacker} - a_1(d_{Attacker}) \\ V_2 = x_{Barrier} - a_2(d_{Barrier}) \\ V_3 = x_{Chaser} - a_3(d_{Chaser}) \\ V_4 = x_{Driver} - a_4(d_{Driver}) \end{cases} \tag{14}$$

$$x(t+1) = \frac{V_1 + V_2 + V_3 + V_4}{4}, \tag{15}$$

where $d_{Attacker}, d_{Barrier}, d_{Chaser}$, and d_{Driver} denote the distance between 4 kinds of chimps and their target in the existing group; $x_{Attacker}, x_{Barrier}, x_{Chaser}$, and x_{Driver} indicate the location vector relative to the prey; V_1, V_2, V_3, and V_4 characterize their location update vector; $x(t+1)$ shows the location of $t+1$ generation chimps; and $a_1 \sim a_4$, $m_1 \sim m_4$, and $c_1 \sim c_4$ denote the coefficient vector. The chimps release hunting responsibility after food satisfaction and scramble to obtain food. These chaotic behaviors assist in preventing the model from becoming trapped in local optima.

$$x_{chimp}(t+1) = \begin{cases} x_{prey}(t) - a \cdot d, \ if \mu < 0.5 \\ Chaotic\_value, \ if \mu \geq 0.5 \end{cases} \tag{16}$$

In Equation (16), μ represents a randomly generated value within [0,1] and $Chaotic\_value$ shows the chaotic mapping.

In ICOA, reverse learning is used to attain the reverse solution of an individual and, later, retained the individual with the higher fitness value to enhance the individual quality of COA and the population diversity. The refraction of light was combined with reverse learning. The refraction angle takes place while attaining the reverse location of the existing individuals, thereby optimizing the generalization capability of the algorithm and extending the search range of the individual. The upper and lower boundaries of the search region are represented as u and l; correspondingly, $\chi \in [u,\ l]$ and O represent the midpoint of the $[u,\ l]$ interval.

$$\begin{cases} \sin\theta_1 = ((u+l)/2 - x)/|PO| \\ \sin\theta_2 = (x/ - (u+l)/2)/|OQ| \end{cases} \tag{17}$$

$$\eta = \frac{\sin\theta_1}{\sin\theta_2}, \tag{18}$$

where η signifies the refractive index. Consider $k = |PO|/|OQ|$; thenm the refraction reverse learning solution was defined:

$$x' = \frac{u+l}{2} + \frac{u+l}{2k\eta} - \frac{x}{k\eta}. \tag{19}$$

The common form of the inverse solution was attained by expanding Equation (19) to n-dimensional space.

$$x_i' = \frac{u_i + l_i}{2} + \frac{u_i + l_i}{2k\eta} - \frac{x_i}{k\eta} \tag{20}$$

In Equation (20), u_i and l_i denote the ith dimension of the upper and lower boundaries, respectively. Thus, the study introduced hyper-parametric ω. According to dissimilar iteration processes, it adaptively adjusts to improve the randomness of the solution enhance the capability of the model with respect to escaping the local optimum.

$$\begin{cases} x_i' = \frac{u_i+l_i}{2} + \frac{u_i+l_i}{2\omega} - \frac{x_i}{\omega} \\ \omega = \frac{\sigma}{2} - \left(\frac{e^{t/T}-1}{e-1}\right)^{\sigma} \end{cases} \tag{21}$$

In Equation (21), T embodies the iteration count and t shows the existing iteration count. σ controls the attenuation rate of ω; the larger the σ, the slower ω decays. By using the greedy approach, individuals with lower fitness value are rejected while individuals with high fitness value are retained after attaining the reverse location of chimps, as follows:

$$x_{update} = max\_fitness\ (x_i, x_i'). \tag{22}$$

The ICOA method derives an FF to achieve high efficiency of classification. It describes a positive integer to portray the better outcomes of the solution. The decline of the classification error rate is considered FF.

$$\begin{aligned} fitness(x_i) &= ClassifierErrorRate(x_i) \\ &= \frac{No.\ of\ misclassified\ samples}{Total\ No.\ of\ samples} * 100 \end{aligned} \tag{23}$$

4. Results and Discussion

The proposed model is simulated using the Python 3.6.5 tool on PC i5-8600k, GeForce 1050Ti 4GB, 16GB RAM, 250GB SSD, and 1TB HDD. The parameter settings are given as follows: learning rate: 0.01; dropout: 0.5; batch size: 5; epoch count: 50; activation: ReLU.

The experimental evaluation of the ICOA-DLVDC technique is performed on two datasets: the VEDAI [22] and ISPRS Postdam [23] datasets. The former dataset includes 3687 images; and the latter dataset has 2244 images. Tables 1 and 2 defined a detailed description of the two datasets. Figure 3 depicts the sample images.

Table 1. Details of VEDAI dataset.

Class	No. of Instances
Car	1340
Truck	300
Van	100
Pickup Car	950
Boat	170
Camping Car	390
Other	200
Plane	47
Tractor	190
Total Instances	3687

Table 2. Details on ISPRS Postdam dataset.

Class	No. of Instances
Car	1990
Truck	33
Van	181
Pickup Car	40
Total Instances	2244

Figure 3. Sample images: (**a**) boat, (**b**) car, (**c**) pickup car, (**d**) airplane.

Figure 4 illustrates the classifier outcomes of the ICOA-DLVDC method under the VEDAI dataset. Figure 4a,b describes the confusion matrix presented by the ICOA-DLVDC technique at 70:30 of the TR set/TS set. The figure denoted that the ICOA-DLVDC method has detected and classified all nine class labels accurately. Similarly, Figure 4c demonstrates the PR examination of the ICOA-DLVDC system. The figure showed that the ICOA-DLVDC method has accomplished maximal PR outcomes under nine classes. Finally, Figure 4d demonstrates the ROC examination of the ICOA-DLVDC method. The figure demonstrates that the ICOA-DLVDC method has resulted in proficient outcomes with the highest ROC values under nine class labels.

In Table 3, the vehicle classification outcomes of the ICOA-DLVDC method on the VEDAI dataset are reported. The table values state that the ICOA-DLVDC technique properly recognized all the vehicle types. With 70% of the TR set, the ICOA-DLVDC technique gains average $accu_y$, $prec_n$, $reca_l$, F_{score}, and MCC of 99.43%, 96.66%, 94.45%, 95.43%, and 95.15% respectively. Moreover, with 30% of the TS set, the ICOA-DLVDC method gains average $accu_y$, $prec_n$, $reca_l$, F_{score}, and MCC of 99.50%, 97.27%, 94.45%, 95.94%, and 95.72%, respectively.

Figure 5 shows the training accuracy TR_accu_y and VL_accu_y of the ICOA-DLVDC method on the VEDAI dataset. The TL_accu_y is determined by the evaluation of the ICOA-DLVDC technique on the TR dataset; whereas the VL_accu_y is computed by evaluating the performance on a separate testing dataset. The outcomes demonstrate that TR_accu_y and VL_accu_y increase with an upsurge in epochs. Thus, the performance of the ICOA-DLVDC method is improved on the TR and TS datasets, with a rise in several epochs.

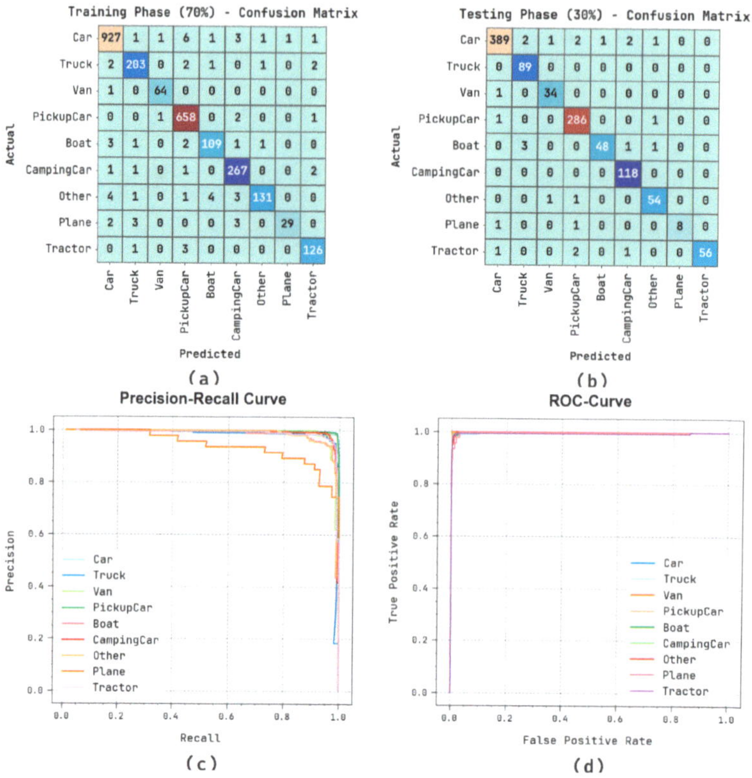

Figure 4. Performance on VEDAI datasets: (**a**,**b**) confusion matrices; (**c**) PR_curve; (**d**) ROC_curve.

Table 3. Vehicle classifier outcome of ICOA-DLVDC technique on VEDAI dataset.

Labels	$Accu_y$	$Prec_n$	$Reca_l$	F_{score}	MCC
Training Phase (70%)					
Car	98.91	98.62	98.41	98.51	97.66
Truck	99.38	96.21	96.21	96.21	95.87
Van	99.88	96.97	98.46	97.71	97.65
Pickup Car	99.26	97.77	99.40	98.58	98.09
Boat	99.46	94.78	93.16	93.97	93.69
Camping Car	99.34	95.70	98.16	96.91	96.56
Other	99.38	97.76	90.97	94.24	93.99
Plane	99.65	96.67	78.38	86.57	86.88
Tractor	99.61	95.45	96.92	96.18	95.98
Average	99.43	96.66	94.45	95.43	95.15

Table 3. *Cont.*

Labels	$Accu_y$	$Prec_n$	$Reca_l$	F_{score}	MCC
Testing Phase (30%)					
Car	98.83	98.98	97.74	98.36	97.45
Truck	99.55	94.68	100.00	97.27	97.06
Van	99.73	94.44	97.14	95.77	95.64
Pickup Car	99.28	97.95	99.31	98.62	98.14
Boat	99.46	97.96	90.57	94.12	93.91
Camping Car	99.64	96.72	100.00	98.33	98.15
Other	99.55	94.74	96.43	95.58	95.34
Plane	99.82	100.00	80.00	88.89	89.36
Tractor	99.64	100.00	93.33	96.55	96.43
Average	99.50	97.27	94.95	95.94	95.72

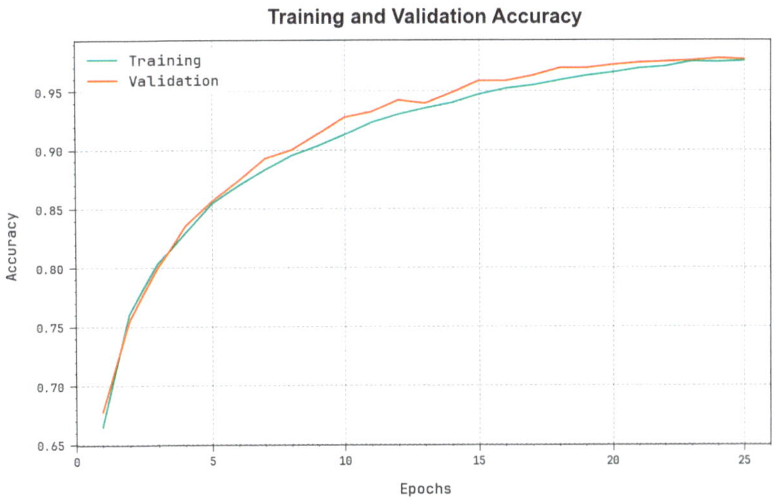

Figure 5. $Accu_y$ curve of the ICOA-DLVDC technique on the VEDAI dataset.

In Figure 6, the TR_loss and VR_loss outcomes of the ICOA-DLVDC method on the VEDAI dataset are shown. The TR_loss defines the error among the predictive performance and original values on the TR data. The VR_loss represents the measure of the performance of the ICOA-DLVDC technique on individual validation data. The results indicate that the TR_loss and VR_loss tend to decrease with rising epochs. They portray the enhanced performance of the ICOA-DLVDC method and its capability to generate accurate classification. The reduced value of TR_loss and VR_loss demonstrates the enhanced performance of the ICOA-DLVDC technique in capturing patterns and relationships.

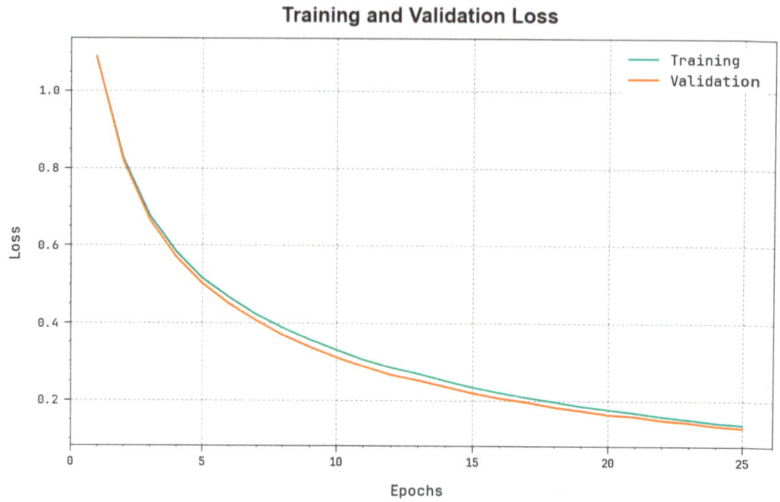

Figure 6. Loss curve of the ICOA-DLVDC technique on the VEDAI dataset.

The comparison study of the ICOA-DLVDC technique with other DL models on the VEDAI dataset is highlighted in Table 4 and Figure 7 [24]. The outcomes show that the ICOA-DLVDC technique accomplishes improved performance with *an* $accu_y$ of 99.50%. On the other hand, the CSOTL-VDCRS, LeNet, AlexNet, and VGG-16 models achieve reduced performance with $accu_y$ of 98.07%, 79.78%, 88.98%, and 94.46%, respectively.

Table 4. $Accu_y$ outcome of the ICOA-DLVDC technique with recent methods on the VEDAI dataset.

VEDAI Dataset	
Methods	**Accuracy (%)**
ICOA-DLVDC	99.50
CSOTL-VDCRS	98.07
LeNet Model	79.74
AlexNet Model	88.98
VGG-16 Model	94.46

Figure 8 illustrates the classifier results of the ICOA-DLVDC technique on the ISPRS Postdam dataset. Figure 8a,b demonstrates the confusion matrix presented by the ICOA-DLVDC system at 70:30 of the TR set/TS set. The figure demonstrates that the ICOA-DLVDC method has detected and classified all four class labels accurately. Similarly, Figure 8c demonstrates the PR examination of the ICOA-DLVDC model. The figure shows that the ICOA-DLVDC technique has accomplished high PR outcomes under four classes. Lastly, Figure 8d elucidates the ROC examination of the ICOA-DLVDC model. The figure shows that the ICOA-DLVDC method has resulted in proficient outcomes, with the highest ROC values under four class labels.

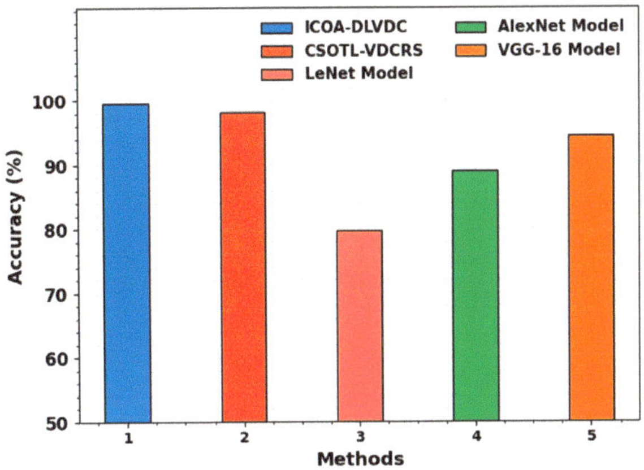

Figure 7. $Accu_y$ outcome of the ICOA-DLVDC technique on the VEDAI dataset.

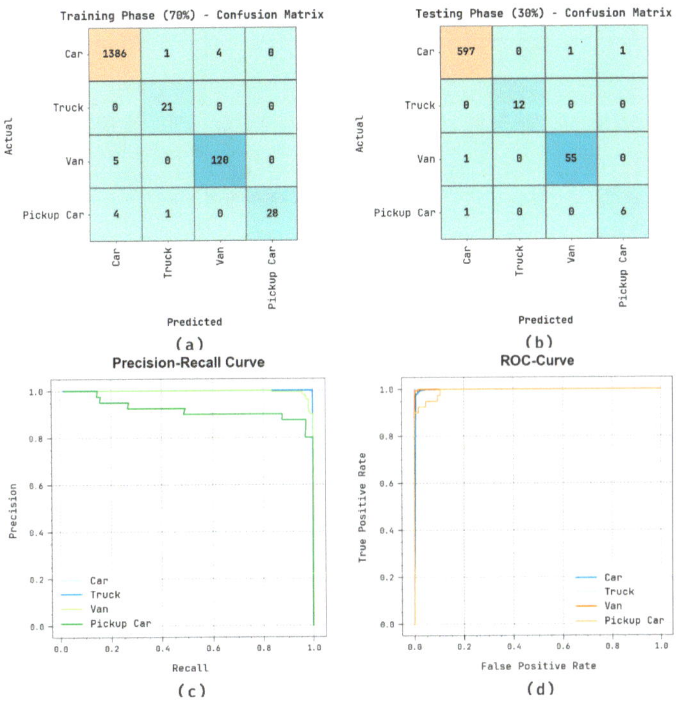

Figure 8. Performance on the ISPRS Postdam dataset: (**a**,**b**) confusion matrices; (**c**) PR_curve; (**d**) ROC_curve.

In Table 5, the vehicle classification outcomes of the ICOA-DLVDC technique on the ISPRS Postdam dataset are reported. The table values stated that the ICOA-DLVDC technique properly recognized all the vehicle types. With 70% of the TR set, the ICOA-DLVDC method gains average $accu_y$, $prec_n$, $recal_l$, F_{score}, and MCC of 99.52%, 96.86%, 95.12%, 95.79%, and 94.77%, respectively. Furthermore, with 30% of the TS set, the ICOA-DLVDC method gains average $accu_y$, $prec_n$, $recal_l$, F_{score}, and MCC of 99.70%, 95.90%, 95.90%, 95.90%, and 95.15%, respectively.

Table 5. Vehicle classifier outcome of the ICOA-DLVDC technique on the ISPRS Postdam dataset.

Labels	$Accu_y$	$Prec_n$	$Reca_l$	F_{score}	MCC
Training Phase (70%)					
Car	99.11	99.35	99.64	99.50	95.55
Truck	99.87	91.30	100.00	95.45	95.49
Van	99.43	96.77	96.00	96.39	96.08
Pickup Car	99.68	100.00	84.85	91.80	91.96
Average	99.52	96.86	95.12	95.79	94.77
Testing Phase (30%)					
Car	99.41	99.67	99.67	99.67	97.00
Truck	100.00	100.00	100.00	100.00	100.00
Van	99.70	98.21	98.21	98.21	98.05
Pickup Car	99.70	85.71	85.71	85.71	85.56
Average	99.70	95.90	95.90	95.90	95.15

Figure 9 shows the training accuracy TR_accu_y and VL_accu_y of the ICOA-DLVDC technique on the ISPRS Postdam dataset. The TL_accu_y is determined by the evaluation of the ICOA-DLVDC technique on the TR dataset; whereas the VL_accu_y is computed by evaluating the performance on a separate testing dataset. The outcomes demonstrate that TR_accu_y and VL_accu_y increase with an upsurge in epochs. As a result, the performance of the ICOA-DLVDC technique is improved on the TR and TS dataset, with a rise in the number of epochs.

Figure 9. $Accu_y$ curve of the ICOA-DLVDC technique on the ISPRS Postdam dataset.

In Figure 10, the TR_loss and VR_loss outcomes of the ICOA-DLVDC technique on ISPRS Postdam dataset are shown. The TR_loss defines the error among the predictive performance and original values on the TR data. The VR_loss represents the measure of the performance of the ICOA-DLVDC technique on individual validation data. The results indicate that the TR_loss and VR_loss tend to decrease with rising epochs. The portray the enhanced performance of the ICOA-DLVDC technique and its capability to generate accurate classification. The reduced value of TR_loss and VR_loss demonstrates the enhanced performance of the ICOA-DLVDC technique in capturing patterns and relationships.

Figure 10. Loss curve of the ICOA-DLVDC technique on the ISPRS Postdam dataset.

The comparison analysis of the ICOA-DLVDC method with other DL techniques [24] on the ISPRS Postdam dataset is highlighted in Table 6 and Figure 11. The outcome specified that the ICOA-DLVDC technique accomplishes improved performance, with an accuracy of 99.70%. On the other hand, the CSOTL-VDCRS, LeNet, AlexNet, and VGG-16 models achieve reduced performance, with accuraciwa of 98.67%, 94.54%, 95.86%, and 89.54%, respectively.

Table 6. $Accu_y$ outcome of ICOA-DLVDC technique with recent methods on ISPRS Postdam dataset.

Methods	Accuracy (%)
ICOA-DLVDC	99.70
CSOTL-VDCRS	98.67
LeNet Model	94.54
AlexNet Model	95.86
VGG-16 Model	89.54

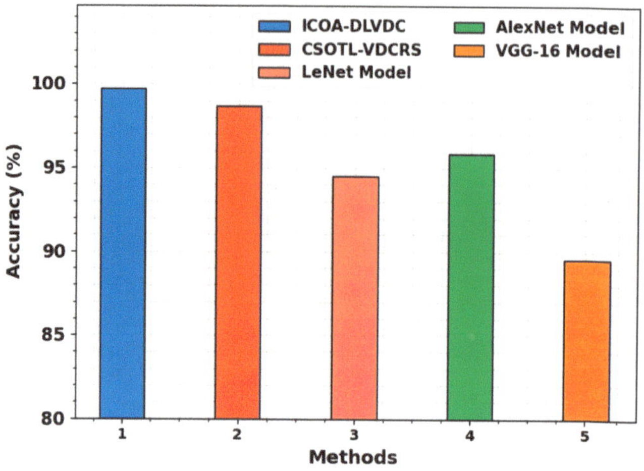

Figure 11. $Accu_y$ outcome of the ICOA-DLVDC technique on the ISPRS Postdam dataset.

5. Conclusions

In this study, we have introduced the ICOA-DLVDC technique for automated vehicle detection and classification on RSI. In the presented ICOA-DLVDC technique, DL-based object detectors and classifiers are applied. The presented ICOA-DLVDC technique involves two phases: EfficientDet-based object detector and ICOA with SAE-based classification. An extensive set of experiments has been conducted to highlight the improved vehicle classification outcomes of the ICOA-DLVDC method. The experimental outcomes demonstrated the remarkable performance of the ICOA-DLVDC technique over other recent approaches, with maximum accuracy of 99.70% and 99.50% on the VEDAI dataset and the ISPRS Postdam dataset, respectively. In the future, we will examine the performance of the ICOA-DLVDC algorithm in different environments, such as day and night times, as well as cloudy and rainy environments. In addition, the computational time of the proposed model can be examined in the future. Moreover, the vehicle detection results can be integrated into geographic information systems (GIS) for better spatial analysis and decision-making. Finally, lightweight models can be developed for edge computing and deployment on resource-constrained devices such as drones and IoT devices.

Author Contributions: Conceptualization, M.A. (Masoud Alajmi) and H.A.; Methodology, M.A. (Masoud Alajmi), H.A., F.A.-M. and K.M.O.; Software, K.M.O.; Validation, K.M.O. and A.S.; Formal analysis, F.A.-M.; Investigation, M.A. (Masoud Alajmi); Data curation, M.A. (Mohammed Aljebreen) and A.S.; Writing—original draft, M.A. (Masoud Alajmi), H.A., F.A.-M. and M.A. (Mohammed Aljebreen); Writing—review & editing, H.A., F.A.-M., M.A. (Mohammed Aljebreen), K.M.O. and A.S.; Visualization, M.A. (Mohammed Aljebreen); Funding acquisition, H.A., F.A.-M. All authors have read and agreed to the published version of the manuscript.

Funding: The authors extend their appreciation to the Deanship of Scientific Research at King Khalid University for funding this work through large group Research Project under grant number (RGP2/35/44). Princess Nourah bint Abdulrahman University Researchers Supporting Project number (PNURSP2023R361), Princess Nourah bint Abdulrahman University, Riyadh, Saudi Arabia. Research Supporting Project number (RSP2023R459), King Saud University, Riyadh, Saudi Arabia. This study is partially funded by the Future University in Egypt (FUE).

Data Availability Statement: Data sharing is not applicable to this article as no datasets were generated during the current study.

Conflicts of Interest: The authors declare that they have no conflict of interest. The manuscript was written through the contributions of all authors. All authors have given approval to the final version of the manuscript.

References

1. Wang, Y.; Peng, F.; Lu, M.; Asif Ikbal, M. Information Extraction of the Vehicle from High-Resolution Remote Sensing Image Based on Convolution Neural Network. *Recent Adv. Electr. Electron. Eng. (Former. Recent Pat. Electr. Electron. Eng.)* **2023**, *16*, 168–177.
2. Anusha, C.; Rupa, C.; Samhitha, G. Region-based detection of ships from remote sensing satellite imagery using deep learning. In Proceedings of the 2nd International Conference on Innovative Practices in Technology and Management (ICIPTM), Gautam Buddha Nagar, India, 23–25 February 2022; IEEE: New York, NY, USA, 2022; Volume 2, pp. 118–122.
3. Chen, Y.; Qin, R.; Zhang, G.; Albanwan, H. Spatial-temporal analysis of traffic patterns during the COVID-19 epidemic by vehicle detection using planet remote-sensing satellite images. *Remote Sens.* **2021**, *13*, 208. [CrossRef]
4. Wang, L.; Shoulin, Y.; Alyami, H.; Laghari, A.A.; Rashid, M.; Almotiri, J.; Alyamani, H.J.; Alturise, F. A novel deep learning—based single shot multibox detector model for object detection in optical remote sensing images. *Geosci. Data J.* **2022**, 1–15. [CrossRef]
5. Ghali, R.; Akhloufi, M.A. Deep Learning Approaches for Wildland Fires Remote Sensing: Classification, Detection, and Segmentation. *Remote Sens.* **2023**, *15*, 1821. [CrossRef]
6. Karnick, S.; Ghalib, M.R.; Shankar, A.; Khapre, S.; Tayubi, I.A. A novel method for vehicle detection in high-resolution aerial remote sensing images using YOLT approach. *Multimed. Tools Appl.* **2022**, *109*, 1–16.
7. Wang, B.; Xu, B. A feature fusion deep-projection convolution neural network for vehicle detection in aerial images. *PLoS ONE* **2021**, *16*, e0250782. [CrossRef]
8. Wang, J.; Teng, X.; Li, Z.; Yu, Q.; Bian, Y.; Wei, J. VSAI: A Multi-View Dataset for Vehicle Detection in Complex Scenarios Using Aerial Images. *Drones* **2022**, *6*, 161. [CrossRef]
9. Safarov, F.; Temurbek, K.; Jamoljon, D.; Temur, O.; Chedjou, J.C.; Abdusalomov, A.B.; Cho, Y.I. Improved Agricultural Field Segmentation in Satellite Imagery Using TL-ResUNet Architecture. *Sensors* **2022**, *22*, 9784. [CrossRef]
10. Momin, M.A.; Junos, M.H.; Mohd Khairuddin, A.S.; Abu Talip, M.S. Lightweight CNN model: Automated vehicle detection in aerial images. *Signal Image Video Process.* **2022**, *17*, 1–9. [CrossRef]
11. Ahmed, I.; Ahmad, M.; Chehri, A.; Hassan, M.M.; Jeon, G. IoT Enabled Deep Learning Based Framework for Multiple Object Detection in Remote Sensing Images. *Remote. Sens.* **2022**, *14*, 4107. [CrossRef]
12. Chen, J.; Hong, H.; Song, B.; Guo, J.; Chen, C.; Xu, J. MDCT: Multi-Kernel Dilated Convolution and Transformer for One-Stage Object Detection of Remote Sensing Images. *Remote. Sens.* **2023**, *15*, 371. [CrossRef]
13. Qiu, Z.; Bai, H.; Chen, T. Special Vehicle Detection from UAV Perspective via YOLO-GNS Based Deep Learning Network. *Drones* **2023**, *7*, 117. [CrossRef]
14. Tian, Z.; Huang, J.; Yang, Y.; Nie, W. KCFS-YOLOv5: A High-Precision Detection Method for Object Detection in Aerial Remote Sensing Images. *Appl. Sci.* **2023**, *13*, 649. [CrossRef]
15. Ye, Y.; Ren, X.; Zhu, B.; Tang, T.; Tan, X.; Gui, Y.; Yao, Q. An Adaptive Attention Fusion Mechanism Convolutional Network for Object Detection in Remote Sensing Images. *Remote. Sens.* **2022**, *14*, 516. [CrossRef]
16. Xiaolin, F.; Fan, H.; Ming, Y.; Tongxin, Z.; Ran, B.; Zenghui, Z.; Zhiyuan, G. Small object detection in remote sensing images based on super-resolution. *Pattern Recognit. Lett.* **2022**, *153*, 107–112. [CrossRef]
17. Javadi, S.; Dahl, M.; Pettersson, M.I. Vehicle Detection in Aerial Images Based on 3D Depth Maps and Deep Neural Networks. *IEEE Access* **2021**, *9*, 8381–8391. [CrossRef]
18. Wu, Y.; Zhang, K.; Wang, J.; Wang, Y.; Wang, Q.; Li, X. GCWNet: A Global Context-Weaving Network for Object Detection in Remote Sensing Images. *IEEE Trans. Geosci. Remote. Sens.* **2022**, *60*, 1–12. [CrossRef]
19. AlDahoul, N.; Karim, H.A.; De Castro, A.; Tan, M.J.T. Localization and classification of space objects using Effi-cientDet detector for space situational awareness. *Sci. Rep.* **2022**, *12*, 21896. [CrossRef]
20. Akila, S.M.; Imanov, E.; Almezhghwi, K. Investigating Beta-Variational Convolutional Autoencoders for the Un-supervised Classification of Chest Pneumonia. *Diagnostics* **2023**, *13*, 2199. [CrossRef]
21. Chen, Q.; He, Q.; Zhang, D. UAV Path Planning Based on an Improved Chimp Optimization Algorithm. *Axioms* **2023**, *12*, 702. [CrossRef]
22. Razakarivony, S.; Jurie, F. Vehicle detection in aerial imagery: A small target detection benchmark. *J. Vis. Commun. Image Represent.* **2016**, *34*, 187–203. [CrossRef]

23. Rottensteiner, F.; Sohn, G.; Jung, J.; Gerke, M.; Baillard, C.; Benitez, S.; Breitkopf, U. The ISPRS Benchmark on Urban Object Classification and 3D Building Reconstruction. *ISPRS Ann. Photogramm. Remote Sens. Spat. Inf. Sci.* **2012**, *1*, 293–298. [CrossRef]
24. Ahmed, M.A.; Althubiti, S.A.; de Albuquerque, V.H.C.; dos Reis, M.C.; Shashidhar, C.; Murthy, T.S.; Lydia, E.L. Fuzzy wavelet neural network driven vehicle detection on remote sensing imagery. *Comput. Electr. Eng.* **2023**, *109*, 108765. [CrossRef]

remote sensing

MDPI

Article

Hybrid Task Cascade-Based Building Extraction Method in Remote Sensing Imagery

Runqin Deng [1], Meng Zhou [1,*], Yinni Huang [1] and Wei Tu [2]

[1] School of Intelligent Systems Engineering, Sun Yat-sen University, Shenzhen 518107, China; dengrq5@mail2.sysu.edu.cn (R.D.); huangyn75@mail2.sysu.edu.cn (Y.H.)
[2] Department of Urban Informatics, School of Architecture and Urban Planning, Shenzhen University, Shenzhen 518060, China; tuwei@szu.edu.cn
* Correspondence: zhoum89@mail.sysu.edu.cn

Abstract: Instance segmentation has been widely applied in building extraction from remote sensing imagery in recent years, and accurate instance segmentation results are crucial for urban planning, construction and management. However, existing methods for building instance segmentation (BSI) still have room for improvement. To achieve better detection accuracy and superior performance, we introduce a Hybrid Task Cascade (HTC)-based building extraction method, which is more tailored to the characteristics of buildings. As opposed to a cascaded improvement that performs the bounding box and mask branch refinement separately, HTC intertwines them in a joint multilevel process. The experimental results also validate its effectiveness. Our approach achieves better detection accuracy compared to mainstream instance segmentation methods on three different building datasets, yielding outcomes that are more in line with the distinctive characteristics of buildings. Furthermore, we evaluate the effectiveness of each module of the HTC for building extraction and analyze the impact of the detection threshold on the model's detection accuracy. Finally, we investigate the generalization ability of the proposed model.

Keywords: deep learning; remote sensing; building extraction; instance segmentation; hybrid task cascade

Citation: Deng, R.; Zhou, M.; Huang, Y.; Tu, W. Hybrid Task Cascade-Based Building Extraction Method in Remote Sensing Imagery. *Remote Sens.* **2023**, *15*, 4907. https://doi.org/10.3390/rs15204907

Academic Editors: Qian Du, Yanni Dong and Xiaochen Yang

Received: 6 September 2023
Revised: 28 September 2023
Accepted: 7 October 2023
Published: 11 October 2023

1. Introduction

The ever-growing cities have witnessed drastic changes and exerted great influence on people's daily lives. Mapping the landscapes of cities is essential for better understanding the urban space and human activities, facilitating more informed decision making in urban policies. The high spatial resolution (HSR) remote sensing imagery contains abundant and detailed land cover information [1]. As one of the main artificial features in a city, the extraction and analysis of the building information provide valuable insights for a wide range of geographic and environmental applications. The automatic building extraction from high-resolution aerial images is currently an active research area and an issue of high importance to many urban scenarios, including disaster assessment, humanitarian aid, change detection in human settlements, urban planning and so on [2–6]. However, for building identification, it remains a challenge to fully leverage the inherent characteristics of building targets and different hierarchical features in building remote sensing images to obtain more accurate building locations.

The appearance of building rooftops in remote sensing images varies due to many factors, including lighting conditions, variety of reflections and diversity of image resolution [7–9]. Moreover, compared with natural images, the spatial scale of high-resolution remote sensing building objects in urban scenes varies greatly, especially for building rooftops in different shapes and sizes. These characteristics raise challenges to the accurate identification of building objects [10]. Previous traditional research usually relied on artificial design features, such as edge, spectral, shape, and texture features to extract buildings from remote sensing imagery [11]. Some scholars also combined these handcrafted features with machine learning methods such as random forest (RF) [12] and support vector machines (SVM) [13] for building detection and classification. However, these approaches are labor-intensive with time-consuming tasks and their performance depends on the low-level hand-engineered local features.

Recent advancements in artificial intelligence (AI) have led to the growing popularity of deep learning algorithms for object detection tasks. In particular, deep Convolutional Neural Networks (CNNs) have been extensively utilized to extract objects from remote sensing images. The deep learning approach is capable of automatic learning and feature extraction from datasets. It also produces more stable results with stronger universality. In recent years, deep learning models have been increasingly adopted for building extraction from remote-sensed imagery. Relevant studies mainly made use of state-of-the-art models, e.g., Faster R-CNN [14], FCN [15], and Mask R-CNN [16], for a wide variety of purposes including object detection, semantic segmentation, and instance segmentation. Among them, Mask R-CNN-based methods have received growing attention in building recognition tasks. This is due to their ability to provide more detailed and accurate localization and segmentation results for the detected targets. More recently, cascade structure networks have been widely applied due to their ability to achieve higher detection accuracy and provide promising alternatives to the existing deep learning models. Moreover, the application of cascade structure networks in the remote sensing image analysis contributes to addressing specific challenges, such as detecting small objects and handling blurred object boundaries in target detection. However, cascade structure models have rarely been used in building object detections from remote sensing imagery. To address the challenges in automatic building extraction, we introduce the hybrid task cascade network (HTC) [17], which combines the advantages of Mask R-CNN and Cascade R-CNN [18] and propose a HTC-based automatic building detection method to extract building information from remote sensing imagery. By incorporating global semantic context (GSC), bounding box information (BBI) and mask information (MI), it can obtain more accurate, regular and smoother results of the building detection. The main contributions of this study are listed as follows:

1. This study introduces the hybrid task cascade network for building detection from remote sensing imagery based on the inherent characteristics of building targets. We also investigate the impact of global semantic features on the performance of building object detection.
2. We compare the influence of the integrated architecture in HTC, which makes the bounding box prediction and mask prediction intertwined, and the mask information flow architecture, which combines mask information at different stages of the cascade structure to improve the accuracy of mask detection.
3. Extensive experiments are conducted on three diverse remote sensing imagery building datasets, including CrowdAI mapping challenge dataset [19], WHU aerial image dataset [20]) and Chinese Typical City Buildings Dataset [21]. The results demonstrate the superiority of the HTC network over the swin-transformer-based method and other existing state-of-the-art (SOTA) segmentation methods.

The remainder of the paper is structured as follows. In Section 2, we provide a review of relevant literature on image segmentation and its applications in remote sensing imagery. Section 3 outlines the details of our HTC-based building extraction method. The experimental setup, including the datasets, evaluation metrics, and selected baseline models, is described in Section 4. In Section 5, we elaborate on the experimental results in detail. Section 6 summarizes the findings, provides discussions, and highlights our future work. Finally, Section 7 concludes the paper.

2. Related Works

This section provides an overview of the current trends in semantic segmentation and instance segmentation using deep learning models, specifically focusing on their application in extracting buildings from aerial and satellite imagery. It also highlights the main contributions of promising CNN-based approaches in this field.

2.1. Encoder–Decoder Structure-Based Semantic Segmentation

In recent years, models based on encoder–decoder structures, such as FCN [15],U-Net [22] and FPN [23], have been widely used for their capability to perform dense predictions on images without using fully connected layers and generate segmentation maps of any size. The commonality among these models is that the encoder performs feature extraction with multi-stage down-sampling. Then, the decoder gradually recovers the size and structure of the image during up-sampling and generates semantic annotations. The fully convolutional network (FCN), proposed by Long et al., is a landmark pixel-based segmentation method. The decoder structure of FCN is the simplest and contains only one deconvolution operation, while U-Net and FPN adopt multiple up-sampling in the decoder structure. With such a U-shape structure, the U-Net and FPN models can extract features of different scales and combine high-level semantic information with low-level geometric details through the fusion of multiscale feature maps. Compared with U-Net which only make predictions in the last layer of the decoder, the feature pyramid network (FPN) makes independent predictions for feature maps at multiple scales and takes the pixel with the highest confidence as the result. All of these encoder–decoder methods aim to obtain richer information and enhance model performance by combining them in different ways.

Many improved encoder–decoder methods have been employed for building extraction from remote sensing images in recent studies due to its ability to obtain more information and better address the scale-variance problems of building objects in remote sensing images. For instance, Shrestha et al. [24] proposed an improved FCN, which used the exponential linear unit (ELU) in place of the commonly used ReLU activation, and applied conditional random fields (CRFs) at the end of the network to reduce the noise and to sharpen the boundary of the buildings. Yang et al. [25] improved the accuracy of building extraction by paying more attention to the edge and using edge information to refine the segmentation results. Tang et al. [26] proposed Capsule–Encoder–Decoder and introduced a vector named capsule to store the characteristics of buildings and their components to improve generalization capabilities on different remote sensing datasets. In [27], three parallel pre-trained ResNet sub-CNNs followed by a fusion operation and a U-shaped deconvolution network were used to learn building features at different scales. It also used the Sobel filters in both the foreground and background as well as an edge detection loss function called the edge constraint loss (ECL) to obtain more precise masks. Huang et al. [9] proposed a novel approach for building extraction by introducing a gated feature labeling unit. This unit reduces unnecessary feature transmission and refines the coarse classification maps at each decoder stage of the fully convolutional network (FCN). Their method leveraged both HSR images and LiDAR point clouds, leading to significant improvements in building the extraction performance. In [28], a non-local block [29] was introduced at the top of the encoder to obtain global information by capturing the dependencies among pixels. Moreover, in the decoder part, they fused the multi-scale features by concatenating different deconvolution layers. In this way, they also achieved good performance. Yuan et al. [30] proposed a multi-scale adaptive semantic segmentation network (MSST-Net) and achieved good performance with the open access WHU building data set. They used a CNN to decode the features at different stages in Swin Transformer [31] and to concatenate the multi-level decoding outputs. Swin Transformer [31], inspired by the attention-based transformers [32] and attached with shifted windows, appeared promising to solve the scale-variance problem in remote sensing images for the powerful ability to catch spatial and global information. Girard et al. [33] trained a semantic segmentation network that aligned predicted frame fields to ground truth contours, in which they utilized the frame fields along with the raster segmentation to aid polygonization methods in resolving ambiguities caused by discrete probability maps.

2.2. Mask R-CNN Structure-Based Instance Segmentation

The instance segmentation approach combines the advantages of both object detection and semantic segmentation. It goes beyond the differentiation of individual objects and aims at also recovering more accurate locations and detailed shapes of objects. Recently, instance segmentation methods based on the Mask R-CNN structure, combined with other feature extraction techniques, have been widely applied to extract ground objects from remote sensing images [2,4,6]. For example, Zhao et al. [34] utilized the Douglas–Peucker (DP) algorithm [35] and Minimum Description Length (MDL) optimization [36] to reduce the number of points in the initial polygons generated by Mask R-CNN by 86%. This simplification process resulted in significantly simplified building polygons compared to the original irregular polygons generated by Mask R-CNN. The regularized polygons obtained through their approach were found to be suitable for various cartographic and engineering applications. By combining Sobel edge detection algorithm and Mask R-CNN, Zhang et al. [37] integrated the artificial edge features and features extracted by RestNet50 to improve the efficiency and accuracy of the building extraction. Fang et al. [38] proposed an attention-based FPN (AFPN), in which channel attention was introduced into each layer of the classical FPN to improve the identification of small buildings. Then, a two-stage coarse-to-fine contour sub-network was designed to refine building instance contours, which adjusted the deviation between the contours obtained by the Sobel operator and the ground truth through a loss function to further improve the contour accuracy. Based on an improved Mask R-CNN model, which used a modified ResNet101 [39] for feature extraction, Han et al. [5] achieved a high building detection accuracy in a manually annotated building dataset. Zhao et al. [40] introduced global context and boundary refinement blocks (BRB) to upgrade feature extraction, and added channel and spatial attention modules to boost the effectiveness of the detection block. They also employed a recurrent neural network (RNN) structure to sequentially decode the vertices of building polygons after the feature extraction stage of Mask R-CNN. In contrast to the traditional approach of converting pixel-level segmentation results into regularized building boundaries, this method directly obtains vector-formatted regularized building outlines. It provides convenience for engineering mapping and other related applications. Chen et al. [3] implemented Swin Transformer to replace the backbone of a current SOTA algorithm, the multiple attending path neural network (MAP-Net) [41], and obtained a more accurate result in the building extraction.

Instance segmentation methods have diversified the applications of building extraction from remote sensing imagery. Chen et al. [2] proposed a building area calculation method based on the number of building pixels, which enables the timely and accurate assessment of the losses caused by natural disasters such as earthquakes and floods. Amo-Boateng et al. [4] implemented Mask R-CNN to detect the rooftops of buildings in a typical rural settlement to estimate the solar generation potential of such areas.

To sum up, there has been a large body of research on the automatic building extraction from remote sensing images with deep learning models. However, few researchers have taken the detection threshold of the model into account to improve its accuracy. Moreover, the fact that building objects in remote sensing images have more regular contour shapes than other targets has been neglected. Based on this characteristic of buildings, we believe that high-quality bounding boxes can have a significant impact on improving the detection accuracy of building targets. In addition, existing models for building extraction from remote sensing images rarely take global semantic information into consideration. Moreover, the constant threshold commonly used for bounding box evaluation often leads to unsatisfactory results. Models with cascade structure, e.g., Cascade R-CNN [18], were proposed to tackle this issue with resampling by cascade regression, albeit with limited improvement. Hybrid Task Cascade (HTC) [17] proposed by Chen et al. provide a viable alternative for instance segmentation which could extract more comprehensive information and thus enable more precise recognition. It has rarely been used for building extraction tasks with an exception of Liu et al. [42]. However, despite the improvements made in the feature extraction backbone network and RPN network in this study, a detailed analysis of the specific impacts of each module in the HTC network for building extraction from remote sensing images was not conducted, e.g., the influence of global semantic information. Furthermore, no relevant analysis was performed regarding the influence of detector thresholds on the model. Finally, they also failed to compare the HTC-based method with other SOTA cascade structure instance segmentation methods and more diverse classes of instance segmentation methods, e.g., the anchor-free instance segmentation method. Nonetheless, HTC could be leveraged for automatically extracting buildings from remote sensing imagery.

3. Methodology

In this study, the Hybrid Task Cascade network (HTC) [17] is introduced to enhance the performance of building extraction from high-resolution remote sensing imagery. The architecture of the proposed network is shown in Figure 1.

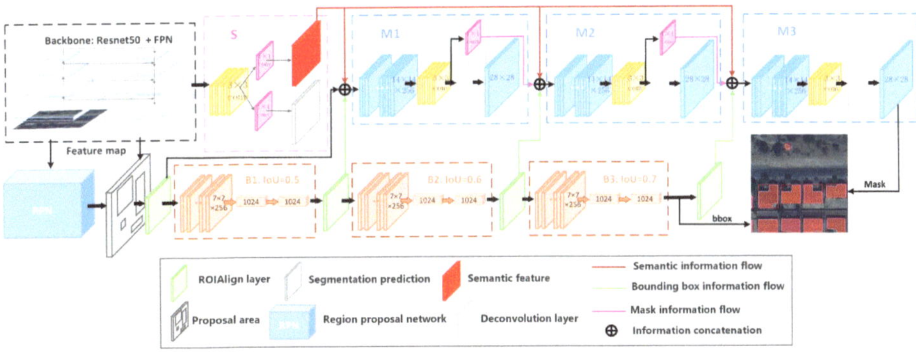

Figure 1. Structure of our HTC building detection model.

Compared to the classical Cascade-Mask-RCNN [43], it exhibits several distinctive characteristics: (1) Instead of performing bounding box regression and mask prediction in parallel, it integrates these tasks. (2) By using a direct path to feed the mask features from the previous stage to the current one, it reinforces the information flow between mask branches. (3) It adopts a fully convolutional branch to explore more global contextual information, fusing it with bbox and mask branches at the same time.

3.1. Network Stucture of HTC

The HTC network primarily consists of a feature extraction network and a three-stage prediction head, mainly focusing on the interaction and fusion of information between the backbone network and the various components. It performs joint processing by interleaving bounding box refinement and masking operations at each stage, as well as full convolutional branching to extract the global semantic context. By incorporating these tasks, the bounding box features, mask features, and global context form a tighter connection, which effectively contributes to the improvement of detection accuracy through the backpropagation.

The following provides a detailed description of the structure of the HTC network. The ResNet50 [39] with the Feature Pyramid Networks (FPN) is first used to extract building rooftop features from input imageries. The extracted feature maps are inputted into the Region Proposal Network (RPN) to extract candidate regions of buildings, and the proposed bounding boxes of the candidate regions are remapped onto the feature maps.

Simultaneously, a semantic segmentation branch is constructed based on the output of the feature pyramid. It performs up-sampling and down-sampling operations on the semantic features extracted at different levels of the feature pyramid, combining high-level features with global information and low-level features with local information. The transformed feature maps from different levels are then fused through element-wise summation. Subsequently, further feature fusion is performed using four 3 × 3 convolutional layers. Then, two 1 × 1 convolutional layers are utilized to obtain global semantic features of buildings and the semantic segmentation results. The semantic segmentation results are compared with the semantic annotations to compute the loss of the branch, while the global semantic features are inputted into each bounding box and mask heads, allowing the model to be more discriminative on the cluttered background. By incorporating global semantic annotations for supervised learning, the model can pay more attention to the recognition of small and medium-sized building objects, leading to higher detection accuracy. Moreover, this approach aligns well with the focus of our building extraction task, for it optimizes the detection threshold progressively and effectively reduces redundant predictions and the overlapping of building targets.

Secondly, the proposal boxes selected by the region proposal network (RPN) are used to extract Regions of Interest (RoIs), and each RoI is pooled into a 7 × 7 feature map using the RoIAlign layer that was the bounding box feature x_1^{box}. Then, a fully connected branch is employed to regress and classify each RoI's bounding box. In the case of traditional Mask-RCNN, following the RPN and RoIAlign layer, the features within the proposal regions are fed into a segmentation head for mask prediction. However, HTC uses the optimized bounding box b_1, which is obtained from the first-stage bounding box head with an Intersection over Union (IoU) threshold of 0.5 for mask prediction. More accurate bounding boxes can provide more precise object positions and boundary information, thereby offering better initialization and localization for the subsequent fully convolutional mask branch. By using accurate bounding boxes as initial regions, the HTC model can achieve more accurate pixel-level segmentation of building objects. Subsequently, the features extracted by backbone network x, along with b_1 and the semantic segmentation features s, are fused and used as inputs to the first stage's mask prediction head M_1 to extract mask information x_1^{mask}. In the mask prediction head M_1, before using the deconvolution operation to obtain the mask prediction m_1, we also learn the intermediate feature m_1^- through four 3 × 3 and a 1 × 1 convolution layers. The mask information m_1^- from the first stage is further fused to the mask branch of the next stage, which allows for interconnection and mutual influence between adjacent stages' mask branches. All features are supervised through backpropagation, leading to more accurate mask segmentation.

At the same time, the optimized bounding boxes b_1 are fed into the bounding box head B_2, with an IoU threshold of 0.6, to further optimize their positions and sizes and extract the features x_2^{box} of the bounding boxes b_2. The mask feature m_1^- from mask head M_1, the bounding box feature x_2^{box} from bounding box head B_2 and the semantic segmentation feature s are fused again and then inputted into the mask prediction head M_2 to obtain the mask feature m_2^-.

Finally, in the third stage, the above operation is repeated. By using the bounding box head B_3 with an IoU threshold of 0.7, much more accurate bounding boxes b_2 can be obtained, together with the mask feature m_2^- and the semantic segmentation feature s to obtain mask information x_3^{mask}. The final mask prediction results m_3 are obtained in the mask head M_3 by a deconvolution operation. The formula of HTC is expressed as:

$$x_i^{bbox} = P(x, b_{i-1}) + P(S(x), b_{i-1}), \tag{1}$$

$$b_i = B_i(x_i^{box}), \tag{2}$$

$$x_i^{mask} = P(x, b_i) + P(S(x), b_i), \tag{3}$$

$$m_i = M_i(\mathcal{F}_i^{mask}, m_{i-1}^-)), \tag{4}$$

In the given expression, we denote x as the features extracted by the backbone network. The variables x_i^{box} and x_i^{mask} represent the box and mask features obtained from x and the input RoIs. The pooling operator $P(\cdot)$, such as RoI Align or ROI pooling, is applied. B_i and M_i refer to the box and mask heads at the i-th stage, while b_i and m_i denote the corresponding box predictions and mask predictions. m_i^- represents the intermediate feature of the i-th mask head. The semantic segmentation head is represented by S, and $\mathcal{F}$ denotes a function that combines the features at the current stage with the preceding one.

3.2. Loss Function

Firstly, the overall loss function $Loss$ is formulated in the form of multi-task learning, which consists of the bounding box regression loss L_{bbox}, mask loss L_{mask} and semantic segmentation L_{seg}. The overall loss function is expressed as follows:

$$Loss = \sum_{i=1}^{3} \alpha_i (L_{bbox}^i + L_{mask}^i) + \beta L_{seg}, \tag{5}$$

Here, we set α_i = [1, 0.5, 0.25] in three stages of the HTC, respectively, and β = 1 by default, which follows the same definition as in HTC [17]. This multi-task learning framework enables the model to jointly optimize these three components, leading to improved performance in various aspects of the task.

Among the loss components, L_{bbox} combines two terms, L_{cls} and L_{reg}, which are used for classification and bounding box regression, respectively. L_{mask} is formulated using the binary cross-entropy form as described in Mask R-CNN [16]. Additionally, in our task, the semantic segmentation loss L_{seg} also appears in the form of binary cross-entropy, serving the purpose of distinguishing the foreground and background. They are given as follows:

$$L_{bbox}^i(c_i, b_i, \hat{c}_i, \hat{b}_i) = L_{cls}(c_i, \hat{c}_i) + L_{reg}(b_i, \hat{b}_i), \tag{6}$$

$$L_{mask}^i(m_i, \hat{m}_i) = BCE(m_i, \hat{m}_i), \tag{7}$$

$$L_{seg}(s, \hat{s}) = BCE(s, \hat{s}), \tag{8}$$

Here, the bbox regression loss L_{reg} is a smoothL1 loss and the classification loss also takes the form of a BCE loss. The bounding box regression loss is given as follows:

$$L_{reg} = \sum SmoothL1(y_i, \hat{y}_i), \tag{9}$$

where $\hat{y}_i$ is the predicted box and y_i is the box label. Finally, the smoothL1 loss and BCE loss is written as follows:

$$L_{SmoothL1}(y, \hat{y}) = \begin{cases} 0.5(\hat{y} - y)^2, & \text{if } |\hat{y} - y| < 1, \\ |\hat{y} - y| - 0.5, & \text{otherwise.} \end{cases} \tag{10}$$

$$L_{BCE}(y, \hat{y}) = -\frac{1}{N} \sum_{i=1}^{N} [y_i \log(\hat{y}_i) + (1 - y_i) \log(1 - \hat{y}_i)], \tag{11}$$

where y_i represents the true label, $\hat{y}_i$ represents the predicted label probability and N is the number of samples.

4. Experiments

In this section, we present our experiment setups, datasets used, evaluation metrics and experimental results.

4.1. Implementation Details

To ensure the consistency of the experimental results, FPN was used in all backbones. The experiments were conducted using the PyTorch framework on a 3060 GPU. A weight decay rate of 0.0001 and a momentum value of 0.9 were applied. The detectors were trained using a batch size of 16 images per GPU for 12 epochs. The initial learning rate was set to 0.02, and it was decreased by a factor of 0.1 after the 8th and 11th epochs, respectively.

4.2. *Datasets*

The proposed method was evaluated on three challenging building instance segmentation datasets: CrowdAI Mapping Challenge Dataset [19], the WHU aerial image dataset [20] and the self-annotated dataset proposed in [21]. Traditional instance annotations are used to supervise the bbox and mask branches, while the semantic branch is supervised by COCO-stuff [44] annotations. By employing the Python COCOAPI, we convert traditional COCO annotations into the COCO-stuff format and perform grayscale transformation to obtain foreground–background semantic annotations.

4.2.1. CrowdAI Mapping Challenge Dataset

The CrowdAI Mapping Challenge dataset is a large-scale remote sensing imagery building dataset published by [19] for the mapping challenge. It is comprised of a total of 280,741 satellite images for training, along with an additional 60,317 images for testing. In academic research, the experiment is commonly conducted with small-scale CrowdAI datasets, which usually comprise 8366 images for training and 1820 images for testing. Each image is presented in JPEG format with a resolution of 300 × 300 pixels. Furthermore, their annotations are provided in the MS-COCO format [45], offering useful information about the objects and their characteristics within the images. The samples displayed in Figure 2 showcase a diverse range of buildings in the CrowdAI dataset, in varying sizes and shapes.

Figure 2. Samples of CrowdAI Mapping Challenge Dataset and corresponding building labels.

4.2.2. WHU Aerial Image Dataset

The WHU aerial image dataset, proposed by Ji et al. [20], consists of JPEG images with a resolution of 512 × 512 pixels. The dataset contains over 187,000 buildings with varying sizes and appearances, captured in Christchurch, New Zealand. It is divided into a training set of 4736 images and a test set of 2416 images. Some samples are illustrated in Figure 3.

Figure 3. Samples of WHU Aerial Image Dataset and building labels.

4.2.3. Chinese Typical City Buildings Dataset

The dataset utilized in this study was provided by Fang et al. [21]. It comprises a total of 7260 tiles, with 5985 images allocated for training purposes and 1275 for testing. The dataset encompasses a diverse range of urban areas in four major Chinese cities: Beijing, Shanghai, Shenzhen and Wuhan. In total, it contains 63,886 instances of buildings. Each image in the dataset is presented in TIF format and possesses dimensions of 500 × 500 pixels, with a spatial resolution of 0.29 m per pixel. The dataset exhibits a wide coverage, encompassing diverse architectural styles with significant variations in terms of shape, size, color and other distinctive features of buildings. Some samples from this building instance segmentation dataset are shown in Figure 4.

Figure 4. Samples of Self-Annotated Dataset and building labels.

4.3. *Evaluation Metrics*

In this study, the proposed method is evaluated using the standard MS COCO metrics [45], which include mean average precision (AP) and mean average recall (AR) at multiple IoU values, e.g., AP, AP_{50}, AP_{75}. Additionally, the mean average precision is computed for objects of different scales, denoted as AP_s, AP_m and AP_l. The IoU is calculated using the equation below:

$$IoU = \frac{intersection}{union} = \frac{S_{pred} \cap S_{gt}}{S_{pred} \cup S_{gt}}, \tag{12}$$

where S_{pred} and S_{gt} represent the prediction results and the ground truth, respectively, and the AP and AR are calculated at 10 IoU overlap thresholds ranging from 0.50 to 0.95 in increments of 0.05, following the evaluation metrics [45]. The formulas are given as follows:

$$AP = \frac{AP_{0.50} + AP_{0.55} + \ldots + AP_{0.90} + AP_{0.95}}{10}, \tag{13}$$

$$AR = 2\int_{0.5}^{1} recall(o)do = \frac{2}{n}\sum_{i=1}^{n} max(IoU(gt_i) - 0.5, 0), \tag{14}$$

In addition, AP_{50} and AP_{75} are AP calculated at IoU thresholds of 0.5 and 0.75, respectively. Between them, the AP_{75} is a much more strict evaluation metric, which can better reflect the position accuracy of the algorithm. Furthermore, the metrics AP_s, AP_m, and AP_l are used to evaluate the performance of the methods on detecting buildings of different sizes. Specifically, AP_s measures the performance in detecting small buildings with an area $S < 32 \times 32$ pixels, AP_m measures the performance in detecting buildings with an area $32 \times 32 < S < 96 \times 96$ pixels, and AP_l measures the performance in detecting large buildings with an area $S > 96 \times 96$ pixels.

4.4. Baseline Methods

In this study, the HTC builing instance segmentation method is compared with five baselines, Mask R-CNN [16], Cascade Mask R-CNN [43], SCNet [46], SOLOv2 [47] and Swin Mask R-CNN [48]. Mask R-CNN is a classic instance segmentation method; recent works showed its popularity for building the instance extraction. Cascade Mask R-CNN effectively integrates cascade structures into instance segmentation, which can obtain better detection accuracy than traditional Mask R-CNN. SCNet, proposed by Vu et al. [46], utilizes stacking and skip connect operations to connect mask branches of the Cascade Mask R-CNN into a sequence of successive convolutional layers, ensuring sample consistency and improving the model training speed and accuracy. It also introduces a feature relay operation to establish a feature prior for bounding box feature to mask prediction. In addition, they introduced a global context branch similar to that in the HTC model for multi-label prediction and fused a global feature with a box and mask branch. SOLOv2 is a powerful, novel and efficient anchor-free instance segmentation model that dynamically segments each instance in an image by decomposing the mask branches into mask kernel prediction and mask feature learning. Swin Mask R-CNN, which introduces Swin Transformer into traditional Mask R-CNN, employs a patch-based approach and the "shift window" conception to effectively capture pixel relationships within each patch while promoting an information exchange between patches. Consequently, this architecture significantly enhances the performance of object detection and instance segmentation tasks and has been frequently used in SOTA methods. All of the baseline models adopt the same settings and use ResNet50 and FPN as the backbone, except for Swin Mask R-CNN, which uses Swin Transformer as the backbone with the window size set to 7 × 7 and transformer blocks with sizes of 2, 2, 6 and 2 in the 4 stages, respectively.

5. Experimental Results

This section presents a comparative analysis of the results obtained from five baseline models and our proposed HTC building detection method, followed by a comparison of their actual performance on the three test sets.

5.1. Results on the CrowdAI Mapping Challenge Dataset

Table 1 presents a comparison of results obtained using different instance segmentation algorithms on the CrowdAI Mapping Challenge Dataset. In general, the HTC-based method demonstrates superior performance across various evaluation metrics, including bounding box and mask accuracy. Our approach exhibits significant improvements in AP and AR for both the bounding box branch and mask branch, surpassing the baseline methods. Specifically, AP is increased by 0.6–3% and 0.3–2.2% in the bounding box and mask branches, respectively, while AR is enhanced by 1.0–3.8% and 0.8–2.9%, respectively. Moreover, for small-sized building instances, the bounding box branch and mask branch achieve an improvement of 0.3–2.6% and 0.4–2.9% in AP_s. For medium-sized buildings, AP_m is enhanced by 0.3–3.1% and 0.1–1.2% in the two branches. Additionally, for large-scale building instances, AP_l shows improvements of 1.7–4.2% and 0.3–4.0% in the bounding box branch and mask branch, respectively.

Table 1. Experimental results on the CrowdAI Mapping Challenge Dataset.

Methods	Bounding Box (%)							Mask (%)						
	AP	AP_{50}	AP_{75}	AP_s	AP_m	AP_l	AR	AP	AP_{50}	AP_{75}	AP_s	AP_m	AP_l	AR
Mask R-CNN	55.7	82.4	64.6	27.1	71.4	69.8	61.1	54.3	82.6	64.4	26.4	69.2	68.5	59.6
Swin Mask R-CNN	56.0	82.5	65.5	28.4	71.4	72.0	61.0	54.9	83.4	64.9	27.7	69.4	71.6	59.8
SOLOv2	—	—	—	—	—	—	—	53.7	82.4	63.0	25.3	69.4	72.0	60.2
Cascade Mask R-CNN	58.1	83.5	66.1	29.4	73.9	71.0	63.5	55.2	83.5	65.0	27.8	69.7	70.1	60.5
SCNet	58.1	83.8	66.1	29.2	74.2	72.3	63.9	55.6	83.7	65.3	27.5	70.3	72.2	61.7
Proposed Method	58.7	84.1	66.6	29.7	74.5	74.0	64.9	55.9	84.2	66.1	28.2	70.4	72.5	62.5

To further illustrate and analyze the results, selected samples are shown in Figure 5, with a focus on the areas outlined by blue circles. Firstly, as illustrated in the first row, Swin Mask R-CNN exhibits an imperfect segmentation of certain building contours. Secondly, in the second row, baseline methods overlook some small-scale buildings and edge targets, whereas our method successfully detects them. Furthermore, in the third rows, baseline methods erroneously detect non-building objects, e.g., vehicles and shadows. Moreover, in the fourth row, compared to the baseline method, our approach extracts building masks with more regular contours and closer approximation to real building outlines. Finally, from the fifth line, we can observe that our method performs well in identifying buildings with interior holes, whereas the baseline methods either fails to recognize the interior holes or produces blurred boundaries. These findings indicate that the proposed method outperforms the baseline methods.

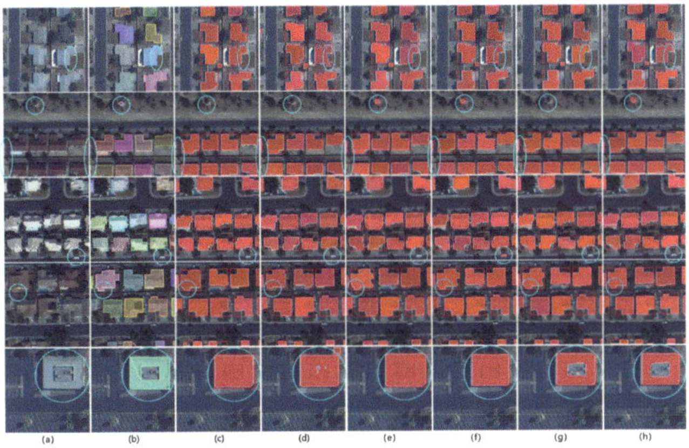

Figure 5. Example of results on the CrowdAI Mapping Challenge Dataset: (**a**) Original image, (**b**) Label, (**c**) Mask R-CNN, (**d**) Swin Mask R-CNN, (**e**) SOLOv2, (**f**) Cascade Mask R-CNN, (**g**) SCNet and (**h**) HTC.

5.2. Results on the WHU Aerial Image Dataset

The experimental results on WHU Aerial Image Dataset are listed in Table 2. Our method achieves a state-of-the-art performance in all evaluation metrics related to bounding box accuracy. However, it slightly falls behind Swin Mask R-CNN in certain evaluation metrics for mask detection accuracy. The proposed method obtains more precise bounding box results than the five baselines, with AP, AP_{50}, AP_{75} and AR values of 66.5%, 85.5%, 75.8% and 70.7%, respectively. Specifically, AP is increased by 0.6–2.8%, while AR is enhanced by 0.8–2.7% compared to the baseline models.

Table 2. Experimental results on the WHU Aerial Image Dataset.

Methods	Bounding Box (%)							Mask (%)						
	AP	AP_{50}	AP_{75}	AP_s	AP_m	AP_l	AR	AP	AP_{50}	AP_{75}	AP_s	AP_m	AP_l	AR
Mask R-CNN	63.7	83.9	73.3	49.3	79.8	73.6	68.0	60.4	84.8	71.4	44.5	77.3	75.2	64.2
Swin Mask R-CNN	63.9	84.8	74.2	50.0	79.5	74.2	68.2	62.7	84.9	73.5	47.2	79.1	78.1	66.5
SOLOv2	—	—	—	—	—	—	—	55.1	83.3	64.3	37.2	74.3	76.8	59.4
Cascade Mask R-CNN	65.9	84.7	74.7	51.2	82.3	76.6	69.9	60.5	84.8	71.5	44.5	77.5	76.7	64.1
SCNet	64.9	85.2	74.3	50.2	81.5	74.7	69.2	60.3	84.8	71.1	44.2	77.4	76.8	64.1
Proposed Method	66.5	85.5	75.8	52.1	82.7	76.5	70.7	61.7	85.5	73.1	46.1	78.3	77.9	65.4

Although our method slightly lags behind Swin Mask R-CNN in several mask accuracy metrics, we can derive valuable insights from the specific experimental results shown in Figure 6 (with a focus on the areas outlined by blue circles). Firstly, compared to the baseline methods, our approach achieves more regular and smoother contour shapes. Secondly, in the third and fourth lines, we can observe that our method demonstrates higher detection accuracy for very small building instances. Furthermore, the precise identification of small interior holes in buildings in the fourth line also demonstrates the superior accuracy of our method in recognizing building outlines. Lastly, in the last row, for different regions of large-scale building instances, the baseline methods exhibit varying degrees of redundant detections. In contrast, our detection method, by gradually increasing the IoU threshold, produces better mask predictions for such buildings, significantly reducing instances of overlapping predicted masks. This characteristic makes our method more suitable for remote sensing image building detection tasks, especially for scenarios with large and unconventionally shaped buildings.

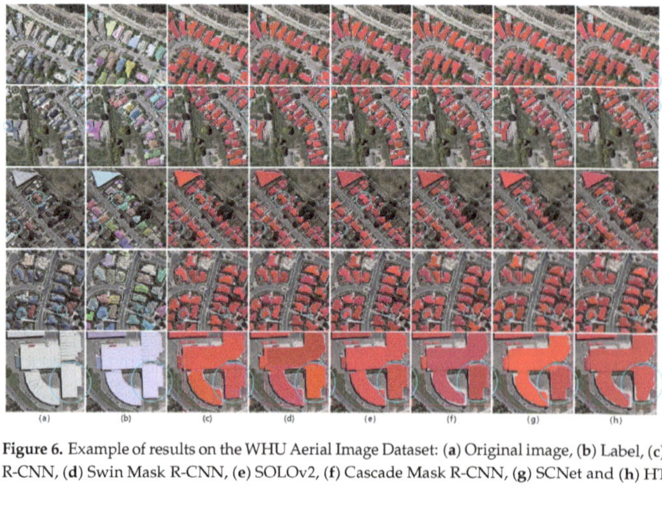

Figure 6. Example of results on the WHU Aerial Image Dataset: (**a**) Original image, (**b**) Label, (**c**) Mask R-CNN, (**d**) Swin Mask R-CNN, (**e**) SOLOv2, (**f**) Cascade Mask R-CNN, (**g**) SCNet and (**h**) HTC.

5.3. Results on the Chinese Typical City Buildings Dataset

The results obtained using different instance segmentation algorithms in this dataset are presented in Table 3. This dataset consists of non-orthophoto images with complex backgrounds, posing challenges in extracting building instances. Moreover, these challenges might have contributed to the relatively lower detection performance of Swin Mask R-CNN. However, the HTC method performs the best in all other evaluation metrics, except for a slight inferiority to SCNet in the metric of mask precision AP_l. Firstly, the bounding box branch exhibits an improvement of 0.8–3.9% in AP, while the mask branch shows an enhancement of 0.3–4.0%. Secondly, the bounding box branch and mask branch demonstrate an increase of 1.9–7.5% and 1.0–6.2% in AR, respectively. Additionally, for small-sized building instances, the bounding box branch and mask branch achieve an improvement of 0.5–2.5% and 0.2–5.1% in AP_s, respectively. Here, we find that SOLOv2 is quite ineffective at detecting small target building objects. For medium-sized buildings, the AP_m in both branches is enhanced by 0.7–3.8% and 0.5–3.6%. Furthermore, when considering large-scale building instances, the bounding box branch demonstrates notable enhancements in AP_l, with improvements ranging from 0.6% to 4.7%. The SCNET achieves nearly the same accuracy in large target building detection. These data highlight the superiority and robustness of the HTC method for building extraction from non-orthophoto remote sensing imagery.

Table 3. Experimental results on the Chinese Typical City Buildings Dataset.

Methods	Bounding Box (%)							Mask (%)						
	AP	AP_{50}	AP_{75}	AP_s	AP_m	AP_l	AR	AP	AP_{50}	AP_{75}	AP_s	AP_m	AP_l	AR
Mask R-CNN	47.8	72.1	52.7	23.2	54.7	58.9	61.7	45.3	71.5	49.7	17.7	51.0	58.3	58.8
Swin Mask R-CNN	46.9	72.1	52.4	22.4	53.9	57.3	59.4	45.1	71.4	49.8	17.3	51.1	57.5	57.3
SOLOv2	—	—	—	—	—	—	—	43.0	69.9	46.9	13.6	49.4	58.7	56.3
Cascade Mask R-CNN	50.0	71.7	55.4	24.0	57.0	61.4	64.3	46.1	71.6	50.9	18.1	51.9	59.2	59.4
SCNet	49.9	72.9	54.6	24.4	56.8	60.7	65.0	46.7	72.7	51.5	18.5	52.5	60.7	61.5
Proposed Method	50.8	73.2	56.0	24.9	57.7	62.0	66.9	47.0	72.8	52.1	18.7	53.0	60.6	62.5

Examples of detailed experimental results, as shown in Figure 7 (with a focus on the areas outlined by blue circles), demonstrate where the HTC method outperforms the baseline methods in building extraction. The HTC method achieves more accurate predictions with fewer errors and clearer building contour extraction. It also reduces the occurrence of repetitive outputs and overlapping objects. From the first, fourth, and fifth rows, we can observe that our method can more accurately identify small target buildings and buildings located at the edges compared to the baseline methods. Meanwhile, from the second and fifth rows, we can observe that our method extracts building masks with more regular and smoother contours. Finally, from the third row, we find that our method can effectively filter out interferences, such as trees, shadows of buildings, and other disturbances, and thus accurately identify building targets and their contours.

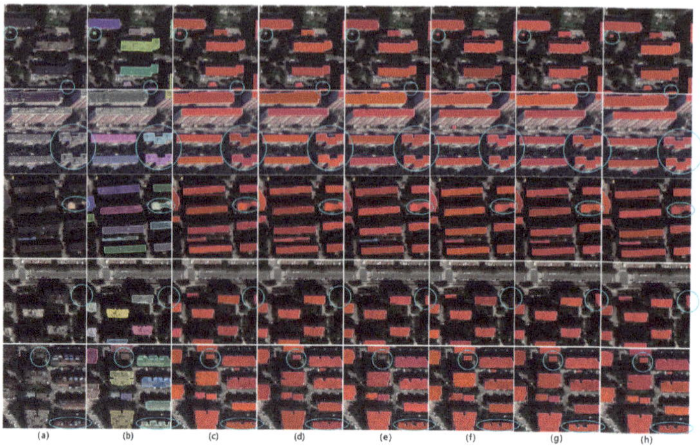

Figure 7. Example of results on Chinese Typical City Buildings Dataset: (**a**) Original image, (**b**) Label, (**c**) Mask R-CNN, (**d**) Swin Mask R-CNN, (**e**) SOLOv2, (**f**) Cascade Mask R-CNN, (**g**) SCNet and (**h**) HTC.

5.4. Ablation Experiments

In order to understand the specific effects of each component of HTC for remote sensing building extraction, we conducted experiments on the CrowdAI Mapping Challenge Dataset and discovered some interesting experimental results.

As shown in Tables 4 and 5, the integrated structure, mask information flow and semantic segmentation branch module all contribute to the improvements in model accuracy compared to the vanilla Cascade Mask R-CNN. However, when each of the three modules is individually excluded, the impact on the final model's detection accuracy is not evident, except for the detection of large-scale building objects. The absence of any single module leads to a decrease in the model's detection accuracy for large-scale building objects AP_l by 1.3–2.3% in the bounding box branch and 0.5–1.5% in the mask branch.

Additionally, it is noteworthy that when only the semantic segmentation branch is used without the integrated structure and mask information flow, the model achieves optimal detection performance on the CrowdAI validation dataset. Compared to HTC, the detection accuracy for small-scale building objects AP_s is improved by 4.9% and 4.7% in the bounding box branch and mask branch, respectively. The detection accuracy for medium-sized buildings AP_m is also improved by 1.1% and 0.9% in the bounding box branch and mask branch, respectively. Although there is a slight decrease of 0.9% and 1.1% in the detection accuracy of large-scale building objects AP_l, overall improvements in AP and AR are observed with only the semantic segmentation module added to Cascade Mask R-CNN, largely due to the more frequent presence of small and medium-sized building objects in the images. While this result may be influenced by the characteristics of the dataset itself and may lack generalizability, it still highlights the importance of global semantic information for the detection of small-scale building objects in remote sensing imagery. Small-size buildings have limited available features, and their semantic information appears in low-level feature maps. As the network deepens, their detailed information may be completely lost. However, HTC addresses this issue by incorporating an additional branch for contextual information utilization in semantic segmentation and utilizing global semantic annotation for supervised learning, thus better preserving the information of small target buildings.

Table 4. Ablation experiments for the bounding box branch on the CrowdAI Mapping Challenge Dataset.

Cascade	Integrated	Mask Info	Semantic	AP	AP_{50}	AP_{75}	AP_s	AP_m	AP_l	AR
✓				58.1	83.5	66.1	29.4	73.9	71.0	63.5
✓	✓	✓		58.6	84.0	66.9	29.8	74.3	72.0	64.9
✓		✓	✓	58.7	84.1	66.6	29.8	74.4	71.7	64.9
✓	✓		✓	58.9	84.1	66.7	30.0	74.7	72.7	65.0
✓			✓	61.4	87.4	69.3	34.6	75.6	73.1	67.9
✓	✓	✓	✓	58.7	84.1	66.6	29.7	74.5	74.0	64.9

Table 5. Ablation experiments for the Mask branch on the CrowdAI Mapping Challenge Dataset.

Cascade	Integrated	Mask Info	Semantic	AP	AP_{50}	AP_{75}	AP_s	AP_m	AP_l	AR
✓				55.2	83.5	65.0	27.8	69.7	70.1	60.5
✓	✓	✓		55.9	84.1	65.9	28.3	70.4	71.1	62.5
✓		✓	✓	55.8	84.2	65.7	28.3	70.3	71.0	62.5
✓	✓		✓	55.8	84.1	65.7	28.3	70.3	72.0	62.3
✓			✓	58.3	87.5	68.4	32.9	71.3	71.4	65.2
✓	✓	✓	✓	55.9	84.2	66.1	28.2	70.4	72.5	62.5

We conducted similar experiments on another two datasets, and the results were largely consistent, except for a slight decrease in model accuracy compared to HTC when using only the semantic segmentation branch module.

5.5. *Effects of different IoU threshold*

The IoU threshold has an impact on the selection of positive and negative samples during the model training process, thereby affecting the training effectiveness of the model. A suitable IoU threshold improves training sample quality and model performance. While 0.5 is commonly used for optimal results in general visual tasks, for building extraction in remote sensing imagery, buildings often have regular and well-defined shapes, leading to higher IoU values for bounding boxes. To validate this hypothesis, we conducted experiments on different datasets. Table 6 shows the results of experiments on the CrowdAI dataset using Mask R-CNN and HTC models, respectively. Increasing the IoU threshold by approximately 0.1 yields slight improvements in key performance metrics, and results in higher AP and AR.

Table 6. Experimental results on the CrowdAI Dataset with different IoU thresholds.

Methods	IoU	Bounding Box (%)							Mask (%)						
		AP	AP_{50}	AP_{75}	AP_s	AP_m	AP_l	AR	AP	AP_{50}	AP_{75}	AP_s	AP_m	AP_l	AR
Mask R-CNN	0.5	55.7	82.4	64.6	27.1	71.4	69.8	61.1	54.3	82.6	64.4	26.4	69.2	68.5	59.6
Mask R-CNN	0.6	56.1	82.1	64.8	27.7	72.0	68.3	61.5	54.4	82.9	64.3	26.7	69.3	68.9	59.9
HTC	(0.5, 0.6, 0.7)	58.7	84.1	66.6	29.7	74.5	74.0	64.9	55.9	84.2	66.1	28.2	70.4	72.5	62.5
HTC	(0.6, 0.7, 0.8)	59.1	83.9	66.9	29.8	75.1	72.7	65.1	56.2	84.1	66.1	28.3	70.9	72.5	62.6

Table 7 shows the results of experiments on the WHU dataset using Mask R-CNN and HTC models, respectively. By appropriately increasing the IoU threshold, we obtained improved results. Specifically, for the HTC model, increasing the detector threshold by 0.1 led to improved precision in terms of AP, AR and detection accuracy for buildings of different scales. Additionally, we conducted an set of experiments with an increased IoU threshold of 0.05 to further validate our hypothesis.

Table 7. Experimental results on the WHU Dataset with different IoU thresholds.

Methods	IoU	Bounding Box (%)							Mask (%)						
		AP	AP_{50}	AP_{75}	AP_s	AP_m	AP_l	AR	AP	AP_{50}	AP_{75}	AP_s	AP_m	AP_l	AR
Mask R-CNN	0.5	63.7	83.9	73.3	49.3	79.8	73.6	68.0	60.4	84.8	71.4	44.5	77.3	75.2	64.2
Mask R-CNN	0.6	64.6	83.8	74.6	50.0	81.4	72.9	68.6	60.9	83.9	72.4	44.7	78.1	75.1	64.3
HTC	(0.5, 0.6, 0.7)	66.5	85.5	75.8	52.1	82.7	76.5	70.7	61.7	85.5	73.1	46.1	78.3	77.9	65.4
HTC	(0.55, 0.65, 0.75)	66.9	85.3	75.6	52.3	83.4	77.0	71.1	62.0	85.5	73.1	46.0	78.9	78.9	65.7
HTC	(0.6, 0.7, 0.8)	67.2	85.1	76.2	52.6	83.9	77.0	71.2	62.4	85.2	73.6	46.5	79.2	78.8	66.0

The findings indicate that appropriately increasing the model's IoU threshold significantly enhances training effectiveness and improves the detection accuracy of the model for automatic building extraction using remote sensing imagery.

5.6. Efficiency and Generalizability Tests

While our method outperforms baseline models in terms of model accuracy, the HTC model is also comparatively more complex with more parameters, which could affect the computational costs. To better evaluate our model, we compared the parameter sizes as well as the detection speed of the proposed HTC-based model with other baseline models on the WHU dataset. As shown in Table 8, while obtaining higher accuracy, our model also sacrifices efficiency to a certain extent. This undoubtedly limits applications of the HTC-based model to scenarios that require fast processing of remote sensing images such as disaster monitoring. To further analyze the model efficiency of HTC-based models, we carried out additional experiments with a modified model structure. We observe that removing the semantic segmentation branch substantially improves the efficiency of HTC by 38.9%. Although from the previous experimental results in Tables 4 and 5 we found that missing global semantic information affects the detection accuracy for large target buildings, the effect on the overall accuracy is modest. Therefore, for application scenarios that require more efficient detection, a modified HTC model without a semantic segmentation branch could be utilized.

Table 8. Parameter sizes and detection speed on the WHU Dataset.

Methods	Parameters (M)	fps (*Frames per Second*)
Mask R-CNN	335	20.64
Cascade Mask R-CNN	587	14.12
Swin Mask R-CNN	542	18.58
SOLOv2	352	15.19
SCNet	720	8.42
HTC	609	8.92
HTC (without semantic branch)	588	12.39

To further validate our model backbone selection, we used four different backbones other than ResNet50 [39] for the experimental comparison on the WHU dataset. The experimental results are shown in Table 9. We found that although ResNet18 [39], with a lower number of network layers, reduces the size and the training time of the model, it sacrifices more in terms of detection accuracy. In addition, deeper networks, ResNet101 [39] and ResNext101 [49], produce slower training without any gain in accuracy. Finally, when we use HRNetV2p [50] as the backbone, which introduce more cross-connections among branches, most of the accuracy metrics are improved, especially for small and medium-sized building targets, although it lags behind in terms of efficiency. It suggests that the method is worth being considered in scenarios where higher detection accuracy is required.

Table 9. Experimental results on HTC with different backbone.

Backbone	Mask (%)							Parameters (M)	fps (*Frames per Second*)
	AP	AP_{50}	AP_{75}	AP_s	AP_m	AP_l	AR		
Resnet-18	59.9	84.5	70.6	43.5	77.0	76.4	64.0	509	9.47
Resnet-50	61.7	85.5	73.1	46.1	78.3	77.9	65.4	609	8.92
Resnet-101	60.3	84.7	71.2	44.2	76.8	77.9	64.2	754	8.16
Resnext-101	61.3	85.0	72.4	45.5	78.2	77.7	64.8	751	8.10
HRNetV2p	62.3	85.9	73.6	46.6	79.3	77.7	65.9	980	7.60

6. Discussion

In this section, we discuss the findings based on our experiment results and discuss our future work.

6.1. Summary of Results

Firstly, our experiments demonstrate the promising capabilities of the HTC model based on the cascaded structure for building extraction from remote sensing imagery. Compared to the selected baseline algorithms, the HTC method achieves higher detection accuracy and generates more regular building contours. In addition, the HTC method integrates the execution of tasks such as the bounding box refinement, masking, and global semantic information extraction. This integration strengthens the information interaction and connection between different branches, ultimately leading to accuracy improvement through backpropagation. This approach has been shown to be effective in enhancing the performance compared to traditional non-hybrid models that handle these tasks separately. The method's accuracy and robustness are validated on three challenging datasets.

Secondly, we show that through cascaded optimization algorithms, our method not only achieves higher recognition accuracy but also reduces the overlap of predicted targets in practical applications. This aligns well with the inherent characteristics of building targets in remote sensing imagery, making the obtained images more suitable for real-world engineering applications. Therefore, the HTC-based method is promising as a suitable alternative for applications in building extraction from remote sensing imagery tasks.

Thirdly, we conduct experiments to evaluate the effectiveness of different modules in the HTC model, revealing their contributions to improving model accuracy and their impacts on recognizing buildings of different scales. Additionally, we find that the utilization of global semantic information plays a vital role in accurately identifying small-sized building targets.

Furthermore, we propose a hypothesis regarding optimal IoU thresholds based on the relatively regular shape of building targets in remote sensing imagery. Our experiments confirm that appropriately increasing the detector's IoU threshold positively affects model accuracy. We believe that this can provide valuable insights for the future application of instance segmentation methods for automatic building detection.

Finally, some of the current limitations of the HTC method are addressed. We have conducted generalization experiments that enable the HTC-based building instance segmentation method to be effectively applied to more scenarios, such as drone disaster assessment (DDS) [51]. In summary, we believe that our proposed method is feasible and has sufficient application prospects.

6.2. Future Work

Our future work will primarily focus on several key directions. Firstly, we aim to investigate the effectiveness of employing additional stages to further enhance the model's accuracy. In a recent study by Wu et al. [52], they proposed an enhanced cascade structure network called Cascade R-CNN++ for high-quality object detection in multi-resolution remote sensing imagery. They found that adding a fourth stage to Cascade R-CNN led to a performance degradation. They attributed this to a potential mismatch between the RoI features and classifier. Additionally, we plan to investigate alternative backbone networks to assess their potential impact on improving HTC's performance.

Secondly, we intend to incorporate building regularization methods to obtain more regular and realistic building contours. Girard et al. [33] proposed a frame-field-based approach for building contour regularization. They trained a deep neural network that aligned a predicted frame field to ground truth contours and utilized the frame field to facilitate polygonization.

Moreover, we aim to explore the applicability of global semantic information in other instance segmentation models. We also extend our efforts to explore how to reduce parameters and improve the speed of model training while preserving globle semantic information.

Finally, we aim to apply our model to real-world scenarios. Experimental results on real-world datasets demonstrate great potential for practical applications, including mapping, urban planning, change detection, and the integration of multi-source geospatial data for urban functional studies [53].

7. Conclusions

This study introduces the HTC model for building extraction from remote sensing imagery and evaluates its effectiveness and robustness on three challenging building extraction datasets. Experimental results demonstrate that the proposed model outperforms existing techniques with higher accuracy and more precise contours. Moreover, the research highlights the positive impact of incorporating global semantic information on efficient and accurate extraction of building targets in remote sensing imagery. Additionally, by considering the inherent characteristics of building targets in remote sensing imagery, the HTC model produces building maps that are more readily applicable.

Finally, through an experimental analysis, we conclude that appropriately increasing the detector's IoU threshold can improve the detection accuracy of building extraction in remote sensing imagery to a certain extent. In summary, our proposed method holds great potential for applications in map-making, urban planning, intelligent urban transportation, geological exploration, urban life quality assessment and beyond. However, alongside these advantages, it is important to address the limitations of large parameter counts and the relatively slow training speed in HTC models, which warrants further investigation in future studies. Our future work will focus on continuing to improve detection accuracy, while reducing time costs and obtaining regular and smooth contours.

Author Contributions: Conceptualization, M.Z. and R.D.; methodology, R.D.; software, R.D.; validation, R.D. and Y.H.; formal analysis, R.D.; investigation, R.D. and M.Z.; resources, Y.H.; data curation, R.D.; writing—original draft preparation, R.D.; writing—review and editing, M.Z. and W.T.; visualization, R.D.; supervision, M.Z. and W.T.; project administration, M.Z. and W.T.; funding acquisition, M.Z. and W.T. All authors have read and agreed to the published version of the manuscript.

Funding: This research was funded by the National Natural Science Foundation of China (42201495), the Guangdong Basic and Applied Basic Research Foundation (2021A1515110049), and Shenzhen Science and Technology Program (JCYJ20220818100200001).

Data Availability Statement: No new data were created or analyzed in this study. Data sharing is not applicable to this article.

Conflicts of Interest: The authors declare no conflict of interest.

References

1. Maggiori, E.; Tarabalka, Y.; Charpiat, G.; Alliez, P. Can semantic labeling methods generalize to any city? the inria aerial image labeling benchmark. In Proceedings of the 2017 IEEE International Geoscience and Remote Sensing Symposium (IGARSS), Fort Worth, TX, USA, 23–28 July 2017; pp. 3226–3229.
2. Chen, J.; Wang, G.; Luo, L.; Gong, W.; Cheng, Z. Building Area Estimation in Drone Aerial Images Based on Mask R-CNN. *IEEE Geosci. Remote Sens. Lett.* **2021**, *18*, 891–894. [CrossRef]
3. Chen, D.; Tu, W.; Cao, R.; Zhang, Y.; He, B.; Wang, C.; Shi, T.; Li, Q. A hierarchical approach for fine-grained urban villages recognition fusing remote and social sensing data. *Int. J. Appl. Earth Obs. Geoinf.* **2022**, *106*, 102661. [CrossRef]

4. Amo-Boateng, M.; Sey, N.E.N.; Amproche, A.A.; Domfeh, M.K. Instance segmentation scheme for roofs in rural areas based on Mask R-CNN. *Egypt. J. Remote Sens. Space Sci.* **2022**, *25*, 569–577. [CrossRef]
5. Han, Q.; Yin, Q.; Zheng, X.; Chen, Z. Remote sensing image building detection method based on Mask R-CNN. *Complex Intell. Syst.* **2021**, *8*, 1847–1855. [CrossRef]
6. Wang, Y.; Li, S.; Teng, F.; Lin, Y.; Wang, M.; Cai, H. Improved mask R-CNN for rural building roof type recognition from uav high-resolution images: A case study in hunan province, China. *Remote Sens.* **2022**, *14*, 265. [CrossRef]
7. Powers, R.P.; Hay, G.J.; Chen, G. How wetland type and area differ through scale: A GEOBIA case study in Alberta's Boreal Plains. *Remote Sens. Environ.* **2012**, *117*, 135–145. [CrossRef]
8. Hu, L.; Zheng, J.; Gao, F. A building extraction method using shadow in high resolution multispectral images. In Proceedings of the 2011 IEEE International Geoscience and Remote Sensing Symposium, Vancouver, BC, Canada, 24–29 July 2011; pp. 1862–1865. [CrossRef]
9. Huang, J.; Zhang, X.; Xin, Q.; Sun, Y.; Zhang, P. Automatic building extraction from high-resolution aerial images and LiDAR data using gated residual refinement network. *ISPRS J. Photogramm. Remote Sens.* **2019**, *151*, 91–105. [CrossRef]
10. Yuan, J.; Cheriyadat, A.M. Learning to count buildings in diverse aerial scenes. In Proceedings of the 22nd ACM SIGSPATIAL International Conference on Advances in Geographic Information Systems, Fort Worth, TX, USA, 4–7 November 2014; Association for Computing Machinery: New York, NY, USA, 2014; pp. 271–280. [CrossRef]
11. Blaschke, T.; Hay, G.J.; Kelly, M.; Lang, S.; Hofmann, P.; Addink, E.; Feitosa, R.Q.; van der Meer, F.; van der Werff, H.; van Coillie, F.; et al. Geographic Object-Based Image Analysis—Towards a new paradigm. *ISPRS J. Photogramm. Remote Sens.* **2014**, *87*, 180–191. [CrossRef]
12. Breiman, L. Random forests. *Mach. Learn.* **2001**, *45*, 5–32. [CrossRef]
13. Noble, W.S. What is a support vector machine? *Nat. Biotechnol.* **2006**, *24*, 1565–1567. [CrossRef]
14. Ren, S.; He, K.; Girshick, R.; Sun, J. Faster r-cnn: Towards real-time object detection with region proposal networks. In Proceedings of the Advances in Neural Information Processing Systems, Montreal, QC, Canada, 7–12 December 2015; Volume 28.
15. Long, J.; Shelhamer, E.; Darrell, T. Fully convolutional networks for semantic segmentation. In Proceedings of the IEEE Conference on Computer Vision and Pattern Recognition, Boston, MA, USA, 7–12 June 2015; pp. 3431–3440.
16. He, K.; Gkioxari, G.; Dollár, P.; Girshick, R. Mask R-CNN. In Proceedings of the IEEE International Conference on Computer Vision, Venice, Italy, 22–29 October 2017.
17. Chen, K.; Pang, J.; Wang, J.; Xiong, Y.; Li, X.; Sun, S.; Feng, W.; Liu, Z.; Shi, J.; Ouyang, W.; et al. Hybrid task cascade for instance segmentation. In Proceedings of the IEEE/CVF Conference on Computer Vision and Pattern Recognition, Long Beach, CA, USA, 16–20 June 2019; pp. 4974–4983.
18. Cai, Z.; Vasconcelos, N. Cascade r-cnn: Delving into high quality object detection. In Proceedings of the IEEE Conference on Computer Vision and Pattern Recognition, Salt Lake City, UT, USA, 18–22 June 2018; pp. 6154–6162.
19. Mohanty, S.P. Crowdai Mapping Challenge 2018: Baseline with Mask RCNN. GitHub Repository. 2018. Available online: https://github.com/crowdai/crowdai-mapping-challenge-mask-rcnn (accessed on 6 October 2023).
20. Ji, S.; Wei, S.; Lu, M. Fully convolutional networks for multisource building extraction from an open aerial and satellite imagery data set. *IEEE Trans. Geosci. Remote Sens.* **2018**, *57*, 574–586. [CrossRef]
21. Fang, F.; Wu, K.; Zheng, D. A dataset of building instances of typical cities in China [DB/OL]. *Sci. Data Bank* **2021**. [CrossRef]
22. Ronneberger, O.; Fischer, P.; Brox, T. U-Net: Convolutional Networks for Biomedical Image Segmentation. In *International Conference on Medical Image Computing and Computer-Assisted Intervention*; Springer: Cham, Switzerland, 2015.
23. Lin, T.Y.; Dollár, P.; Girshick, R.; He, K.; Hariharan, B.; Belongie, S. Feature pyramid networks for object detection. In Proceedings of the IEEE Conference on Computer Vision and Pattern Recognition, Honolulu, HI, USA, 21–26 July 2017; pp. 2117–2125.
24. Shrestha, S.; Vanneschi, L. Improved fully convolutional network with conditional random fields for building extraction. *Remote Sens.* **2018**, *10*, 1135. [CrossRef]
25. Yang, G.; Zhang, Q.; Zhang, G. EANet: Edge-aware network for the extraction of buildings from aerial images. *Remote Sens.* **2020**, *12*, 2161. [CrossRef]
26. Tang, Z.; Chen, C.Y.C.; Jiang, C.; Zhang, D.; Luo, W.; Hong, Z.; Sun, H. Capsule–Encoder–Decoder: A Method for Generalizable Building Extraction from Remote Sensing Images. *Remote Sens.* **2022**, *14*, 1235. [CrossRef]
27. Liu, Y.; Chen, D.; Ma, A.; Zhong, Y.; Fang, F.; Xu, K. Multiscale U-Shaped CNN Building Instance Extraction Framework with Edge Constraint for High-Spatial-Resolution Remote Sensing Imagery. *IEEE Trans. Geosci. Remote Sens.* **2021**, *59*, 6106–6120. [CrossRef]
28. Ma, J.; Wu, L.; Tang, X.; Liu, F.; Zhang, X.; Jiao, L. Building extraction of aerial images by a global and multi-scale encoder-decoder network. *Remote Sens.* **2020**, *12*, 2350. [CrossRef]
29. Wang, X.; Girshick, R.; Gupta, A.; He, K. Non-local neural networks. In Proceedings of the IEEE Conference on Computer Vision and Pattern Recognition, Salt Lake City, UT, USA, 18–22 June 2018; pp. 7794–7803.
30. Yuan, W.; Xu, W. Msst-net: A multi-scale adaptive network for building extraction from remote sensing images based on swin transformer. *Remote Sens.* **2021**, *13*, 4743. [CrossRef]
31. Liu, Z.; Lin, Y.; Cao, Y.; Hu, H.; Wei, Y.; Zhang, Z.; Lin, S.; Guo, B. Swin transformer: Hierarchical vision transformer using shifted windows. In Proceedings of the IEEE/CVF International Conference on Computer Vision, Montreal, QC, Canada, 10–16 October 2021; pp. 10012–10022.

32. Dosovitskiy, A.; Beyer, L.; Kolesnikov, A.; Weissenborn, D.; Zhai, X.; Unterthiner, T.; Dehghani, M.; Minderer, M.; Heigold, G.; Gelly, S.; et al. An Image is Worth 16x16 Words: Transformers for Image Recognition at Scale. *arXiv* **2020**, arXiv:2010.11929.
33. Girard, N.; Smirnov, D.; Solomon, J.; Tarabalka, Y. Polygonal building extraction by frame field learning. In Proceedings of the IEEE/CVF Conference on Computer Vision and Pattern Recognition, Virtual, 19–25 June 2021; pp. 5891–5900.
34. Zhao, K.; Kang, J.; Jung, J.; Sohn, G. Building extraction from satellite images using mask R-CNN with building boundary regularization. In Proceedings of the IEEE Conference on Computer Vision and Pattern Recognition Workshops, Salt Lake City, UT, USA, 18–22 June 2018; pp. 247–251.
35. Douglas, D.H.; Peucker, T.K. Algorithms for the reduction of the number of points required to represent a digitized line or its caricature. *Cartogr. Int. J. Geogr. Inf. Geovis.* **1973**, *10*, 112–122. [CrossRef]
36. Sohn, G.; Jwa, Y.; Jung, J.; Kim, H. An implicit regularization for 3D building rooftop modeling using airborne lidar data. *ISPRS Ann. Photogramm. Remote Sens. Spat. Inf. Sci.* **2012**, *1*, 305–310. [CrossRef]
37. Zhang, L.; Wu, J.; Fan, Y.; Gao, H.; Shao, Y. An efficient building extraction method from high spatial resolution remote sensing images based on improved mask R-CNN. *Sensors* **2020**, *20*, 1465. [CrossRef] [PubMed]
38. Fang, F.; Wu, K.; Liu, Y.; Li, S.; Wan, B.; Chen, Y.; Zheng, D. A coarse-to-fine contour optimization network for extracting building instances from high-resolution remote sensing imagery. *Remote Sens.* **2021**, *13*, 3814. [CrossRef]
39. He, K.; Zhang, X.; Ren, S.; Sun, J. Deep residual learning for image recognition. In Proceedings of the IEEE Conference on Computer Vision and Pattern Recognition, Las Vegas, NV, USA, 27–30 June 2016; pp. 770–778.
40. Zhao, W.; Persello, C.; Stein, A. Building outline delineation: From aerial images to polygons with an improved end-to-end learning framework. *ISPRS J. Photogramm. Remote Sens.* **2021**, *175*, 119–131. [CrossRef]
41. Zhu, Q.; Liao, C.; Hu, H.; Mei, X.; Li, H. MAP-Net: Multi Attending Path Neural Network for Building Footprint Extraction from Remote Sensed Imagery. *IEEE Trans. Geosci. Remote Sens.* **2020**, *59*, 6169–6181. [CrossRef]
42. Liu, X.; Chen, Y.; Wei, M.; Wang, C.; Goncalves, W.N.; Marcato, J.; Li, J. Building Instance Extraction Method Based on Improved Hybrid Task Cascade. *IEEE Geosci. Remote Sens. Lett.* **2022**, *19*, 3002005. [CrossRef]
43. Cai, Z.; Vasconcelos, N. Cascade R-CNN: High quality object detection and instance segmentation. *IEEE Trans. Pattern Anal. Mach. Intell.* **2021**, *43*, 1483–1498. [CrossRef]
44. Caesar, H.; Uijlings, J.; Ferrari, V. Coco-stuff: Thing and stuff classes in context. In Proceedings of the Conference on Computer Vision and Pattern Recognition, Salt Lake City, UT, USA, 18–22 June 2018; pp. 1209–1218.
45. Lin, T.Y.; Maire, M.; Belongie, S.; Bourdev, L.; Girshick, R.; Hays, J.; Perona, P.; Ramanan, D.; Zitnick, C.L.; Dollár, P. Microsoft COCO: Common Objects in Context. In *European Conference on Computer Vision*; Springer: Cham, Switzerland, 2014.
46. Vu, T.; Kang, H.; Yoo, C.D. Scnet: Training inference sample consistency for instance segmentation. In Proceedings of the AAAI Conference on Artificial Intelligence, Virtual, 2–9 February 2021; Volume 35, pp. 2701–2709.
47. Wang, X.; Zhang, R.; Kong, T.; Li, L.; Shen, C. Solov2: Dynamic and fast instance segmentation. *Adv. Neural Inf. Process. Syst.* **2020**, *33*, 17721–17732.
48. Fu, R.; He, J.; Liu, G.; Li, W.; Mao, J.; He, M.; Lin, Y. Fast seismic landslide detection based on improved mask R-CNN. *Remote Sens.* **2022**, *14*, 3928. [CrossRef]
49. Xie, S.; Girshick, R.; Dollár, P.; Tu, Z.; He, K. Aggregated residual transformations for deep neural networks. In Proceedings of the IEEE Conference on Computer Vision and Pattern Recognition, Honolulu, HI, USA, 21–26 July 2017; pp. 1492–1500.
50. Wang, J.; Sun, K.; Cheng, T.; Jiang, B.; Deng, C.; Zhao, Y.; Liu, D.; Mu, Y.; Tan, M.; Wang, X.; et al. Deep high-resolution representation learning for visual recognition. *IEEE Trans. Pattern Anal. Mach. Intell.* **2020**, *43*, 3349–3364. [CrossRef]
51. Xu, C.; Wang, G.; Yan, S.; Yu, J.; Zhang, B.; Dai, S.; Li, Y.; Xu, L. Fast vehicle and pedestrian detection using improved Mask R-CNN. *Math. Probl. Eng.* **2020**, *2020*, 5761414. [CrossRef]
52. Wu, B.; Shen, Y.; Guo, S.; Chen, J.; Sun, L.; Li, H.; Ao, Y. High Quality Object Detection for Multiresolution Remote Sensing Imagery Using Cascaded Multi-Stage Detectors. *Remote Sens.* **2022**, *14*, 2091. [CrossRef]
53. Zhang, Y.; Li, Q.; Tu, W.; Mai, K.; Yao, Y.; Chen, Y. Functional urban land use recognition integrating multi-source geospatial data and cross-correlations. *Comput. Environ. Urban Syst.* **2019**, *78*, 101374. [CrossRef]

remote sensing

Article

An Infrared Maritime Small Target Detection Algorithm Based on Semantic, Detail, and Edge Multidimensional Information Fusion

Jiping Yao, Shanzhu Xiao *, Qiuqun Deng, Gongjian Wen, Huamin Tao and Jinming Du

National Key Laboratory of Science and Technology on Automatic Target Recognition, Collage of Electronic Science and Technology, National University of Defense Technology, Changsha 410073, China; yaojiping@nudt.edu.cn (J.Y.)
* Correspondence: xiaoshanzhu@nudt.edu.cn

Abstract: The infrared small target detection technology has a wide range of applications in maritime defense warning and maritime border reconnaissance, especially in the maritime and sky scenes for detecting potential terrorist attacks and monitoring maritime borders. However, due to the weak nature of infrared targets and the presence of background interferences such as wave reflections and islands in maritime scenes, targets are easily submerged in the background, making small infrared targets hard to detect. We propose the multidimensional information fusion network(MIFNet) that can learn more information from limited data and achieve more accurate target segmentation. The multidimensional information fusion module calculates semantic information through the attention mechanism and fuses it with detailed information and edge information, enabling the network to achieve more accurate target position detection and avoid detecting one target as multiple ones, especially in high-precision scenes such as maritime target detection, thus effectively improving the accuracy and reliability of detection. Moreover, experiments on our constructed dataset for small infrared targets in maritime scenes demonstrate that our algorithm has advantages over other state-of-the-art algorithms, with an IoU of 79.09%, nIoU of 79.43%, F1 score of 87.88%, and AuC of 95.96%.

Keywords: infrared maritime images; small infrared target detection; feature fusion; gradient information; attention mechanism

Citation: Yao, J.; Xiao, S.; Deng, Q.; Wen, G.; Tao, H.; Du, J. An Infrared Maritime Small Target Detection Algorithm Based on Semantic, Detail, and Edge Multidimensional Information Fusion. *Remote Sens.* **2023**, *15*, 4909. https://doi.org/10.3390/rs15204909

Academic Editors: Yanni Dong, Xiaochen Yang and Qian Du

Received: 29 August 2023
Revised: 5 October 2023
Accepted: 9 October 2023
Published: 11 October 2023

1. Introduction

The automatic monitoring and detection of targets on the sea surface have significant scientific research and practical application significance in maintaining national sovereignty and safeguarding maritime rights and interests. Compared to radar imaging methods, infrared imaging is a passive imaging method with strong resistance to smoke interference, longer detection distance, and wider temporal applicability. Infrared small target detection technology has been widely used in fields such as maritime defense and maritime surveillance in sea-sky scenes.

However, in general, targets are far away from the observation equipment, and infrared small targets occupy very few pixels in the image, appearing as patches or even dots, lacking effective shape features [1], as well as lacking texture, color, and shape features of common objects [2]. At the same time, in practical applications, sea surface scenes are complex and often contain static or slowly varying clutter, such as sea-sky lines, cloud clusters, and islands, as well as dynamic clutter, such as fish-scale light and sun glint. In harsh environments, strong wave clutter may also occur, making targets easily submerged in complex backgrounds. Based on these circumstances, infrared small target detection has great application value and research value but faces significant challenges.

Infrared small target detection technology can be divided into Track Before Detect (TBD) [3–6] and Detect Before Track (DBT) according to the order of utilizing prior information [7]. TBD can handle low signal-to-noise ratio situations but generally requires high computational complexity, making it difficult to detect high-speed targets. On the other hand, DBT algorithms generally have lower complexity and can meet real-time requirements, making them widely deployed on hardware platforms and widely applied in various fields. The quality of DBT algorithms is determined by single-frame infrared small target detection algorithms, making them a research hotspot. Currently, mainstream single-frame infrared small target detection algorithms can be divided into deep learning-based algorithms and traditional feature extraction-based algorithms. Traditional algorithms can be further divided into filter-based algorithms [8,9], human vision-inspired algorithms [10,11], and optimization-based methods [12–14]. Filter-based algorithms assume that there is a contrast difference between the target and the background, making high-contrast regions more likely to be targeted. In optimization-based methods, the target class is treated as a sparse matrix, while the background class is treated as a low-rank matrix. The goal is to continuously optimize the separation of low-rank and sparse matrices to achieve target detection. These methods heavily rely on handcrafted features, and their performance weakens when the pre-designed features do not match the actual scenarios.

Unlike traditional algorithms that rely on handcrafted features, deep learning methods rely on a large amount of data to learn target features [15–17]. It has achieved remarkable results in terms of speed and accuracy in object detection on general datasets. Dai et al. proposed the ACM [18], which combines high-level semantic information with low-level details and uses the U-net and FPN network structures suitable for extracting multiscale features to detect small infrared targets. Although these methods have achieved significant achievements, they still face some challenges when applied to the detection of small infrared targets in marine and aerial scenes [19]. Due to the sensitivity of the data, efficiently utilizing the limited data to ensure the effectiveness of the algorithm in similar scenarios becomes a key task. Therefore, we must delve into the image information that can effectively describe the targets. More importantly, the targets and backgrounds in the marine environment are complex and diverse. The direct application of models trained on general object detection datasets or specific non-marine small object detection datasets to marine target detection still needs to be verified for its effectiveness.

This paper explores an infrared small target detection algorithm in marine and aerial scenes that utilizes edge information, detail information, and semantic information for multidimensional information fusion. Specifically, it consists of three information processing channels. Firstly, the input infrared image is processed by a multiscale object detection FPN network to extract semantic information. Then, the input infrared image is simultaneously processed by the edge information extraction module and the detail information extraction module to extract edge information and detail information, respectively. As shown in Figure 1, the three types of information, with semantic information as the main component, detail information, and edge information as the auxiliary, are fused and processed by the FCN module to obtain the detection results. Additionally, considering that the weights of different layers in the fusion process of FPN cross-level connections are different, we design a multiscale information fusion module to address this issue. Experimental results on our dataset demonstrate that our algorithm outperforms state-of-the-art algorithms in terms of IoU, nIoU, F1, and AuC metrics.

The main contributions of this study are as follows:

a. We propose an infrared small target detection model that demonstrates excellent performance on a dataset specifically designed for infrared small targets in maritime and aerial scenes. We introduce an edge information extraction module, which not only compensates for the loss of target information caused by downsampling but also provides edge information to enable more precise target detection;

b. We draw inspiration from the deeplab network structure and introduce shallow feature maps with richer detailed information in the last stage to further reduce the loss caused by downsampling;
c. We propose a fusion mechanism that combines semantic information with detail and edge information. This mechanism first extracts semantic information from the FPN baseline network and then organically integrates all three components using an attention mechanism;
d. Experimental results on the dataset compared with other state-of-the-art algorithms demonstrate the excellent performance of our algorithm. It can effectively extract and learn the features of the targets.

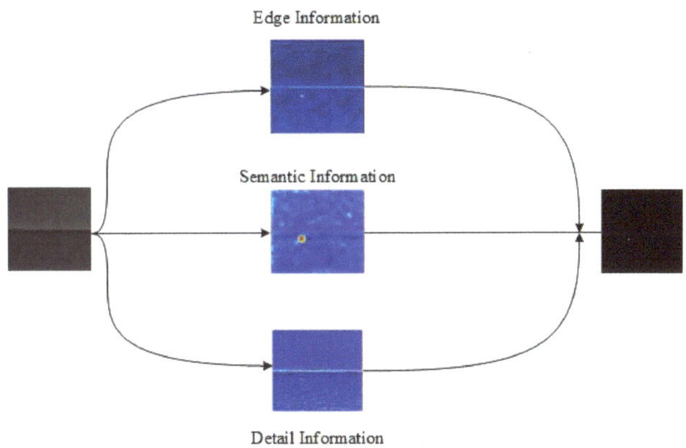

Figure 1. Main motivation.

The organizational structure of the entire article is as follows: In Section 2, the framework of the network and specific algorithm details are presented. Section 3 delves into our experimental details, results, and a comparative analysis of our algorithm against others. Section 4 comprehensively discusses and analyzes the experimental results. Finally, in Section 5, a comprehensive summary of the article is provided.

2. Materials and Methods

2.1. Related Work

2.1.1. Infrared Small Target Detection

Presently, the categorization of infrared small target detection algorithms in the context of maritime and sky scenes can be divided into single-frame-based detection methods and multi-frame-based detection methods. In the realm of multi-frame methods, approaches encompass particle filtering [5], Markov random fields [6], pipeline filtering [8], and dynamic programming [9], among others. Traditional single-frame-based infrared target detection algorithms can be further grouped into filter-based methods, visual saliency-based methods, and matrix decomposition-based methods.

Spatial filtering methods enhance the signal-to-noise ratio of infrared targets and suppress background noise by constructing filters. Notable algorithms include the Tophat [8] filter, Maxmean [9] filter, etc. These methods offer high real-time processing speed and simplicity. However, when encountering complex cloud layers or sea clutter, they might diverge from the underlying algorithm assumptions, leading to elevated false alarms.

Saliency detection methods emphasize the differences in grayscale values between objects and backgrounds akin to spatial filtering. This category includes Gaussian difference filters, Gaussian Laplacian filters, Gabor filters, second-order directional derivative filters, and more. In 2013, Chen et al. proposed the Local Contrast Model (LCM) [10] based on the contrast differences between an image and its neighborhood. Although LCM is straightforward and effective, its effectiveness is limited to relatively bright targets against a background, and it might inadvertently enhance brighter noise, thereby resulting in a high false alarm rate. Subsequently, Wei et al. extended LCM to the MPCM model [11], which can detect both bright and dark targets but lacks adaptive threshold selection. Nevertheless, in complex scenarios, the prevalence of background interference often leads to high false alarm rates in these methods.

In recent years, low-rank sparse decomposition has garnered significant attention in the field of infrared small target detection due to its promising background suppression capabilities. IPI [12] is an exemplary model that treats targets as sparse matrices and backgrounds as low-rank components, constructing target patch models using local patches. Given that small targets occupy only a fraction of the entire image, this sparse hypothesis is suitable for diverse scenes. However, IPI preserves strong edge features of targets, resulting in time-consuming processing. Many researchers assume that targets come from multiple subspaces, leading to approaches like LRR [20], SMSL [21], and SRWS [13]. Zhang et al. [14] introduced the Partial Sum of the Tensor Nuclear Norm (PSTNN) model, employing the tensor nuclear norm and weighted L1 norm to suppress background effectively while retaining targets.

With the advancement of deep learning, researchers have increasingly turned to data-driven methods to address infrared weak small target detection challenges. Prominent algorithms include Fast RCNN [22–24], YOLO [25–30], as well as the Transformer initially applied in natural language processing. Unet and FPN are common network models in the field of infrared target detection, integrating high-level semantic information and low-level structural details. In 2021, Dai et al. introduced the ACM model [18], which combines non-local attention mechanisms with bidirectional pathways for cross-layer fusion. ISNet [19] employs Taylor finite difference edge blocks and bidirectional attention aggregation modules to ensure precise edge capture. AGPC [31] integrates an Attention-Guided Context Block (AGCB) to discern pixel correlations within and across distinct scales, and it further incorporates a Context Pyramid Module (CPM) for multiscale context information fusion. Li et al. [32] proposed a Dense Nested Attention Network (DNANet), where tri-directional interaction modules are densely nested, enhancing various depths of the UNet architecture. This augmentation significantly bolsters the model's performance in small target detection. However, the application of deep learning in this domain remains in the developmental phase due to the inherent sensitivity of infrared data.

2.1.2. Attention Mechanism

Attention mechanisms in deep learning refer to methods that redirect focus toward crucial regions within an image while disregarding irrelevant portions [33]. Attention mechanisms can be conceptualized as dynamic selection processes. These processes adaptively weight features based on their significance within the input, thus facilitating the generation of weighted features. Due to its excellent performance, attention mechanisms have been widely applied in neural networks, and various attention mechanisms are constantly evolving, such as SE-net [34], CBAM [35], EMANet [36], CCNet [37], and HamNet [38]. SEnet generates optimal feature maps through squeeze and excitation. CBAM enhances feature maps from both channel-wise and spatial-wise perspectives using channel attention modules and spatial attention modules. As the name suggests, the channel attention mechanism computes the significance level of each channel, making it commonly applied within convolutional neural networks. Among the established channel attention methods, the SENet model stands out. By learning inter-channel relationships, SENet enhances the network's expressive capability in feature representation, subsequently boosting model performance. The spatial attention mechanism shares a similar essence with the channel attention mechanism, with the former concentrating on capturing spatial importance by introducing attention modules. This enables the model to adaptively learn attention weights for distinct regions. This approach allows the model to emphasize crucial image areas while disregarding less significant regions. Notably, the Convolutional Block Attention Module (CBAM) is a prime example, seamlessly integrating channel and spatial attention. CBAM aims to augment the convolutional neural network's capacity to focus on image content. EMANet approaches self-attention from an Expectation Maximization (EM) perspective. It introduces EM attention, utilizing the EM algorithm to derive a concise set of basis vectors rather than utilizing all points as reconstruction bases. In the context of CCNet, self-attention operations are conceptualized as graph convolutions. Instead of densely connected graphs handled by self-attention, CCNet introduces sparsely connected graphs. To achieve this, it introduces cross attention, involving both row and column attention mechanisms to capture global information comprehensively.

2.1.3. Edge Information

Edge features in images have always been the focus of scholars' attention [39]. Recently, researchers have been exploring the integration of edge information to tackle challenges in semantic segmentation. For instance, RPCNet [40] introduced an iterative pyramid context module, merging semantic edge detection and semantic segmentation within a unified multi-task learning framework. GSCNN [41] integrates shape flow to explicitly extract edge information, embedding it into the regular flow features. DFN [42] devised a depth-supervised edge network for enhancing the prediction of semantic edge. However, works such as [41,43] often solely utilize the deepest layer features for representing both semantic segmentation heads and edge detection heads, overlooking the potential of hierarchical convolution features at different stages. Considering that high-level features aid in classifying object categories and low-level features preserve fine image structures, it is imperative to comprehensively exploit hierarchical features across multiple stages to enhance both semantic segmentation and boundary detection.

2.2. *Method*

In this section, we describe MIFNet in detail. The following subsections detail the overall architecture, edge information extractor, multiscale information fusion module, multiple information fusion module, and training loss function of the proposed MIFNet.

2.2.1. Overall Architecture

The overall framework of the infrared maritime small target detection network based on edge information assistance is shown in Figure 2. Firstly, in one branch, the infrared image is fed into the FPN backbone network, while in another branch, the infrared image is input into the edge information extraction module. At the same time, the infrared image is fed into the detail module to enhance the extraction of detailed information. The main network adopts the mainstream FPN architecture, with Resnet-20 as the backbone network. The detailed structure of the baseline network is shown in Table 1. Features are extracted from different layers to obtain x2, x3, and x4. The input infrared image undergoes stage one processing to obtain feature map x2. X2 is then fused with x6 through the MSF (multiscale information fusion module) module. x2 undergoes stage two processing to obtain feature maps x3 and x4, which are then fused into x5 through the MSF module. The MSF module is detailed in Section 2.1.3. In the MSF module, high-level feature maps with rich semantic information are processed by the Channel Attention and Pixel Attention modules to generate a weight that guides the fusion of different scale information contained in high-level and low-level feature maps extracted from the previous stage. Additionally, high-level feature maps with semantic information processed by the Channel Attention module can make the feature maps pay more attention to regions of interest in images.

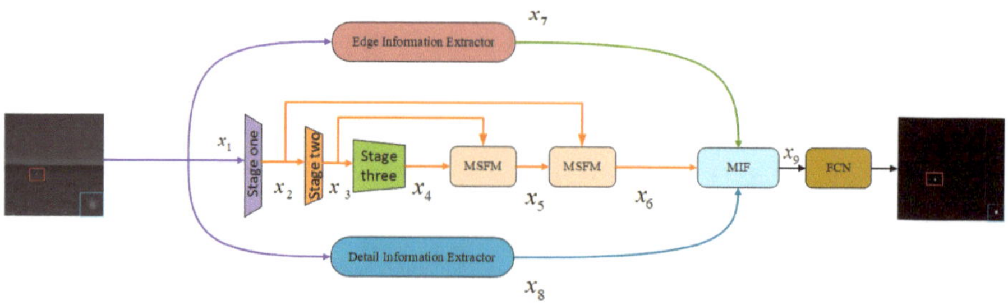

Figure 2. Overall Architecture.

Table 1. BNet backbones.

Stage	Output	Backbone
Stage one	$H \times W$	$\begin{bmatrix} 3 \times 3\text{conv}, 16 \\ 3 \times 3\text{conv}, 16 \end{bmatrix} \times b$
Stage two	$\frac{H}{2} \times \frac{W}{2}$	$\begin{bmatrix} 3 \times 3\text{conv}, 32 \\ 3 \times 3\text{conv}, 32 \end{bmatrix} \times b$
Stage three	$\frac{H}{4} \times \frac{W}{4}$	$\begin{bmatrix} 3 \times 3\text{conv}, 64 \\ 3 \times 3\text{conv}, 64 \end{bmatrix} \times b$

In addition, the input infrared image is processed by the edge information extraction module and the detail information extraction module, respectively, to obtain the edge and detail information of the infrared image. Finally, in the decision stage, the fusion is performed by the multidimensional information fusion module, and the binary image containing the target position information is obtained through the FCN module. Additionally, the input infrared image is individually processed by the edge information extraction module and the detail information extraction module, respectively, to obtain the edge information and the detail information of the infrared image. Then, in the decision stage, the fusion is performed by the multidimensional information fusion module, and the binary image containing the target position information is obtained through the FCN module. The pseudocode for the MIFNet is presented in Algorithm 1.

Algorithm 1: The Method Processing of an Image

Input: An Infrared Image
begin
Do abstract feature extraction
$X_2 = X_1 \otimes Resnet1$
$X_3 = X_2 \otimes Resnet2$
$X_4 = X_3 \otimes Resnet3$
$X_7 = X_1 \otimes SConv$
$X_8 = X_1 \otimes BConv$
End
Do FPN feature fusion
$X_5 = MSFM(X_4, X_3)$
$X_6 = MSFM(X_5, X_2)$
End
Do Multiple dimension information fusion
$X_9 = MIF(X_6, X_7, X_8)$
$result = FCN(X_9)$
End
Output: Binary Mask Image

MIFNet adopts FPN as the network framework and uses Resnet-20 as the backbone network to extract semantic features, as shown in Table 1. As illustrated in Figure 2, when an infrared image with a height of H and width of W is input into the network, it goes through the detail information extraction module, edge information extraction module, and semantic information extraction module separately. After processing and fusion, it outputs a binary mask image with target value 1. Due to its ability to simultaneously fuse high-level features with low resolution and high semantic information, as well as low-level features with high resolution but low semantic information, FPN has been widely applied in the field of object detection. In FPN, the high-resolution information from lower layers is transmitted to higher layers through cross-layer connections, enabling the network to integrate features at different scales and reduce information loss during downsampling processes.

2.2.2. Edge Information Extractor

During the downsampling process, the resolution of the image gradually decreases. We observed that edge information can be combined with deep learning methods to complement the loss of image resolution.

For all input infrared images, the processing flow of the edge information branch is as follows:

$$x_{i.v} = Conv2d(x_i, kernel_v) \tag{1}$$

$$x_{i,h} = Conv2d(x_h, kernel_h) \tag{2}$$

where $Conv2d$ represents the two-dimensional convolution operation, x_i is the i-th input channel, and $kernel_h$ and $kernel_v$ are the vertical and horizontal convolution kernels. In this paper, the configuration of the convolution kernel is as shown in Equation (1):

$$kernel_h = \begin{bmatrix} 0 & -1 & 0 \\ 0 & 0 & 0 \\ 0 & 1 & 0 \end{bmatrix} \tag{3}$$

$$kernel_v = \begin{bmatrix} 0 & -1 & 0 \\ 0 & 0 & 0 \\ 0 & 1 & 0 \end{bmatrix} \tag{4}$$

After calculating the horizontal and vertical gradients separately, the final gradient is calculated using the formula shown in Equation (2):

$$x_i = \sqrt{x_{i,v}^2 + x_{i,h}^2} \tag{5}$$

2.2.3. Multiscale Information Fusion Module

In the semantic information extraction module, we use cross-layer connections to fuse feature maps of different scales. Inspired by the non-local idea [18], we employ a non-modulation demodulation fusion mechanism to fuse information from different layers. Inspired by BiseNet [44] and DMANet [45], we adopt a two-branch architecture to use semantic information. MSF (Multiscale Information Fusion) module consists of Channel Attention and Pixel Attention, and the final fusion is performed by elementwise addition. The structure of Module MSFM is depicted in Figure 3.

Compared to other fusion modules, our MSFM module tends to focus more on semantic information, with shallow-level details serving as supplementary components. Due to the typically small proportion of infrared targets in images, they often lack detailed information and texture. Therefore, we have devised the Channel Attention mechanism to enhance targets by reinforcing channel-related information, which enhances the network's overall attention toward the targets. As shown in the diagram, the formula for Channel Attention can be summarized as follows:

$$MSFM(F_D, F_S) = (\alpha \otimes (CA(F_D))) \oplus (((1-\alpha) \otimes (F_S) \otimes CA(X_l))) \quad (6)$$

$$F_M = CA(F_D) == \sigma(\mathcal{B}(\mathbf{W}_2\delta(\mathcal{B}(\mathbf{W}_1 F_D)))) \quad (7)$$

$$\alpha = PA(F_M) = \sigma((\mathbf{W}\delta(\beta(\mathbf{W}F_M)))) \quad (8)$$

$\sigma, \delta, \beta, W, \oplus, \otimes$ represents the Sigmoid function, Rectified Linear Unit(ReLU), Batch Normalization(BN), Convolution function, add function, and element-wise multiplication, respectively. F_D represents the deep feature map containing semantic information, F_S represents the shallow feature map containing detail information, *CA* represents the channel attention mechanism, *PA* represents the pixel attention mechanism, α represents the dynamic weights obtained by fusing guidance information extracted from the deep feature map, and F_M represents F_D after channel attention.

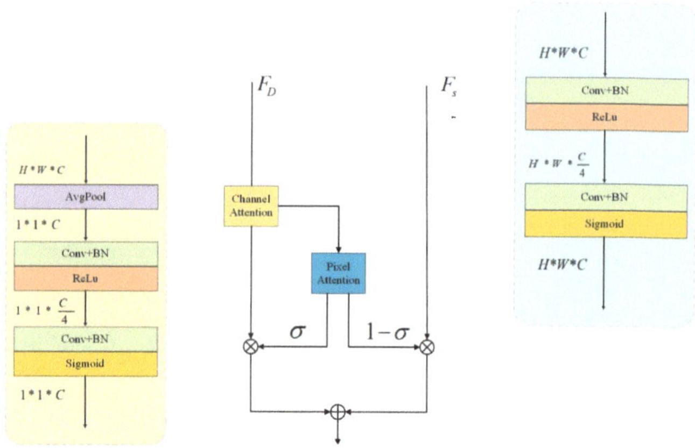

Figure 3. Multiscale Information Fusion Module.

2.2.4. Multiple Information Fusion Module

The MIF (Multiple Information Fusion) module is designed to guide the fusion of edge features and enrich detailed information using semantic information. As shown in the figure, the context branch contains rich semantic information but lacks specific spatial and geometric information. On the other hand, the detailed branch preserves relatively rich spatial details. Additionally, due to the differences between infrared targets and the environment, we introduce edge information as a supplement. The structural layout of Module MIF aligns with the visual representation presented in Figure 4.

The extracted semantic information is guided by weights obtained through global average pooling, convolution blocks, batch normalization, and the Sigmoid function. Based on these guided weights, different weights are assigned and multiplied into the edge information feature map and detail information feature map. The formula is as follows:

$$MIF(F_B, F_S, F_D) = Conv((\alpha \otimes F_B) \oplus (((1-\alpha) \otimes (F_D)))) \quad (9)$$

$$\alpha = SW(F_S) = \sigma((\beta(\mathbf{W}f_D))) \quad (10)$$

The formula for average pooling is as follows:

$$f_D = \mathcal{F}_{ap}(F_D) = \frac{1}{H \times W} \sum_{i=1,j=1}^{H,W} F[:, i, j] \tag{11}$$

σ, β, W, $\oplus$, $\otimes$ represents the Sigmoid function, Batch Normalization(BN), Convolution function, add function, and element-wise multiplication, respectively. F_B represents edge information, F_S represents semantic information, F_D represents detail information, SW represents the mechanism for generating semantic information guided weights, and α represents dynamic weights fused from the guided information extracted from semantic information.

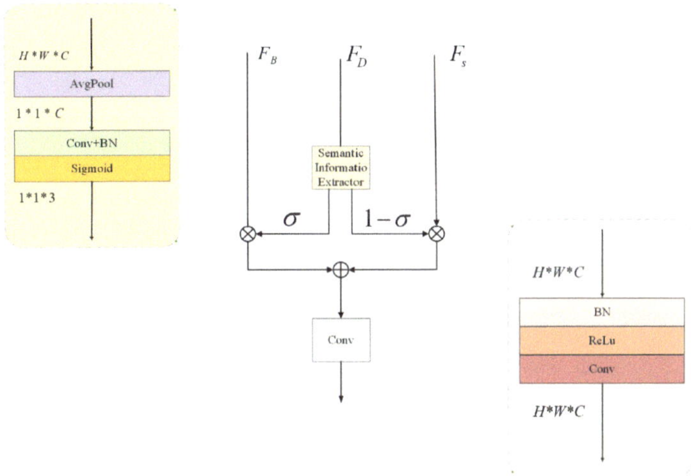

Figure 4. Multiple dimension Fusion Module.

2.2.5. Loss Function

Due to the issue of positive-negative sample imbalance between infrared small targets and background, we employ a soft margin loss function during the network training process. The formula for this loss function is defined as follows:

$$\mathcal{L}_{\text{soff}-\text{IoU}}(x, s) = \frac{\sum_{i,j} x_{i,j} \cdot s_{i,j}}{\sum_{i,j} s_{i,j} + x_{i,j} - x_{i,j} \cdot s_{i,j}} \tag{12}$$

$s \in \mathbb{R}^{H \times W}$ represents the predicted target score map, while $x \in \mathbb{R}^{H \times W}$ represents the annotated ground truth target image.

3. Results

In this section, a series of experiments were conducted to validate the effectiveness of our method. Firstly, we briefly introduced the evaluation metrics used to validate the algorithm. Then, the dataset used for validation was described. Subsequently, the algorithm was tested qualitatively and quantitatively. Finally, a series of ablation experiments were conducted to demonstrate the effectiveness of the added module. In the following equation, N represents the total number of samples, and TP, FP, TN, FN, T, and P, respectively, represent true positives, false positives, true negatives, false negatives, true positives, and positives.

3.1. Evaluation Metrics

(1) Intersection over Union (IoU) is a pixel-level evaluation metric used to assess the contour description capability of an algorithm [18]. It is calculated as the ratio of the intersection area between the predicted region and the ground truth label to the union area of the two, as shown below:

$$IoU = \frac{TP}{T + P - TP} \tag{13}$$

(2) Normalized Intersection over Union (nIoU) is a metric specifically designed to evaluate the performance of infrared small target detection [18], aiming to avoid the influence of large target segmentation on the evaluation metric. It is defined as follows:

$$nIoU = \frac{1}{N}\sum_{i}^{N}\frac{TP[i]}{T[i] + P[i] - TP[i]} \tag{14}$$

(3) The Receiver Operating Characteristic (ROC) curve is used to describe the relationship between the True Positive Rate (*TPR*) and the False Positive Rate (*FPR*) [19].

$$TPR = \frac{TP}{TP + FN} \quad FPR = \frac{FP}{FP + TN} \tag{15}$$

(4) The F1-measure [46] is a key metric used to evaluate algorithm performance. It calculates the harmonic mean of precision and recall, effectively capturing both dimensions. This metric encompasses the inherent trade-off relationship between precision and recall, thus forming a comprehensive performance evaluation scale that balances the two key evaluation metrics.

$$precision = \frac{TP}{TP + FP} \quad recall = \frac{FP}{TP + FN} \tag{16}$$

$$Fmeasure = \frac{2precsion \times recall}{precsion + recall} \tag{17}$$

3.2. Experiment Settings and Dataset

Dataset: In order to compare infrared small target detection algorithms in maritime scenes, we have constructed an infrared ship target dataset in maritime scenes. This dataset consists of six scenes, and the detailed information of the dataset is shown in Table 2. The total number of images used for training is 210, and there are 215 images used for testing. The dataset images have a size of 640 × 512 and overall appear dark, with small targets located in complex backgrounds. The backgrounds are blurry and primarily consist of sea clutter, islands, and other interference.

Table 2. Information of the test Dataset.

	Target Size	Target Category	Background Type
(a)	5 × 7	One small target with low local contrast	Calm sea
(b)	4 × 8, 5 × 6	Two small targets with low local contrast	Floating interface
(c)	5 × 5, 5 × 6, 5 × 7, 5 × 8	Three small targets with low local contrast	Calm sea
(d)	9 × 9, 5 × 6, 5 × 6	Three small targets with low local contrast	Wave clutter
(e)	7 × 7, 3 × 6, 9 × 9, 4 × 8, 4 × 7	Four small targets with low local contrast	Wave clutter
(f)	4 × 6, 4 × 5, 5 × 6	Three small targets with low local contrast	Dynamic camera

Experimental Details: ACMUNet [18], ACMFPN [18], AGPCNet [33], and our method are trained on an NVIDIA GTX 1080Ti with 12 GB of memory. We use Python as the programming language, Pycharm version 2022 as the editor, and PyTorch 1.8.0 as the deep learning framework. We choose Adagrad as the optimizer with a learning rate update strategy of CosineAnnealingLR. The initial learning rate is set to 0.05, the batch size is set to 4, and the total number of training epochs is set to 200. Traditional algorithms such as Tophat [8], Maxmean [9], MPCM [11], IPI [12], SRWS [13], and PSTNN [14] are implemented on an Intel i7-12700 CPU with 32 GB of memory using Matlab R2021b. The specific experimental parameters for traditional algorithms are shown in Table 3 [47].

Table 3. Computational complexity and running time of deep learning algorithms.

	ACMFPN	ACMUnet	AGPC	Ours
FLOPs	564.537 M	1.003 G	86.362 G	2.013 G
Params	386.615 K	519.271 K	12.360 M	397.666 K

3.3. *Equations Comparision to the State-of-the-Art Method*

To validate the effectiveness of our method, we compare our method with other advanced traditional and deep learning algorithms in both qualitative and quantitative aspects. The results are shown in Table 4, Figures 5–10.

Table 4. Parameter settings.

Method	Hyper-Parameters Settings
MaxMean	$\text{Patchsize}_{median} = 3\times 3$
Tophat	Patch size $= 3\times 3$
MPCM	window size = $\{3, 5, 7, 9\}$
PTSNN	Patch size : 40, Slide step : 40, $\lambda = 0.6/\sqrt{\max(n_1, n_2) * n_3}$
IPI	Patch size : 50×50, Slide $= 10$, $\lambda = L/\sqrt{\max(m,n)}, L = 4.5,, \varepsilon = 10^{-7}$
SRWS	Patch size : 50, Slide step : 10, $\beta = 1/\sqrt{\max(m,n)}$

Quantitative results: Table 5 presents the results of different algorithms. It is evident that our algorithm performs among the top algorithms tested and achieves the best detection results. From the quantitative results, we can observe that deep learning algorithms outperform traditional algorithms. The reasons for the suboptimal performance of the traditional algorithm are as follows:

1. Reliance on Manual Feature Design: Most traditional algorithms heavily rely on manually designed features. However, in maritime and aerial scenes, interference from factors such as reflected light from waves and islands can render the designed features incapable of meeting the requirements for distinguishing targets from the background. As a result, the comparative metrics are lower.

2. Fragmented Results in Visual Analysis: By combining the visualized results for analysis, it can be observed that the traditional detection algorithm exhibits a fragmented phenomenon in the results for ships, failing to recognize ships as a whole entity. This may result in lower detection metrics.

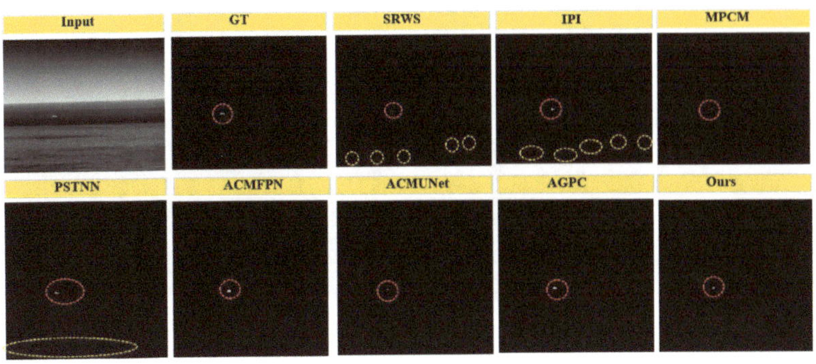

Figure 5. Detection results of infrared small targets in Scene 1 using different detection methods.

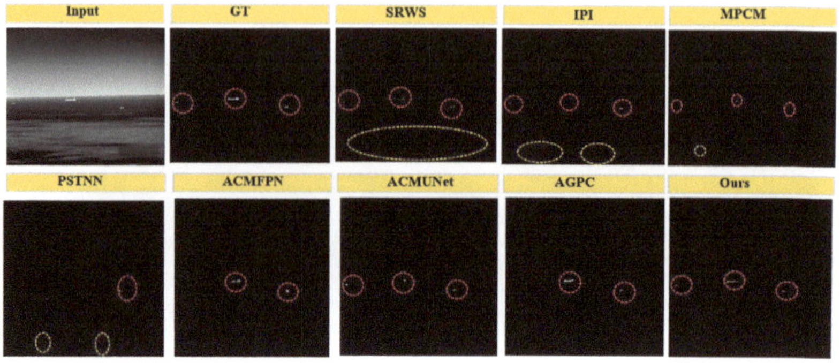

Figure 6. Detection results of infrared small targets in Scene 2 using different detection methods.

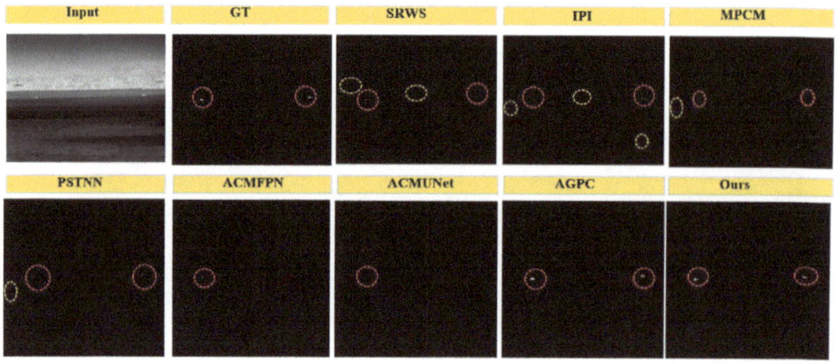

Figure 7. Detection results of infrared small targets in Scene 3 using different detection methods.

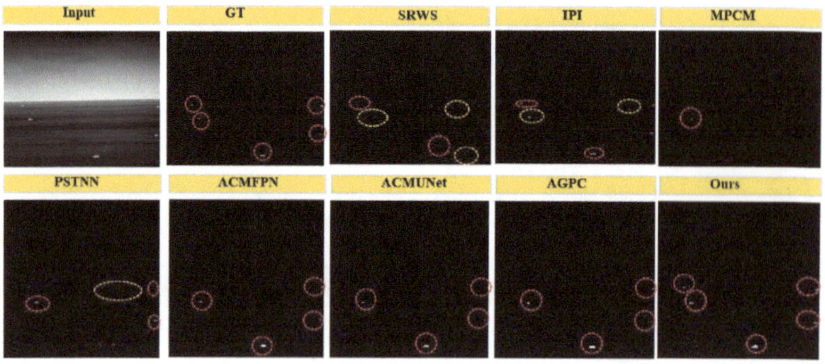

Figure 8. Detection results of infrared small targets in Scene 4 using different detection methods.

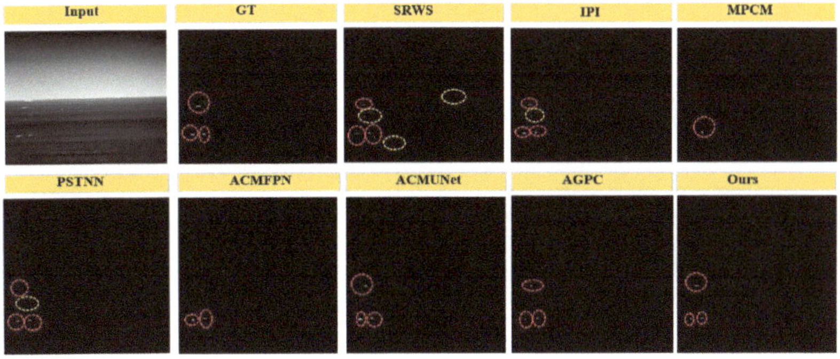

Figure 9. Detection results of infrared small targets in Scene 5 using different detection methods.

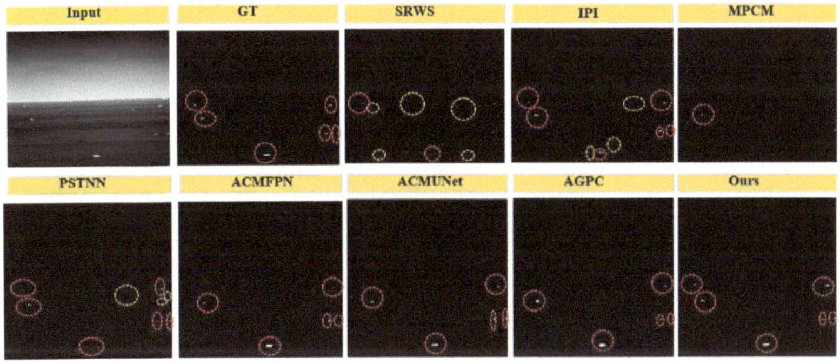

Figure 10. Detection results of infrared small targets in Scene 6 using different detection methods.

Table 5. Comparison with other state-of-the-art methods.

Method	IoU	nIoU	F1	AuC
Maxmean	0.12	3.54	0.23	54.45
Tophat	26.33	26.7	41.69	63.44
MPCM	11.58	12.49	20.75	55.80
IPI	48.05	48.17	64.91	77.23
PSTNN	43.51	44.22	60.64	74.52
SRWS	26.39	27.71	41.76	26.39
ACMFPN	72.66	72.97	84.19	95.75
ACMUnet	72.55	73.24	83.3	95.50
AGPC	77.61	78.13	83.39	95.58
Ours	79.09	79.43	87.88	95.96

Qualitative results: Figures 5–10 present the qualitative results of different detection methods on the infrared small target dataset in maritime and aerial scenes. The true targets are indicated by red bounding boxes in the images, while the false alarms are indicated by yellow bounding boxes. By comparing the detection results of each algorithm with the ground truth (GT) of the input image, it can be observed that the algorithms have missed detections.

As shown in Figure 5, the traditional algorithm SRWS fails to detect the targets, while Tophat, Maxmean, and PSTNN are able to detect the targets but with a large number of false alarms. On the other hand, deep learning algorithms such as ACMFPN, ACMUnet, AGPC, MPCM, and our algorithm successfully detect the targets. In Figure 6, SRWS, IPI, MPCM, PSTNN, ACMFPN, ACMUnet, and our algorithm are able to detect the targets. However, SRWS, IPI, and PSTNN algorithms have a large number of false alarms and can only detect the edges of the targets. In this case, both ACMFPN and AGPC algorithms have some missed detections., while ACMFPN can only detect the edges of the targets. ACMUnet and our algorithm successfully detect the targets, but ACMUnet's detection is incomplete. In Figure 7, SRWS, IPI, MPCM, PSTNN, ACMFPN, ACMUnet, and our algorithm are able to detect the two targets successfully but with a large number of false alarms. As for the deep learning detection results, ACMFPN and ACMUnet have some missed detections, while AGPC and our algorithm successfully detect the targets. In Figure 8, SRWS, IPI, MPCM, PSTNN, ACMUnet, AGPC, GNet, and other deep learning algorithms are able to detect the targets, but with some missed detections, while GNet successfully detects all the targets. SRWS, IPI, and PSTNN also have relatively high false alarm rates. In Figure 9, SRWS, IPI, PSTNN, ACMUnet, AGPC, GNet, and other algorithms are able to detect the targets, but SRWS, IPI, and PSTNN have relatively high false alarm rates, and ACMFPN has some missed detections. In Figure 10, SRWS, IPI, PSTNN, ACMFPN, ACMUnet, AGPC, GNet, and other algorithms are able to detect the targets, but SRWS, IPI, and PSTNN have relatively high false alarm rates, while ACMFPN, ACMUnet, and AGPC have some missed detections.

Figure 11 shows the ROC curves of the detection results obtained by different testing algorithms on the dataset. The larger the area under the ROC curve (AUC), the better the algorithm's performance. It can be observed that deep learning algorithms demonstrate better performance on the ROC curves compared to traditional algorithms. Additionally, our algorithm achieves top performance in terms of the AuC metric among the tested algorithms. The 3D visualization of the detection results simultaneously presents a more vivid display of the superior performance of our proposed algorithm (Appendix A Figure A1).

3.4. Ablation Study

In this phase, we designed a series of experiments to validate the potential benefits of the proposed network modules and design choices, ensuring the rationality of the contributions made by the components in our proposed model.

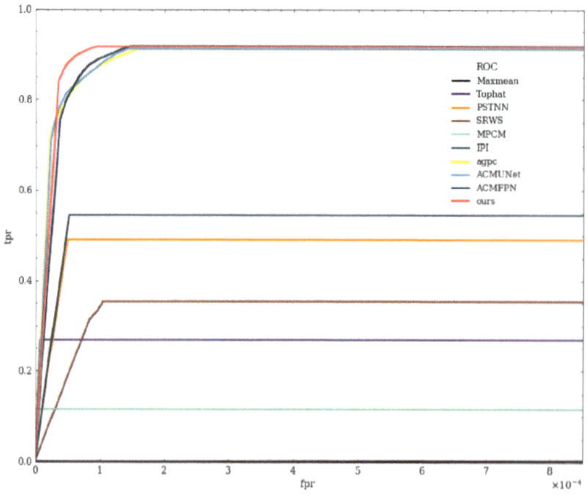

Figure 11. ROC curve.

1. Influence of edge information.

As shown in Table 6, incorporating edge information leads to an improvement of 1.8% and 2.19% in network IoU and nIoU, respectively. Since infrared small targets often occupy fewer pixels in the image, it is crucial to preserve and utilize the effective features of small targets in the network. Therefore, on the one hand, in order to provide the model with more dimensional information, and on the other hand, to compensate for the information loss caused by the downsampling process, we introduce edge information. Experimental results demonstrate that the inclusion of edge information enhances the network's performance and provides more features of infrared small targets.

Table 6. Ablation study on the module.

Method	IoU	nIoU
Baseline	72.66	72.97
Baseline + fusion	76.19	76.28
Baseline + sobel + fusion	77.99	78.47
Baseline + x1 + sobel + fusion	79.09	79.43

2. Influence of detailed information.

With the increase in network depth and the improvement in downsampling frequency, the resolution of the image decreases. In order to alleviate the information loss caused by downsampling operations, we introduce the finer details present in the shallow feature maps during the final fusion stage. As shown in Table 6, the introduction of detail information leads to an increase of 0.2% and 0.12% in network IoU and nIoU, respectively. The experimental results effectively demonstrate that the introduction of detail information in the final stage preserves some original features of the infrared small targets and compensates for the information loss caused by downsampling, thereby improving the network's performance.

3. Influence of Fusion module

As shown in Table 6, the introduction of the context interaction module resulted in an improvement of 3.53% and 3.31% in network IoU and nIoU, respectively. Compared to the fusion module in ACM, we focused more on extracting semantic information from high-level feature maps and guiding the fusion of high-level feature maps with semantic information and shallow-level feature maps with detailed information through the generation of weights. Additionally, an adversarial approach was employed to achieve an effective balance in the fusion. Furthermore, the Channel Attention mechanism was used to optimize the high-level feature maps. Experimental results on the infrared small target dataset in the maritime and sky scenes demonstrated that this fusion approach outperformed the fusion approach in ACM.

4. Influence of depth of layers

As shown in Table 7, we also considered the impact of network depth on the algorithm performance. When the number of Resnet blocks in the network was reduced from 4 to 3, IoU and nIoU decreased by 0.8% and 0.65%, respectively. Similarly, when the number of Resnet blocks in the network was reduced from 3 to 2, IoU and nIoU decreased by 0.32% and 0.38%, respectively. Furthermore, reducing the number of Resnet blocks in the network from 2 to 1 resulted in a decrease of 1.3% and 1.31% in IoU and nIoU, respectively. It can be observed that as the network depth decreases, the detection accuracy of the network also decreases.

Table 7. Ablation study on the depth of the network.

Depth	IoU	nIoU
1	76.60	77.09
2	77.90	78.40
3	78.22	78.78
4	79.09	79.43

4. Discussion

Comparing the visualization and quantitative results obtained by our proposed algorithm with other advanced algorithms, it is inevitable for model-driven algorithms relying on manually designed features to have false alarms and missed detections when facing complex background changes. Compared with other data-driven algorithms of the same type, our algorithm has stronger robustness and finer detection results. Meanwhile, the ablation experiments have demonstrated the effectiveness of the multiscale feature fusion module and multidimensional feature fusion module, which effectively utilizes multidimensional information in the process of target detection. In the future, we will explore the field of multidimensional information fusion, including how to represent each dimension of information and fusion methods.

5. Conclusions

This paper proposes an infrared maritime small target detection algorithm in maritime scenes. To overcome the issue of information loss during the typical downsampling process, we choose to simultaneously fuse detailed information and edge information. By comparing our algorithm with other state-of-the-art methods on the infrared small target detection dataset we established in maritime scenes, our algorithm performs at the top. Our algorithm achieves IoU, nIoU, F1, and AuC scores of 79.09%, 79.43%, 87.88%, and 95.96%, respectively. Visualizing the detection results in different scenes reveals that our algorithm has a lower false negative rate compared to other algorithms, indicating that our network can extract more effective information from the targets. Additionally, the results of the ablation experiments demonstrate that the adoption of attention mechanism-based cross-layer connection information fusion and the introduction of edge information and detail information contribute to improving the detection performance of the algorithm. In future work, we will explore the integration of different forms of information with deep learning techniques to address the issue of infrared small target detection. This article did not consider the algorithm performance on embedded platforms such as DSP, FPGA, ARM, etc., and the algorithm performance on embedded platforms will be further investigated in the future.

Author Contributions: Conceptualization, J.Y.; methodology, J.Y.; software, J.Y. and J.D.; validation, J.Y. and J.D.; formal analysis, J.Y., Q.D. and G.W.; investigation, J.Y. and S.X.; resources, J.Y.; data curation, J.Y. and J.D.; writing—original draft preparation, J.Y.; writing—review and editing, J.Y., S.X. and Q.D.; visualization, J.Y. and S.X.; supervision, G.W. and S.X.; project administration, H.T.; funding acquisition, H.T. All authors have read and agreed to the published version of the manuscript.

Funding: This work was supported by the National Natural Science Foundation of China (61921001).

Data Availability Statement: Not applicable.

Conflicts of Interest: The authors declare no conflict of interest.

Appendix A. Three-Dimensional Visualization Results of Different Methods on 6 Test Images

Figure A1 presents the 3D intensity visualization results of different testing algorithms, and it can be observed that our algorithm's detection results are closer to the ground truth.

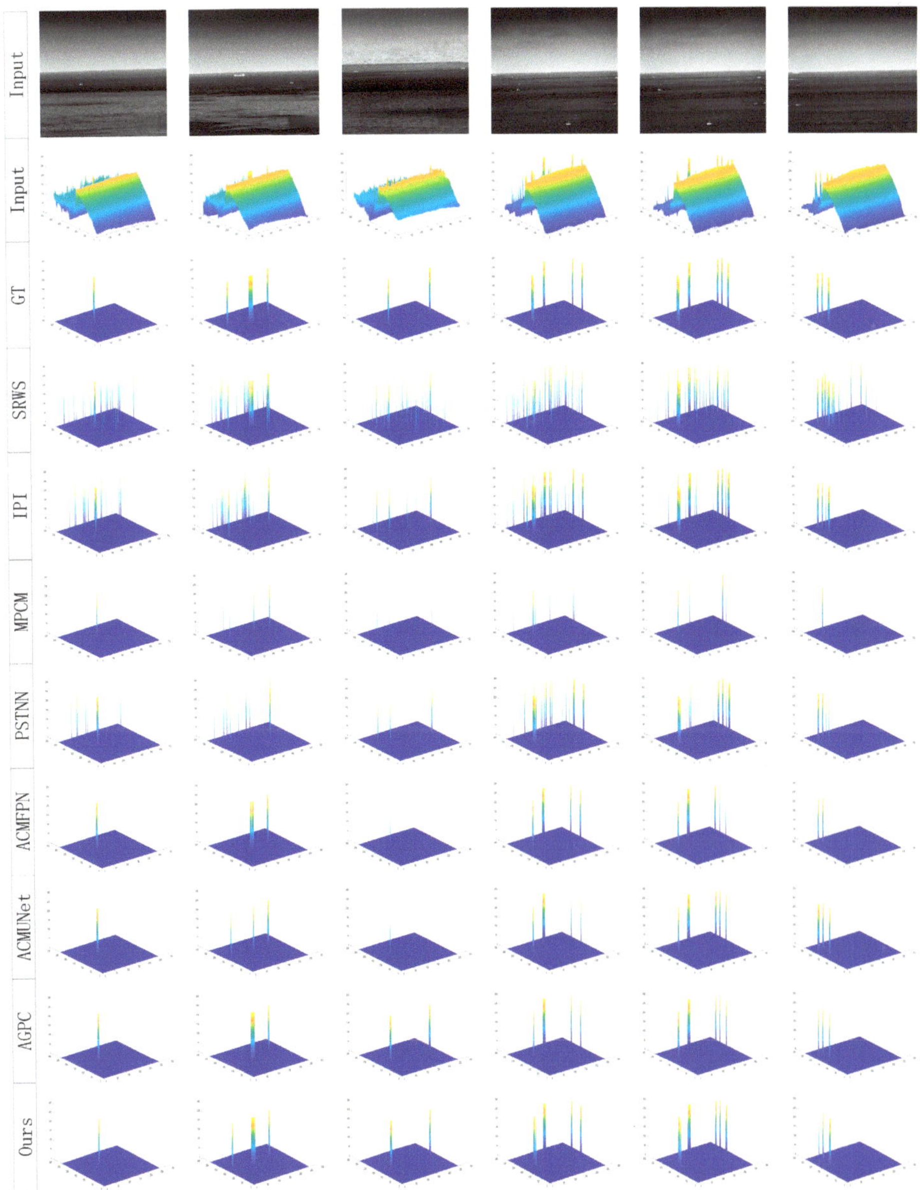

Figure A1. Three-dimensional visualization results of different methods on 6 test images.

References

1. Yang, C.; Ma, J.; Qi, S.; Tian, J.; Zheng, S.; Tian, X. Directional support value of gaussian transformation for infrared small target detection. *Appl. Opt.* **2015**, *54*, 2255–2265. [CrossRef] [PubMed]
2. Qi, H.; Mo, B.; Liu, F.; He, Y.; Liu, S. Small infrared target detection utilizing local region similarity difference map. *Infared Phys. Technol.* **2015**, *71*, 131–139. [CrossRef]
3. Pak, J. Visual odometry particle filter for improving accuracy of visual object trackers. *Electron. Lett.* **2020**, *56*, 884–887. [CrossRef]
4. Lin, G.; Fan, W. Unsupervised video object segmentation based on mixture models and saliency detection. *Neural Process. Lett.* **2020**, *51*, 657–674. [CrossRef]
5. Li, B.; Xu, Z.; Zhang, J.; Wang, X.; Fan, X. Dim-Small Target Detection Based on Adaptive Pipeline Filtering. *Math. Probl. Eng.* **2020**, *2020*, 8234349. [CrossRef]
6. Fu, J.; Zhang, H.; Luo, W.; Gao, X. Dynamic Programming Ring for Point Target Detection. *Appl. Sci.* **2022**, *12*, 1151. [CrossRef]
7. Zhao, M.; Li, W.; Li, L.; Hu, J.; Ma, P.; Tao, R. Single-Frame Infrared Small-Target Detection: A Survey. *IEEE Geosci. Remote Sens. Mag.* **2022**, *10*, 87–119. [CrossRef]
8. Rivest, J.F.; Fortin, R. Detection of dim targets in digital infrared imagery by morphological image processing. *Opt. Eng.* **1996**, *35*, 1886–1893. [CrossRef]
9. Deshpande, S.D.; Er, M.H.; Venkateswarlu, R.; Chan, P. Maxmean and max-median filters for detection of small targets. In *Signal and Data Processing of Small Targets*; International Society for Optics and Photonics: Bellingham, WA, USA, 1999; Volume 3809, pp. 74–83.
10. Chen, C.P.; Li, H.; Wei, Y.; Xia, T.; Tang, Y.Y. A local contrast method for small infrared target detection. *IEEE Trans. Geosci. Remote Sens.* **2013**, *52*, 574–581. [CrossRef]
11. Wei, Y.; You, X.; Li, H. Multiscale patch-based contrast measure for small infrared target detection. *Pattern Recognit.* **2016**, *58*, 216–226. [CrossRef]
12. Gao, C.; Meng, D.; Yang, Y.; Wang, Y.; Zhou, X.; Hauptmann, A.G. Infrared patch-image model for small target detection in a single image. *IEEE Trans. Image Process.* **2013**, *22*, 4996–5009. [CrossRef] [PubMed]
13. Zhang, T.; Peng, Z.; Wu, H.; He, Y.; Li, C.; Yang, C. Infrared small target detection via self-regularized weighted sparse model. *Neurocomputing* **2021**, *420*, 124–148. [CrossRef]
14. Zhang, L.; Peng, Z. Infrared small target detection based on partial sum of the tensor nuclear norm. *Remote Sens.* **2019**, *11*, 382. [CrossRef]
15. Haq, M.A.; Hassine, S.B.H.; Malebary, S.J.; Othman, H.A.; Tag-Eldin, E.M. 3D-cnnhsr: A 3-dimensional convolutional neural network for hyperspectral super-resolution. *Comput. Syst. Sci. Eng.* **2023**, *47*, 2689–2705.
16. Haq, M.A. CNN based automated weed detection system using uav imagery. *Comput. Syst. Sci. Eng.* **2022**, *42*, 837–849.
17. Stupariu, M.-S.; Cushman, S.A.; Pleşoianu, A.-I.; Pătru-Stupariu, I.; Fuerst, C. Machine learning in landscape ecological analysis: A review of recent approaches. *Landsc. Ecol.* **2022**, *37*, 1227–1250. [CrossRef]
18. Dai, Y.; Wu, Y.; Zhou, F.; Barnard, K. Asymmetric contextual modulation for infrared small target detection. In Proceedings of the IEEE/CVF Winter Conference on Applications of Computer Vision, Virtual, 5–9 January 2021; pp. 950–959.
19. Zhang, M.; Zhang, R.; Yang, Y.; Bai, H.; Zhang, J.; Guo, J. ISNet: Shape matters for infrared small target detection. In Proceedings of the IEEE/CVF Conference on Computer Vision and Pattern Recognition (CVPR), New Orleans, LA, USA, 18–24 June 2022; pp. 877–886.
20. Li, M.; He, Y.J.; Zhang, J. Small infrared target detection based on low-rank representation. In *Image and Graphics*; Springer: Berlin/Heidelberg, Germany, 2015; pp. 393–401.
21. Wang, X.; Peng, Z.; Kong, D.; He, Y. Infrared dim and small target detection based on stable multisubspace learning in heterogeneous scene. *IEEE Trans. Geosci. Remote Sens.* **2017**, *55*, 5481–5493. [CrossRef]
22. Girshick, R.; Donahue, J.; Darrell, T. Rich feature hierarchies for accurate object detection and semantic segmentation. In Proceedings of the IEEE Conference on Computer Vision and Pattern Recognition, Columbus, OH, USA, 23–28 June 2014; pp. 580–587.
23. Girshick, R. Fast r-cnn. In Proceedings of the IEEE International Conference on Computer Vision, Santiago, Chile, 7–13 December 2015; pp. 1440–1448.
24. He, K.; Gkioxari, G.; Dollár, P. Mask r-cnn. In Proceedings of the IEEE International Conference on Computer Vision, Venice, Italy, 22–29 October 2017; pp. 2961–2969.
25. Redmon, J.; Divvala, S.; Girshick, R. You only look once: Unified, real-time object detection. In Proceedings of the IEEE Conference on Computer Vision and Pattern Recognition, Las Vegas, NV, USA, 27–30 June 2016; pp. 779–788.
26. Redmon, J.; Farhadi, A. YOLO9000: Better, faster, stronger. In Proceedings of the IEEE Conference on Computer Vision and Pattern Recognition, Honolulu, HI, USA, 21–26 July 2017; pp. 7263–7271.
27. Redmon, J.; Farhadi, A. Yolov3: An incremental improvement. *arXiv* **2018**, arXiv:1804.02767.
28. Bochkovskiy, A.; Wang, C.Y.; Liao, H.Y.M. Yolov4: Optimal speed and accuracy of object detection. *arXiv* **2020**, arXiv:2004.10934.
29. Li, C.; Li, L.; Jiang, H. YOLOv6: A single-stage object detection framework for industrial applications. *arXiv* **2022**, arXiv:2209.02976.
30. Wang, C.Y.; Bochkovskiy, A.; Liao, H.Y.M. YOLOv7: Trainable bag-of-freebies sets new state-of-the-art for real-time object detectors. *arXiv* **2022**, arXiv:2207.02696.

31. Zhang, T.; Li, L.; Cao, S.; Pu, T.; Peng, Z. Attention-Guided Pyramid Context Networks for Detecting Infrared Small Target Under Complex Background. *IEEE Trans. Aerosp. Electron. Syst.* **2023**, *59*, 4250–4261. [CrossRef]
32. Li, B.; Xiao, C.; Wang, L.; Wang, Y.; Lin, Z.; Li, M.; An, W.; Guo, Y. Dense Nested Attention Network for Infrared Small Target Detection. *IEEE Trans. Image Process.* **2023**, *32*, 1745–1758. [CrossRef] [PubMed]
33. Lee, M. The geometry of feature space in deep learning models: A holistic perspective and comprehensive review. *Mathematics* **2023**, *11*, 2375. [CrossRef]
34. Hu, J.; Shen, L.; Sun, G. Squeeze-and-excitation neworks. In Proceedings of the IEEE Conference on Computer Vision and Pattern Recognition (CVPR), Salt Lake City, UT, USA, 18–22 June 2018; pp. 7132–7141.
35. Woo, S.; Park, J.; Lee, J.; Kweon, I.S. Cbam: Convolutional block attention module. In Proceedings of the European Conference on Computer Vision (ECCV), Munich, Germany, 8–14 September 2018; pp. 3–19.
36. Li, X.; Zhong, Z.; Wu, J.; Yang, Y.; Lin, Z.; Liu, H. Expectation-maximization attention networks for semantic segmentation. In Proceedings of the IEEE/CVF International Conference on Computer Vision, Seoul, Republic of Korea, 27 October–2 November 2019.
37. Huang, Z.; Wang, X.; Wei, Y.; Huang, L.; Shi, H.; Liu, W.; Huang, T.S. Ccnet: Criss-cross attention for semantic segmentation. In Proceedings of the 2019 IEEE/CVF International Conference on Computer Vision (ICCV), Seoul, Republic of Korea, 27 October–2 November 2020.
38. Abdulqader, D.A.; Hathal, M.S.; Mahmmod, B.M.; Abdulhussain, S.H.; Al-Jumeily, D. Plain, edge, and texture detection based on orthogonal moment. *IEEE Access* **2022**, *10*, 114455–114468. [CrossRef]
39. Geng, Z.; Guo, M.-H.; Chen, H.; Li, X.; Wei, K.; Lin, Z. Is attention better than matrix decomposition? In Proceedings of the International Conference on Learning Representations, Virtual, 3–7 May 2021.
40. Zhen, M.; Wang, J.; Zhou, L.; Li, S.; Shen, T.; Shang, J.; Fang, T.; Quan, L. Joint semantic segmentation and boundary detectionusing iterative pyramid contexts. In Proceedings of the IEEE/CVF Conference on Computer Vision and Pattern Recognition (CVPR), Seattle, WA, USA, 13–19 June 2020; pp. 13666–13675.
41. Takikawa, T.; Acuna, D.; Jampani, V.; Fidler, S. Gated-SCNN: Gatedshape CNNs for semantic segmentation. In Proceedings of the IEEE/CVF International Conference on Computer Vision (ICCV), Seoul, Republic of Korea, 27 October 2019–2 November 2019; pp. 5229–5238.
42. Yu, C.; Wang, J.; Peng, C.; Gao, C.; Yu, G.; Sang, N. Learninga discriminative feature network for semantic segmentation. In Proceedings of the IEEE/CVF Conference on Computer Vision and Pattern Recognition (CVPR), Salt Lake City, UT, USA, 18–23 June 2018; pp. 1857–1866.
43. Yuan, Y.; Xie, J.; Chen, X.; Wang, J. SegFix: Model-agnostic boundary refinement for segmentation. In Proceedings of the European Conference on Computer Vision (ECCV), Glasgow, UK, 23–28 August 2020; pp. 489–506.
44. Yu, C.; Wang, J.; Peng, C.; Gao, C.; Yu, G.; Sang, N. BiSeNet: Bilateral Segmentation Network for Real-Time Semantic Segmentation. In *Computer Vision—ECCV 2018, Proceedings of the 15th European Conference, Munich, Germany, 8–14 September 2018*; Ferrari, V., Hebert, M., Sminchisescu, C., Weiss, Y., Eds.; Springer International Publishing: Cham, Switzerland, 2018; pp. 334–349.
45. Sun, X.; Xie, Y.; Jiang, L.; Cao, Y.; Liu, B. DMA-Net: DeepLab With Multi-Scale Attention for Pavement Crack Segmentation. *IEEE Trans. Intell. Transport. Syst.* **2022**, *23*, 18392–18403. [CrossRef]
46. Faisal, M.M.; Mohammed, M.S.; Abduljabar, A.M.; Abdulhussain, S.H.; Mahmmod, B.M.; Khan, W.; Hussain, A. Object detection and distance measurement using AI. In Proceedings of the 2021 14th International Conference on Developments in eSystems Engineering (DeSE), Sharjah, United Arab Emirates, 7–10 December 2021; pp. 559–565.
47. Lv, G.; Dong, L.; Liang, J.; Xu, W. Novel Asymmetric Pyramid Aggregation Network for Infrared Dim and Small Target Detection. *Remote Sens.* **2022**, *14*, 5643. [CrossRef]

remote sensing

MDPI

Article

A Multi-Stream Attention-Aware Convolutional Neural Network: Monitoring of Sand and Dust Storms from Ordinary Urban Surveillance Cameras

Xing Wang [1,2,3], Zhengwei Yang [1,*], Huihui Feng [4], Jiuwei Zhao [5], Shuaiyi Shi [6] and Lu Cheng [7]

1 School of Atmosphere Science, Nanjing University, Nanjing 210023, China; wangxing@nju.edu.cn
2 Key Laboratory of Meteorological Disaster (KLME), Ministry of Education & Collaborative Innovation Center on Forecast and Evaluation of Meteorological Disasters (CIC-FEMD), Nanjing University of Information Science & Technology, Nanjing 210044, China
3 School of Computer Engineering, Nanjing Institute of Technology, Nanjing 211167, China
4 School of Geosciences and Info-Physics, Central South University, Changsha 410083, China; hhfeng@csu.edu.cn
5 School of Atmospheric Sciences, Nanjing University of Information Science and Technology, Nanjing 210044, China; jiuwei@nuist.edu.cn
6 State Key Laboratory of Remote Sensing Science, Aerospace Information Research Institute, Chinese Academy of Sciences, Beijing 100094, China; shisy01@radi.ac.cn
7 School of Geography Science and Geomatics, Suzhou University of Science and Technology, Suzhou 215009, China; lucheng@usts.edu.cn
* Correspondence: yang.zhengwei@nju.edu.cn; Tel.: +86-136-5519-2849

Abstract: Sand and dust storm (SDS) weather has caused several severe hazards in many regions worldwide, e.g., environmental pollution, traffic disruptions, and human casualties. Widespread surveillance cameras show great potential for high spatiotemporal resolution SDS observation. This study explores the possibility of employing the surveillance camera as an alternative SDS monitor. Based on SDS image feature analysis, a Multi-Stream Attention-aware Convolutional Neural Network (MA-CNN), which learns SDS image features at different scales through a multi-stream structure and employs an attention mechanism to enhance the detection performance, is constructed for an accurate SDS observation task. Moreover, a dataset with 13,216 images was built to train and test the MA-CNN. Eighteen algorithms, including nine well-known deep learning models and their variants built on an attention mechanism, were used for comparison. The experimental results showed that the MA-CNN achieved an accuracy performance of 0.857 on the training dataset, while this value changed to 0.945, 0.919, and 0.953 in three different real-world scenarios, which is the optimal performance among the compared algorithms. Therefore, surveillance camera-based monitors can effectively observe the occurrence of SDS disasters and provide valuable supplements to existing SDS observation networks.

Keywords: sand and dust storm; surveillance camera; deep learning; attention mechanism

Citation: Wang, X.; Yang, Z.; Feng, H.; Zhao, J.; Shi, S.; Cheng, L. A Multi-Stream Attention-Aware Convolutional Neural Network: Monitoring of Sand and Dust Storms from Ordinary Urban Surveillance Cameras. *Remote Sens.* **2023**, *15*, 5227. https://doi.org/10.3390/rs15215227

Academic Editor: Yuriy Kuleshov

Received: 26 September 2023
Revised: 26 October 2023
Accepted: 28 October 2023
Published: 3 November 2023

1. Introduction

Sand and dust storms (SDS) are typical weather disasters caused by strong and turbulent winds entraining dust particles into the air, severely affecting the environment, agriculture, industry, and human health for a long time [1]. Central Asia, Northern Africa, the Middle East, North America, and Australia are considered the primary sources of SDSs worldwide during the spring, winter, and early summer [2]. Since the extensive losses and damage caused by SDS events, the early warning and monitoring of SDS have attracted considerable attention. As early as 2007, the World Meteorological Organization launched the Sand and Dust Storm Warning Advisory and Assessment System (https://public.wmo.int/en/our-mandate/focus-areas/environment/SDS/warnings) (accessed on 1 January 2023) to enhance the ability of global countries to deliver timely and quality

SDS forecasts, observations, information, and knowledge to human beings [3]. After decades of efforts, several technologies or tools are available and can be classified as ground- and space-based monitoring technologies [4]. However, the rapid and small-scale characteristics of some SDS events pose challenges to the current monitoring system:

(1) Traditional ground-based methods, like lookout towers, are considered more objective and directed observations. With advances in industrial manufacturing and sensing technologies, ground radars, lidars, and wireless sensor networks become essential options for SDS monitoring tasks while being irreplaceable parts in the calibration and integration of remote sensing-based measurements [2,5,6]. Limited by cost, operating requirements, and deployment conditions, these ground-based observations usually cannot be deployed at a high density over a large spatial scale, resulting in a lack of spatial representation of observation results.

(2) Space-based technologies mainly refer to satellite-based remote sensing methods, which are important in monitoring, assessing, and predicting SDS events. The well-known Moderate Resolution Imaging Spectroradiometer (MODIS) and Total Ozone Mapping Spectrometer (TOMS) satellites [7] enable passive detection of the occurrence and transportation of an SDS on the vertical distribution, while Cloud-Aerosol Lidar and Infrared Pathfinder Satellite Observation (CALIPSO) satellites [8] can actively sense the vertical profile of cloud aerosols and inverse the vertical information of SDSs. However, the satellite is more suitable for large-scale and long-term SDS monitoring tasks. With the enhancement of satellite imagery resolution and the number of satellites, the problem of insufficient spatial and temporal resolution has been gradually alleviated. However, rapid sensing of small-scale dust storm events still needs to be improved.

The ground camera enables continuous recording of changes in the scene and rapid feedback on the emergence of SDSs, which researchers have noticed for its high temporal resolution observation advantage [5]. For example, Bukhari [9] attempted to use ground cameras to capture the startup phase of large-scale SDSs. In addition, some researchers have used what the cameras capture as evidence of dust storm events: Narasimhan and Nayar [10] analyzed the visual manifestations of haze and fog weather closer to SDSs and gave an essential reference for the visual-based SDS events identification and images reconstruction during SDS conditions; Chavez Jr et al. [11] deployed some cameras to take photos when the winds exceeded a pre-selected threshold to augment remote sensing satellites in order to enhance their capability of detecting SDSs; Dagsson-Waldhauserova et al. [12] combined images captured by surveillance cameras and portable dust measurement instruments to identify the extent, magnitude, and grain size characteristics of SDS events in southwestern Iceland; Urban et al. [13] used the images captured by ground remote cameras to estimate the amount and frequency of dust emissions from a setting in the Mojave Desert, USA. Although the above three works manually reviewing the imagery are practical, visually interpreting imagery is laborious, repetitive, and time-consuming, which could be more friendly to the operators, especially for long-term observations/images. To achieve automatic monitoring purposes, Gutierrez [6] proposed a feed-forward neural network for the automatic classification of SDS events from camera frames.

Generally, SDS videos/images captured by purposely placed surveillance cameras provide redundant ground-based observations serving to correct SDS information sensed by other means, such as space-based sensing results, and the potential of the surveillance camera network as a stand-alone SDS monitoring or early warning system needs to be further exploited. According to the authors, the reasons for this are the following: (1) The limited field of view dictates that ground cameras can detect SDSs at a small scale, but it is difficult to describe the full extent of large-scale SDSs. However, large-area deployment of cameras is costly, which is a bottleneck problem for ground camera-based SDS monitoring networks. (2) Existing ground camera-based SDS monitoring efforts are inadequately automated and still require high labor costs, while the complex and dynamic urban surveillance scenarios present further challenges to the accuracy of SDS detection algorithms, which require the development of algorithms specifically for SDS monitoring from surveillance cameras.

As the smart city develops, widespread surveillance cameras have become essential for traffic, security, emergency management, etc. According to the survey by Comparitech, approximately 770 million surveillance cameras are deployed worldwide (https://www.comparitech.com/) (accessed on 1 January 2023). Such a high-density surveillance camera network provides the hardware basis for high spatial resolution SDS observation; meanwhile, 4G/5G communication technology provides the software basis for high temporal resolution SDS video transmission. Therefore, in theory, surveillance cameras have great potential for high spatiotemporal resolution SDS event monitoring. On the other hand, with the development of deep learning technology, CNN-based algorithms have achieved satisfactory performance in object detection tasks, which brings new opportunities to surveillance camera-based SDS detectors. In such a context, we used deep learning to build a surveillance camera-based SDS monitor. The main contributions of this study are as follows:

(1) The designed monitoring system is deployed on existing urban surveillance resources rather than specifically deploying cameras, thus having higher reliability and obvious low-cost advantages. To the best of our knowledge, prior efforts did not use ordinary urban surveillance cameras for such a task.
(2) A novel Multi-Stream Attention-aware Convolutional Neural Network (MA-CNN) is proposed, which learns SDS image features at different scales through a multi-stream structure and employs the attention mechanism to achieve satisfactory performance for accurate and automatic SDS identification from complex and dynamic surveillance scenarios.
(3) For deep learning model training and testing, sand and dust storm image (SDSI), a new dataset consisting of 13,216 images (6578 SDS images and 6638 non-SDS images taken from scenes similar to SDS scenarios), was constructed.

The rest of this paper is organized as follows: Following this introduction, we detail the proposed MA-CNN in Section 2; we discuss the experimental results in Section 3 and conclude in Section 4.

2. Methodology

Intuitively, SDS monitoring via surveillance cameras can be considered a typical object detection issue in computer vision, i.e., detecting the occurrence of SDSs from surveillance videos, which can be divided into two parts: selecting image features and constructing the detection algorithm. Both of the above two steps contribute to SDS detection performance.

2.1. Image Features Selection

In meteorology, the definition of SDS weather levels is mainly based on horizontal visibility while referring to wind strength during SDS occurrences. A higher SDS level is associated with higher wind speed and lower visibility, and vice versa. It was observed that there were differences in the impact of different levels of SDS on surveillance images/videos.

For a lower-level SDS event, similar to fog, smog, and haze weather events, its appearance commonly causes a significant reduction in the sharpness and contrast of images captured by outdoor surveillance cameras, resulting in color shifting and missing detail problems [14]. The degradation effect caused by these weather events can be generically expressed as follows [10,15,16]:

$$Intensity(x) = J(x)t(x) + A(1 - t(x)) \tag{1}$$

where $Intensity(x)$ is the observed intensity, $J(x)$ is the scene radiance, $A(\bullet)$ is the global atmospheric light, and $t(x)$ is the medium transmission describing the portion of the light that is not scattered and reaches the surveillance camera.

Previous studies have pointed out that the radius of particles in SDSs is close to 25 μm, which is much larger than that in haze (0.01–1 μm) and fog (1–10 μm) conditions [10,17]. Since particles of different sizes show varying reflection interactions with visible light, the degradation effects of SDS and the other three weather conditions differ slightly. Nevertheless, the generic model (i.e., Equation (1)) still works, so images taken in these environments are hard to discriminate.

Visually, these weather events change the color of the surveillance image. Researchers were quick to build single or multiple color channels (classical ones like dark channels [18]) that are sensitive to weather conditions, thus establishing a prior or assumption in terms of color space for fog, haze, smog, and SDS removal and image enhancement tasks. Single channels of RGB [19,20], HSV [21], and YIQ [22] color spaces are popular options. As shown in Figure 1, surveillance images captured under haze, fog, smog, and SDS weather conditions were randomly selected. Their effects on the surveillance images in HSV color space were compared. Figure 1 illustrates that the listed weather events blurred the natural images. Moreover, the Hue color channel exhibits a single-peak pattern in the weather-influenced images and can be considered as a color prior or assumption of these blurred images. To date, image deblurring algorithms still favor this strategy highly and have made significant progress [17,19,23,24].

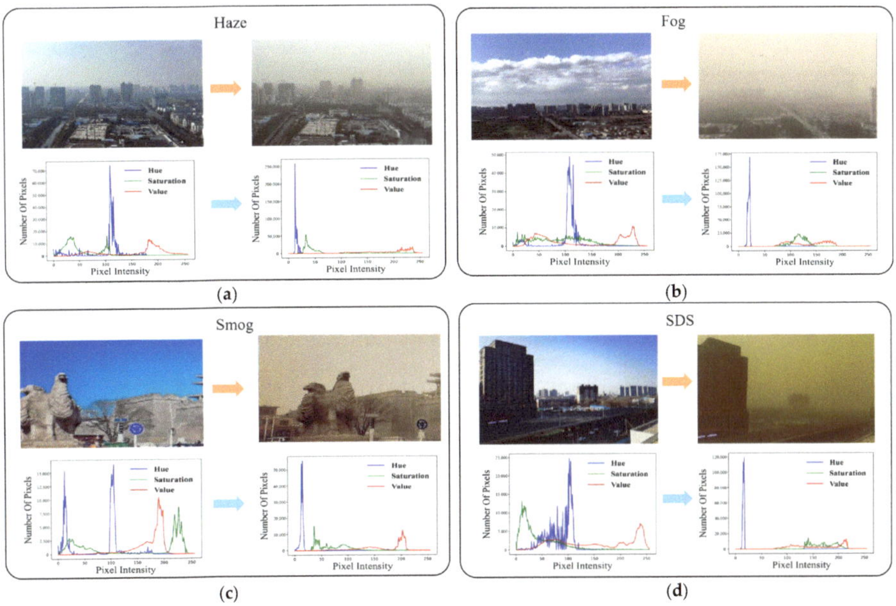

Figure 1. Comparison of HSV histogram of different weather conditions. (**a**) Haze; (**b**) Fog; (**c**) Smog; (**d**) SDS.

In addition to the images taken during the above weather, we found that similar surveillance images also have single peaks in the Hue channel, as shown in Figure 2, making it more challenging to accurately distinguish SDSs in surveillance images.

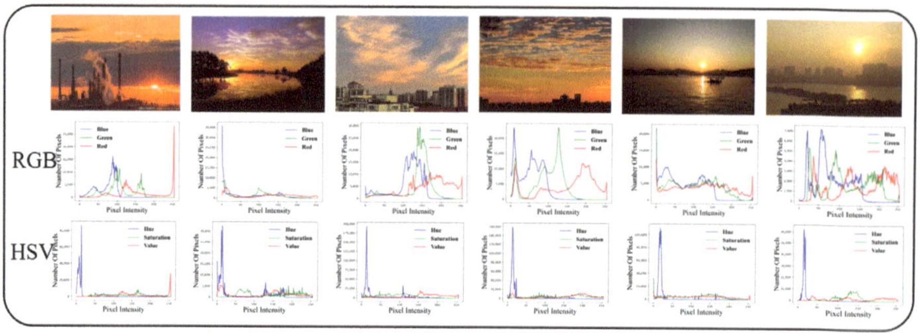

Figure 2. Comparison of HSV histogram of similar surveillance scenarios.

Essentially, the purpose of the deblurring studies is to realize image restoration, which solves the problem of inconsistency between SDS images and the corresponding natural images. In contrast, our study focuses on monitoring the appearance of SDS via surveillance cameras. The crucial point is accurately distinguishing those surveillance scenarios that are similar to SDS. That is, sharing similar image characteristics with fog, haze, smog, etc., makes it more challenging to accurately distinguish SDS visually.

High-level SDS usually move fast together with low visibility. Here, a surveillance video recording the complete process of SDS from generation to transiting is collected, whose key frames are shown in Figure 3. While still using the image HSV color space as an example, it can be observed that as the SDS moves and the proportion of sand in the image changes, accompanied by changes in the characteristics of the HSV color space (for example, in the first two images the single-peak pattern of the Hue channel is not apparent). Therefore, the image features of high-level SDS events are dynamically changing, resulting in the fixed manual image features needing to be more manageable to describe thoroughly the randomly occurring SDS events in nature.

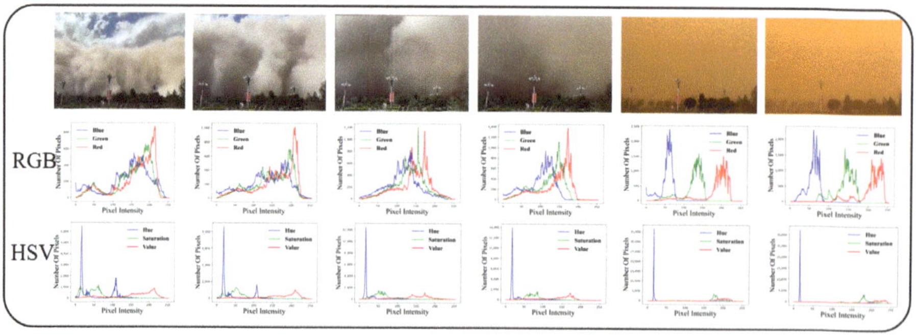

Figure 3. Comparison of HSV histogram of moving SDS events.

Given the dynamic behavior of SDSs and the high similarity of image features with relevant scenarios, constructing image features based on specific channels for a robust and accurate description of SDSs poses significant challenges. Inspired by [17], we consider the three channels (R, G, B) to be related and not independent. Therefore, taking the original image as an inseparable indivisibility in the form of a tensor for the input of the following detection algorithm.

2.2. Multi-Stream Attention-Aware CNN Network

As described in Section 2.1, how to detect the occurrence of SDSs from surveillance images, mainly to distinguish SDS events from similar scenarios, becomes a fundamental problem for a detector to solve. With the development of deep learning technology, CNN-based algorithms have achieved satisfactory performance in object detection and classification tasks, bringing new opportunities to construct surveillance camera-based SDS detectors. On the other hand, inspired by the biological systems of humans that tend to focus on the distinctive parts when processing large amounts of information, the attention mechanism has the advantage of solving the problem of information overload and improving the accuracy of results [25,26]. As one of the latest advancements in deep learning, the attention mechanism has become an essential tool in deep learning [27,28].

In this section, the attention mechanism is employed to locate the most salient components of the feature maps in CNNs and to remove the redundancy. Then, a Multi-Stream Attention-aware CNN Network (MA-CNN) is proposed for SDS detection from surveillance images, as shown in Figure 4. In MA-CNNs, the input data are simultaneously fed into multiple streams with different scales. In each stream, the features acquired by the stacked 2D CNN layers are fed into the spatial attention layer to acquire more focused SDS-sensitive features. Then the two layers are concatenated as input to a deeper layer.

Considering a CNN with L layers, the hidden state of layer l is represented as H_l, where $l\epsilon\{1,2\ldots,L\}$. In this notation, the input image can be defined as H_1. The convolutional layer consists of two parts learnable parameters, that is, the weight matrix W_l that connects layer l and layer $l-1$ and the bias term vector b_l. Hence, each neuron in convolutional layer l is only connected with a local region on H_{l-1} and W_l is shared among all spatial locations. In order to improve translation invariance and representation capability, convolutional layers are interleaved with point-wise nonlinearity (i.e., Leaky ReLU, whose parameter alpha = 0.1 [29]) and nonlinear down-sampling operation (i.e., 2D max pooling). The feature map of l can be obtained by the following:

$$F_1 = Max\_pooling(Leaky\_ReLU(H_{l-1} \ * W_l + b_l)),\ l = 1,2,\ldots,L \tag{2}$$

where $*$ denotes the 2D convolution operation.

As shown in Figure 4, in order to enable the proposed network to learn richer semantic representations of the input image progressively, the number of channels in the outputs of successive blocks gradually increases as $64 \rightarrow 128 \rightarrow 256 \rightarrow 512$, and the kernel size of the connects layer gradually increases as $3 \times 3 \rightarrow 5 \times 5 \rightarrow 7 \times 7$. We integrate an attention mechanism [30] into each stream. As described in Section 3.1, considering the high similarity between SDSs and other monitoring scenarios (e.g., fog, smog, and haze), we introduce spatial attention mechanism to adaptively bootstrap features related to the key SDS-relevant features and pay less attention to those less feature-rich regions. The features obtained from the spatial attention layer and the last CNN layer are then concatenated as the final vectors of each stream. The details of the spatial attention mechanism are shown in Figure 5.

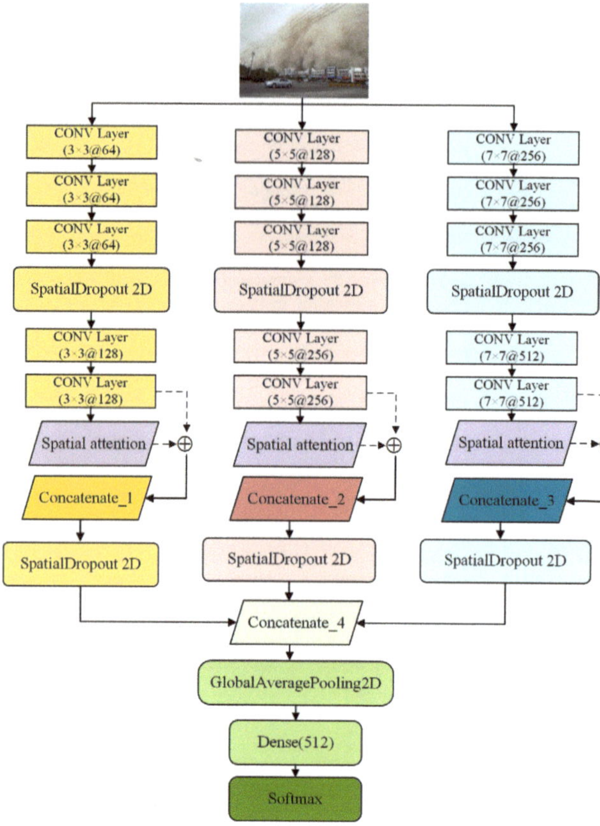

Figure 4. The architecture of an MA-CNN. (⊕ is the feature concatenate operation).

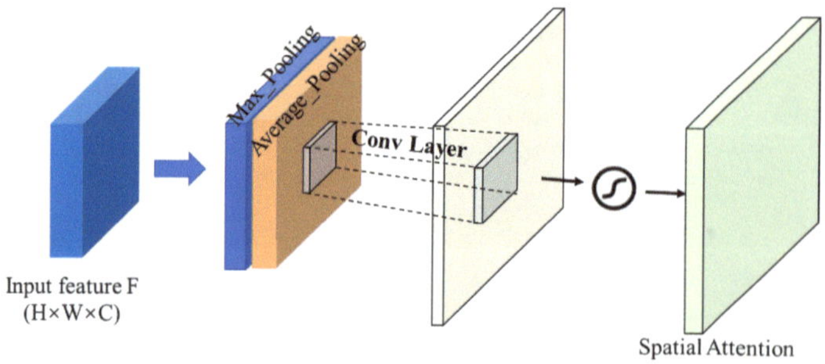

Figure 5. The architecture of the spatial attention mechanism.

For an input feature F with a size of H × W × C, firstly, the spatial information of a 2D feature map is generated by the average pooling feature ($F_{avg} \epsilon R^{l\times H\times W}$) and max pooling feature ($F_{max} \epsilon R^{l\times H\times W}$), respectively. Then, F_{avg} and F_{max} are concatenated and convolved by a standard convolution layer, producing a 2D spatial attention feature map $M_F(S') \epsilon R^{H\times W}$. The spatial attention is computed as follows:

$$M_F(S') = \delta\left(f^{7\times7}([avg\_pooling(F); \max\_pooling(F)])\right) = \delta\left(f^{7\times7}([F_{avg}; F_{avg}])\right) \quad (3)$$

where $\delta(\bullet)$ denotes the sigmoid function and $f^{7\times7}(\bullet)$ means the convolution operation with the filter size of 7 × 7.

Moreover, to mitigate overfitting, the 0.5 via spatialdropout2D [31] is added to each stream's second and last layers. Then, the feature responses from three streams are concatenated and fed into a GlobalAveragePooling2D output layer to produce the final vectors. Lastly, a fully connected layer with a size of 512 and a Softmax classifier can be utilized for a two-class classification.

2.3. *Experiment Setup*

2.3.1. Experimental Environment

Our experiments were performed on a workstation with Ubuntu 11.2.0 (Linux 5.15.0-25-generic) for the operating system. More specifications are as follows:

- 2× Intel Xeon Silver 4216 CPU@2.10 GHz (32 cores);
- 8×NVIDIA GEFORCE GTX2080Ti graphics cards equipped with 11 GB GDDR6 memory;
- 188 GB RAM;
- Python 3.9.16;
- TensorFlow 2.4.1, Scikit-learn 1.2.1, and Keras 2.4.3 libraries;
- CUDA 11.8 and CUDNN 8.

2.3.2. Evaluation Metrics

The performance of SDS detection algorithms is evaluated by the weighted Precision, weighted Recall, and weighted F1 score, which are calculated as follows:

$$\text{Precision} = \sum_{i=1}^{n} \frac{\text{TP}_i}{(\text{TP}_i + \text{FP}_i)} \times \text{r}_i \quad (4)$$

$$\text{Recall} = \sum_{i=1}^{n} \frac{\text{TP}_i}{(\text{TP}_i + \text{FN}_i)} \times \text{r}_i \quad (5)$$

$$\text{F1\_score} = \sum_{i=1}^{n} 2 \times \frac{\text{precision}_i \cdot \text{recall}_i}{\text{precision}_i + \text{recall}_i} \times \text{r}_i \quad (6)$$

where TP is the number of true positives; FP is the number of false positives; FN is the number of true negatives; i is the class index; and r_i is the ratio between the number of samples of class i and the total number of samples in all classes.

2.3.3. Dataset Building

For the training and testing of the deep learning model, we constructed a dataset called sand and dust storm image (SDSI). As described in Section 3.1, surveillance images obtained between smog, fog, and haze weather conditions and SDSs are easily confused. Therefore, fog and haze images were collected as negative samples. Furthermore, it was observed that scenes such as sunset, sunrise, rainy weather, and smoke in everyday life also have a relatively high similarity to SDS, as shown in Figure 6d–g; thus, images of these scenes were also collected as part of the negative samples.

Figure 6. Surveillance scenarios obtained in the SDSI dataset. (**a**) Sandstorm; (**b**) Haze; (**c**) Fog; (**d**) Sunrise; (**e**) Sunset; (**f**) Smoke; (**g**) Cloudy.

To ensure the number and diversity of the constructed SDSI dataset, images searched from the Internet and captured by surveillance cameras were jointly used. Finally, the positive samples (labeled as "Sandstorm") comprised 6578 images of SDSs taken in various scenarios. In comparison, 6638 images with similar scenarios to SDS conditions were taken to form the negative samples (labeled as "Others"). After that, the obtained images were split into training, validation, and test datasets, as shown in Table 1.

Table 1. Definition of Training-Test-Validation Split.

	Training Dataset	Validation Dataset	Test Dataset	Total
Other	4431	1029	1178	6638
Sandstorm	4425	1028	1125	6578
Total	8856	2057	2303	13,216

2.3.4. Model Training

Employing accuracy as a metric, all algorithm/model training is performed by minimizing the categorical cross-entropy loss with the Adam optimizer ($\beta_1 = 0.99$, $\beta_2 = 0.999$). Moreover, the learning rate is set to 0.0001 and the batch size is set to 16. We trained each algorithm from 100 to 300 epochs until a stable result is obtained. The models were evaluated with a K-fold cross-validation scheme in our SDSI dataset (K = 10).

3. Results

3.1. Experiments in SDSI Dataset

To further evaluate the effectiveness of our proposed CNNs, nine well-known and commonly used deep learning models for object detection tasks, i.e., VGG16 and VGG19 [32], NasNetMobile [33], Xception [34], ResNet50 [35], Mobile Net and Mobile Net V2 [36], InceptionV3 [37], and DenseNet121 [38], are selected for comparison. The performance of different algorithms in the SDSI dataset (test dataset) is shown in Table 2.

Table 2. Performance of different algorithms in the SDSI dataset.

	Precision	**F1_Score**	**Recall**	**Parameters**
VGG16	0.830	0.864	0.858	14,715,714
VGG19	0.834	0.829	0.830	20,025,410
NasNetMobile	0.778	0.833	0.820	4,271,830
Xception	0.771	0.852	0.834	20,865,578
ResNet50	0.736	0.633	0.677	23,591,810
Mobile Net V1	0.693	0.814	0.774	3,230,914
Mobile Net V2	0.788	0.854	0.840	2,260,546
InceptionV3	0.800	0.850	0.840	21,806,882
DenseNet121	0.802	0.867	0.855	7,039,554
MA-CNN (Ours)	**0.857**	**0.868**	**0.866**	28,171,314

Note: The best results are highlighted in bold.

Evaluated on the SDSI dataset, the proposed MA-CNN achieves state-of-the-art performance regarding Precision (0.857), F1 _Score (0.868), and Recall (0.866), as shown in Table 2. In general, MA-CNN shows significant improvement in precision than the compared methods. VGG16 and DenseNet121 performed closer to the proposed method in the F1 _Score and Recall metrics but with approximately 2.7% and 5.5% less than the proposed method in precision, respectively.

Considering that the spatial attention mechanism has good generality for existing deep learning networks, it was added before the output layer of the selected nine comparison algorithms. The spatial attention layer is added before the first fully connected layer for VGG16, VGG19, ResNet50, and Mobile Net V1. In contrast, for InceptionV3, Xception, NasNetMobile, and DenseNet121, the spatial attention layer is added after the last concated layer. For Mobile Net V2, the spatial attention layer is added after the last bottleneck). Taking VGG16 as an example, the structure of VGG16 with the spatial attention mechanism is labeled as "VGG16-A". Analogously, nine other comparison algorithms with the attention mechanism were obtained. The performance of different algorithms with spatial attention mechanisms in the SDSI dataset (test dataset) is presented in Table 3.

Table 3. Performance of different algorithms in the SDSI dataset after spatial attention mechanisms were added.

	Precision	**F1_Score**	**Recall**	**Parameters**
VGG16	0.812	0.855	0.847	14,781,924
VGG19	0.821	0.847	0.842	20,091,620
NasNetMobile	0.801	0.861	0.861	4,551,802
Xception	0.774	0.844	0.828	21,916,458
ResNet50	0.704	0.642	0.670	24,642,690
Mobile Net V1	0.799	0.854	0.852	3,494,210
Mobile Net V2	0.732	0.822	0.796	2,671,586
InceptionV3	0.786	0.835	0.824	22,857,762
DenseNet121	0.790	0.858	0.845	7,039,554
MA-CNN (Ours)	**0.857**	**0.868**	**0.866**	28,171,314

Note: The best results are highlighted in bold.

Table 3 suggests that our proposed algorithm achieves optimal performance after adding the attention mechanism. Comparing Tables 2 and 3, the performance of VGG16, Mobile Net V2, Inception V3, and DenseNet121 algorithms decreased after the attention layers were added. In summary, the attention mechanism is unsuitable for all the comparing deep learning models when detecting SDS via the SDSI dataset. However, Tables 2 and 3 indicate that our proposed algorithm contains the maximum number of parameters compared to the other listed algorithms, resulting in the MA-CNN model taking more extended time in the training process.

3.2. Experiments in Real-World Scenarios

Next, we evaluate the performance of the different SDS monitor algorithms in a real-world surveillance scenario. As shown in Figure 7, three surveillance videos that recorded the SDS from its appearance in the distance to its passage through the filming site were selected. Each video was fed into the SDS monitor algorithms in the form of a single frame. In the end, scenario_1 produced 375 images, while that of scenario_2 and scenario_3 were 241 and 282, respectively.

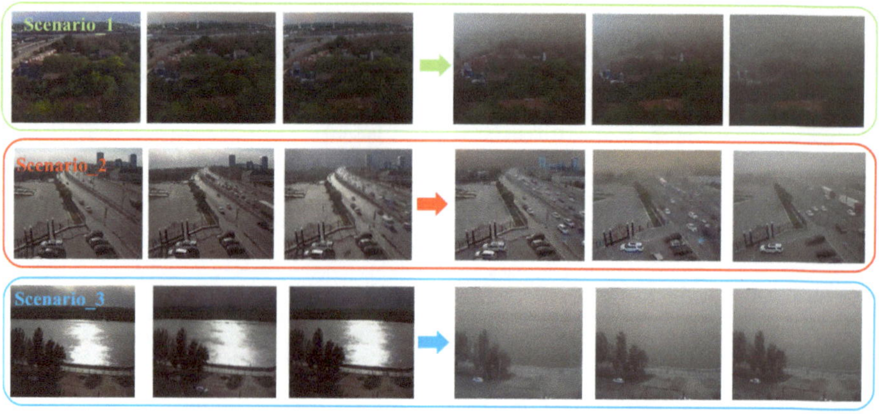

Figure 7. Three real-world SDS events recorded by surveillance cameras.

The experiment results of the three abovementioned real-world surveillance scenarios are reported in Tables 4–6.

Table 4 shows that, in surveillance scenario_1, the proposed algorithm achieves a Precision of 0.945 and an F1 _Score of 0.967, which is the highest among the listed algorithms; although NasNet has the highest Recall of 0.957, this is only 0.001 higher than that of our proposed MA-CNN; Table 5 illustrates that the MA-CNN achieved the best performance in surveillance scenario_2 with a Precision, F1 _Score, and Recall of 0.919, 0.936, and 0.934, respectively; Table 6 demonstrates that in surveillance scenario_3, DenseNET achieves the maximum values of F1 _Score and Recall, 0.971 and 0.955, which are 0.004 and 0.009 higher than the algorithm proposed in this paper, respectively; however, in terms of accuracy, the algorithm in this paper achieves the optimal performance of 0.953, which is 0.031 higher than DenseNet. The proposed MA-CNN performs stable and well for SDS monitoring in the above three real-world surveillance scenarios and can be considered the optimal model.

Table 4. Performance of different algorithms in real-world surveillance scenario_1.

	Precision	F1_Score	Recall
VGG16	0.86	0.883	0.902
VGG19	0.891	0.912	0.925
NasNetMobile	0.925	0.936	**0.957**
Xception	0.887	0.892	0.865
ResNet50	0.785	0.832	0.804
Mobile Net V1	0.742	0.788	0.769
Mobile Net V2	0.924	0.939	0.94
InceptionV3	0.879	0.886	0.897
DenseNet121	0.897	0.913	0.944
MA-CNN (Ours)	**0.945**	**0.967**	0.956

Note: The best results are highlighted in bold.

Table 5. Performance of different algorithms in real-world surveillance scenario_2.

	Precision	F1_Score	Recall
VGG16	0.823	0.835	0.870
VGG19	0.856	0.894	0.908
NasNetMobile	0.842	0.851	0.850
Xception	0.822	0.817	0.834
ResNet50	0.771	0.820	0.799
Mobile Net V1	0.692	0.722	0.743
Mobile Net V2	0.870	0.895	0.913
InceptionV3	0.887	0.913	0.920
DenseNet121	0.906	0.923	0.926
MA-CNN (Ours)	**0.919**	**0.936**	**0.934**

Note: The best results are highlighted in bold.

Table 6. Performance of different algorithms in real-world surveillance scenario_3.

	Precision	F1_Score	Recall
VGG16	0.764	0.792	0.800
VGG19	0.902	0.933	0.941
NasNetMobile	0.910	0.934	0.922
Xception	0.861	0.823	0.842
ResNet50	0.845	0.883	0.871
Mobile Net V1	0.797	0.811	0.826
Mobile Net V2	0.853	0.892	0.887
InceptionV3	0.915	0.935	0.946
DenseNet121	0.922	**0.971**	**0.957**
MA-CNN (Ours)	**0.953**	0.967	0.948

Note: The best results are highlighted in bold.

After that, the performance of different algorithms with spatial attention mechanisms in the three real-world surveillance scenarios was also used for comparison, as presented in Tables 7–9.

Tables 7–9 show that the MA-CNN still achieves optimal performance compared to other algorithms with spatial attention mechanisms. In addition, the accuracy of some comparison algorithms after the spatial attention mechanism was added shows a certain degree of degradation in real-world surveillance scenarios; for example, in surveillance scenarios 2 and 3, the Resnet50 algorithm shows a degradation in accuracy of 0.029 and 0.028 (please see Table 5, Table 6, Table 8 and Table 9). This phenomenon also occurs on the SDSI dataset, further indicating that the spatial attention mechanism is not suitable to be added to all deep learning algorithms.

Table 7. Performance of different algorithms in real-world surveillance scenario_1 after spatial attention mechanisms were added.

	Precision	F1_Score	Recall
VGG16	0.875	0.891	0.884
VGG19	0.912	0.918	0.914
NasNetMobile	0.864	0.859	0.862
Xception	0.824	0.83	0.832
ResNet50	0.812	0.809	0.822
Mobile Net V1	0.787	0.789	0.792
Mobile Net V2	0.933	0.944	0.938
InceptionV3	0.902	0.911	0.907
DenseNet121	0.919	0.923	0.954
MA-CNN (Ours)	**0.945**	**0.967**	**0.956**

Note: The best results are highlighted in bold.

Table 8. Performance of different algorithms in real-world surveillance scenario_2 after spatial attention mechanisms were added.

	Precision	F1_Score	Recall
VGG16	0.852	0.859	0.862
VGG19	0.873	0.889	0.878
NasNetMobile	0.792	0.783	0.801
Xception	0.754	0.753	0.759
ResNet50	0.742	0.732	0.753
Mobile Net V1	0.725	0.737	0.734
Mobile Net V2	0.798	0.803	0.808
InceptionV3	0.907	0.915	0.918
DenseNet121	0.916	0.917	0.917
MA-CNN (Ours)	**0.919**	**0.936**	**0.934**

Note: The best results are highlighted in bold.

Table 9. Performance of different algorithms in real-world surveillance scenario_3 after spatial attention mechanisms were added.

	Precision	F1_Score	Recall
VGG16	0.814	0.807	0.815
VGG19	0.917	0.923	0.921
NasNetMobile	0.922	0.913	0.917
Xception	0.87	0.863	0.872
ResNet50	0.817	0.823	0.821
Mobile Net V1	0.825	0.818	0.834
Mobile Net V2	0.914	0.925	0.917
InceptionV3	0.929	0.933	0.952
DenseNet121	0.928	0.944	**0.953**
MA-CNN (Ours)	**0.953**	**0.967**	0.948

Note: The best results are highlighted in bold.

For a more intuitive understanding of the SDS recognition pattern of the proposed algorithm, we compare the attention feature maps extracted by different deep learning algorithms shown in Tables A1 and A2 (Please see Appendix A). In Tables A1 and A2, the first row show four SDS figures randomly selected from surveillance scenario_1, and from the second row onwards, the corresponding attention maps extracted by different algorithms are presented.

In the first two images, the SDS appears in the upper area of the surveillance images, correctly identified by most comparison algorithms. For the last two images, when the SDS is diffused throughout the image, more significant uncertainty emerges in the performance of different comparison algorithms. In contrast, Tables A1 and A2 demonstrate that, in the four selected images, the attention feature maps of our MA-CNN method can match well with the changes of the pixels blurred by SDS, which indicates that the MA-CNN can identify and localize the SDSs that appear in the surveillance scenarios well. The visualization results further indicate that the proposed MA-CNN algorithm is more suitable for SDS monitoring via surveillance cameras.

4. Discussion

Compared to existing studies, the main contribution of our study is the use of surveillance cameras for SDS monitoring tasks, which sheds light on building a low-cost, high temporal and spatial resolution SDS monitoring network based on existing urban surveillance resources. However, some shortcomings need to be addressed.

(1) Although the proposed MA-CNN model brings an improvement in SDS accuracy, it suffers from a long computational delay due to a relatively large number of parameters (as presented in Tables 2 and 3). Taking the experimental platform shown in Section 2.3.1 as an example, the MA-CNN algorithm costs 0.78 s to judge whether an SDS appears in a surveillance image with a resolution of 1920 × 1080, whereas VGG16, VGG19, NasNetMobile, Xception, ResNet50, Mobile Net V1, Mobile Net V2, InceptionV3, and DenseNet121 take 0.39 s, 0.52 s, 0.27 s, 0.54 s, 0.65 s, 0.25 s, 0.21 s, 0.62 s, and 0.44 s, respectively. In practical applications, increasing the computational capability of the hardware devices or adopting a distributed computing manner are alternative solutions to reduce the abovementioned latency.

(2) In this study, we detect SDSs in visible light surveillance video captured during the daytime, while in low-light scenarios such as nighttime, it is difficult to capture the appearance of an SDS in visible light surveillance video, and the MA-CNN method will on not work. Extensive research has shown that ordinary surveillance cameras can have "night vision" through near-infrared (NIR) video to perceive low-light scenarios [39], which provides the opportunity to monitor SDSs at night. Research on NIR video-based SDS monitoring, thereby constructing an all-weather observation system, will be the focus of the next step.

(3) Rapid changes over time are an important distinction between SDSs and similar weather events (e.g., fog, haze, and smoke). Understanding SDSs from both temporal and spatial dimensions is an effective strategy to improve SDS monitoring accuracy. The proposed MA-CNN algorithm can mine the image features of SDSs from the spatial dimension, and the patterns of SDSs in the temporal dimension are not utilized. The main reason is that few dust storm videos can be collected currently, making it challenging to build a deep learning dataset for describing an SDS's spatial and temporal features. Therefore, more SDS and similar weather video data should be collected in the future, which is the basis for the study of more accurate SDS monitoring methods.

5. Conclusions

In this paper, we attempt to build urban surveillance cameras as SDS monitors. After analyzing the image features of SDSs, RGB images were selected as the input of the recognition algorithm; attentive mechanisms were introduced, and a deep learning model named MA-CNN was proposed for the accurate discrimination of SDSs via surveillance images. In addition, the SDSI dataset was constructed for the training and testing of the model. The experimental results on the SDSI dataset show that compared with 18 related deep learning algorithms, the MA-CNN model achieves the best performance with a precision, F1_score, and recall of 0.857, 0.868, and 0.866, respectively. Moreover, the performance on three real-world surveillance scenarios also demonstrates that the MA-CNN can effectively identify an SDS's occurrence in surveillance images. The constructed SDS monitor can be deployed on existing urban surveillance resources, which has the advantage of a low cost and provides a promising means for high spatiotemporal resolution SDS observation.

However, the monitors proposed in this study can only determine the occurrence of SDS events, resulting in limited practical value. Quantifying SDS levels from surveillance images or videos will be our focus in the future.

Author Contributions: Conceptualization, X.W. and Z.Y.; methodology, X.W.; software, X.W., L.C. and S.S; validation, X.W., Z.Y. and J.Z.; formal analysis, X.W.; investigation, X.W.; resources, X.W.; data curation, X.W.; writing—original draft preparation, X.W.; writing—review and editing, X.W. and H.F.; visualization, X.W. and S.S.; supervision, X.W. and H.F.; project administration, X.W., J.Z. and S.S.; funding acquisition, X.W., J.Z. and S.S. All authors have read and agreed to the published version of the manuscript.

Funding: This research was funded by the National Natural Science Foundation of China (NSFC) (No. 42105022 and 42005104) and the Joint Open Project of KLME & CIC-FEMD, NUIST (No. KLME202312).

Data Availability Statement: SDSI dataset IS available at this site (https://pan.baidu.com/s/1eRAxM0SVtsa_SdwfJfYhXw, accessed on 1 December 2023).

Conflicts of Interest: The authors declare no conflict of interest.

Appendix A

Table A1. Attention feature maps extracted by different deep learning algorithms.

Images from scenario_1				
VGG16	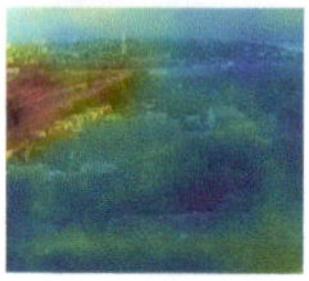		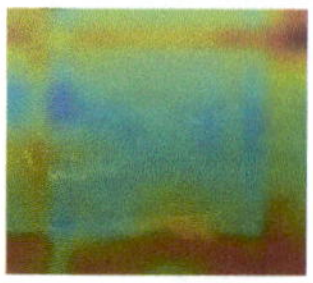	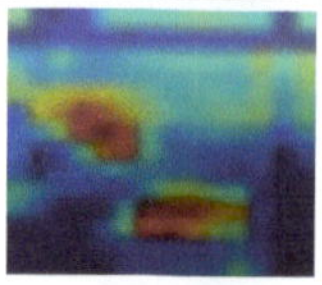
VGG19			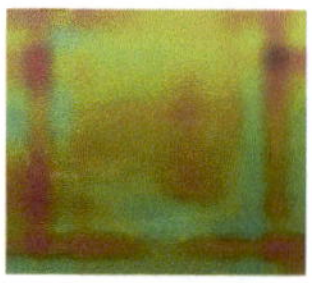	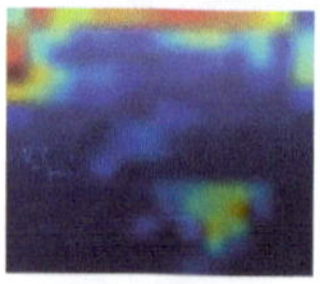
NasNet				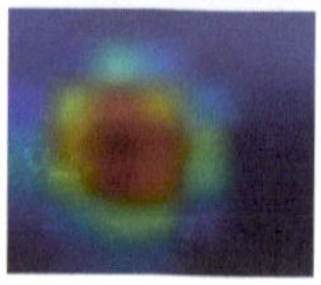

Table A1. *Cont.*

Xception		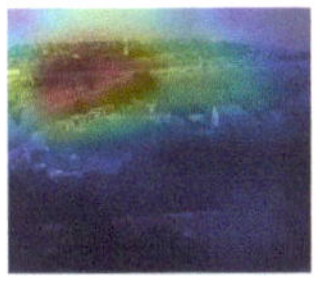	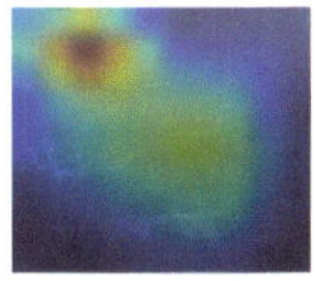	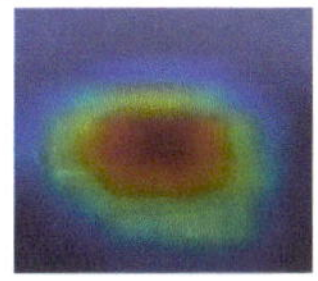
ResNet50	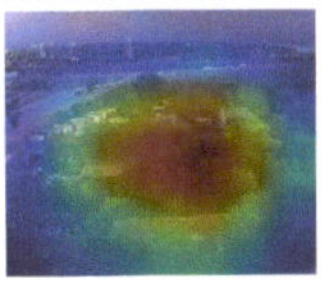	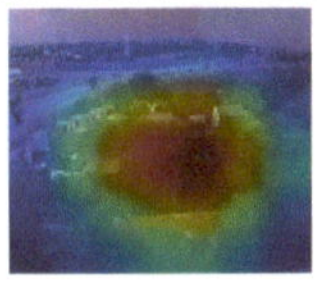	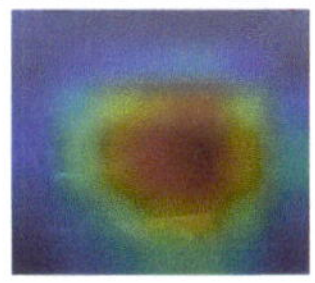	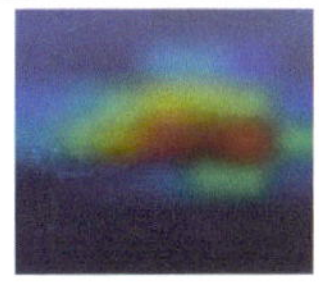
Mobile Net V1			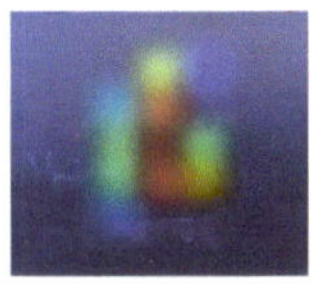	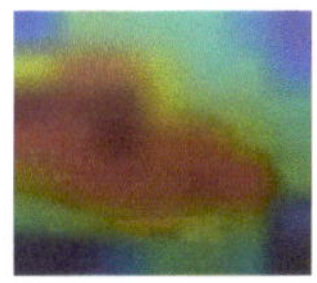
MobileNet V2	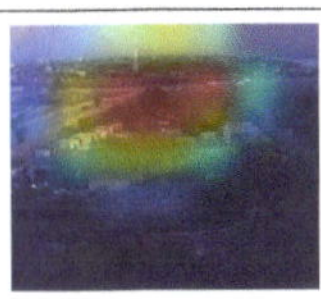			
InceptionV3			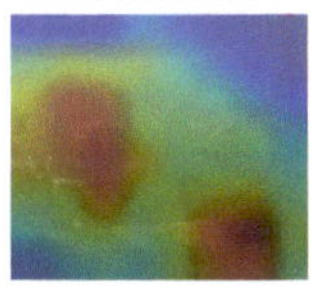	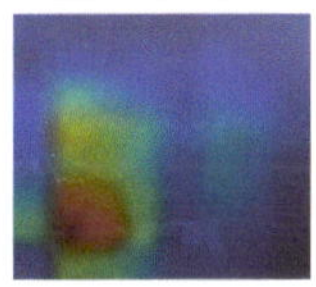
DenseNet121		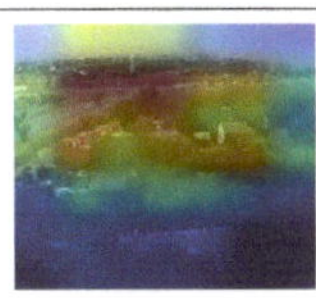	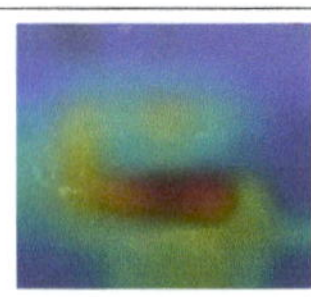	
MA-CNN (Ours)				
		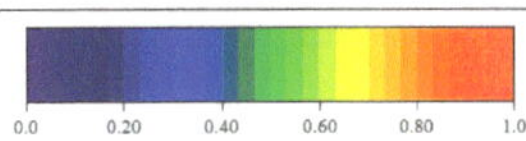		

Table A2. Attention feature maps extracted by different deep learning algorithms after attention mechanism.

Images from scenario_1				
VGG16	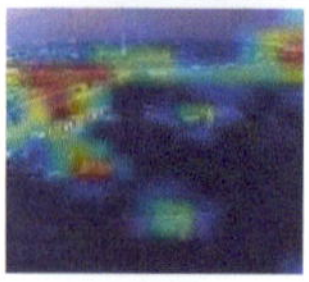	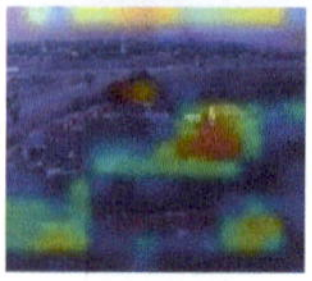	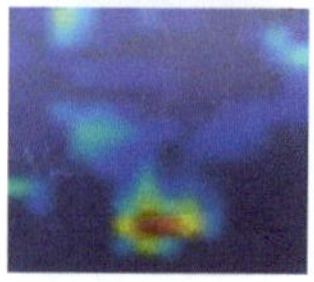	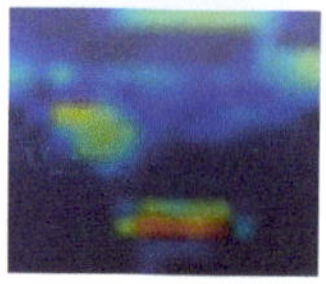
VGG19				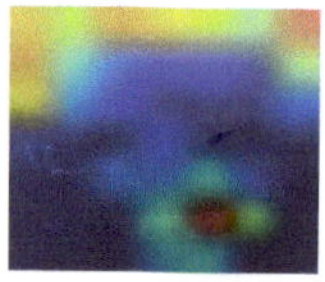
NasNet				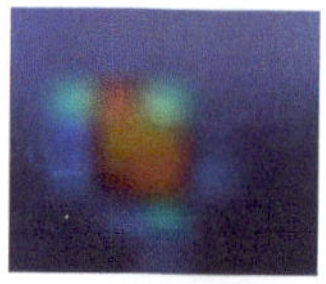
Xception			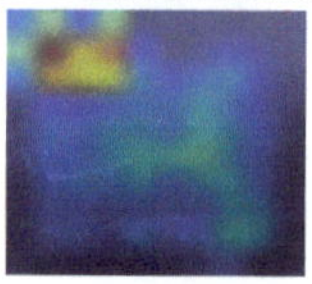	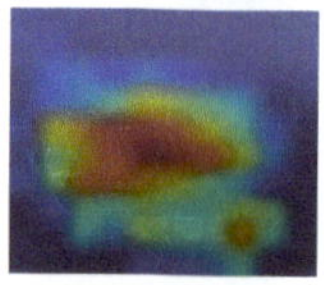
ResNet50			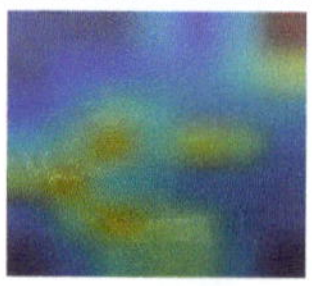	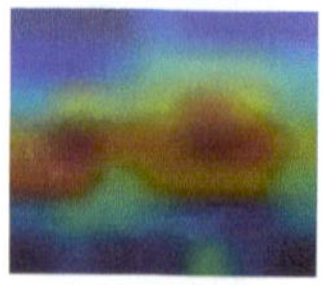
Mobile Net V1	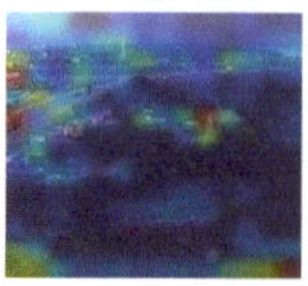	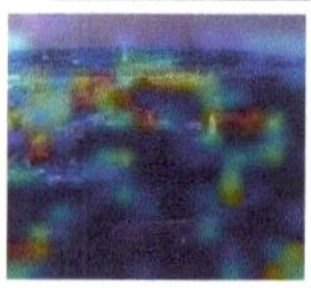	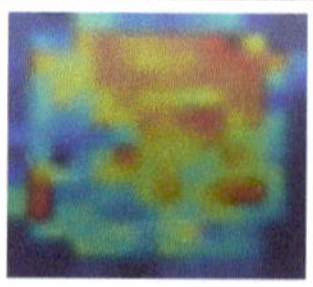	

Table A2. *Cont.*

MobileNet V2	
InceptionV3	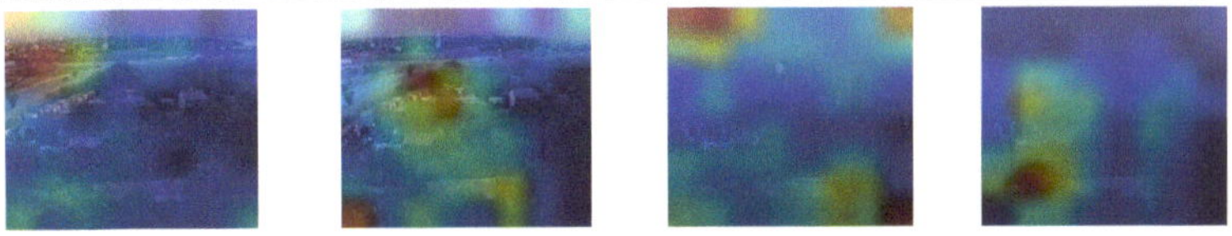
DenseNet121	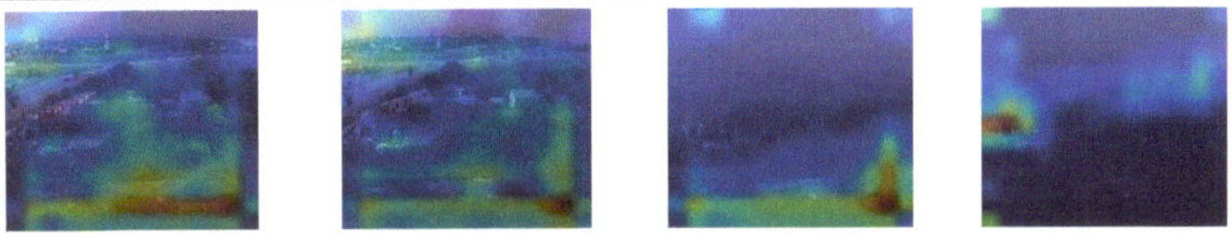
MA-CNN (Ours)	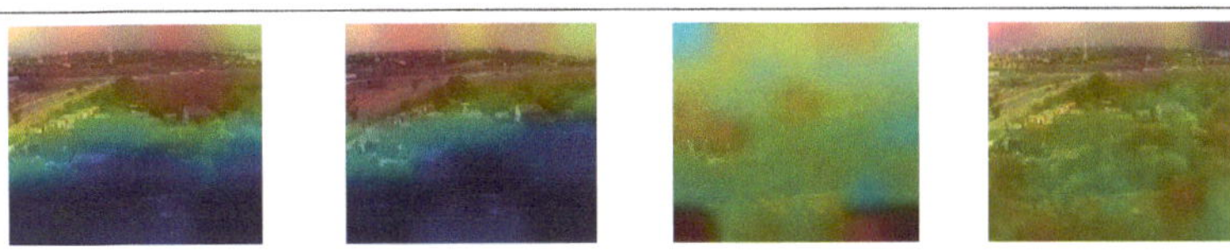
	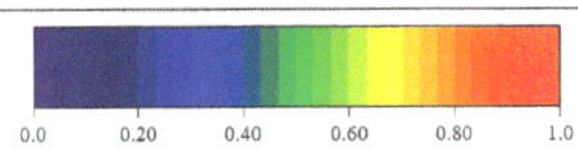

References

1. Shepherd, G.; Terradellas, E.; Baklanov, A.; Kang, U.; Sprigg, W.; Nickovic, S.; Boloorani, A.D.; Al-Dousari, A.; Basart, S.; Benedetti, A. *Global Assessment of Sand and Dust Storms*; United Nations Environment Programme: Nairobi, Kenya, 2016.
2. Jiao, P.; Wang, J.; Chen, X.; Ruan, J.; Ye, X.; Alavi, A.H. Next-Generation Remote Sensing and Prediction of Sand and Dust Storms: State-of-the-Art and Future Trends. *Int. J. Remote Sens.* **2021**, *42*, 5277–5316.
3. Nickovic, S.; Agulló, E.C.; Baldasano, J.M.; Terradellas, E.; Nakazawa, T.; Baklanov, A. *Sand and Dust Storm Warning Advisory and Assessment System (Sds-Was) Science and Implementation Plan: 2015–2020*; World Meteorological Organization: Geneva, Switzerland, 2015.
4. Behzad, R.; Barati, S.; Goshtasb, H.; Gachpaz, S.; Ramezani, J.; Sarkheil, H. Sand and Dust Storm Sources Identification: A Remote Sensing Approach. *Ecol. Indic.* **2020**, *112*, 106099.
5. Muhammad, A.; Sheltami, T.R.; Mouftah, H.T. A Review of Techniques and Technologies for Sand and Dust Storm Detection. *Rev. Environ. Sci. Bio Technol.* **2012**, *11*, 305–322.
6. Gutierrez, J. Automated Detection of Dust Storms from Ground-Based Weather Station Imagery Using Neural Network Classification. Ph.D. Thesis, New Mexico State University, Las Cruces, NM, USA, 2020.
7. McPeters, R.D. *Nimbus-7 Total Ozone Mapping Spectrometer (Toms) Data Products User's Guide*; National Aeronautics and Space Administration, Scientific and Technical: Washington, DC, USA, 1996; Volume 1384.
8. Sassen, K.; Wang, Z.; Liu, D. Global Distribution of Cirrus Clouds from Cloudsat/Cloud-Aerosol Lidar and Infrared Pathfinder Satellite Observations (Calipso) Measurements. *J. Geophys. Res. Atmos.* **2008**, *113*, D00A12.
9. Bukhari, S.A. Depicting Dust and Sand Storm Signatures through the Means of Satellite Images and Ground-Based Observations for Saudi Arabia. PhD Thesis, University of Colorado at Boulder, Boulder, CO, USA, 1993.
10. Narasimhan, S.G.; Nayar, S.K. Vision and the Atmosphere. *Int. J. Comput. Vis.* **2002**, *48*, 233. [CrossRef]
11. Jr, C.; Pat, S.; Mackinnon, D.J.; Reynolds, R.L.; Velasco, M. Monitoring Dust Storms and Mapping Landscape Vulnerability to Wind Erosion Using Satellite and Ground-Based Digital Images. *Arid. Lands. Newsl.* **2002**, *51*, 1–8.

12. Dagsson-Waldhauserova, P.; Magnusdottir, A.Ö.; Olafsson, H.; Arnalds, O. The Spatial Variation of Dust Particulate Matter Concentrations During Two Icelandic Dust Storms in 2015. *Atmosphere* **2016**, *7*, 77. [CrossRef]
13. Urban, F.E.; Goldstein, H.L.; Fulton, R.; Reynolds, R.L. Unseen Dust Emission and Global Dust Abundance: Documenting Dust Emission from the Mojave Desert (USA) by Daily Remote Camera Imagery and Wind-Erosion Measurements. *J. Geophys. Res. Atmos.* **2018**, *123*, 8735–8753.
14. Abdulameer, F.S. Using Color Spaces Hsv, Yiq and Comparison in Analysis Hazy Image Quality. *Adv. Phys. Theor. Appl* **2019**, *76*, 15–23.
15. Fattal, R. Single Image Dehazing. *ACM Trans. Graph. (TOG)* **2008**, *27*, 1–9. [CrossRef]
16. Narasimhan, S.G.; Nayar, S.K. Chromatic Framework for Vision in Bad Weather. In Proceedings of the IEEE Conference on Computer Vision and Pattern Recognition, CVPR 2000 (Cat. No. PR00662), Hilton Head, SC, USA, 15 June 2000.
17. Xu, G.; Wang, X.; Xu, X. Single Image Enhancement in Sandstorm Weather Via Tensor Least Square. *IEEE/CAA J. Autom. Sin.* **2020**, *7*, 1649–1661. [CrossRef]
18. Sun, K.J.; Tang, X. Single Image Haze Removal Using Dark Channel Prior. *IEEE Trans. Pattern Anal. Mach. Intell.* **2010**, *33*, 2341–2353.
19. Cheng, Y.; Jia, Z.; Lai, H.; Yang, J.; Kasabov, N.K. Blue Channel and Fusion for Sandstorm Image Enhancement. *IEEE Access* **2020**, *8*, 66931–66940. [CrossRef]
20. Shi, F.; Jia, Z.; Lai, H.; Kasabov, N.K.; Song, S.; Wang, J. Sand-Dust Image Enhancement Based on Light Attenuation and Transmission Compensation. *Multimed. Tools Appl.* **2023**, *82*, 7055–7077. [CrossRef]
21. Fu, X.; Huang, Y.; Zeng, D.; Zhang, X.-P.; Ding, X. A Fusion-Based Enhancing Approach for Single Sandstorm Image. In Proceedings of the 2014 IEEE 16th International Workshop on Multimedia Signal Processing (MMSP), Jakarta, Indonesia, 22–24 September 2014.
22. Lee, H.S. Efficient Sandstorm Image Enhancement Using the Normalized Eigenvalue and Adaptive Dark Channel Prior. *Technologies* **2021**, *9*, 101. [CrossRef]
23. Ding, B.; Zhang, R.; Xu, L.; Cheng, H. Sand-Dust Image Restoration Based on Gray Compensation and Feature Fusion. *Acta Armamentarii* **2022**. [CrossRef]
24. Gao, G.; Lai, H.; Liu, Y.; Wang, L.; Jia, Z. Sandstorm Image Enhancement Based on Yuv Space. *Optik* **2021**, *226*, 165659. [CrossRef]
25. Brauwers, G.; Frasincar, F. A General Survey on Attention Mechanisms in Deep Learning. *IEEE Trans. Knowl. Data Eng.* **2021**, *35*, 3279–3298.
26. Hassanin, M.; Anwar, S.; Radwan, I.; Khan, F.S.; Mian, A. Visual Attention Methods in Deep Learning: An in-Depth Survey. *arXiv* **2022**, arXiv:2204.07756.
27. Niu, Z.; Zhong, G.; Yu, H. A Review on the Attention Mechanism of Deep Learning. *Neurocomputing* **2021**, *452*, 48–62.
28. Hafiz, M.A.; Parah, S.A.; Bhat, R.U.A. Attention Mechanisms and Deep Learning for Machine Vision: A Survey of the State of the Art. *arXiv* **2021**, arXiv:2106.07550.
29. Maas, A.L.; Hannun, A.Y.; Ng, A.Y. Rectifier Nonlinearities Improve Neural Network Acoustic Models. In Proceedings of the 30th International Conference on Machine Learning, Atlanta, GA, USA, 16–21 June 2013.
30. Woo, S.; Park, J.; Lee, J.-Y.; Kweon, I.S. Cbam: Convolutional Block Attention Module. In Proceedings of the European Conference on Computer Vision (ECCV), Munich, Germany, 8–14 September 2018.
31. Tompson, J.; Goroshin, R.; Jain, A.; LeCun, Y.; Bregler, C. Efficient Object Localization Using Convolutional Networks. In Proceedings of the IEEE Conference on Computer Vision and Pattern Recognition, Boston, MA, USA, 7–12 June 2015.
32. Simonyan, K.; Zisserman, A. Very Deep Convolutional Networks for Large-Scale Image Recognition. *arXiv* **2014**, arXiv:1409.1556.
33. Zoph, B.; Vasudevan, V.; Shlens, J.; Le, Q.V. Learning Transferable Architectures for Scalable Image Recognition. In Proceedings of the IEEE Conference on Computer Vision and Pattern Recognition, Salt Lake City, UT, USA, 18–23 June 2018.
34. Chollet, F. Xception: Deep Learning with Depthwise Separable Convolutions. In Proceedings of the IEEE Conference on Computer Vision and Pattern Recognition, Honolulu, HI, USA, 21–26 July 2017.
35. Zhang, K.X.; Ren, S.; Sun, J. Deep Residual Learning for Image Recognition. In Proceedings of the IEEE Conference on Computer Vision and Pattern Recognition, Las Vegas, NV, USA, 27–30 June 2016.
36. Howard, A.G.; Zhu, M.; Chen, B.; Kalenichenko, D.; Wang, W.; Weyand, T.; Andreetto, M.; Adam, H. Mobilenets: Efficient Convolutional Neural Networks for Mobile Vision Applications. *arXiv* **2017**, arXiv:1704.04861.
37. Szegedy, C.; Vanhoucke, V.; Ioffe, S.; Shlens, J.; Wojna, Z. Rethinking the Inception Architecture for Computer Vision. In Proceedings of the IEEE Conference on Computer Vision and Pattern Recognition, Las Vegas, NV, USA, 27–30 June 2016.
38. Huang, G.; Liu, Z.; Van Der Maaten, L.; Weinberger, K.Q. Densely Connected Convolutional Networks. In Proceedings of the IEEE Conference on Computer Vision and Pattern Recognition, Honolulu, HI, USA, 21–26 July 2017.
39. Wang, X.; Wang, M.; Liu, X.; Zhu, L.; Shi, S.; Glade, T.; Chen, M.; Xie, Y.; Wu, Y.; He, Y. Near-Infrared Surveillance Video-Based Rain Gauge. *J. Hydrol.* **2023**, *618*, 129173. [CrossRef]

remote sensing

MDPI

Article

A Lightweight Man-Overboard Detection and Tracking Model Using Aerial Images for Maritime Search and Rescue

Yijian Zhang, Qianyi Tao and Yong Yin *

Navigation College, Dalian Maritime University, Dalian 116026, China
* Correspondence: bushyin@dlmu.edu.cn

Abstract: Unmanned rescue systems have become an efficient means of executing maritime search and rescue operations, ensuring the safety of rescue personnel. Unmanned aerial vehicles (UAVs), due to their agility and portability, are well-suited for these missions. In this context, we introduce a lightweight detection model, YOLOv7-FSB, and its integration with ByteTrack for real-time detection and tracking of individuals in maritime distress situations. YOLOv7-FSB is our lightweight detection model, designed to optimize the use of computational resources on UAVs. It comprises several key components: FSNet serves as the backbone network, reducing redundant computations and memory access to enhance the overall efficiency. The SP-ELAN module is introduced to ensure operational speed while improving feature extraction capabilities. We have also enhanced the feature pyramid structure, making it highly effective for locating individuals in distress within aerial images captured by UAVs. By integrating this lightweight model with ByteTrack, we have created a system that improves detection accuracy from 86.9% to 89.2% while maintaining a detection speed similar to YOLOv7-tiny. Additionally, our approach achieves a MOTA of 85.5% and a tracking speed of 82.7 frames per second, meeting the demanding requirements of maritime search and rescue missions.

Keywords: unmanned aerial vehicles (UAVs); object tracking; object detection; tracking by detection; maritime search and rescue

Citation: Zhang, Y.; Tao, Q.; Yin, Y. A Lightweight Man-Overboard Detection and Tracking Model Using Aerial Images for Maritime Search and Rescue. *Remote Sens.* **2024**, *16*, 165. https://doi.org/10.3390/rs16010165

Academic Editors: Qian Du, Yanni Dong and Xiaochen Yang

Received: 26 October 2023
Revised: 27 December 2023
Accepted: 28 December 2023
Published: 30 December 2023

1. Introduction

In recent years, there has been a surge in maritime accidents worldwide, resulting in significant human and economic losses. Since 2014, the number of maritime accidents has steadily increased, averaging 2647 casualties and incidents annually [1]. Maritime search and rescue (SAR) operations play a vital role in national emergency response systems, with the primary challenge being swiftly and accurately locating and identifying objects at sea. The emergence of unmanned aerial vehicle (UAV) technology has revolutionized various fields, including robotics, security surveillance, intelligent transportation, wildlife conservation, and geospatial information [2–7]. Owing to their agility, portability, and aerial accessibility, UAVs offer rapid deployment, high data capacity, and outstanding spatial resolution, making them particularly effective for executing maritime SAR missions [8].

However, identifying individuals in maritime distress within UAV imagery presents unique challenges due to their small scale. Operator fatigue and distractions can also lead to missed sightings of man-overboard situations in aerial images. The application of deep learning-based object detection algorithms has significantly improved this situation. UAVs equipped with such algorithms can rapidly cover extensive maritime areas while promptly identifying individuals in distress, ensuring that they can be rescued faster, thereby increasing the probability of survival. Nonetheless, UAV image object detectors must strike a balance between lightweight design and real-time processing due to hardware limitations and specific application scenarios. This study focuses on enhancing the efficiency of identifying man-overboard situations in UAV imagery by addressing these challenges.

Several scholars have demonstrated the exceptional performance of the You Only Look Once (YOLO) series in the field of object detection, particularly in UAV-based detection systems [9–11]. Therefore, we have selected YOLOv7-tiny [12] as the UAV visual framework for maritime search and rescue operations involving man-overboard situations.

Our primary contributions to this work are as follows:

- The introduction of a lightweight object detector, "YOLOv7-FSB," tailored to efficiently identify man-overboard situations for rescue missions.
- To address the challenge of small objects in images, we propose a lightweight backbone network named "FSNet" to enhance the model's global information perception and improve its robustness.
- In response to the limited feature information of submerged individuals in aerial images, we introduce the module design called "SP-ELAN." This module integrates channel reconstruction units, reducing feature redundancy and enhancing algorithm efficiency.
- To enhance the utilization efficiency of feature information for submerged individuals in aerial images, we implement an improved Bidirectional Feature Pyramid Network. This network performs a bidirectional fusion of features related to submerged individuals in aerial images, integrating both local and global features.
- The selection of ByteTrack as the tracking component, combined with the improved detection algorithm, demonstrates real-time detection and tracking capabilities through experimentation.

The remainder of this paper is organized as follows: Section 2 reviews related research, Section 3 introduces the algorithm for identifying maritime man-overboard situations based on aerial imagery, and detailed information about the experiments and analysis is provided in Section 4. In Section 5, conclusions are drawn, and future research directions are outlined.

2. Relate Works

2.1. Object Detection

In the early stages of computer vision, conventional object detection algorithms heavily relied on intricately handcrafted feature engineering. These methods required the meticulous design of complex feature representations and the incorporation of various acceleration techniques, all within an era characterized by a lack of truly effective image representations [13–15]. In subsequent years, progress in object detection research experienced a significant slowdown as the performance of handcrafted features reached a saturation point. However, a pivotal turning point occurred in 2014 when convolutional neural networks (CNNs) emerged as a solution to this challenge [16]. CNN-based object detection methods can be broadly categorized into one- and two-stage detectors.

2.1.1. Two-Stage Detectors

Two-stage algorithms typically follow a sequence where they first identify candidate regions within an image and then proceed to classify and precisely locate the target within these regions. This approach has seen significant evolution in the field of object detection. The journey of two-stage object detection algorithms can be traced back to the pioneering work of Girshick et al. [17], who introduced the region-based CNN (R-CNN) algorithm. R-CNN marked the initial steps in employing deep learning for object detection. It laid the foundational concepts for subsequent CNN-based object detection techniques. The field continued to advance with the work of He et al. [18], who made notable improvements to the CNN architecture. They introduced spatial pyramid pooling (SPP) layers, a critical innovation that led to the development of the faster SPP-Net detection algorithm. This introduction of SPP layers significantly improved the speed and efficiency of object detection compared to the earlier R-CNN model. Building on this progress, Girshick further refined the methodology with the introduction of the Fast R-CNN model [19]. Fast R-CNN streamlined the detection process by discerning all potential candidate boxes directly from the feature maps extracted from the image. This innovation not only enhanced the efficiency of training and detection but also marked a leap forward compared to R-CNN. The field reached new heights with the presentation of the Faster R-CNN algorithm by Ren et al. [20]. Faster R-CNN introduced a two-stage detection approach, further optimizing training and detection speeds when compared to the Fast R-CNN model.

2.1.2. One-Stage Detectors

On the other hand, one-stage detectors, in contrast to two-stage detectors, rely on a global regression-based classification approach to directly predict the position and category of a target. These one-stage methods are well suited for real-time object detection, where speed is of utmost importance. Several influential algorithms have been developed in this category, each with its own set of features and trade-offs. One notable algorithm is the "You Only Look Once" detection algorithm introduced by Redmon et al. [21]. YOLO employs a neural network to predict both the positions and categories of objects within images. While it excels in real-time detection, it tends to exhibit lower accuracy compared to some two-stage detectors. Liu et al. introduced the Single-Shot MultiBox Detector (SSD) algorithm [22], an improved iteration of YOLO. SSD utilizes the VGG-16 [23] deep convolutional neural network for extracting multiscale feature maps and directly outputs target positions. This refinement aims to strike a balance between speed and accuracy. The YOLO series, pioneered by Redmon, continued to evolve with YOLOv3 [24], which is characterized by high accuracy and speed. YOLOv4, introduced in 2020, incorporated CSPDarkNet-53 as the backbone network, further enhancing performance. YOLOv5 emphasizes engineering considerations, prioritizing model flexibility and ease of deployment, although it may involve certain performance trade-offs. Li et al. introduced YOLOv6 [25], which incorporates the RepBlock module inspired by the renowned RepVGG network [26], significantly enhancing training and inference speeds. YOLOv7 [12] introduces trainable freebies, reparameterizes the model structure, and integrates "expansion" and "compound scaling" techniques to enhance inference speed and accuracy. YOLOv8 builds upon YOLOv5 by incorporating the latest idea of decoupled detection heads. Additionally, it enhances the backbone network and feature pyramid, resulting in improved performance. Furthermore, Lin et al. [27] proposed an innovative "focal loss" method to address class imbalance issues in dense object detectors. This innovation enabled the creation of a high-speed, high-precision single-stage detector named RetinaNet. Tan et al. [28] introduced a composite scale transformation method for object detection models, achieving superior results under different computational resource constraints compared to existing models.

2.1.3. Lightweight Detectors

To meet the demands of effective detection in industrial applications with limited memory and computational resources, researchers have introduced a range of lightweight object detection algorithms. These algorithms are crucial for scenarios where efficiency and real-time processing are paramount. One noteworthy approach was presented by Huang et al. [29] with their novel 3FL-Net. This algorithm significantly enhances the performance of lightweight object detectors, particularly under adverse weather conditions. 3FL-Net achieves this by closely integrating various components, including feature enhancement, feature extraction, feature adaptation, and lightweight detection subnetworks. It is a valuable solution for industrial applications that require reliable object detection in challenging environmental conditions. Wu et al. [30] introduced a lightweight backbone network and an efficient feature fusion network tailored for road damage object detection. They leveraged lightweight modules (LWCs), optimized attention mechanisms, and activation functions to build a system that efficiently identifies road damage, even with limited computational resources. Pang et al. [31] took a unique approach by designing structural reparameterization blocks (SRB) at the network module level. This innovation is particularly aimed at improving inference accuracy, especially for applications like satellite orbit calculations where precision is critical. Liu et al. [32] addressed the challenges of lightweight object detection with their structurally reparameterized LightNet design. This approach not only improves feature extraction capabilities but also reduces model inference complexity. Furthermore, they employed knowledge distillation techniques to meet the accuracy and real-time requirements of industrial defect detection. These lightweight object detection algorithms are invaluable for a wide range of industrial applications where resource constraints, real-time processing, and the need for efficient yet accurate object detection are key considerations.

2.2. Object Tracking

Target tracking is a vital computer vision task that involves the real-time localization and tracking of specific objects within video sequences. This task comes with the added challenge of not only tracking objects but also preserving their identities over time. Unlike object detection, target tracking focuses on maintaining object identity throughout a sequence of frames. In the realm of multi-object tracking (MOT), the task typically involves two crucial components: a detection module responsible for identifying objects in each frame and a data association module that links the detections across frames. With advancements in object detection technology, "tracking by detection" has become a predominant framework in MOT. Two widely recognized tracking-by-detection frameworks are ByteTrack [33] and DeepSORT [34]. DeepSORT is a robust tracking framework that relies on the Kalman filter for object tracking. It consists of two essential components: SORT [35], which utilizes Kalman filtering for tracking, and a re-identification module that employs deep neural networks to extract unique features. DeepSORT's straightforward yet effective architecture makes it a standout performer, particularly in high-frame-rate scenarios. On the other hand, ByteTrack excels in the data association part. It introduces a simple and efficient data association method known as BYTE. BYTE leverages the similarity between detection boxes and tracking trajectories. It retains high-scoring detection results while effectively eliminating background noise from low-scoring detections. This approach significantly improves the consistency of object tracking.

In summary, deep-learning-based object detection and tracking algorithms play a pivotal role in the field of computer vision. These algorithms enable the efficient utilization of UAVs for various applications. In our specific context, we combine the improved version of YOLOv7-tiny, which strikes a balance between accuracy and speed, with the straightforward and efficient ByteTrack framework for detecting and tracking individuals in maritime distress in aerial imagery. This combination leverages lightweight visual models, meeting the dual requirements of speed and accuracy in search and rescue operations using UAVs.

3. Materials and Methods

YOLOv7-tiny, like other detectors in the YOLO series, features a structured architecture comprising a backbone, neck, and head. The structure of YOLOv7-tiny is illustrated in Figure 1.

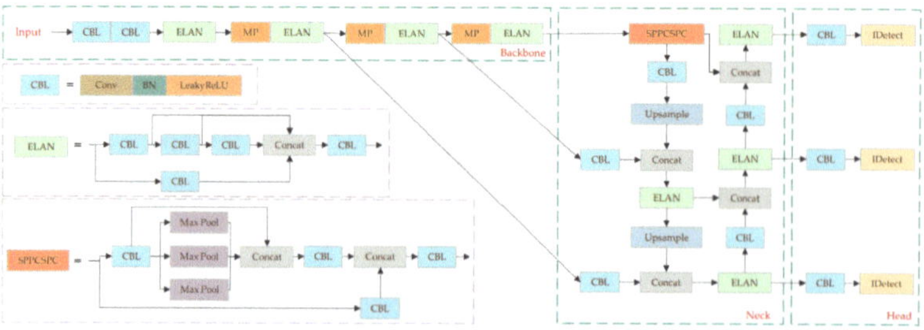

Figure 1. The general framework of YOLOv7-tiny.

The backbone of YOLOv7-tiny includes multiple CBL modules, ELAN structures, and MP convolution layers. Each CBL module consists of convolution layers, batch normalization layers, and an activation function. YOLOv7-tiny employs LeakyReLU as its activation function, which is an evolution from the ReLU (Rectified Linear Unit) activation function. The ELAN structure enhances the network's learning capacity by controlling the shortest and longest gradient paths, effectively extracting features. The downsampling structure is designed with MP convolution layers, incorporating both max-pooling and convolution layers to facilitate parallel feature extraction and compression.

The neck of YOLOv7-tiny adheres to the PANet structure from YOLOv5. It involves operations for upsampling and downsampling features of different scales obtained from the backbone. This enables feature fusion, enhancing the model's ability to capture information across scales. The neck comprises modules such as SPPCSPC and ELAN-Y. The SPPCSP structure connects to the final layer of the backbone, introducing a substantial residual branch to optimize feature extraction, reduce computational load, and expand the receptive field.

The head module of YOLOv7-tiny encompasses detection heads operating at various scales, including large, medium, and small dimensions. These heads serve as the network's classifier and regressor, managing classification and regression tasks for object detection. Ultimately, this module achieves the dual objectives of object classification and precise localization in the context of object detection.

YOLOv7 encompasses GIoU, DIoU, CIoU, and ∝IoU. Notably, ∝IoU is well suited for training sets with small samples, but given the abundant number of images in our dataset, ∝IoU is not aligned with our circumstances. CIoU outperforms GIoU and DIoU by considering geometric parameters such as overlapping area, center-point distance, and aspect ratio. It effectively addresses issues related to inaccurate convergence and slow convergence speed. Hence, we employ CIoU as the model's loss function to ensure accurate and rapid convergence.

Building upon this model, we introduce YOLOv7-FSB, tailored for real-time detection and tracking of individuals in maritime distress within aerial imagery captured by UAVs.

3.1. Improvement-Based FSNet

In aerial images captured by drones, the targets are often small, placing higher demands on the backbone network of a model. However, the onboard computational resources are limited, necessitating a lightweight yet powerful feature extraction backbone network. Therefore, we propose an innovative lightweight backbone network named FSNet, specifically designed for efficiently extracting features relevant to individuals in distress in aerial images captured by unmanned aerial vehicles.

FasterNet [36], a crucial component of FSNet, comprises an inverted residual block that includes a partial convolution (PConv) layer, two 1×1 convolution layers, batch normalization, and ReLU activation. PConv has proven to be highly effective in extracting spatial features of individuals in maritime distress within aerial imagery. It not only reduces redundant computations but also minimizes memory access. Traditionally, lightweight model design focused on reducing floating-point operations (FLOPs). However, it was observed that simply reducing FLOPs did not significantly improve speed, as memory access often became the bottleneck. The introduction of PConv effectively addresses this issue.

The Simple Attention Module (SimAM) [37], incorporated into FSNet, introduces three-dimensional attention weights that consider both spatial and channel dimensions. Unlike previous attention modules, SimAM excels in capturing both channel and spatial features, offering enhanced flexibility and modularity without the need for additional parameters in the base network.

As illustrated in Figure 2, each FSNetBlock in FSNet cleverly integrates FasterNet with the SimAM attention module. Within an FSNetBlock, a combination of 3×3 PConv and 1×1 Conv is employed, where the use of 1×1 convolutional layers reduces the parameter count, accelerates the training speed, and enhances the model's nonlinear fitting capability. However, these layers exhibit a limited receptive field, impeding the acquisition of global features. Leveraging the lightweight attention mechanism of the SimAM module addresses this issue, resulting in an improved receptive field for the model. The FSNet lightweight backbone network is ultimately composed of multiple FSNetBlocks. This design significantly enhances the extraction of relevant features for individuals in distress at sea, enabling FSNet to efficiently and rapidly extract these crucial features from aerial images.

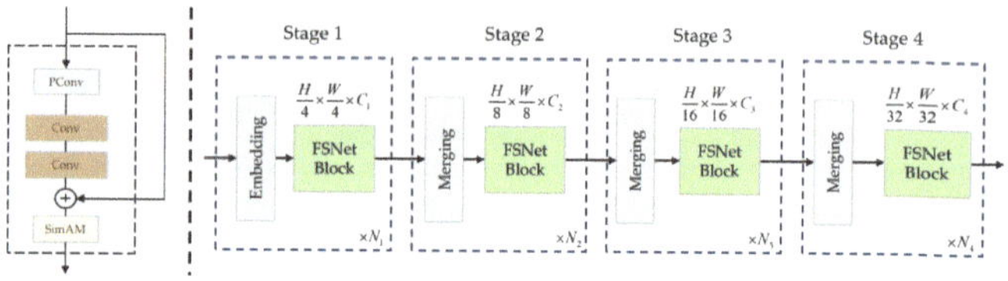

Figure 2. The left side is the FSNetBlock structure, and the right side is the structure and details of FSNet.

FSNet offers several advantages, including reduced parameters, lower computational requirements, and superior feature extraction efficiency. These attributes are particularly relevant for real-time feature extraction in the context of maritime distress detection using UAV-captured aerial imagery.

3.2. Improvement-Based SP-ELAN

When detecting individuals in aerial images, the submerged person is often only partially visible above the water surface, limiting the available features in the image. This poses a greater demand for the feature extraction capabilities of the model. To enhance feature extraction in computer vision applications within the context of aerial image analysis, we drew inspiration from ScConv [38] and PConv, leading to the development of the SP-ELAN module.

ScConv, following a split–transform–merge strategy, extracts features from multiple parallel branches with distinct roles and concatenates these outputs for the final result. It divides the input into two parts, each processed by branches dedicated to extracting diverse types of contextual information. One branch involves the adaptive calibration of input features through convolution filters, facilitating communication between the filters, while the other branch preserves the original spatial context. PConv, on the other hand, judiciously applies regular convolution to a subset of input channels, leaving the rest untouched. This approach optimizes computational resources and enhances the capability for extracting spatial features from aerial imagery.

As depicted in Figure 3, the SP-ELAN module leverages the advantages of ScConv and PConv. Specifically, we strategically replace certain Convs in the original ELAN with PConv, and the fused results undergo adaptive calibration through ScConv to yield the final output. This approach effectively integrates self-calibration with original spatial context information, generating highly discriminative output content while simplifying parameters. Such integration significantly enhances the precision of feature extraction, positioning it as a valuable complement to computer vision applications.

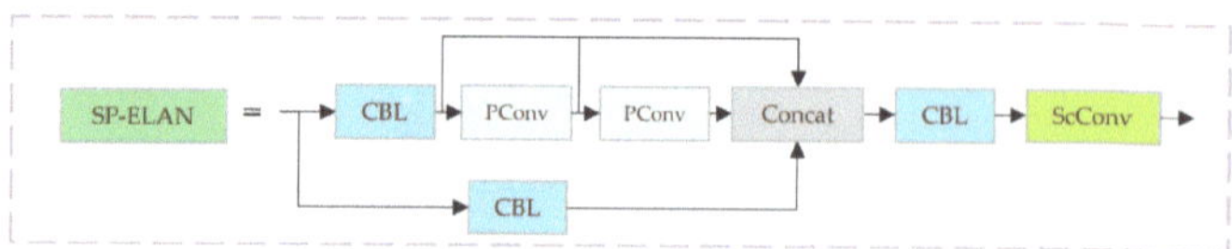

Figure 3. The structure of SP-ELAN.

3.3. Improvement-Based BiFPN-S

As a critical component bridging the gap between the backbone and the head, the neck plays a pivotal role in processing and amalgamating features extracted from the main trunk to better suit the requirements of object detection tasks. While the Feature Pyramid Network (FPN) has been a fundamental element in recognizing objects of varying sizes, its traditional top–down structure is limited by the unidirectional flow of information. To address this limitation, the Path Aggregation Network (PAN) introduced a bottom–up aggregation path, which improved accuracy but added complexity in terms of parameters and computational requirements.

In aerial images, the feature information of man-overboard situations holds particular significance. Traditional methods of feature fusion face challenges such as limitations in unidirectional information flow and high search costs. To establish a lightweight feature pyramid that strikes a balance between efficiency and accuracy, this paper introduces an enhanced version of the Bidirectional Feature Pyramid Network (BiFPN) [28], referred to as BiFPN-S in this paper. The original concept of BiFPN aimed to enhance pathways through efficient bidirectional cross-scale connections and weighted feature fusion. However, higher-level feature layers offer limited assistance in detecting small targets in aerial images. To address this issue and achieve the goal of eliminating redundant feature extraction calculations without compromising the model's ability to extract features related to man-overboard situations, we propose BiFPN-S.

In BiFPN-S, fusion is conducted as follows:

$$O = \sum_i \frac{w_i}{\varepsilon + \sum_j w_j} \cdot I_i \quad (1)$$

Here, a ReLU activation function is added after each set of weights, w_i, to ensure $w_i \geq 0$, $\varepsilon = 0.0001$ is introduced to avoid numerical instability. Taking the fifth layer as an example:

$$P_5^{td} = Conv\left(\frac{w_1 \cdot P_5^{in} + w_2 \cdot \mathrm{Re}\,size\left(P_6^{in}\right)}{w_1 + w_2 + \varepsilon}\right) \quad (2)$$

$$P_5^{out} = Conv\left(\frac{w_1' \cdot P_5^{in} + w_2' \cdot P_5^{td} + w_3' \cdot \mathrm{Re}\,size\left(P_5^{out}\right)}{w_1' + w_2' + w_3' + \varepsilon}\right) \quad (3)$$

where P_5^{td} represents the intermediate feature of level 5 in the top–down path, and P_5^{out} is the output feature of level 5 in the bottom–up path. In this context, $Conv(\cdot)$ denotes depthwise separable convolution, with batch normalization and activation functions added after each convolution. In our optimized version, we reduce the number of feature extraction layers, specifically eliminating the highest-level feature extraction layer, resulting in a lightweight model, as depicted in Figure 4. BiFPN-S provides an effective solution to enhance feature extraction in drone imagery without introducing unnecessary complexity.

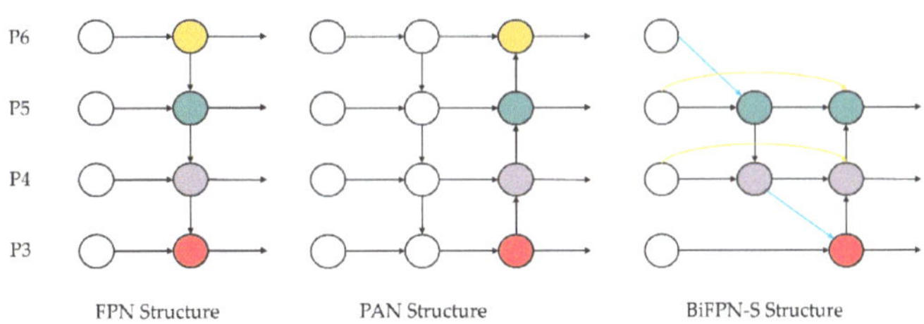

Figure 4. The details of BiFPN-S.

3.4. Tracking Model

Building upon our well-optimized object detection network, the next crucial step involves selecting an appropriate tracking method. In our evaluation, ByteTrack stands out as an exceptional choice, demonstrating outstanding performance and offering a streamlined solution for practical applications.

ByteTrack's approach places a strong emphasis on low-scoring detection boxes. It efficiently reassigns these low-scoring boxes for matching with previously unassociated tracking trajectories once the high-scoring boxes have been matched. Furthermore, when ByteTrack encounters detection boxes with sufficiently high scores but cannot find a match, it initiates the creation of new tracking trajectories. During data matching, ByteTrack has a low dependency on ReID features for appearance similarity calculations. This is a crucial advantage, particularly in scenarios such as aerial images depicting individuals in water rescue situations, where the features available for identifying distressed individuals are notably limited.

Considering the specific challenges posed by water rescue scenarios in aerial images and the emphasis on efficient tracking, we have chosen ByteTrack as our tracking model. It aligns well with our objectives, and as a tracking-by-detection method, ByteTrack's tracking effectiveness is highly contingent on the detector's performance. When the detector performs well, it yields favorable tracking results.

In the future, we will refer to the initials FSNet, SP-ELAN, and BiFPN-S as our method, denoted as 'YOLOv7-FSB.' In summary, to achieve efficient and effective detection and tracking of individuals in water rescue scenarios, we leverage the optimized object detection algorithm presented in this paper. This involves the seamless integration of YOLOv7-FSB with ByteTrack to accomplish the tracking task. The collaboration between these two components results in a lightweight and efficient model, perfectly suited for the demanding requirements of water rescue scenarios.

4. Experiments

4.1. Dataset

To validate the enhanced performance of YOLOv7-FSB, we conducted experiments using a meticulously curated dataset that combines selected MOBDrone [39] and SeeDronesSea [40] datasets.

The MOBDrone dataset comprises 49 videos captured with a DJI FC2 camera mounted on a Phantom 6310 Pro V4 drone. These videos portray various scenarios simulating individuals falling into water, encompassing both conscious and unconscious individuals, as well as other objects. This dataset has been post-processed to a resolution of 1080p. Professional annotators manually labeled the bounding boxes for objects falling into five categories: person, boat, surfboard, wood, and lifebuoy. The SeaDronesSee dataset showcases a diverse range of situations, with altitudes spanning from 5 m to 260 m and varying viewing angles from 0° to 90°. Each frame is accompanied by the pertinent altitude, angle, and other metadata. This dataset is captured using multiple cameras, providing a wide range of scenarios. The dataset included annotations covering various categories, such as swimmers, boats, jet skis, life-saving equipment, and buoys.

MOBDrone focuses on individuals in maritime man-overboard situations who are not wearing life jackets, whereas SeaDronesSee encompasses a wide range of scenarios related to an entire rescue process. These datasets have been amalgamated and processed for joint utilization in validating the proposed methodology. By combining the scenarios from MOBDrone and the diverse conditions from SeaDronesSee, we achieve a more comprehensive assessment of the methodology's performance in maritime search and rescue scenarios.

4.2. Experimental Setup

This research was conducted using the Linux operating system. The configuration included an Intel(R) Xeon(R) Gold 6338 CPU @ 2.00 GHz with a minimum clock frequency of 0.8 GHz, complemented by 512 GB of memory. We harnessed the power of the NVIDIA

A100-PCIE-40GB graphics processing unit (GPU) with 40 GB of memory capacity. To leverage GPU acceleration, the system ran on CUDA 11.7, and we primarily employed PyTorch 2.0.1 as the deep learning framework.

The interplay between software and hardware is crucial for model performance, and equally significant in our experiments are the hyperparameter settings. Model performance and speed are significantly impacted by image size. The step size for each parameter update is determined by the learning rate, which should be carefully adjusted according to the specific problem and the model at hand. To facilitate improved model convergence, modifications to the learning rate decay frequency are necessary. The batch size should be adjusted to an optimal value to maximize memory utilization. Increasing the number of workers is required to expedite the data preparation process, but it should be carried out cautiously to prevent memory leaks. Finally, the overall duration of model training should be determined by selecting an appropriate number of epochs based on the task at hand and the complexity of the model. The experimental settings are summarized in Table 1.

Table 1. Configuration.

Configuration	Name	Type
Hardware	CPU	Intel(R) Xeon(R) Gold 6338 CPU @ 2.00 GHz
	GPU	NVIDIA A100-PCIE-40GB
	Memory	512 GB
Software	CUDA	11.7
	PyTorch	2.0.1.
Hyperparameters	Image Size	1280 × 1280
	Learning Rate	0.01
	Learning Rate Decay Frequency	0.1
	Batch Size	64
	Workers	32
	Maximum Training Epochs	100

4.3. *Evaluation Metrics*

Confusion matrices are fundamental tools in the assessment of deep learning models, particularly in the context of computer vision. They offer a comprehensive means of evaluating a model's performance by quantifying correct and incorrect predictions for each class.

Typically, a confusion matrix comprises four quadrants: true positives (TP), false positives (FP), true negatives (TN), and false negatives (FN). True positives represent instances where the model accurately predicts positive cases. False positives refer to cases where the model incorrectly classifies negatives as positives. True negatives represent accurate predictions of negatives, while false negatives signify incorrect classifications of positives as negatives.

To provide a more comprehensive and intuitive evaluation of target detection and tracking algorithms, advanced evaluation metrics have been developed based on the foundation established by confusion matrices. These metrics significantly enhance our ability to understand and assess the performance of these algorithms.

4.3.1. Detection Evaluation Metrics

The evaluation of object detection algorithms hinges on two critical aspects: accurate object localization and correct object classification. In the context of deep learning for maritime applications involving target detection and tracking, evaluation metrics play a pivotal role in assessing the performance of detection models. IoU is a simple measure, but it focuses solely on overlap areas, disregarding considerations of object size and shape. While accuracy is easy to comprehend, it can be misleading in unbalanced datasets. Precision emphasizes correct positive predictions while ignoring false positives, and recall prioritizes correct positive detections while disregarding false negatives, making both less suitable for unbalanced data. AP calculates single-class accuracy, and mean average precision (mAP) aggregates multiple-class AP.

The primary evaluation metrics based on these criteria are presented in Table 2. In the "Note" column of the table, "↑" indicates that a higher value corresponds to better model performance, while "↓" indicates that a lower value signifies better performance. "Perfect" denotes the theoretical value for optimal performance.

Table 2. Object detection evaluation metric.

Metric	Description	Formula	Note
IoU	The degree of overlap between ground truth, B_{GT}, and predicted value, B_p.	$\frac{area(B_p \cap B_{GT})}{area(B_p \cup B_{GT})}$	↑
Precision	The proportion of true positive samples among the predicted positive samples.	$\frac{TP}{TP+FP}$	↑
Recall	The proportion of detected regions in true positive samples	$\frac{TP}{TP+FN}$	↑
AP	The area enclosed by the precision, P, and recall curve, R, with the coordinate axes, where $r_1, r_2, \cdots, r_n$ represents the recall rate corresponding to different precision interpolations.	$\sum_{i=1}^{n-1} (r_{i+1} - r_i) p_{\text{interp}}(r_{i+1})$	↑
mAP	The mean average precision is the average of the average precisions for each class, where K represents the total number of target classes.	$\frac{\sum_{i=1}^{K} AP_i}{K}$	↑

4.3.2. Tracking Evaluation Metrics

Evaluating the performance of object tracking algorithms requires consideration of two fundamental principles. Firstly, it involves a thorough examination of the algorithm's ability to accurately locate target positions. Secondly, it evaluates the algorithm's effectiveness in maintaining the individual identities of each target over time. Based on these principles, several advanced evaluation metrics have been developed to provide a comprehensive assessment.

Bernardin and Stiefelhagen introduced CLEAR MOT [41], which serves as a metric for measuring the tracking model's localization accuracy and association matching capabilities. Lisanti et al. proposed the ID Score [42], focusing on the stability and durability of the tracker's object tracking.

Table 3 summarizes these evaluation metrics.

Table 3. Object tracking evaluation metric.

Metric	Description	Formula	Note
ID_SW	The number of times the identity of the target changes.	/	↓
MOTA	The ability of a tracking algorithm to detect targets and maintain their trajectories is measured. Here, t represents the video frame corresponding to the t moment in the sequence.	$1 - \frac{\sum_t (FN_t + FP_t + ID\_SW_t)}{\sum_t GT_t}$	↑

4.4. Detection Results Discussion

4.4.1. Activation Function Comparison

We conducted a performance comparison of YOLOv7-tiny using two different activation functions: LeakyReLU and SiLU. SiLU is the default activation function used in YOLOv5 and YOLOv7, and it is recognized as an improved version of ReLU with a smoother behavior near zero. The experimental results, as shown in Table 4, reveal that LeakyReLU outperforms SiLU as an activation function. Based on these results, we have selected LeakyReLU as the activation function for our model.

Table 4. Results of activation function comparison test.

Detector	Precision (%)	Recall (%)	mAP (%)
YOLOv7_SiLU	88.2	83.4	86.4
YOLOv7_ReLU	88.8	85.0	86.9

4.4.2. Comparison of FasterNet and FSNet Performance

To validate the effectiveness of FSNet, we conducted a performance comparison between FSNet and FasterNet as backbone network models. The experimental results, as presented in Table 5, demonstrate that FSNet, being an enhanced version of FasterNet, excels in the detection of individuals in maritime distress in aerial imagery. This superiority is attributed to FSNet's reduction in redundant computations and memory access while maintaining efficient computational processes and excellent feature extraction capabilities.

Table 5. Results of the backbone comparison test.

Detector	Precision (%)	Recall (%)	mAP (%)
YOLOv7_FasterNet	88.9	84.4	87.1
YOLOv7_FSNet	89.8	85.1	87.3

4.4.3. Ablation Test Comparison

To analyze the effectiveness of different methods, a series of ablation experiments were designed. Each set of experiments employed the same dataset, training parameters, and methods for training. YOLOv7 served as the baseline for the initial experiments. Subsequent experiments introduced the following enhancements: FSNet was integrated into YOLOv7-A, YOLOv7-B used SP-ELAN to improve accuracy, YOLOv7-C combined with BiFPN-S, YOLOv7-D integrated both FSNet and BiFPN-S, YOLOv7-E combined SP-ELAN and BiFPN-S, YOLOv7-F combined FSNet and SP-ELAN, and YOLOv7-FSB synergistically utilized all three methods.

As depicted in Figure 5 and detailed in Table 6, it is evident that, compared to the original YOLOv7-tiny model, the detection speed of each modular method has not been significantly affected, while detection accuracy has improved. This underscores the effectiveness of the lightweight techniques employed in this study for personnel overboard detection. Bold font indicates the best-performing results, and the same applies throughout.

- The analysis of experiments 1–3 demonstrates that each method, when added, enhances both the model's detection accuracy and speed compared to the baseline YOLOv7-tiny model. Among these methods, FSNet stands out with the highest speed of 98.1 frames per second and an mAP of 87.3%. The introduction of SP-ELAN effectively integrates self-calibration and original spatial context information, making more efficient use of the device's computational capabilities, resulting in an mAP of 87.9% and a speed of 96.2 frames per second. The inclusion of BiFPN-S strengthens the model's capability for extracting features of individuals in maritime distress and improves the inference speed, leading to an mAP of 87.5% and a speed of 92 frames per second.
- Experiments 4–5 reveal that, in comparison to YOLOv7, the combination of FSNet with SP-ELAN and BiFPN-S individually boosts the mAP by 1.3%, 1.8%, and 1.5%, respectively. The integration of all three modules contributes to improved model performance while maintaining model detection speed, resulting in a more lightweight model without sacrificing accuracy.

- Based on the summarized optimization methods, we developed a lightweight search and rescue algorithm tailored for personnel overboard scenarios. This algorithm combines FSNet as the backbone network, integrates SP-ELAN for model enhancement, and incorporates BiFPN-S for feature fusion. The proposed method maintains the same outstanding detection speed as the original YOLOv7-tiny model while enhancing the mAP by 2.3%.

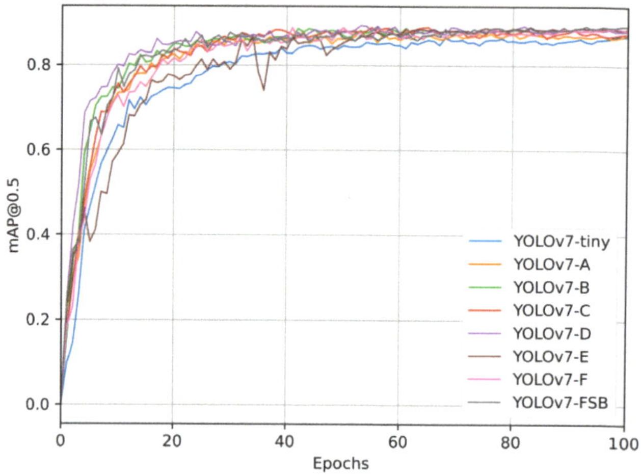

Figure 5. The average precision across various experiments.

Table 6. Results of ablation experiments with different methods.

Detector	FSNet	SP-ELAN	BiFPN-S	mAP (%)	Parameters/M	FPS
YOLOv7-tiny				86.9	6.02	95.3
YOLOv7_A	✓			87.3	**5.64**	**98.1**
YOLOv7_B		✓		87.9	6.18	96.2
YOLOv7_C			✓	87.5	6.02	92
YOLOv7_D	✓		✓	88.2	5.66	94.5
YOLOv7_E		✓	✓	88.7	6.19	93.1
YOLOv7_F	✓	✓		88.4	5.81	97.4
YOLOv7-FSB	✓	✓	✓	**89.2**	5.82	96.5

4.4.4. Comparison of Different Object Detection Models

To further analyze the performance of the lightweight algorithm in detecting individuals in maritime distress in aerial imagery, a comparison was conducted on the test dataset between YOLOv7-FSB and other networks, including YOLOv8n, YOLOv7-tiny, YOLOv5s, RetinaNet, SSD, and EfficientDet. Given that two-stage detection models have longer inference times, which may not meet real-time requirements, we opted to compare them with faster single-stage detectors. The recognition results of each network model are presented in Table 7.

When comparing the lightweight model YOLOv7-FSB to other models, the results indicate significant differences. The SSD model experiences a 47.6% decrease in speed, a parameter increase of 17.8 million, and a 22.5% drop in the mAP. The RetinaNet model shows a 28.3% reduction in speed, a parameter increase of 30.5 million, and a 21.1% decrease in the mAP. The EfficientDet model's speed decreases by 51.1%, with a parameter increase of 14.7 million and a 36.1% drop in the mAP. The YOLOv5s model's speed drops by 12.9%, with a parameter increase of 1.21 million and a 4.1% reduction in the mAP. The YOLOv7-tiny model experiences a negligible 0.01% reduction in speed, an increase of 0.2 million parameters, and a 2.3% decrease in the mAP.

Table 7. Comparison of detection performance for different methods.

Detector	mAP (%)	Parameters/M	FPS
SSD	66.7	23.6	50.6
RetinaNet	68.1	36.3	69.2
EfficientDet	53.1	20.5	47.2
YOLOv5s	85.1	7.03	84
YOLOv7-tiny	86.9	6.02	95.3
YOLOv8n	85.3	**3.1**	96.1
YOLOv7-FSB	**89.2**	5.82	**96.5**

By integrating FSNet, our model achieves heightened non-linear fitting capabilities, improved receptive fields, and an optimized parameter count and training speed. The incorporation of the SP-ELAN module adeptly allocates computational resources, seamlessly integrating self-calibration and spatial contextual information. Even with streamlined parameters, the model sustains a high degree of discriminative power in its outputs. With support from the BiFPN-S module, the model reinforces pathways through efficient bidirectional cross-scale connections and weighted feature fusion, ensuring a delicate balance between enhanced performance and swift inference speeds. In the context of personnel overboard detection and tracking, our proposed YOLOv7-FSB method surpasses all other algorithms in terms of detection accuracy and speed, making it a practical choice for real-world applications. These results underscore the effectiveness of the lightweight techniques employed in this study, enabling real-time detection tasks based on UAV imagery of individuals in maritime distress.

4.4.5. Results and Visualization

To enhance transparency and facilitate a more intuitive evaluation and comparison of the proposed small-target detection methods, we incorporated the Grad-CAM (Gradient-weighted Class Activation Mapping) technique [43]. This approach visualizes the heat maps corresponding to detected objects, providing a visual representation of the network's focus areas. Grad-CAM computes the gradients of the target class output based on the final convolutional layer's feature maps. Subsequently, these gradients are leveraged to perform a weighted summation, resulting in activation maps that emphasize regions of interest. The visualization of these attention regions is crucial for understanding the network's decision-making process, highlighting areas where the network is most confident about object detection or areas with high activation values.

Figure 6 illustrates Grad-CAM images for YOLOv7-tiny and YOLOv7-FSB across various scenarios, while Figure 7 showcases corresponding Grad-CAM images for different enhancement methods in the same scene. In these images, brighter regions denote specific areas prioritized by the network. The enhanced models exhibit remarkable feature extraction capabilities, particularly in recognizing individuals in distress at sea and mitigating the impact of noise on the model.

As a result, the YOLOv7-FSB model demonstrates outstanding performance in the tasks of locating and rescuing individuals in maritime distress. We have employed this model as the detector in our tracking-by-detection visual framework.

4.5. Tracking Results Discussion

4.5.1. Comparison of Different Tracking Models

To validate the performance of the improved YOLOv7-FSB model when combined with ByteTrack and DeepSORT separately, we conducted experiments using three sets of image sequences. The results are presented in Table 8.

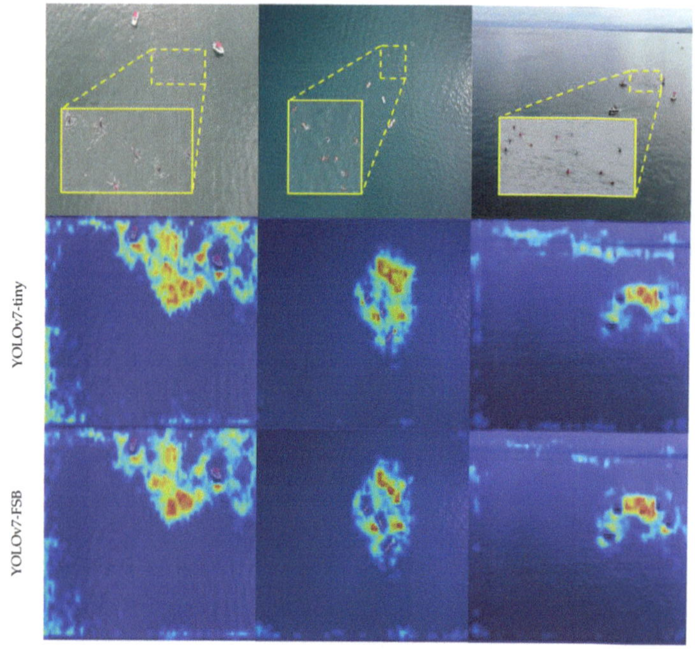

Figure 6. Grad-CAM diagrams of YOLOv7-tiny and YOLOv7-FSB.

Table 8. Comparison of tracking performance for different tracking methods.

Video Sequence	Tracker	MOTA (%)	ID Switch	FPS
Sequence 1	ByteTrack	83.4%	26	82
	DeepSORT	68.5%	68	23.6
Sequence 2	ByteTrack	87.6%	18	83.4
	DeepSORT	82.3%	26	24.5
Sequence 3	ByteTrack	85.4%	34	82.7
	DeepSORT	76.8%	84	21.7

In sequences 1 and 3, ByteTrack outperformed DeepSORT in terms of MOTA by 14.9% and 8.7%, respectively. Throughout the entire tracking process, the number of ID switches when using ByteTrack was significantly lower than when using DeepSORT. Trackers need to run in tandem with the detector, and the number of frames per second achieved using the ByteTrack algorithm was around 82. In sequence 2, ByteTrack's MOTA was 5.3% higher than DeepSORT's, with a similar number of ID switches. Typically, real-time monitoring requires processing at least 30 frames per second, ensuring the system can keep up with the flow of data. When combining the improved YOLOv7-FSC model with ByteTrack for validating video sequences, the average processing speed reached approximately 82.7 frames per second. Based on the data in the table, it is evident that ByteTrack, which relies on motion features alone, is more suitable for tracking maritime individuals in distress, offering significant advantages in terms of tracking efficiency and ID switching reduction.

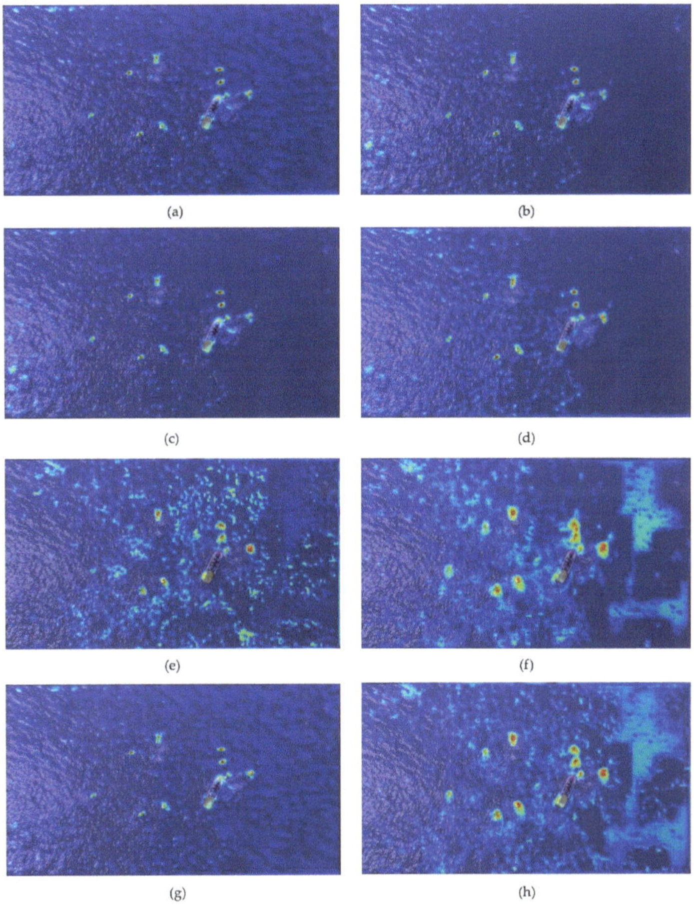

Figure 7. Grad-CAM diagram: (a) YOLOv7_A; (b) YOLOv7_B; (c) YOLOv7_C; (d) YOLOv7_D; (e) YOLOv7_E; (f) YOLOv7_F; (g) YOLOv7-tiny; (h) YOLOv7-FSB.

4.5.2. Comparison of Different Detection Models

To assess the impact of the improved YOLOv7-FSB model on tracking results, we conducted experiments by combining YOLOv7-tiny and YOLOv7-FSB separately with ByteTrack. The results are displayed in Table 9.

Across all the image sequences, when using YOLOv7-FSB as the detector for tracking, the number of ID switches was higher compared to when using YOLOv7-tiny as the detector. However, the MOTA value when combining YOLOv7-FSB with ByteTrack was higher than when using YOLOv7-tiny with ByteTrack. Typically, an increase in ID switches leads to a decrease in MOTA because it indicates a less stable tracking process. The situation where both the number of ID switches and the MOTA values increase is due to YOLOv7-FSB having fewer instances of missed detections, meaning it provides better object detection capabilities compared to YOLOv7-tiny. This further validates the effectiveness of the YOLOv7-FSB model proposed in this paper.

Table 9. Comparison of tracking performance for different detection methods.

Video Sequence	Tracker	MOTA (%)	ID Switch	FPS
Sequence 1	YOLOv7-tiny	76.3%	21	80.5
	YOLOv7-FSB	83.4%	26	82
Sequence 2	YOLOv7-tiny	81.1%	14	81.9
	YOLOv7-FSB	87.6%	18	83.4
Sequence 3	YOLOv7-tiny	79.3%	28	82.1
	YOLOv7-FSB	85.4%	34	82.7

In conclusion, the lightweight solution presented in this paper, YOLOv7-FSC, reduces the number of parameters and computations, allowing it to overcome the constraints of UAV computing resources. When detecting individuals in maritime distress in aerial images, this algorithm achieves detection speeds comparable to YOLOv7-tiny while significantly improving detection accuracy. Combining YOLOv7-FSC with ByteTrack results in excellent tracking performance, meeting the practical engineering requirements for UAV-based search and rescue operations. The model proposed in this article can find people who have fallen into water faster and improve the possibility of their rescue, which is of great significance.

4.6. Future Research Directions

In the future, our research will continue to focus on improving detector accuracy by exploring multi-sensor fusion techniques. We aim to integrate data from multiple sources, such as visible light, thermal imaging, near-infrared, etc., to enhance the system's detection robustness. Additionally, we plan to refine the architecture and expand functionalities to create a more powerful and comprehensive solution.

5. Conclusions

This study presents YOLOv7-FSC, a novel algorithm designed for real-time detection and tracking of individuals in maritime distress at sea when they fall overboard. The algorithm is built upon the YOLOv7-tiny framework with a focus on lightweight design to reduce detector size, maintain recognition speed, and enhance detection accuracy, thereby ensuring real-time recognition of individuals in maritime distress in aerial images. YOLOv7-FSB, the proposed model, employs FSNet as the backbone network, reducing redundant computations and memory access. The SP-ELAN module optimizes device computational capabilities. Additionally, the enhanced feature pyramid structure, BiFPN-S, bolsters its feature extraction capability and inference speed. To validate the effectiveness of YOLOv7-FSC, rigorous testing was conducted using datasets selected from MOBDrone and SeaDronesSee as benchmarks. The testing included ablation experiments and comparative trials. Subsequently, we combined the lightweight YOLOv7-FSC model with ByteTrack as a detection-based tracker, ensuring that the tracking performance meets the real-time and accuracy requirements for detecting and tracking individuals in maritime distress in aerial images. The visual model proposed in this paper can accurately perform real-time detection and tracking tasks, offering a suitable technological solution for large-scale and rapid search and rescue operations for individuals in maritime distress.

Author Contributions: Conceptualization, Y.Z., Y.Y. and Q.T.; formal analysis, Y.Z. and Y.Y.; funding acquisition, Y.Y. and Q.T.; investigation, Y.Z. and Q.T.; methodology, Y.Z., Y.Y. and Q.T.; project administration, Y.Z.; software, Y.Z. and Q.T.; supervision, Y.Y.; validation, Y.Z. and Y.Y.; writing the original draft, Y.Z. and Q.T.; writing—review and editing, Y.Z. and Y.Y. All authors have read and agreed to the published version of the manuscript.

Funding: This work is supported by the Ship Maneuvering Simulation in Yunnan Inland Navigation, grant number 851333J; the National Key R&D Program of China, grant number 2022YFB4300803; the National Key R&D Program of China, grant number 2022YFB4301402; and the Liaoning Provincial Science and Technology Plan (Key) project, grant number 2022JH1/10800096.

Data Availability Statement: Data are contained within the article.

Conflicts of Interest: The authors declare no conflict of interest.

References

1. European Maritime Safety Agency (EMSA). Annual Overview of Marine Casualties and Incidents. 2022. Available online: https://emsa.europa.eu/csn-menu/items.html?cid=14&id=4867 (accessed on 30 November 2022).
2. Tomic, T.; Schmid, K.; Lutz, P.; Domel, A.; Kassecker, M.; Mair, E.; Grixa, I.; Ruess, F.; Suppa, M.; Burschka, D. Toward a Fully Autonomous UAV: Research Platform for Indoor and Outdoor Urban Search and Rescue. *IEEE Robot. Automat. Mag.* **2012**, *19*, 46–56. [CrossRef]
3. Manyam, S.G.; Rasmussen, S.; Casbeer, D.W.; Kalyanam, K.; Manickam, S. Multi-UAV Routing for Persistent Intelligence Surveillance & Reconnaissance Missions. In Proceedings of the 2017 International Conference on Unmanned Aircraft Systems (ICUAS), Miami, FL, USA, 13–16 June 2017; pp. 573–580.
4. Jung, S.; Hwang, S.; Shin, H.; Shim, D.H. Perception, Guidance, and Navigation for Indoor Autonomous Drone Racing Using Deep Learning. *IEEE Robot. Autom. Lett.* **2018**, *3*, 2539–2544. [CrossRef]
5. Ammar, A.; Koubaa, A.; Ahmed, M.; Saad, A.; Benjdira, B. Vehicle Detection from Aerial Images Using Deep Learning: A Comparative Study. *Electronics* **2021**, *10*, 820. [CrossRef]
6. Yao, H.; Qin, R.; Chen, X. Unmanned Aerial Vehicle for Remote Sensing Applications—A Review. *Remote Sens.* **2019**, *11*, 1443. [CrossRef]
7. Ravindran, R.; Santora, M.J.; Jamali, M.M. Multi-Object Detection and Tracking, Based on DNN, for Autonomous Vehicles: A Review. *IEEE Sens. J.* **2021**, *21*, 5668–5677. [CrossRef]
8. Yang, T.; Jiang, Z.; Sun, R.; Cheng, N.; Feng, H. Maritime Search and Rescue Based on Group Mobile Computing for Unmanned Aerial Vehicles and Unmanned Surface Vehicles. *IEEE Trans. Ind. Inf.* **2020**, *16*, 7700–7708. [CrossRef]
9. Bomantara, Y.A.; Mustafa, H.; Bartholomeus, H.; Kooistra, L. Detection of Artificial Seed-like Objects from UAV Imagery. *Remote Sens.* **2023**, *15*, 1637. [CrossRef]
10. Zhao, X.; Xia, Y.; Zhang, W.; Zheng, C.; Zhang, Z. YOLO-ViT-Based Method for Unmanned Aerial Vehicle Infrared Vehicle Target Detection. *Remote Sens.* **2023**, *15*, 3778. [CrossRef]
11. Wang, Y.; Zou, H.; Yin, M.; Zhang, X. SMFF-YOLO: A Scale-Adaptive YOLO Algorithm with Multi-Level Feature Fusion for Object Detection in UAV Scenes. *Remote Sens.* **2023**, *15*, 4580. [CrossRef]
12. Wang, C.-Y.; Bochkovskiy, A.; Liao, H.-Y.M. YOLOv7: Trainable Bag-of-Freebies Sets New State-of-the-Art for Real-Time Object Detectors 2022. *arXiv* **2022**, arXiv:2207.02696.
13. Viola, P.; Jones, M. Rapid Object Detection Using a Boosted Cascade of Simple Features. In Proceedings of the 2001 IEEE Computer Society Conference on Computer Vision and Pattern Recognition, CVPR 2001, Kauai, HI, USA, 8–14 December 2001; Volume 1, pp. I-511–I-518.
14. Dalal, N.; Triggs, B. Histograms of Oriented Gradients for Human Detection. In Proceedings of the 2005 IEEE Computer Society Conference on Computer Vision and Pattern Recognition (CVPR'05), San Diego, CA, USA, 20–25 June 2005; Volume 1, pp. 886–893.
15. Felzenszwalb, P.; McAllester, D.; Ramanan, D. A Discriminatively Trained, Multiscale, Deformable Part Model. In Proceedings of the 2008 IEEE Conference on Computer Vision and Pattern Recognition, Anchorage, AK, USA, 23–28 June 2008; pp. 1–8.
16. LeCun, Y.; Bengio, Y.; Hinton, G. Deep Learning. *Nature* **2015**, *521*, 436–444. [CrossRef] [PubMed]
17. Girshick, R.; Donahue, J.; Darrell, T.; Malik, J. Rich Feature Hierarchies for Accurate Object Detection and Semantic Segmentation. In Proceedings of the 2014 IEEE Conference on Computer Vision and Pattern Recognition, Columbus, OH, USA, 23–28 June 2014; pp. 580–587.
18. He, K.; Zhang, X.; Ren, S.; Sun, J. Spatial Pyramid Pooling in Deep Convolutional Networks for Visual Recognition. *IEEE Trans. Pattern Anal. Mach. Intell.* **2015**, *37*, 1904–1916. [CrossRef] [PubMed]
19. Girshick, R. Fast R-CNN. In Proceedings of the 2015 IEEE International Conference on Computer Vision (ICCV), Santiago, Chile, 7–13 December 2015; pp. 1440–1448.
20. Ren, S.; He, K.; Girshick, R.; Sun, J. Faster R-CNN: Towards Real-Time Object Detection with Region Proposal Networks. *IEEE Trans. Pattern Anal. Mach. Intell.* **2017**, *39*, 1137–1149. [CrossRef] [PubMed]

21. Redmon, J.; Divvala, S.; Girshick, R.; Farhadi, A. You Only Look Once: Unified, Real-Time Object Detection. In Proceedings of the 2016 IEEE Conference on Computer Vision and Pattern Recognition (CVPR), Las Vegas, NV, USA, 26 June–1 July 2016; pp. 779–788.
22. Liu, W.; Anguelov, D.; Erhan, D.; Szegedy, C.; Reed, S.; Fu, C.-Y.; Berg, A.C. SSD: Single Shot MultiBox Detector. In *Computer Vision—ECCV 2016*; Leibe, B., Matas, J., Sebe, N., Welling, M., Eds.; Lecture Notes in Computer Science; Springer International Publishing: Cham, Switzerland, 2016; Volume 9905, pp. 21–37, ISBN 978-3-319-46447-3.
23. Simonyan, K.; Zisserman, A. Very Deep Convolutional Networks for Large-Scale Image Recognition 2015. *arXiv* **2014**, arXiv:1409.1556.
24. Redmon, J.; Farhadi, A. YOLOv3: An Incremental Improvement 2018. *arXiv* **2018**, arXiv:1804.02767.
25. Li, C.; Li, L.; Jiang, H.; Weng, K.; Geng, Y.; Li, L.; Ke, Z.; Li, Q.; Cheng, M.; Nie, W.; et al. YOLOv6: A Single-Stage Object Detection Framework for Industrial Applications 2022. *arXiv* **2022**, arXiv:2209.02976.
26. Ding, X.; Zhang, X.; Ma, N.; Han, J.; Ding, G.; Sun, J. RepVGG: Making VGG-Style ConvNets Great Again. In Proceedings of the 2021 IEEE/CVF Conference on Computer Vision and Pattern Recognition (CVPR), Nashville, TN, USA, 20–25 June 2021; pp. 13728–13737.
27. Lin, T.-Y.; Goyal, P.; Girshick, R.; He, K.; Dollar, P. Focal Loss for Dense Object Detection. *IEEE Trans. Pattern Anal. Mach. Intell.* **2020**, *42*, 318–327. [CrossRef]
28. Tan, M.; Pang, R.; Le, Q.V. EfficientDet: Scalable and Efficient Object Detection 2020. *arXiv* **2019**, arXiv:1911.09070.
29. Huang, S.-C.; Jaw, D.-W.; Hoang, Q.-V.; Le, T.-H. 3FL-Net: An Efficient Approach for Improving Performance of Lightweight Detectors in Rainy Weather Conditions. *IEEE Trans. Intell. Transport. Syst.* **2023**, *24*, 4293–4305. [CrossRef]
30. Wu, C.; Ye, M.; Zhang, J.; Ma, Y. YOLO-LWNet: A Lightweight Road Damage Object Detection Network for Mobile Terminal Devices. *Sensors* **2023**, *23*, 3268. [CrossRef]
31. Pang, Y.; Zhang, Y.; Kong, Q.; Wang, Y.; Chen, B.; Cao, X. SOCDet: A Lightweight and Accurate Oriented Object Detection Network for Satellite On-Orbit Computing. *IEEE Trans. Geosci. Remote Sens.* **2023**, *61*, 5608115. [CrossRef]
32. Liu, J.; Li, H.; Zuo, F.; Zhao, Z.; Lu, S. KD-LightNet: A Lightweight Network Based on Knowledge Distillation for Industrial Defect Detection. *IEEE Trans. Instrum. Meas.* **2023**, *72*, 3525713. [CrossRef]
33. Zhang, Y.; Sun, P.; Jiang, Y.; Yu, D.; Weng, F.; Yuan, Z.; Luo, P.; Liu, W.; Wang, X. ByteTrack: Multi-Object Tracking by Associating Every Detection Box. In *Computer Vision—ECCV 2022*; Avidan, S., Brostow, G., Cissé, M., Farinella, G.M., Hassner, T., Eds.; Lecture Notes in Computer Science; Springer Nature Switzerland: Cham, Switzerland, 2022; Volume 13682, pp. 1–21, ISBN 978-3-031-20046-5.
34. Wojke, N.; Bewley, A.; Paulus, D. Simple Online and Realtime Tracking with a Deep Association Metric. In Proceedings of the 2017 IEEE International Conference on Image Processing (ICIP), Beijing, China, 17–20 September 2017; pp. 3645–3649.
35. Bewley, A.; Ge, Z.; Ott, L.; Ramos, F.; Upcroft, B. Simple Online and Realtime Tracking. In Proceedings of the 2016 IEEE International Conference on Image Processing (ICIP), Phoenix, AZ, USA, 25–28 September 2016; pp. 3464–3468.
36. Chen, J.; Kao, S.; He, H.; Zhuo, W.; Wen, S.; Lee, C.-H.; Chan, S.-H.G. Run, Don't Walk: Chasing Higher FLOPS for Faster Neural Networks. In Proceedings of the 2023 IEEE/CVF Conference on Computer Vision and Pattern Recognition (CVPR), Vancouver, BC, Canada, 18–22 June 2023; pp. 12021–12031.
37. Yang, L.; Zhang, R.-Y.; Li, L.; Xie, X. SimAM: A Simple, Parameter-Free Attention Module for Convolutional Neural Networks. *arXiv* **2021**, arXiv:2110.06534.
38. Li, J.; Wen, Y.; He, L. SCConv: Spatial and Channel Reconstruction Convolution for Feature Redundancy. In Proceedings of the 2023 IEEE/CVF Conference on Computer Vision and Pattern Recognition (CVPR), Vancouver, BC, Canada, 18–22 June 2023; pp. 6153–6162.
39. Cafarelli, D.; Ciampi, L.; Vadicamo, L.; Gennaro, C.; Berton, A.; Paterni, M.; Benvenuti, C.; Passera, M.; Falchi, F. MOBDrone: A Drone Video Dataset for Man OverBoard Rescue. In *Image Analysis and Processing—ICIAP 2022*; Sclaroff, S., Distante, C., Leo, M., Farinella, G.M., Tombari, F., Eds.; Lecture Notes in Computer Science; Springer International Publishing: Cham, Switzerland, 2022; Volume 13232, pp. 633–644, ISBN 978-3-031-06429-6.
40. Varga, L.A.; Kiefer, B.; Messmer, M.; Zell, A. SeaDronesSee: A Maritime Benchmark for Detecting Humans in Open Water. In Proceedings of the 2022 IEEE/CVF Winter Conference on Applications of Computer Vision (WACV), Waikoloa, HI, USA, 3–8 January 2022; pp. 3686–3696.
41. Bernardin, K.; Stiefelhagen, R. Evaluating Multiple Object Tracking Performance: The CLEAR MOT Metrics. *EURASIP J. Image Video Process.* **2008**, *2008*, 246309. [CrossRef]
42. Ristani, E.; Solera, F.; Zou, R.; Cucchiara, R.; Tomasi, C. Performance Measures and a Data Set for Multi-Target, Multi-Camera Tracking. In *Computer Vision—ECCV 2016 Workshops*; Hua, G., Jégou, H., Eds.; Lecture Notes in Computer Science; Springer International Publishing: Cham, Switzerland, 2016; Volume 9914, pp. 17–35, ISBN 978-3-319-48880-6.
43. Selvaraju, R.R.; Cogswell, M.; Das, A.; Vedantam, R.; Parikh, D.; Batra, D. Grad-CAM: Visual Explanations from Deep Networks via Gradient-Based Localization. *Int. J. Comput. Vis.* **2020**, *128*, 336–359. [CrossRef]

remote sensing

MDPI

Article

Sensing and Deep CNN-Assisted Semi-Blind Detection for Multi-User Massive MIMO Communications

Fengxia Han [1,2], Jin Zeng [1,*], Le Zheng [3,4], Hongming Zhang [5] and Jianhui Wang [6]

1 School of Software Engineering, Tongji University, Shanghai 201804, China; fengxiahan@tongji.edu.cn
2 Engineering Research Center of Key Software Technologies for Smart City Perception and Planning, Ministry of Education, Shanghai 200003, China
3 School of Information and Electronics, Beijing Institute of Technology, Beijing 100081, China; le.zheng.cn@gmail.com
4 Beijing Institute of Technology Chongqing Innovation Center, Chongqing 401120, China
5 School of Information and Communication Engineering, Beijing University of Posts and Telecommunications, Beijing 100876, China; zhanghm@bupt.edu.cn
6 School of Electronic and Information Engineering, Tongji University, Shanghai 201804, China; 1952934@tongji.edu.cn
* Correspondence: zengjin@tongji.edu.cn

Abstract: Attaining precise target detection and channel measurements are critical for guiding beamforming optimization and data demodulation in massive multiple-input multiple-output (MIMO) communication systems with hybrid structures, which requires large pilot overhead as well as substantial computational complexity. With benefits from the powerful detection characteristics of MIMO radar, we aim for designing a novel sensing-assisted semi-blind detection scheme in this paper, where both the inherent low-rankness of signal matrix and the essential knowledge about geometric environments are fully exploited under a designated cooperative manner. Specifically, to efficiently recover the channel factorizations via the formulated low-rank matrix completion problem, a low-complexity iterative algorithm stemming from the alternating steepest descent (ASD) method is adopted to obtain the solutions in case of unknown noise statistics. Moreover, we take one step forward by employing the denoising convolutional neural network (DnCNN) to preprocess the received signals due to its favorable performance of handling Gaussian denoising. The overall paradigm of our proposed scheme consists of three stages, namely (1) target parameter sensing, (2) communication signal denoising and (3) semi-blind detection refinement. Simulation results show that significant estimation gains can be achieved by the proposed scheme with reduced training overhead in a variety of system settings.

Keywords: sensing-assisted communication; deep CNN; semi-blind detection; target AoA/AoD estimation; low-rank matrix completion

Citation: Han, F.; Zeng, J.; Zheng, L.; Zhang, H.; Wang, J. Sensing and Deep CNN-Assisted Semi-Blind Detection for Multi-User Massive MIMO Communications. *Remote Sens.* **2024**, *16*, 247. https://doi.org/10.3390/rs16020247

Academic Editors: Qian Du, Yanni Dong and Xiaochen Yang

Received: 14 December 2023
Revised: 1 January 2024
Accepted: 5 January 2024
Published: 8 January 2024

1. Introduction

With the vigorous development of the wireless communication industry and the deep integration with intelligent information processing technology, various remote sensing applications, e.g., vehicle-to-everything, smart manufacturing and environmental monitoring, come to the force, which puts forward increasing demand for network capacity and spectrum utilization [1–3]. As an important solution to support ultra-high transmission rate and intelligence requirements of the future mobile communications, massive multiple-input multiple-output (MIMO) inherently allows for simultaneous serving of multiple users by deploying large-scale antennas at a base station (BS) and facilitates the generation of directed beams towards target users, thereby expanding connected density and reducing the mutual interference in scenarios involving multiple users [4,5]. To obtain larger antenna gains and alleviate the pressure of spectrum scarcity, exploring novel transceiver design and frame structure operating at higher frequencies, such as the millimeter wave (mmWave) band [6,7], paves the way to guarantee reliable communications while enabling satisfactory user experiences.

1.1. Background

Considering that major obstacles preventing the deployment of massive MIMO are related to huge hardware costs and power consumptions imposed on the excessive number of radio frequency (RF) chains, hybrid analog/digital structures have been utilized to canonically route the RF chains via a well-designed phase-shifter network [8–10]. Furthermore, to compensate for prominent path loss encountered in mmWave channels and mitigate severe interference among multiple users, optimization on hybrid beamforming is expected to strike a favorable tradeoff between cost and rate [11–13]. One of the critical premises for obtaining the optimal beamforming is to acquire channel state information. Nevertheless, accurate channel estimation is challenging due to increased dimensionality of signal matrix and the application of hybrid architecture that limits the available observations. In this way, developing an efficient receiver scheme to enhance the transmission performance is essential in hybrid massive MIMO systems.

Recently, based on the fact that radar and communication systems share similar underline signal processing and gradually work in approximate frequency bands, there emerges a technological trend to integrate both functionalities into a single system [14–16], which offers an exciting opportunity to implement radar sensing by inherent wireless infrastructures. Given the potential ability of rapid and wide-range detection, it would be beneficial to leverage the radar sensing to fully extract geometric information from the surrounding environment [17,18] and thus assist in obtaining more precise parameter estimations. Toward this end, the coexistence of radar and communication components will constitute the primary architecture of this paper for enhanced performance gains.

1.2. Problem Being Addressed

Against the above background, as the dimensionality of antennas and carrier frequencies proliferate, the task of effectively balancing detection performance with computational cost becomes increasingly arduous. Consequently, the critical problem to be addressed is to propose novel solutions for comprehensive extraction of valuable information from the received signals in hybrid massive MIMO communication systems, thereby contributing to more reliable and cost-effective channel estimators.

1.3. Existing Solutions

In the literature, various channel estimators have been conducted for guiding the follow-up decoding and beamforming operations in hybrid massive MIMO systems. Among them, many techniques refer to the non-blind detections, which perform explicit channel estimation basically by transmitting and dealing with pilot symbols [19–21]. However, as the number of antennas increases, the pilot overhead could be enlarged accordingly and makes a negative impact on spectral efficiency. To address such issue, several research works began to focus on leveraging the sparse representation exhibited in delay-domain or angle-domain of mmWave channels [22–24], based on which a number of compressed-sensing (CS) tools have been adopted to figure out the formulated sparse recovery problem. Considering the grid-based CS estimator may lead to effects of basis mismatch, another gridless method appeared to construct continuous parameter domain by employing the atomic norm minimization [25]. Moreover, in order to avoid substantial computational complexity caused by high-dimensional optimizations and exploit more information from the payload data, our prior work [26] has proposed a semi-blind detection scheme for a hybrid massive MIMO system over frequency-selective fading channels, which utilized the uplink data as virtual pilots to improve the accuracy of channel estimation and data detection. However, the existing research has not sufficiently capitalized on the intrinsic features of massive MIMO systems and realistic channel characteristics, which constitute a significant challenge for achieving better tradeoff between training overhead and detection performance.

In addition to searching for rich scattering assumptions at higher frequencies, the low-rankness of massive MIMO systems has also been widely considered. In [27], according to the fact that both the number of BS antennas and the number of symbols in one coherence block are typically larger than the number of active users, the joint channel and data estimation was modeled as a low-rank matrix completion problem, for which two iterative algorithms were put forward to effectively recover the original channel and data. Moreover, schemes exploiting both the low-rank property and the knowledge of array responses have been proposed in [28–30], where the initial channel estimates provided by solutions to an inductive matrix completion would be further refined through multiple stages. Without loss of generality, most of the existing methods need to recover the matrix factorizations following an alternating minimization procedure [31–33]. Adhering to the existing methodologies, the intricate matrix multiplication or matrix inversion would induce unacceptable computational costs, especially when generalized to large-scale problems. As an effort to relieve the computational burden, an alternating steepest descent (ASD) algorithm was introduced in [34] to replace the least squares (LS) subproblem solutions with exact line-search updates. In view of the simplicity and moderate accuracy in the presence of non-negligible noise, we are inspired to tailor the ASD algorithm for estimating channels in hybrid massive MIMO systems.

Instead of simply obeying stochastic distributions, more realistic channels consisting of multiple propagation paths usually display a certain geometric structure. Since the process of communication channel estimation is similar to the radar target detection to some extent, research on making full use of sensing parameters to support communication functionality has attracted growing attention [35–37]. For example, by combining both functionalities of phased-MIMO radar and hybrid communication in the mmWave band, a novel strategy for joint target detection and channel estimation was proposed in [35]. To reduce the spectral resources as well as hardware cost, the employment of radar sensing for assisting the prediction of motion parameters in vehicular communication networks was investigated in [36]. Moreover, a robust MIMO-radar-aided channel estimation scheme deployed in multi-user (MU) MIMO communication systems has proven to be advantageous in improving the estimation accuracy with fewer training overhead [37]. Given that channel estimation may suffer from severe performance degradation in low signal-to-noise ratio (SNR) regions due to conceivable noise enhancement along with the LS-based estimation process, a supervised deep-learning (DL)-based signal denoiser was also involved in [37] to eliminate the noise on received signals before estimating the channel gains. However, training a neural network with high-dimensional signals and large datasets seems not tolerable in terms of complexity and latency [38]. Therefore, developing a cost-effective signal denoising module to make the state-of-the-art intelligent algorithms practically implementable for massive MIMO channel estimation is worth exploring.

1.4. Motivations and Objectives

A summary of the existing literature is provided in Table 1. The motivation for this paper stems from the notable limitations in current receiver techniques for hybrid massive MIMO communication systems. As one can see, challenges of large pilot overhead and substantial computational complexity will arise due to increased dimensionality and complicated signal nature in hybrid architectures, which restrict observation capabilities. Additionally, the growing demand for employing radar sensing in extracting environment information motivates us to seek advisable coordination between sensing and communication modules. In light of these challenges, the objective of this paper is to develop an efficient detection scheme tailored for hybrid massive MIMO systems with the aid of radar sensing, aimed at enhancing estimation performance and alleviating computational burdens.

Table 1. Existing works comparison on receiver techniques.

Existing Works	System Models	Receiver Techniques	Limitations or Features
[19–21]	mmWave MIMO communications	non-blind detection based on pilots only	Pilot overhead can be enlarged greatly with massive antennas.
[22–24]	mmWave MIMO communications	CS leveraging the channel sparsity	Gird-based estimator may lead to effects of basis mismatch.
[25]	mmWave MIMO communications	atomic norm minimization	High-dimensional optimizations may cause substantial computational complexity.
[26]	mmWave MIMO communications	semi-blind detection aided with payload data	Neglect of geometric structure of channel distributions.
[27]	massive MIMO communications	LS leveraging the channel low-rankness	Underexploitation of realistic channel characteristic with multiple propagation paths.
[28–30]	mmWave MIMO communications	multiple stages exploiting both the low-rankness and sparsity	High computational complexity when generalized to large-scale problems.
[35]	joint radar and communications	multiple signal classification (MUSIC) for angle estimation	Joint signal processing strategy can simultaneously detect targets while estimating the communication channel.
[36]	joint radar and communications	maximum likelihood (ML) estimator	Sensing parameter estimation can promote the dynamic topology construction of surrounding environments.
[37]	joint radar and communications	LS detection vs. DL-based denoiser	Efficient estimation with fewer training resources by eliminating noise before recovering channels.

1.5. Our Proposal and Contributions

In this paper, we propose a sensing and deep CNN-assisted semi-blind detection refinement scheme for a hybrid massive MU-MIMO system, where different functional modules, i.e., radar and communication, are integrated into the BS by partitioning multiple antennas. Specifically, both the low-rankness and geometric characteristic of high-frequency channels are fully exploited in a progressive manner. To reduce the large training overhead and substantial computational complexity, a novel transmission frame and an efficient receiver scheme are presented that can be split into three stages, namely (1) target parameter sensing, (2) communication signal denoising and (3) semi-blind detection refinement. Below, we crisply summarize the main contributions in this paper:

- A novel time-division duplex (TDD) transmission frame capable of coordinating the radar and communication operations is designed, based on which the root multiple signal classification (MUSIC) algorithm is firstly applied to environment sensing and then the extracted target angle information is utilized for refining the subsequent communication detection results.
- A generic representation for analog combining with phase shifters is considered, and the signal recovery problem is transformed into a low-rank matrix completion. To obtain the matrix factorizations with lower complexity, an iterative algorithm modified from ASD is proposed without any prior knowledge of noise statistics. In addition, different from the conventional pilot-only method, the semi-blind detection scheme is employed with reduced training overhead.
- A pre-trained denoising convolutional neural network (DnCNN) is adopted to preprocess the received signals before performing the semi-blind detection, which attempts

to handle Gaussian noise removal with unknown noise level and shows powerful ability of improved accuracy especially in low SNR regions.

The remainder of this paper is organized as follows. In Section 2, the system model is presented and the conventional pilot-only estimation method is reviewed. By transforming the signal recovery problem into a low-rank matrix completion, our proposed sensing and deep CNN-assisted semi-blind detection scheme is detailed with three stages in Section 3. Simulation results are provided in Section 4. Finally, conclusions are given in Section 5.

Notations: The list of acronyms used in this paper is summarized in Table 2. Unless otherwise specified, bold uppercase letters are used to represent matrices, bold lowercase letters are used to represent vectors and scalars are denoted by normal font. $\mathbf{A}(i,:)$ and $\mathbf{A}(:,j)$ denote the i-th row and j-th column of $\mathbf{A}$, respectively. $A_{i,j}$ denotes the (i,j)-th entry of $\mathbf{A}$. $\mathbf{A}^T$, $\mathbf{A}^H$ and $\mathbf{A}^*$ denote the operations of transpose, Hermitian transpose and element-wise conjugate, respectively. $\text{diag}(\mathbf{a})$ stands for a diagonal matrix with diagonal elements given by the vector $\mathbf{a}$. $\mathbb{E}$ and $\text{vec}(\cdot)$ represent the expectation operation and vectorization operation, respectively.

Table 2. List of acronyms.

Abbreviations	Expansion
ALS	Alternating Least Squares
AoA	Angle-of-Arrival
AoD	Angle-of-Departure
BS	Base Station
CNN	Convolutional Neural Network
CS	Compressed Sensing
DnCNN	Denoising Convolutional Neural Network
DL	Deep Learning
LMMSE	Linear Minimum Mean Squared Error
LoS	Line-of-Sight
LS	Least Squares
mmWave	Millimeter Wave
MIMO	Multiple-Input Multiple-Output
MU-MIMO	Multi-User MIMO
MUSIC	Multiple Signal Classification
NMSE	Normalized Mean Squared Error
RF	Radio Frequency
RMSE	Root Mean Squared Error
SNR	Signal-to-Noise Ratio
SVD	Singular Value Decomposition
TDD	Time-Division Duplex
UE	User Equipment
ULA	Uniform Linear Array

2. System Model and Background

In this paper, we consider a sensing-assisted hybrid massive MU-MIMO communication system, as shown in Figure 1, where K single-antenna users are served by a BS with two modules: one for the uplink communication and the other for the target sensing. Similarly as introduced in [37], the system operates in TDD mode and the multi-functional transmission frame structure is designed as shown in Figure 2, which consists of two parts. In the former part, the sensing module transmits probing signals and receives the echoes to detect the target directions. In the latter part, the communication module receives the pilots and payload data to complete the channel estimation and recover the uplink signals.

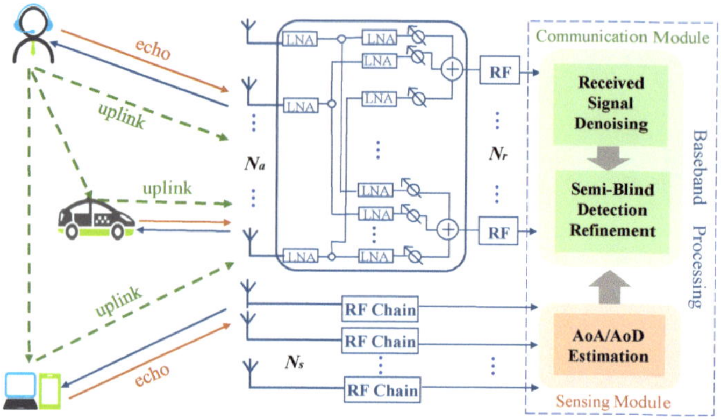

Figure 1. Sensing-assisted hybrid massive MU-MIMO communication system. The blue solid lines denote the probing signal sent by the sensing module. The red solid lines denote the corresponding echoes reflected by target users. The green dashed lines denote the uplink communication signals.

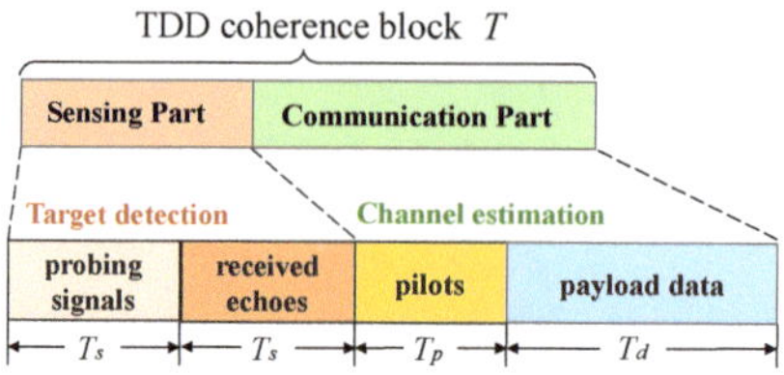

Figure 2. The structure of multi-functional transmission frame. T_s denotes the duration of transmitted waveform by radar module. T_p and T_d denote the duration of uplink pilots and payload data, respectively.

Suppose the uniform linear arrays (ULAs) are employed and are parallel with the users such that the angle-of-arrival (AoA) is identical to the angle-of-departure (AoD). Given the massive MIMO scenario and asymptotical orthogonality of steering vectors, we assume the sensing signals are emitted through line-of-sight (LoS) channels. Meanwhile, for the uplink communication, the transmitted signals from each single-antenna user will be scattered by the others around it when regarding them as point targets [36]. In addition, we consider the block-fading channel model, whose coefficients stay constant during the coherence block T.

Specifically, we first introduce the sensing model for assisting the AoA/AoDs estimation of target users. Then, the uplink communication model is presented, based on which the channel estimation problem for the hybrid massive MU-MIMO system is formulated. Finally, the conventional linear minimum mean squared error (LMMSE) estimator for solving the corresponding problem is reviewed.

2.1. Sensing Model

For the sensing module, we consider the mono-static radar case, where both the transmit and receiver arrays are equipped with N_s antenna elements and we assume there is sufficient isolation between them. Let $\mathbf{X}_s \in \mathbb{C}^{N_s \times T_s}$ denote the probing signal sent by the BS, then the corresponding echoes reflected by target users can be expressed as [35]

$$\mathbf{Y}_s = \sum_{k=1}^{K} \beta_k \mathbf{b}(\theta_k)\mathbf{b}^H(\theta_k)\mathbf{X}_s + \mathbf{N}_s, \tag{1}$$

where $\mathbf{N}_s \sim \mathcal{CN}(0, \sigma_s^2 \mathbf{I})$ represents the additive white Gaussian noise plus interference, $\beta_k \sim \mathcal{CN}(0, \sigma_{\beta_k}^2)$ denotes the reflection coefficient of the k-th target user, θ_k denotes the k-th user's azimuth angle relative to the BS and $\mathbf{b}(\theta_k) \in \mathbb{C}^{N_s \times 1}$ represents the steering vector satisfying

$$\mathbf{b}(\theta_k) = \frac{1}{\sqrt{N_s}}[1, e^{j\frac{2\pi}{\lambda}d\sin(\theta_k)}, \ldots, e^{j\frac{2\pi}{\lambda}d(N_s-1)\sin(\theta_k)}]^T, \tag{2}$$

where λ and d represent the signal wavelength and inter-element spacing, respectively. Generally, we set $d = \lambda/2$.

For notational brevity, we arrange all steering vectors into a uniform matrix $\mathbf{B}(\boldsymbol{\Theta}) = [\mathbf{b}(\theta_1), \ldots, \mathbf{b}(\theta_K)] \in \mathbb{C}^{N_s \times K}$ with $\boldsymbol{\Theta} = \{\theta_1, \ldots, \theta_K\}$ and set $\boldsymbol{\Lambda}_s = \mathrm{diag}(\beta_1, \ldots, \beta_K)$. Then, we rewrite the sensing signal model in (1) as a matrix form given by

$$\mathbf{Y}_s = \mathbf{B}(\boldsymbol{\Theta})\boldsymbol{\Lambda}_s \mathbf{B}^H(\boldsymbol{\Theta})\mathbf{X}_s + \mathbf{N}_s. \tag{3}$$

2.2. Uplink Communication Model

For the uplink communication module, suppose the BS adopts a hybrid analog–digital architecture with N_a antennas and N_r ($N_r < N_a$) RF chains to combine the incoming signals as depicted in Figure 1. Let $x_k[t]$ be the t-th transmitted symbol from user k with unit average power. Then, the received signals across N_a BS antennas is given by

$$\mathbf{r}_c[t] = \sum_{k=1}^{K} \mathbf{h}_k x_k[t] + \mathbf{n}_c[t], \quad 1 \le t \le T_c, \tag{4}$$

where $\mathbf{n}_c[t] \sim \mathcal{CN}(\mathbf{0}, \sigma_c^2 \mathbf{I})$ and $\mathbf{h}_k \in \mathbb{C}^{N_a \times 1}$ denotes the channel between user k and the BS.

Considering the hybrid arrays deployed, $\mathbf{r}_c[t]$ will further pass through the RF phase shifter network. In particular, let $\mathbf{W}[t] \in \mathbb{C}^{N_r \times N_a}$ denote the analog combining matrix, then the signal model combined in the RF band is consequently formulated as

$$\mathbf{y}_c[t] = \mathbf{W}[t]\mathbf{H}\mathbf{x}_c[t] + \mathbf{W}[t]\mathbf{n}_c[t], \tag{5}$$

where $\mathbf{y}_c[t] \in \mathbb{C}^{N_r \times 1}$ indicates the received signal after passing through the analog structure, $\mathbf{x}_c[t] \in \mathbb{C}^{K \times 1}$ denotes the transmitted signal vector from K users at time slot t and $\mathbf{H} \triangleq [\mathbf{h}_1, \ldots, \mathbf{h}_K] \in \mathbb{C}^{N_a \times K}$ denotes the channel matrix between K users and the BS.

Next, we elaborate on the high-frequency channel model containing the intrinsic geometric structure. Instead of simply modeling the channels by stochastic distributions as [27], we adopt the scattering model here to describe the geometric environment over which the communication takes place. Suppose the users equipped with single-antenna are randomly distributed within communication distance, then the transmitted signal from each user will be scattered by its surrounding users. In this case, according to the widely used Saleh–Valenzuela channel model as presented in [7], the multi-path channel vector between user k and the BS can be provided as

$$\mathbf{h}_k = \sqrt{\frac{N_a}{K}} \sum_{l=1}^{K} \alpha_{l,k} \mathbf{a}(\theta_{l,k}), \tag{6}$$

where $\alpha_{l,k}$ and $\theta_{l,k}$ denote the complex scattering coefficient and the azimuth angle relative to the BS for the l-th path of user k, respectively. Specifically, the path $l = k$ contributes to the direct path from user k to the BS [36]. In addition, $\mathbf{a}(\theta_{l,k}) \in \mathbb{C}^{N_a \times 1}$ denotes the steering vector expressed as

$$\mathbf{a}(\theta_{l,k}) = \frac{1}{\sqrt{N_a}}[1, e^{j\frac{2\pi}{\lambda}d\sin(\theta_{l,k})}, \ldots, e^{j\frac{2\pi}{\lambda}d(N_a-1)\sin(\theta_{l,k})}]^T. \tag{7}$$

By referring to (5)–(6), we arrive at the input–output relationship for uplink communication signals, which is given by

$$\mathbf{y}_c[t] = \mathbf{W}[t]\mathbf{A}(\boldsymbol{\Theta})\mathbf{G}\mathbf{x}_c[t] + \mathbf{v}[t], \tag{8}$$

where $\mathbf{v}[t] \triangleq \mathbf{W}[t]\mathbf{n}_c[t]$ and $\mathbf{v}[t] \sim \mathcal{CN}(\mathbf{0}, \sigma_c^2\mathbf{W}[t]\mathbf{W}[t]^H)$, $\mathbf{A}(\boldsymbol{\Theta}) = [\mathbf{a}(\theta_1), \ldots, \mathbf{a}(\theta_K)] \in \mathbb{C}^{N_a \times K}$ corresponds to the steering vector matrix and $\mathbf{G} \in \mathbb{C}^{K \times K}$ merges the channel path gains and scattering coefficients that satisfies

$$g_{i,j} = \sqrt{\frac{N_a}{K}}\alpha_{l,k}, \ \forall i = l, j = k, i, j \leq K. \tag{9}$$

According to the communication signal model in (8), even with sufficient resolution for target sensing parameters, i.e., the geometric pattern of the channel matrix $\mathbf{A}(\boldsymbol{\Theta})$ is known to the BS, the stochastic channel gains of different paths are still unable to infer. Therefore, a certain amount of pilots is required for completing the channel estimation.

2.3. The Conventional LMMSE Estimator

The conventional pilot-only estimation methods may cause large training overhead, especially with massive MIMO scenarios. To be specific, let $\mathbf{X}_p \triangleq [\mathbf{x}_c[1], \ldots, \mathbf{x}_c[T_p]] \in \mathbb{C}^{K \times T_p}$ denote the transmitted pilots, where T_p indicates the number of pilots. Then, the signal model of (8) can be transformed as [27]

$$\mathbf{y}_c[t] = \left(\mathbf{x}_p^T[t] \otimes \mathbf{W}[t]\right)(\mathbf{I}_K \otimes \mathbf{A}(\boldsymbol{\Theta}))\mathbf{g} + \mathbf{v}[t], \tag{10}$$

where $\mathbf{g} \triangleq \mathrm{vec}(\mathbf{G}) \in \mathbb{C}^{KK \times 1}$ and $\otimes$ denotes the Kronecker product.

In terms of whether the azimuth angles information is known or unknown to the BS, we define the channel counterparts to be estimated as $\mathbf{g}$ and $\mathbf{h}_u \triangleq (\mathbf{I}_K \otimes \mathbf{A}(\boldsymbol{\Theta}))\mathbf{g}$, respectively. By further defining $\mathbf{y}_p \triangleq [\mathbf{y}_c^T[1], \ldots, \mathbf{y}_c^T[T_p]]^T \in \mathbb{C}^{N_r T_p \times 1}$, $\mathbf{v}_p \triangleq [\mathbf{v}^T[1], \ldots, \mathbf{v}^T[T_p]]^T \in \mathbb{C}^{N_r T_p \times 1}$ and $\mathbf{W}_p \triangleq [\mathbf{x}_p^T[1] \otimes \mathbf{W}[1]; \ldots; \mathbf{x}_p^T[T_p] \otimes \mathbf{W}[T_p]] \in \mathbb{C}^{N_r T_p \times N_a K}$, the LMMSE estimates of the uplink channels can be given by [39]

$$\hat{\mathbf{h}}_u = (\mathbf{W}_p^H\mathbf{W}_p + \sigma_c^2\mathbf{I})^{-1}\mathbf{W}_p^H\mathbf{y}_p, \tag{11}$$

and

$$\hat{\mathbf{g}} = \left(\mathbf{A}_\mathbf{I}^H(\boldsymbol{\Theta})\mathbf{W}_p^H\mathbf{W}_p\mathbf{A}_\mathbf{I}(\boldsymbol{\Theta}) + \sigma_c^2\mathbf{I}\right)^{-1}\mathbf{A}_\mathbf{I}^H(\boldsymbol{\Theta})\mathbf{W}_p^H\mathbf{y}_p, \tag{12}$$

for which $\mathbf{A}_\mathbf{I}(\boldsymbol{\Theta}) = \mathbf{I} \otimes \mathbf{A}(\boldsymbol{\Theta})$ and we suppose $\mathbf{W}[t]\mathbf{W}[t]^H = \mathbf{I}$ according to popular choices of analog combining matrix.

Upon determining the estimated channels, let $\mathbf{X}_d \triangleq [\mathbf{x}_c[T_p+1], \ldots, \mathbf{x}_c[T_c]] \in \mathbb{C}^{K \times T_d}$ with $T_d = T_c - T_p$ indicating the length of payload data, then the LMMSE-based data detection can be performed by [39]

$$\hat{\mathbf{x}}_d[t] = \left((\mathbf{W}[t]\widehat{\mathbf{H}})^H\mathbf{W}[t]\widehat{\mathbf{H}} + \sigma_c^2\mathbf{I}\right)^{-1}(\mathbf{W}[t]\widehat{\mathbf{H}})^H\mathbf{y}_c[t], \tag{13}$$

where $\widehat{\mathbf{H}} = \mathrm{vec}^{-1}(\hat{\mathbf{h}}_u)$ or $\widehat{\mathbf{H}} = \mathrm{vec}^{-1}(\mathbf{A}_\mathbf{I}(\boldsymbol{\Theta})\hat{\mathbf{g}})$ depends on whether the azimuth angles of target users are already known to the BS.

From the above discussions, we can conclude that for the conventional pilot-only estimator, the number of transmitted pilots and the variance of noise have significant influences on the accuracy of channel estimation. To be specific, for the most common case without knowledge of geometric information [27], since there are involved $N_a \times K$ unknowns in $\widehat{\mathbf{H}}$, the number of required pilots needs to scale with $T_p \propto N_a K/N_r$, which may occupy a large portion of transmission frame given the massive receive antennas. For the case where the channel estimation is conducted after resolving the AoA/AoDs [37], although the number of required pilots can be reduced to $T_p \propto KK/N_r$, the accuracy of the LMMSE estimator will be greatly affected by non-negligible noise even with super-resolution of angle detections [40]. Therefore, to reduce the large training overhead while maintaining acceptable estimation accuracy, we will propose a sensing and deep denoising-assisted semi-blind detection scheme to improve the performance achieved by the existing methods.

3. Proposed Sensing and Deep CNN-Assisted Semi-Blind Detection Scheme

In this section, we introduce the proposed scheme to recover the signals formulated in Section 2. Based on the transmission frame structure designed in Figure 2, the proposed estimator will successively address the AoAs estimation, low-rank matrix completion and semi-blind detection refinement to exploit both the low-rank property and the geometric knowledge of channels in hybrid massive MIMO systems. In particular, by integrating the low-complexity spectrum estimation algorithm and high-efficiency matrix completion algorithm, MIMO radar's rapid detection characteristics can be fully utilized to assist in obtaining more accurate channel estimation as well as taking up less training overhead.

3.1. Target Parameter Sensing

Considering the potential superiority of MIMO radar in achieving high-resolution angular estimations of target users, especially in identifying user positions, scattering paths and channel scenarios, the primary task of the sensing module is to detect and acquire the geometric directions of target users.

As shown in Figure 2, the angular estimation phase at the sensing module consists of two parts, during which the BS firstly sends probing signals and then receives the echo signals. Suppose the azimuth angles of K users are randomly distributed as $\mathbf{\Theta} \sim \mathrm{U}[-\frac{\pi}{2}, \frac{\pi}{2}]$. To achieve the optimal estimation performance, we assume the probing signals sent by different antennas are mutually orthogonal such that the energy is evenly dispersed at each angle, i.e.,

$$\mathbf{R}_s = \frac{1}{T_s}\mathbf{X}_s\mathbf{X}_s^H = \frac{P_s}{N_a}\mathbf{I}_{N_s}, \tag{14}$$

where P_s denotes the total energy of the BS. Moreover, under the assumption that the additive white Gaussian noise is unrelated to the probing signals and the averaged received power at each antenna being fixed as one, we can obtain the covariance matrix of the received echoes as

$$\mathbf{R}_{Y_s} = \frac{1}{T_s}\mathbf{Y}_s\mathbf{Y}_s^H = \mathbf{B}(\mathbf{\Theta})\mathbf{R}_{\tilde{s}}\mathbf{B}^H(\mathbf{\Theta}) + \sigma_s^2\mathbf{I}_{N_a}, \tag{15}$$

where

$$\mathbf{R}_{\tilde{s}} = \frac{1}{T_s}\mathbf{\Lambda}_s\mathbf{B}^H(\mathbf{\Theta})\mathbf{B}(\mathbf{\Theta})\mathbf{\Lambda}_s^*. \tag{16}$$

To recover the AoAs of different target users, subspace decomposition-based methods can be invoked to separate the effective signals from background noise. Specifically, the signal subspace contains information from different directions, while the noise subspace comprises unrelated noise information. Then, following the standard subspace decomposition process [41], we present the signal and noise subspaces by taking eigenvalue decomposition of (15) as follows

$$\mathbf{R}_{Y_s} = (\mathbf{U}_s, \mathbf{U}_n)\mathbf{\Lambda}_r(\mathbf{U}_s, \mathbf{U}_n)^H, \tag{17}$$

where $\mathbf{\Lambda}_r \in \mathbb{C}^{N_s \times N_s}$ is a diagonal matrix with eigenvalues in descending order and $\mathbf{U}_s \in \mathbb{C}^{N_s \times K}$ and $\mathbf{U}_n \in \mathbb{C}^{N_s \times (N_s - K)}$ correspond to eigenvector matrices that constitute the signal and noise subspaces, respectively. It can be seen that the columns of $\mathbf{B}(\mathbf{\Theta})$ and $\mathbf{U}_s$ span the same subspace. Without loss of generality, given the orthogonal relationship between the signal and noise subspaces, the pseudo-spectrum used for locating the AoAs of K target users can be readily formulated as

$$P_a(\theta) = \frac{1}{|\mathbf{b}^H(\theta)\mathbf{U}_n\mathbf{U}_n^H\mathbf{b}(\theta)|}. \tag{18}$$

Instead of applying the classical approach that finds the angles at peaks through tedious spectrum search, we here employ the root-MUSIC approach, which transforms the problem of spectral peak detection into polynomial root finding, thus eliminating the dependence on a predefined angular grid. As a result, the proposed approach can avoid the grid mismatch phenomenon induced by superficial quantization and identify distinct source signals even when their angular differences are minimal, which allows for super-resolution angular estimation with low complexity. Specifically, according to [30], we rewrite the denominator of (18) as

$$\mathbf{b}^H(\theta)\mathbf{U}_n\mathbf{U}_n^H\mathbf{b}(\theta) = \sum_{n=0}^{N_s-1}\sum_{m=0}^{N_s-1} e^{-j\xi m \sin\theta} C_{m,n} e^{j\xi n \sin\theta} = \sum_{l=-N_s+1}^{N_s-1} C_l e^{-j\xi l \sin\theta}, \tag{19}$$

where $\xi = \frac{2\pi d}{\lambda}$, $\mathbf{C} = \mathbf{U}_n\mathbf{U}_n^H$ represents the correlation matrix with C_l denoting the sum of elements along its l-th diagonal. Let $z \triangleq e^{-j\xi \sin\theta}$, then we can further simplify (19) to a polynomial

$$f(z) = \sum_{l=-N_s+1}^{N_s-1} C_l z^l. \tag{20}$$

Then, the $2(N_s - 1)$ roots of $f(z)$ exactly correspond to the poles of the MUSIC spectrum given by

$$z_i = |z_i| e^{j\arg(z_i)}, \ i = 1, \ldots, 2(N_s - 1), \tag{21}$$

K roots out of which lying closest to the unit circle are chosen to yield the AoAs as

$$\hat{\theta}_i = \sin^{-1}\left(-\frac{\arg(z_i)}{\xi}\right), \ i = 1, \ldots, K, \tag{22}$$

where $\arg(z_i)$ denotes the phase angle of z_i.

3.2. *Low-Rank Matrix Completion*

When the sensing module obtains the knowledge of all the AoAs of target users, the angle information will be transferred to the communication module, which is devoted to identifying the coefficients of different scattering paths by using the uplink pilot and payload data. For the sake of saving the training overhead while making full use of available information at hand, we consider to explore the low-rankness of the massive MIMO system and then recover the uplink channels based on low-rank matrix completion.

Before constructing the signal recovery problem, we show a generic model of the analog combining matrix and discuss how the transmitted signals from K users are related to the received signals at the BS. Specifically, given the fully-connected hybrid structure as presented in Figure 1 such that each RF chain is connected to all antennas, various realizations of the phase shifter with constant modulus can be modeled by $\omega e^{j\phi}$ and then make up the combining matrix $\mathbf{W}[t]$. To this end, we can regard the generation of analog combining matrix at each time slot as randomly selecting N_r rows from a DFT matrix $\mathbf{F} \in \mathbb{C}^{N_a \times N_a}$ with elements given by [42]

$$F_{m,n} = \frac{1}{\sqrt{N_a}} e^{-\frac{j2mn\pi}{N_a}}, \quad m,n = 0,\ldots,N_a-1. \tag{23}$$

According to the signal model formulated in (5), by further defining $\widetilde{\mathbf{H}} \triangleq \mathbf{FH} \in \mathbb{C}^{N_a \times K}$ and $\mathbf{z}[t] \triangleq \widetilde{\mathbf{H}}\mathbf{x}_c[t] \in \mathbb{C}^{N_a \times 1}$, we can find that the received signal $\mathbf{y}_c[t]$ is exactly composed of N_r elements from $\mathbf{z}[t]$. Then, let $\Omega \in \mathbb{C}^{N_r \times T_c}$ represent the index set indicating each independent selection of $\mathbf{W}[t]$ during the communication period, and denote $\mathbf{Y}_c^{\Omega} \in \mathbb{C}^{N_a \times T_c}$ as the observation matrix with entries in Ω given by $\{\mathbf{y}_c[t]\}_{t=1}^{T_c}$ and otherwise zero, then the received signals can be recast as a matrix form

$$\mathbf{Y}_c^{\Omega} = \mathcal{P}_{\Omega}(\mathbf{Z}+\mathbf{N}), \tag{24}$$

where $\mathbf{Z} \triangleq [\mathbf{z}[1],\ldots,\mathbf{z}[T_c]] \in \mathbb{C}^{N_a \times T_c}$, $\mathbf{N} \sim \mathcal{CN}(\mathbf{0},\sigma_c^2\mathbf{I})$ and $\mathcal{P}_{\Omega}(\cdot)$ represents a linear operation that preserves the entries in Ω while filling them not in Ω with zero.

Based on (24), we plausibly view $\mathbf{Y}_c^{\Omega}$ as an incomplete observation of the matrix $\mathbf{Z}$ corrupted by the Gaussian noise $\mathbf{N}$, where $\mathbf{Z} = \mathbf{HX}_c$ with $\mathbf{X}_c \triangleq [\mathbf{x}_c[1],\ldots,\mathbf{x}_c[T_c]] \in \mathbb{C}^{K \times T_c}$. Noting the fact that in massive MIMO scenarios, the number of users is normally much smaller than the number of BS antennas and the length of coherence block, i.e., $N_a, T_c \gg K$, the matrix $\mathbf{Z}$ displays an inherent low-rank property, i.e., $\text{rank}[\mathbf{Z}] \leq K \ll \min\{N_a, T_c\}$. Moreover, since multiplying by unitary matrix does not change the distributions, $\widetilde{\mathbf{H}}$ and $\mathbf{H}$ are interchangeable for the channel estimation. Consequently, as derived in [27], the channel and data estimation is formulated as the following low-rank matrix completion problem

$$\min_{\mathbf{U}\in\mathbb{C}^{N_a \times K}, \mathbf{V}\in\mathbb{C}^{K \times T_c}} \text{rank}[\mathbf{UV}], \qquad s.t. \quad \mathbf{Y}_c^{\Omega} = \mathcal{P}_{\Omega}(\mathbf{UV}+\mathbf{N}), \tag{25}$$

where $\mathbf{U}$ and $\mathbf{V}$ are related to the solutions of channel and data estimates $\widehat{\mathbf{H}}$ and $\widehat{\mathbf{X}}_c$, respectively.

Different from the LMMSE estimator, which requires channel and noise second order statistics, we here propose to employ a low-complexity algorithm invoked in [34] to solve (25) without any prior information and exclusively based on matrix-wise updates. In this vein, considering the original rank-minimization problem is non-convex and generally NP-hard, an approximate relaxation scheme such as Frobenious norm minimization comes to the scene, which suggests to recover the matrix factorizations from

$$(\mathbf{U}^*, \mathbf{V}^*) = \arg\min_{\substack{\mathbf{U}\in\mathbb{C}^{N_a \times K} \\ \mathbf{V}\in\mathbb{C}^{K \times T_c}}} \frac{1}{2}\left\|\mathbf{Y}_c^{\Omega} - \mathcal{P}_{\Omega}(\mathbf{UV})\right\|_F^2. \tag{26}$$

To solve the above optimization problem, alternative minimization techniques, i.e., successively finding out $\mathbf{U}$ and $\mathbf{V}$ with the minimized residual, have been widely used due to their simplicity and flexibility. As an effort to circumvent the computational burden, a better choice called ASD algorithm incorporates a simple line-search to update solutions instead of dealing with the least squares subproblems. To be specific, let $\nabla f_{\mathbf{V}}(\mathbf{U})$ and $\nabla f_{\mathbf{U}}(\mathbf{V})$ define the gradient of the function $f(\mathbf{U},\mathbf{V}) = ||\mathbf{Y}_c^{\Omega} - \mathcal{P}_{\Omega}(\mathbf{UV})||_F^2$ with respect to $\mathbf{U}$ and $\mathbf{V}$, respectively, the stepsizes for updating solutions along the steepest gradient descent direction can be explicitly computed as [34]

$$t_{\mathbf{U}} = \frac{||\nabla f_{\mathbf{V}}(\mathbf{U})||_F^2}{||\mathcal{P}_{\Omega}(\nabla f_{\mathbf{V}}(\mathbf{U})\mathbf{V})||_F^2}, \tag{27}$$

and

$$t_{\mathbf{V}} = \frac{||\nabla f_{\mathbf{U}}(\mathbf{V})||_F^2}{||\mathcal{P}_{\Omega}(\mathbf{U}\nabla f_{\mathbf{U}}(\mathbf{V}))||_F^2}, \tag{28}$$

where

$$\nabla f_{\mathbf{V}}(\mathbf{U}) = -(\mathbf{Y}_c^{\Omega} - \mathcal{P}_{\Omega}(\mathbf{UV}))\mathbf{V}^H, \tag{29}$$

$$\nabla f_{\mathbf{U}}(\mathbf{V}) = -\mathbf{U}^H(\mathbf{Y}_c^{\Omega} - \mathcal{P}_{\Omega}(\mathbf{UV})). \tag{30}$$

For initializations, the intuition behind (8) tells that the random realization of the channel mainly depends on the distribution of $\mathbf{G}$ assuming perfect estimates of angles. In view of this, given the angle information $\widehat{\mathbf{\Theta}}$ estimated by Section 3.1, we initialize $\mathbf{U}_0 = \mathbf{A}(\widehat{\mathbf{\Theta}})\mathbf{G}_0 \in \mathbb{C}^{N_a \times K}$ and $\mathbf{V}_0 \in \mathbb{C}^{K \times T_c}$ with elements of $\mathbf{G}_0 \in \mathbb{C}^{K \times K}$ and $\mathbf{V}_0$ drawing from i.i.d. Gaussian random variables with zero mean and unit variance. Then, the key idea of our proposed algorithm is to alternatively update $\mathbf{U}$ and $\mathbf{V}$ until the termination criteria are reached. Specifically, for the $(t+1)$-th iteration, after obtaining the steepest descent stepsizes and gradient descent directions following (27)–(30), we propose to update $\mathbf{U}_t$ to $\mathbf{U}_{t+1}$ in a two-step way, i.e.,

$$\mathbf{U}_{t+1}^{'} = \mathbf{U}_t - t_{\mathbf{U}_t}\nabla f_{\mathbf{V}_t}(\mathbf{U}_t), \;\; \mathbf{U}_{t+1} = \mathbf{A}(\widehat{\mathbf{\Theta}})\mathbf{G}_{t+1}, \tag{31}$$

where

$$\mathbf{G}_{t+1} = (\mathbf{A}^H(\widehat{\mathbf{\Theta}})\mathbf{A}(\widehat{\mathbf{\Theta}}))^{-1}\mathbf{A}^H(\widehat{\mathbf{\Theta}})\mathbf{U}_{t+1}^{'}. \tag{32}$$

As a similar procedure, $\mathbf{V}_t$ will be updated to $\mathbf{V}_{t+1}$ by

$$\mathbf{V}_{t+1} = \mathbf{V}_t - t_{\mathbf{V}_t}\nabla f_{\mathbf{U}_{t+1}}(\mathbf{V}_t), \tag{33}$$

where

$$t_{\mathbf{V}_t} = \frac{||\nabla f_{\mathbf{U}_{t+1}}(\mathbf{V}_t)||_F^2}{||\mathcal{P}_{\Omega}(\mathbf{U}_{t+1}\nabla f_{\mathbf{U}_{t+1}}(\mathbf{V}_t))||_F^2}. \tag{34}$$

Note that obtaining the gradient and stepsize will involve the product of a residual matrix and a $N_a \times K$ or $K \times T_c$ matrix. However, such operation only implements once at the beginning of each iteration and then can be efficiently used to update variables. Moreover, since the above approach replaces the least square solutions with an exact line-search step and thus avoids complex operations of matrix inversions, the computational cost for each iteration is much smaller than that of alternating least squares (ALS) and the conventional singular value decomposition (SVD) methods.

3.3. *Refined Semi-Blind Detection*

3.3.1. Ambiguity Removal

Although the above approach is able to provide solutions to (26) directly from the received observation $\mathbf{Y}_c^{\Omega}$, the non-uniqueness of matrix factorization may lead to great estimation bias. More clearly, suppose the initial channel and data estimates obtained from the ASD iterations are denoted by $\widehat{\mathbf{U}}$ and $\widehat{\mathbf{V}}$, respectively. If there exists an invertible matrix $\mathbf{\Sigma} \in \mathbb{C}^{K \times K}$ satisfying $\mathbf{HX}_c = \widehat{\mathbf{U}}\mathbf{\Sigma}\mathbf{\Sigma}^{-1}\widehat{\mathbf{V}}$, then solving (26) would not guarantee a unique recovery for $\mathbf{H}$ and $\mathbf{X}$. Therefore, a certain amount of pilots are required to resolve the ambiguity.

On this basis, considering the number of elements that need to be determined is $K \times K$, we develop a pilot and data placement as $\mathbf{X}_c \equiv [\mathbf{X}_p, \mathbf{X}_d]$, where $\mathbf{X}_p \in \mathbb{C}^{K \times T_p}$ with $T_p \propto K$. Then, we have

$$\widehat{\mathbf{V}} = \mathbf{\Sigma}\mathbf{X}_c = \mathbf{\Sigma}[\mathbf{X}_p, \mathbf{X}_d]. \tag{35}$$

Due to the fact that pilot sequences are usually selected to be orthogonal, i.e., $\mathbf{X}_p\mathbf{X}_p^H = T_p \cdot \mathbf{I}_K$, we can obtain $\mathbf{\Sigma}$ as

$$\mathbf{\Sigma} = \frac{\widehat{\mathbf{V}}(:, 1 \sim T_p) \cdot \mathbf{X}_p^H}{T_p}, \tag{36}$$

where $\widehat{\mathbf{V}}(:, i \sim j)$ refers to the sub-matrix consisting of i-th to j-th column vectors of $\widehat{\mathbf{V}}$. Upon resolving $\boldsymbol{\Sigma}$, the data and channel estimation can be accordingly given by [26]

$$\widehat{\mathbf{X}}_d = \boldsymbol{\Sigma}^{-1}\widehat{\mathbf{V}}(:, T_p + 1 \sim T_c), \tag{37}$$

and

$$\widehat{\mathbf{H}} = \widehat{\mathbf{U}}\widehat{\mathbf{V}}(:, 1 \sim T_p)\mathbf{X}_p^H / T_p. \tag{38}$$

The process of our proposed sensing-assisted semi-blind detection scheme is detailed in Algorithm 1. By carefully inspecting the workflow of the proposed approach, we can observe that the resulting estimation is not directly obtained from matrix factorization solutions; indispensable pilots are needed to facilitate the overall estimation as shown in Line 14, which explains the origin of the name "semi-blind detection". The advantage of employing semi-blind detection lies in the fact that while the conventional non-blind detection implements channel estimation based on pilots only as displayed in (11)–(12), semi-blind detection can reduce the number of required pilots from an order of N_a to K, naturally leading to the consideration of its application in large-scale MIMO systems.

Since the core idea of semi-blind detection hinges on full exploitation of prior information carried by both known pilots and estimated payload data, the efficiency of semi-blind detection mainly depends on the efficiency of the preceding matrix completion processes as illustrated in Line 4–13. As the scale of antennas or users increases, some computational delays may arise. Fortunately, instead of adopting the conventional LMMSE/ALS, the proposed ASD also offers advantages in terms of cost-effectiveness, which will be validated by experimental results in Section 4.

Algorithm 1 The proposed sensing-assisted semi-blind detection with reduced pilot overhead

1: **Input** $\mathbf{A}(\widehat{\boldsymbol{\Theta}}) \in \mathbb{C}^{N_a \times K}$, $\mathbf{Y}_c^{\Omega} \in \mathbb{C}^{N_a \times T_c}$, $\mathbf{X}_p \in \mathbb{C}^{K \times T_p}$
2: Randomly initialize $\mathbf{G}_0 \in \mathbb{C}^{K \times K}$ and $\mathbf{V}_0 \in \mathbb{C}^{K \times T_c}$.
3: Set $\mathbf{U}_0 = \mathbf{A}(\widehat{\boldsymbol{\Theta}})\mathbf{G}_0$ and $t = 0$.
4: **repeat**
5: **Step (1)** Update $\mathbf{U}_t$ to $\mathbf{U}_{t+1}$ as follows
6: Compute $\nabla f_{\mathbf{V}}(\mathbf{U}) = -(\mathbf{Y}_c^{\Omega} - \mathcal{P}_{\Omega}(\mathbf{U}\mathbf{V}))\mathbf{V}^H$ and $t_{\mathbf{U}} = ||\nabla f_{\mathbf{V}}(\mathbf{U})||_F^2 / ||\mathcal{P}_{\Omega}(\nabla f_{\mathbf{V}}(\mathbf{U})\mathbf{V})||_F^2$.
7: $\mathbf{U}'_{t+1} \leftarrow \mathbf{U}_t - t_{\mathbf{U}_t}\nabla f_{\mathbf{V}_t}(\mathbf{U}_t)$.
8: $\mathbf{U}_{t+1} \leftarrow \mathbf{A}(\widehat{\boldsymbol{\Theta}})(\mathbf{A}^H(\widehat{\boldsymbol{\Theta}})\mathbf{A}(\widehat{\boldsymbol{\Theta}}))^{-1}\mathbf{A}^H(\widehat{\boldsymbol{\Theta}})\mathbf{U}'_{t+1}$.
9: **Step (2)** Update $\mathbf{V}_t$ to $\mathbf{V}_{t+1}$ as follows
10: Compute $\nabla f_{\mathbf{U}}(\mathbf{V}) = -\mathbf{U}^H(\mathbf{Y}_c^{\Omega} - \mathcal{P}_{\Omega}(\mathbf{U}\mathbf{V}))$ and $t_{\mathbf{V}} = ||\nabla f_{\mathbf{U}}(\mathbf{V})||_F^2 / ||\mathcal{P}_{\Omega}(\mathbf{U}\nabla f_{\mathbf{U}}(\mathbf{V}))||_F^2$.
11: $\mathbf{V}_{t+1} \leftarrow \mathbf{V}_t - t_{\mathbf{V}_t}\nabla f_{\mathbf{U}_{t+1}}(\mathbf{V}_t)$.
12: **Step (3)** Update t to $t+1$.
13: **until** Convergence
14: Get $\boldsymbol{\Sigma} \leftarrow \widehat{\mathbf{V}}(:, 1 \sim T_p) \cdot \mathbf{X}_p^H / T_p$.
15: **Output** $\widehat{\mathbf{H}} = \widehat{\mathbf{U}}\boldsymbol{\Sigma}$

3.3.2. Signal Denoising

Given that the estimation accuracy of the semi-blind method partially relies on assumptions about the signal structure, it may suffer from pilot contamination and exhibit performance degradation, especially in the presence of high noise levels. In this way, it may be beneficial to firstly recover the latent clean signal from the noisy observations before performing the semi-blind detection for enhancing the reliability of payload data estimates.

With the recent development of deep learning, approaches based on CNNs have achieved remarkable success in the task of signal denoising, among which DnCNN formulated in [43] provides the state-of-the-art performance. Due to the integration of residual learning and batch normalization in the network architecture design, DnCNN has been validated to be highly effective in removing the Gaussian-type noise, which is exactly suitable for our considered signal models, especially at high noise levels. Specifically, the residual learning makes the inputs to each layer Gaussian-like distributed and less correlated with the image content, therefore avoid oversmoothing at high noise level. Furthermore, batch normalization alleviates the internal covariate shift by incorporating a normalization step and a re-scaling step before the nonlinear layer, which makes the model less sensitive to the noise level and generalizes well to inputs with strong noise. Moreover, DnCNN is shown to be able to extend to general denoising tasks, for example, the blind Gaussian denoising, which means the capacity of signal denoising without being informed of the noise level. Based on the above reasons, we propose to apply the DnCNN as a preprocessing procedure before performing the semi-blind detection, which aims to denoise the received signal $\mathbf{Y}_c \in \mathbb{C}^{N_r \times T_c}$ and expects better estimation accuracy in noisy environments.

Specifically, we choose to use the pre-trained model provided in [43] for Gaussian denoising, which takes the noisy signals as the input and generates the residual noise as the output. By subtracting the residual from the input, we then obtain the denoised signal $\mathbf{Y}_c'$. It is worth noting that the model is without fine-tuning on our dataset because our signals exhibit similar distributions with the Gaussian noise, which can be seen by comparing Figure 3d,e, making it difficult for the DnCNN model to distinguish our signal from the noise. In this way, if the model is trained using our dataset, it tends to overestimate the noise level and over-smooth the input signal. On the contrary, the pre-trained model is trained with natural images with distinct distribution from the Gaussian noise, which can be seen by comparing Figure 3d,f, making it easier to implement proper denoising strength and avoid over-smoothing the input signal. Therefore, the pre-trained model is adopted, which not only saves the effort of the time-consuming training process but also enhances the denoising accuracy.

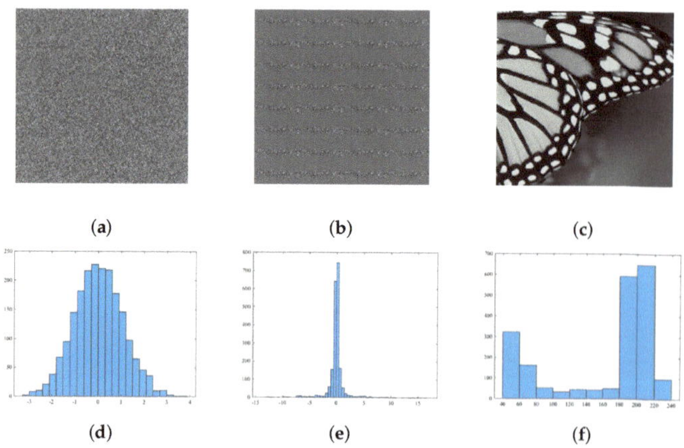

Figure 3. Comparisons among sample signals of (**a**) Gaussian noise, (**b**) our signal and (**c**) natural image, with corresponding histograms in (**d**), (**e**), (**f**), respectively. (**e**) exhibits similar distribution with (**d**), while (**f**) is distinct from (**d**).

The overall framework of our proposed sensing and deep CNN-assisted semi-blind detection scheme for hybrid massive MU-MIMO systems is summarized in Figure 4, which can be divided into three stages, i.e., target parameter sensing, communication signal denoising and semi-blind detection refinement. In the sensing stage, the BS detects and acquires the azimuth angles of target user equipments (UEs) by sending probing signals and receiving echoes. Then, the uplink communication signals transmitted by UEs are processed in the RF band and denoised by using the pre-trained DnCNN model. Finally, based on the estimated angles and denoised signals, the refined semi-blind detection will be performed at the communication module.

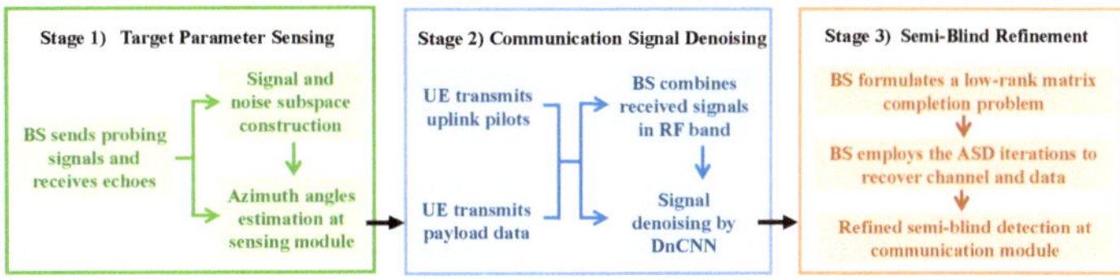

Figure 4. The overall framework of the proposed sensing and deep CNN-assisted semi-blind detection.

4. Simulation Results

Simulation results are provided in this section to demonstrate the superior performance of the proposed sensing and deep CNN-assisted semi-blind detection scheme. We consider a hybrid massive MU-MIMO system as given in Figure 1, where $K = 6$ single-antenna users tend to communicate with the BS-integrated radar and communication modules. For the sensing module, the transmit and receive arrays of radar are equipped with $N_s = 64$, and the length of snapshots are set to be $T_s \geq N_s$. For the communication module, the receiver is equipped with $N_a = 64$ antennas and $N_r = 32$ chains. Here, we assume the pilot sequences are mutually orthogonal with the pilot length satisfying $T_p \geq K$. In addition, we choose the payload data independently from the QPSK constellation with unit average power. Moreover, all the scattering and reflection coefficients are supposed to obey the standard complex Gaussian distributions.

4.1. The Proposed Transmission Frame

We firstly compare the channel estimation performance of adopting the multi-functional transmission frame as shown in Figure 2 with some existing schemes, i.e., the conventional pilot-only estimator [39] and the original semi-blind estimator relying solely on the communication module without aid of radar sensing [27]. Specifically, the channel estimation performance is evaluated by the normalized mean squared error (NMSE) defined as NMSE $= \mathbb{E}[||\widehat{\mathbf{H}} - \mathbf{H}||_{\mathcal{F}}^2]/\mathbb{E}[||\mathbf{H}||_{\mathcal{F}}^2]$. For fair comparison, we assume the process of solving the low-rank matrix completion in Algorithm 1 is temporarily replaced by the same ALS iteration as derived in [27], and the noise variance is approximated according to SNR $= 10\lg(\bar{P}/\sigma^2)$ with $\bar{P}$ being the average received power. The other configurations for the transmission frame are set as $T_s = 100$, $T_p = 2K$ and $T_d = 100$. Unless otherwise stated, each simulation result is obtained through Monte-Carlo with 1000 independent channel realizations. Figure 5 depicts the NMSE performance versus SNR by using different transmission schemes. We can observe that the proposed sensing-assisted scheme shows significant performance gains especially with low SNRs compared to the alternative estimators, which thanks to that more geometric information about the scattering environments can be gleaned from the target sensing stage, while the other two schemes neglect the structural characteristics of the multi-path channel models.

To further validate the superiority of utilizing radar sensing function to yield the target angles rather than extracting the path angles from the roughly estimated channels as conducted in [30], which designed a three-stage estimator (denoted by "TSTE") and suggested to obtain the knowledge of array responses directly from an initial low-rank channel estimate, Figure 6 illustrates the root mean squared error (RMSE) defined as $\mathrm{RMSE} = 10\lg\sqrt{\mathbb{E}[\sum_{k=1}^{K}||\hat{\theta}_k - \theta_k||_2^2/K]}$ to measure the performance of AoAs resolutions of target users by using different detection schemes. Since the estimation deviation of the subspace decomposition-based method mainly depends on the dimension of received covariance matrix and some preset parameters such as the array spacing, we consider two settings of d/λ. It can be seen that the proposed sensing-assisted detection scheme achieves better resolutions, especially when the inter-element spacing is larger. Moreover, it is noteworthy that the RMSE under our proposed scheme would improve as the SNR increases, while that obtained by the TSTE estimator only fluctuates slightly versus different SNR, which is reasonable because their path angles are learned from the low-rank matrix solutions instead of straight from the received echoes, and the noise components have been removed before constructing the signal subspaces.

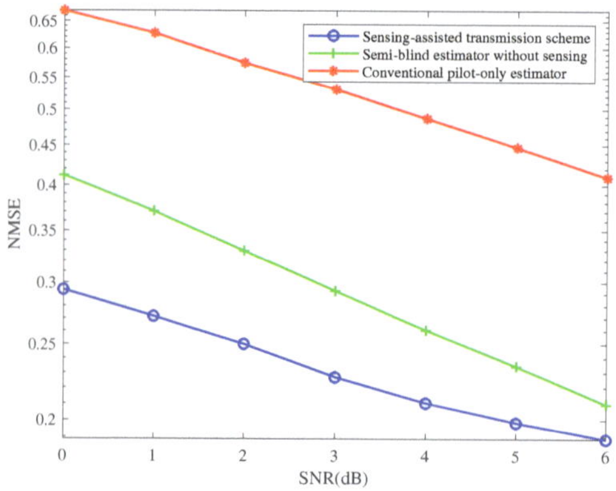

Figure 5. The NMSE performance versus SNR by using different transmission schemes.

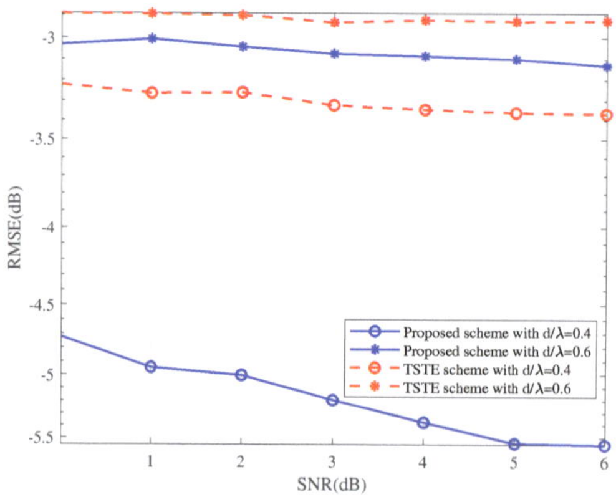

Figure 6. The detection performance of AoAs resolutions under different schemes.

4.2. Refined Semi-Blind Detection

Next, we illustrate the performance gain of our proposed Algorithm 1. To show the advantage of performing semi-blind detection at the receiver rather than the LS/LMMSE-based detection [37], suppose the comparison experiment is set as performing the LMMSE channel estimation after obtaining the knowledge of target angles from the sensing module. Considering two settings of pilot length, $T_p = K$ and $T_p = 1.5K$, Figure 7 plots the NMSE performance of channel estimation with different receiver schemes. From Figure 7, we observe that the NMSE performance of Algorithm 1 outperforms the LMMSE method even if the sensing module is involved, which indicates the necessity of exploiting low-rank characteristics in massive MIMO systems to improve the estimation accuracy with relatively lower pilots. In addition, we can note that the NMSE performance of Algorithm 1 improves as the length of data sequences T_d increases, while the performance of LMMSE remains almost the same in all settings. The reason is that by incorporating data into the decomposition process of received signals, Algorithm 1 can take full advantage of payload data as "virtual pilots" to glean more useful information over the conventional pilot-only method.

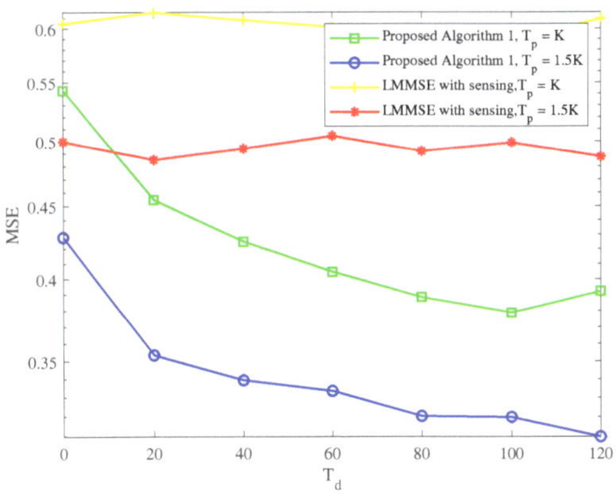

Figure 7. The NMSE performance versus length of data sequences under different receiver schemes with SNR = 2dB.

We then characterize the performances of Algorithm 1 in terms of the computational efficiency and the estimation accuracy without any prior knowledge of channel or noise statistics. By fixing $T_p = K$ and $T_d = 100$, Figure 8 depicts the convergence rates by employing different types of iterations, where "R-ALS iteration" refers to the regularized alternating least squares approach proposed by [27] for solving the formulated low-rank matrix completion problem (26). As shown in Figure 8, both the ASD iterations presented in Algorithm 1 and the R-ALS iterations will converge within limited iterations, and their convergence behaviors are similar for different SNR levels. Although the convergence rate of R-ALS iterations seems faster that that of ASD iterations, the calculations of $\mathbf{U}_{t+1}$ and $\mathbf{V}_{t+1}$ in Algorithm 1 avoid two matrix inversion operations for each iteration, which would otherwise involve additional complexity with cubic order of the matrix dimension, i.e., $\mathcal{O}(N_a^3)$. Therefore, the total computational overhead incurred by Algorithm 1 can be relatively reduced especially with high-dimensional matrices.

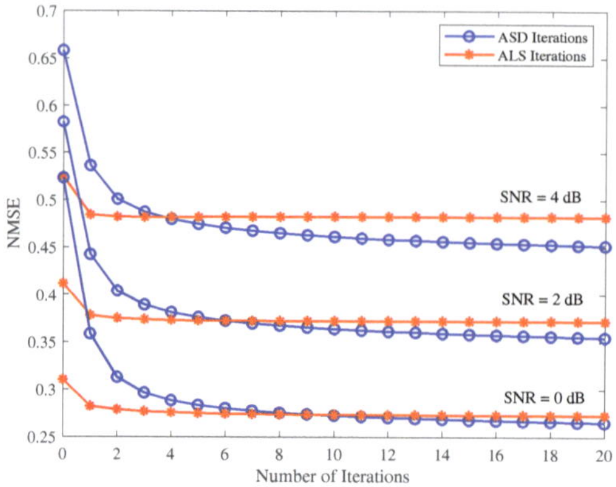

Figure 8. Convergence rates of adopting different iterative algorithms, where NMSE $= \mathbb{E}[||\widehat{\mathbf{Z}} - \mathbf{Z}||_{\mathcal{F}}^2]/\mathbb{E}[||\mathbf{Z}||_{\mathcal{F}}^2]$ with $\widehat{\mathbf{Z}} = \widehat{\mathbf{H}}\widehat{\mathbf{X}}_c$.

To further validate the computational complexity of the proposed ASD approach, we conduct more experiments compared with the conventional LMMSE and R-ALS, focusing on their execution time and memory consumption. Following the principles of controlled variable experimental methodology, and with reference to the results illustrated in Figures 7 and 8, we initially quantify the averaged computational cost for all three approaches, under the conditions of SNR $= 2$ dB and the number of iterations set to 6. Subsequently, given that the LMMSE approach would induce a single execution at the receiver adhered to Equations (12) and (13) without convergence issues, we only record the time and memory expenditures for the proposed ASD and R-ALS iterations by establishing the termination criterion as $\epsilon = ||\hat{\mathbf{H}}_t - \hat{\mathbf{H}}_{t-1}||_{\mathcal{F}}^2/||\hat{\mathbf{H}}_t||_{\mathcal{F}}^2 < 0.01$. The outcomes are presented in Table 3, from which we can observe that the proposed ASD approach exhibits lower time and memory consumption compared to its existing alternatives, thereby demonstrating the advantages in reduced complexity and higher computational efficiency.

Table 3. Comparisons on computational cost by using different approaches.

Approaches	Averaged Computational Cost		Total Cost until Convergence	
performance	memory	time	memory	time
proposed ASD	0.216 Kb	3.11 ms	10.22 Kb	22.80 ms
R-ALS	8.644 Kb	43.22 ms	46.61 Kb	86.96 ms
LMMSE	0.636 Kb	3.50 ms	/	/

In order to visualize the influence with unknown prior knowledge on the estimation accuracy, Figure 9 compares the NMSE performance versus different SNRs by adopting Algorithm 1 and the R-ALS-based estimator, which mimics and modifies the process in [27] given the target angles. For the former scenarios without available noise, the maximum iteration number is set to be $t_{\max} = 20$; while for the latter, the regularization parameters used to incorporate the prior information are set to be consistent with the noise variance. It can be seen that benefiting from taking noise effect into account, the R-ALS estimator performs better with low SNR values, which is owing to the non-negligible noise that imposes great challenges on accurate estimations in such SNR ranges. Nevertheless, the performance differences between two algorithms become narrow as the SNR increases, and Algorithm 1 would even achieve better performance under the condition of higher SNRs. By combining the results in Figure 8 and Table 3, we can conclude that our proposed Algorithm 1 is a good choice for refining semi-blind detection due to its comparable performance and lower complexity.

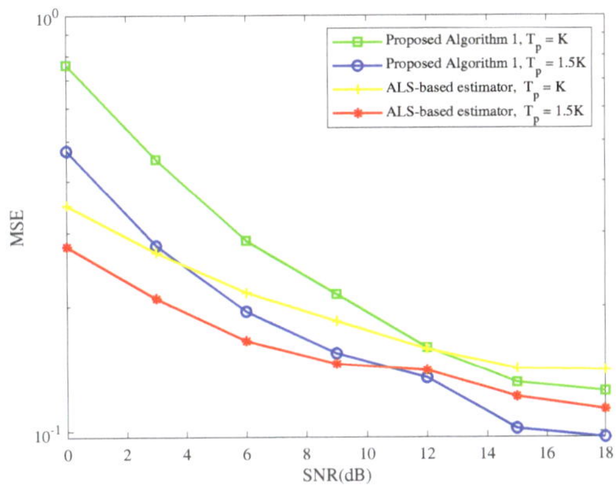

Figure 9. The NMSE performance versus SNR by adopting different iterative algorithms.

4.3. Signal Denoising by DnCNN

Finally, we confirm the effectiveness of signal denoising by DnCNN before performing semi-blind detection in low SNR regions. To evaluate the denoising performance of adopting different DnCNN models for our received signal dataset, we select the pre-trained DnCNN models in [43] for Gaussian denoising with specific noise levels (which are referred to as DnCNN-15 and DnCNN-25) and for blind Gaussian denoising (which is referred to as DnCNN-B) as comparisons. During the denoising phase, we successively obtain the output $\mathbf{Y}_{denoise}$ as the denoised data of the input $\mathbf{Y}_{noisy} = \mathbf{Y}_c$. Suppose the clean signal that refers to $\mathbf{Y}_{noiseless}$ is related to the combined results of $\mathbf{HX}_c$, then the following performance metrics include NMSE and NMSE-Y defined as NMSE-Y $= \mathbb{E}[||\mathbf{Y}_{denoise} - \mathbf{Y}_{noiseless}||^2_{\mathcal{F}}]/\mathbb{E}[||\mathbf{Y}_{noiseless}||^2_{\mathcal{F}}]$, which are used to measure the accuracy of channel estimation and signal denoising, respectively.

The averaged NMSE and NMSE-Y results over 1000 samples by adopting different signal denoising models are shown in Table 4, where the best results for each SNR value are highlighted in bold. As one can see, all the selected DnCNN models outperform the competing method without signal denoising when SNR is below -2 dB due to the consideration of noise influence in low SNR regions. In contrast, as the noise level decreases, both DnCNN-25 and DnCNN-B produce better results on NMSE-Y while fail to reach higher estimation accuracy, which is intuitively reasonable because the models with strong denoising power are likely to distort the original signal structures. It is pleasant to find that the model DnCNN-15, which is trained with lower noise level, can not only yield the best denoising performance on most of the discussed SNRs but also generate better channel estimates, which indicates the feasibility of deploying the signal denoising before performing the semi-blind detection.

Table 4. Performance comparison by adopting different signal denoising models.

SNR Value	−5 dB		−4 dB		−3 dB	
Selected Model	NMSE-Y	NMSE	NMSE-Y	NMSE	NMSE-Y	NMSE
Without DnCNN	3.1727	1.6561	2.5430	1.2967	1.9962	0.9956
DnCNN-15	1.5223	1.0362	1.1352	**0.8159**	**0.8320**	**0.6679**
DnCNN-25	**0.8566**	**0.8875**	**0.8618**	0.8951	0.8888	0.9199
DnCNN-B	0.9424	0.9635	0.9265	0.9488	0.8888	0.9042
SNR Value	**−2 dB**		**−1 dB**		**0 dB**	
Selected Model	NMSE-Y	NMSE	NMSE-Y	NMSE	NMSE-Y	NMSE
Without DnCNN	1.6025	0.8733	1.2425	0.7477	0.9951	0.6051
DnCNN-15	**0.6335**	**0.6242**	**0.4998**	**0.5947**	**0.4312**	**0.5720**
DnCNN-25	0.9070	0.9363	0.9314	0.9552	0.9356	0.9610
DnCNN-B	0.9350	0.8631	0.8218	0.8875	0.7700	0.8672

5. Conclusions

In this paper, we have proposed a sensing and deep CNN-assisted semi-blind detection refinement scheme for uplink hybrid massive MU-MIMO systems. The main idea is to extract the geometric structure corresponding to the communication channels via radar sensing, based on which both the transmitted pilot and payload data are contributed to yield the refined estimates. We have treated the signal recovery problem as acquiring factorization solutions to a low-rank matrix completion, for which an iterative algorithm modified from ASD is presented without any prior knowledge of noise statistics. Specifically, a pre-trained DnCNN has been adopted for signal denoising to alleviate the noise enhancement along with the LS-based estimation process, especially in low SNR regions. The overall approach has been carried out in three progressive stages, i.e., target parameter sensing, communication signal denoising and semi-blind detection refinement. Consequently, simulation results have been provided to validate the superior performance of our proposed scheme. It is seen that the utilization of co-designed multi-functional transmission frame and the refined semi-blind receiver could achieve better estimation accuracy with reduced training overhead, which facilitates the availability of sensing-assisted communications.

Despite the advancements introduced by the proposed architecture, focusing on leveraging radar sensing capabilities to enhance communication gains, there still remain several limitations requiring further investigation. Instead of relying on the coexistence and information sharing between different modules, future work could develop more powerful signal processing techniques to integrate radar and communication modules into a unified hardware platform and decongest the RF environment by using the same signal for both functionalities. In such a case, devising sophisticated signal models that harmoniously incorporate the distinctions between LoS and multi-path channels might unveil new dimensions of system performance.

Another crucial aspect that needs to be considered in future work involves the time-varying property of propagation paths under user mobility scenarios, including time-delay parameters and Doppler shift effects. This aspect necessitates the BS to pre-compensate for these effects before demodulating the received signals for most robust detection performance. Toward this end, we plan to explore the time-varying channel tracking algorithms and compensation methods to mitigate the adverse effects of temporal fluctuations and phase deviations on estimation errors, which should be of great importance to cater to real-time applications.

Author Contributions: Conceptualization, F.H. and J.Z.; methodology, F.H. and L.Z.; software, H.Z. and J.W.; validation, J.Z. and J.W.; formal analysis, H.Z. and L.Z.; investigation, F.H.; resources, L.Z. and H.Z.; data curation, J.Z. and J.W.; writing—original draft preparation, F.H.; writing—review and editing, L.Z. and H.Z.; visualization, J.W.; supervision, J.Z.; project administration, F.H.; funding acquisition, F.H. All authors have read and agreed to the published version of the manuscript.

Funding: This research was funded in part by the National Natural Science Foundation of China under Grants 62201388 and 62201389, in part by the Natural Science Foundation of Shanghai under Grant 22ZR1463400 and in part by the Fundamental Research Funds for the Central Universities under Grant 2022-5-YB-01.

Data Availability Statement: The data are not publicly available due to privacy.

Conflicts of Interest: The authors declare no conflicts of interest.

References

1. Tong, W.; Li, G.Y. Nine challenges in artificial intelligence and wireless communications for 6G. *IEEE Wirel. Commun.* **2022**, *29*, 140–145. [CrossRef]
2. Li, X.; Liu, F.; Zhou, Z.; Zhu, G.; Wang, S.; Huang, K.; Gong, Y. Integrated sensing, communication, and computation over-the-Air: MIMO beamforming design. *IEEE Trans. Wirel. Commun.* **2023**, *22*, 5383–5398. [CrossRef]
3. Koubaa, A.; Ammar, A.; Abdelkader, M.; Alhabashi, Y.; Ghouti, L. AERO: AI-enabled remote sensing observation with onboard edge computing in UAVs. *Remote Sens.* **2023**, *15*, 1873. [CrossRef]
4. Zhu, Y.; Guo, H.; Lau, V.K.N. Bayesian channel estimation in multi-user massive MIMO with extremely large antenna array. *IEEE Trans. Signal Process.* **2021**, *69*, 5463–5478. [CrossRef]
5. Guo, J.; Wen, C.-K.; Jin, S.; Li, G.Y. Overview of deep learning-based CSI feedback in massive MIMO systems. *IEEE Trans. Commun.* **2022**, *70*, 8017–8045. [CrossRef]
6. Zhang, R.; Cheng, L.; Wang, S.; Lou, Y.; Wu, W.; Ng, D.W.K. Tensor decomposition-based channel estimation for hybrid mmWave massive MIMO in high-mobility scenarios. *IEEE Trans. Commun.* **2022**, *70*, 6325–6340. [CrossRef]
7. Wei, X.; Hu, C.; Dai, L. Deep learning for beamspace channel estimation in millimeter-wave massive MIMO systems. *IEEE Trans. Commun.* **2021**, *69*, 182–193. [CrossRef]
8. Nguyen, N.T.; Ma, M.; Lavi, O.; Shlezinger, N.; Eldar, Y.C.; Swindlehurst, A.L.; Juntti, M. Deep unfolding hybrid beamforming designs for THz massive MIMO systems. *IEEE Trans. Signal Process.* **2023**, *71*, 3788–3804. [CrossRef]
9. Mirzaei, J.; ShahbazPanahi, S.; Sohrabi, F.; Adve, R. Hybrid analog and digital beamforming design for channel estimation in correlated massive MIMO systems. *IEEE Trans. Signal Process.* **2021**, *69*, 5784–5800. [CrossRef]
10. Wei, X.; Dai, L. Channel estimation for extremely large-scale massive MIMO: Far-field, near-field, or hybrid-field? *IEEE Commun. Lett.* **2022**, *26*, 177–181. [CrossRef]
11. Chafaa, I.; Negrel, R.; Belmega, E.V.; Debbah, M. Self-supervised deep learning for mmWave beam steering exploiting sub-6 GHz channels. *IEEE Trans. Wirel. Commun.* **2022**, *21*, 8803–8816. [CrossRef]
12. Wu, K.; Zhang, J.A.; Huang, X.; Guo, Y.J.; Hanzo, L. Simultaneous beam and user selection for the beamspace mmWave/THz massive MIMO downlink. *IEEE Trans. Commun.* **2023**, *71*, 1785–1797. [CrossRef]
13. Cheng, Z.; Liao, B. QoS-aware hybrid beamforming and DOA estimation in multi-carrier dual-function radar-communication systems. *IEEE J. Sel. Areas Commun.* **2022**, *40*, 1890–1905. [CrossRef]
14. Liu, F.; Cui, Y.; Masouros, C.; Xu, J.; Han, T.X.; Eldar, Y.C.; Buzzi, S. Integrated sensing and communications: Toward dual-functional wireless networks for 6G and beyond. *IEEE J. Sel. Areas Commun.* **2022**, *40*, 1728–1767. [CrossRef]
15. Cui, Y.; Liu, F.; Jing, X.; Mu, J. Integrating sensing and communications for ubiquitous IoT: Applications, trends, and challenges. *IEEE Netw.* **2021**, *35*, 158–167. [CrossRef]
16. Wang, J.; Varshney, N.; Gentile, C.; Blandino, S.; Chuang, J.; Golmie, N. Integrated Sensing and Communication: Enabling Techniques, Applications, Tools and Data Sets, Standardization, and Future Directions. *IEEE Internet Things J.* **2022**, *9*, 23416–23440. [CrossRef]
17. Graff, A.; Chen, Y.; González-Prelcic, N.; Shimizu, T. Deep learning-based link configuration for radar-aided multiuser mmWave vehicle-to-infrastructure communication. *IEEE Trans. Veh. Technol.* **2023**, *72*, 7454–7468. [CrossRef]
18. Wang, X.; Zhai, W.; Zhang, X.; Wang, X.; Amin, M.G. Enhanced automotive sensing assisted by joint communication and cognitive sparse MIMO radar. *IEEE Trans. Aerosp. Electron. Syst.* **2023**, *59*, 4782–4799. [CrossRef]
19. Hassan, K.; Masarra, M.; Zwingelstein, M.; Dayoub, I. Channel estimation techniques for millimeter-wave communication systems: Achievements and challenges. *IEEE Open J. Commun. Soc.* **2020**, *1*, 1336–1363. [CrossRef]
20. Zhang, R.; Yang, L.; Tang, M.; Tan, W.; Zhao, J. Channel estimation for mmWave massive MIMO systems with mixed-ADC architecture. *IEEE Open J. Commun. Soc.* **2023**, *4*, 606–613. [CrossRef]
21. Wang, Y.; Zhang, Y.; Tian, Z.; Leus, G.; Zhang, G. Super-resolution channel estimation for arbitrary arrays in hybrid millimeter-wave massive MIMO systems. *IEEE J. Sel. Top. Signal Process.* **2019**, *13*, 947–960. [CrossRef]
22. Fan, D.; Gao, F.; Liu, Y.; Deng, Y.; Wang, G.; Zhong, Z.; Nallanathan, A. Angle domain channel estimation in hybrid millimeter wave massive MIMO systems. *IEEE Trans. Wirel. Commun.* **2018**, *17*, 8165–8179. [CrossRef]
23. Zhou, L.; Dai, J.; Xu, W.; Chang, C. Sparse channel estimation for intelligent reflecting surface assisted massive MIMO systems. *IEEE Trans. Green Commun. Netw.* **2022**, *6*, 208–220. [CrossRef]
24. Wang, Y.; Qi, C.; Li, P.; Lu, Z.; Lu, P. Channel estimation for wideband mmWave MIMO OFDM system exploiting block sparsity. *IEEE Commun. Lett.* **2022**, *26*, 897–901. [CrossRef]

25. Chu, H.; Zheng, L.; Wang, X. Semi-blind millimeter-wave channel estimation using atomic norm minimization. *IEEE Commun. Lett.* **2018**, *22*, 2535–2538. [CrossRef]
26. Han, F.; Wang, X.; Deng, H. A very-low pilot scheme for mmWave hybrid massive MIMO-OFDM systems. *IEEE Wirel. Commun. Lett.* **2021**, *10*, 2061–2064. [CrossRef]
27. Liang, S.; Wang, X.; Ping, L. Semi-blind detection in hybrid massive MIMO systems via low-rank matrix completion. *IEEE Trans. Wirel. Commun.* **2019**, *18*, 5242–5254. [CrossRef]
28. Li, X.; Fang, J.; Li, H.; Wang, P. Millimeter wave channel estimation via exploiting joint sparse and low-rank structures. *IEEE Trans. Wirel. Commun.* **2018**, *17*, 1123–1133. [CrossRef]
29. Vlachos, E.; Alexandropoulos, G.C.; Thompson, J. Massive MIMO channel estimation for millimeter wave systems via matrix completion. *IEEE Signal Process. Lett.* **2018**, *25*, 1675–1679. [CrossRef]
30. Masood, K.F.; Hu, R.; Tong, J.; Xi, J.; Guo, Q.; Yu, Y. A low-complexity three-stage estimator for low-rank mmWave channels. *IEEE Trans. Veh. Technol.* **2021**, *70*, 5920–5931. [CrossRef]
31. Nguyen, L.T.; Kim, J.; Shim, B. Low-rank matrix completion: A contemporary survey. *IEEE Access* **2019**, *7*, 94215–94237. [CrossRef]
32. Giampouras, P.V.; Rontogiannis, A.A.; Koutroumbas, K.D. Alternating iteratively reweighted least squares minimization for low-rank matrix factorization. *IEEE Trans. Signal Process.* **2019**, *67*, 490–503. [CrossRef]
33. Chi, Y.; Lu, Y.M.; Chen, Y. Nonconvex optimization meets low-rank matrix factorization: An overview. *IEEE Trans. Signal Process.* **2019**, *67*, 5239–5269. [CrossRef]
34. Tanner, J.; Wei, K. Low rank matrix completion by alternating steepest descent methods. *Appl. Comput. Harmon. Anal.* **2016**, *40*, 417–429. [CrossRef]
35. Liu, F.; Masouros, C.; Petropulu, A.P.; Griffiths, H.; Hanzo, L. Joint radar and communication design: Applications, state-of-the-art, and the road ahead. *IEEE Trans. Commun.* **2020**, *68*, 3834–3862. [CrossRef]
36. Yuan, W.; Wei, Z.; Li, S.; Yuan, J.; Ng, D.W.K. Integrated sensing and communication-assisted orthogonal time frequency space transmission for vehicular networks. *IEEE J. Sel. Top. Signal Process.* **2021**, *15*, 1515–1528. [CrossRef]
37. Huang, S.; Zhang, M.; Gao, Y.; Feng, Z. MIMO radar aided mmWave time-varying channel estimation in MU-MIMO V2X communications. *IEEE Trans. Wirel. Commun.* **2021**, *20*, 7581–7594. [CrossRef]
38. He, H.; Wen, C.-K.; Jin, S.; Li, G.Y. Deep learning-based channel estimation for beamspace mmWave massive MIMO systems. *IEEE Wirel. Commun. Lett.* **2018**, *7*, 852–855. [CrossRef]
39. Kay, S.M. *Fundamentals of Statistical Signal Processing, Volume III: Practical Algorithm Development*; Prentice Hall: Upper Saddle River, NJ, USA, 2013.
40. Gizzini, A.K.; Chafii, M.; Nimr, A.; Fettweis, G. Enhancing least square channel estimation using deep learning. In Proceedings of the 2020 IEEE 91st Vehicular Technology Conference: VTC2020-Spring, Antwerp, Belgium, 25–28 May 2020; pp. 1–5.
41. Schmidt, R. Multiple emitter location and signal parameter estimation. *IEEE Trans. Antennas Propag.* **1986**, *34*, 276–280. [CrossRef]
42. Liu, A.; Lau, V. Phase only RF precoding for massive MIMO systems with limited RF chains. *IEEE Trans. Signal Process.* **2014**, *62*, 4505–4515. [CrossRef]
43. Zhang, K.; Zuo, W.; Chen, Y.; Meng, D.; Zhang, L. Beyond a gaussian denoiser: Residual learning of deep CNN for image denoising. *IEEE Trans. Image Process.* **2017**, *26*, 3142–3155. [CrossRef] [PubMed]

remote sensing

MDPI

Article

Ship Target Detection in Optical Remote Sensing Images Based on E2YOLOX-VFL

Qichang Zhao [1,2], Yiquan Wu [1,*] and Yubin Yuan [1]

1 School of Electronic Information Engineering, Nanjing University of Aeronautics and Astronautics, Nanjing 211106, China; zhao031111@nuaa.edu.cn (Q.Z.); harley_yuan@nuaa.edu.cn (Y.Y.)
2 Satellite General Department, Shanghai Institute of Satellite Engineering, Shanghai 201109, China
* Correspondence: imagestrong@nuaa.edu.cn

Abstract: In this research, E2YOLOX-VFL is proposed as a novel approach to address the challenges of optical image multi-scale ship detection and recognition in complex maritime and land backgrounds. Firstly, the typical anchor-free network YOLOX is utilized as the baseline network for ship detection. Secondly, the Efficient Channel Attention module is incorporated into the YOLOX Backbone network to enhance the model's capability to extract information from objects of different scales, such as large, medium, and small, thus improving ship detection performance in complex backgrounds. Thirdly, we propose the Efficient Force-IoU (EFIoU) Loss function as a replacement for the Intersection over Union (IoU) Loss, addressing the issue whereby IoU Loss only considers the intersection and union between the ground truth boxes and the predicted boxes, without taking into account the size and position of targets. This also considers the disadvantageous effects of low-quality samples, resulting in inaccuracies in measuring target similarity, and improves the regression performance of the algorithm. Fourthly, the confidence loss function is improved. Specifically, Varifocal Loss is employed instead of CE Loss, effectively handling the positive and negative sample imbalance, challenging samples, and class imbalance, enhancing the overall detection performance of the model. Then, we propose Balanced Gaussian NMS (BG-NMS) to solve the problem of missed detection caused by the occlusion of dense targets. Finally, the E2YOLOX-VFL algorithm is tested on the HRSC2016 dataset, achieving a 9.28% improvement in mAP compared to the baseline YOLOX algorithm. Moreover, the detection performance using BG-NMS is also analyzed, and the experimental results validate the effectiveness of the E2YOLOX-VFL algorithm.

Keywords: ship; detection; classification; deep learning; YOLOX; optical remote sensing

Citation: Zhao, Q.; Wu, Y.; Yuan, Y. Ship Target Detection in Optical Remote Sensing Images Based on E2YOLOX-VFL. *Remote Sens.* **2024**, *16*, 340. https://doi.org/10.3390/rs16020340

Academic Editors: Qian Du, Yanni Dong and Xiaochen Yang

Received: 18 November 2023
Revised: 31 December 2023
Accepted: 9 January 2024
Published: 15 January 2024

1. Introduction

Remote sensing images are obtained by sensors carried by satellites, aircraft, and other platforms, allowing for observation of the Earth's surface. One notable characteristic is their ability to capture large-scale data in a short period of time. Remote sensing images typically include information about objects such as rivers, farmland, forests, roads, cars, airplanes, and ships. Utilizing remote sensing images for object detection is an important application area. Ships hold significant importance, as they are crucial targets at sea, and automated ship detection and recognition play a vital role in gaining maritime control. Therefore, in-depth research on ship detection and recognition is essential.

Currently, remote sensing images employed for ship detection include Synthetic Aperture Radar (SAR) images, visual images, infrared images, and hyperspectral images. Among these, SAR and visual images are the most widely employed. Although SAR has the advantage of being able to see through clouds, SAR images lack visual intuitiveness and cannot provide detailed information such as ship material attributes [1–3]. Visual images obtained through optical remote-sensing technology offer a more realistic representation of color information of targets and backgrounds. They exhibit prominent structural features, making them unmatched in the domain of ship detection and classification in maritime regions compared to other image sources [4].

The detection and classification of ship targets in optical remote sensing images can be categorized into conventional approaches and deep learning-based techniques.

1. Conventional detection approaches often involve multiple steps and have limited accuracy. For instance, in [5], ship detection is achieved using image edge intensity, which is effective in distinguishing between ships and nonships and obtains a 95% detection accuracy. He et al. [6] utilized texture fractal dimension and gap features for ship detection. This method effectively eliminates the impact of sea clutter on the detection results, achieving good detection results under different signal-to-noise ratios. Furthermore, it is not affected by occlusion and rotation of moving targets, demonstrating better robustness compared to traditional methods based on target shape and point-line structures. In [7], utilizing the differences in multi-scale fractal features between ship targets and background images, the interference of sea and sky backgrounds on ship detection has been effectively eliminated. The experimental results demonstrate that this method can accurately and effectively detect targets, with a 97.1% detection rate and a 9.5% false alarm rate. Li et al. [8] used an image-based maritime ship detection method based on fuzzy theory, effectively eliminating false targets and efficiently and accurately detecting ship targets in images. Its Figure of Merit (FoM) is 25% higher than the CFAR algorithm. Zhao et al. [9] proposed a ship detection method combining multi-scale visual saliency, capable of successfully detecting ships of different sizes and orientations while overcoming interference from complex backgrounds. This method achieved a detection rate of 93% and a false alarm rate of 4%. Wang et al. [10] employed maximum symmetric surround saliency detection for initial candidate region extraction and combined ship target geometric features for candidate region pre-filtering, achieving ship detection even in complex sea backgrounds. The detection accuracy is 97.2%, and the detection performance is better than most ship detection algorithms. Though these methods make use of the strong perception ability of the human visual system for nearby objects, they face challenges in selecting salient features due to factors such as atmospheric conditions, marine environments, and imaging performance.
2. In contrast, deep learning methods have gained considerable momentum due to their advantage in extracting diverse deep features containing semantic information from labeled data, thereby achieving higher accuracy compared to traditional methods. Deep learning algorithms can generally be divided into two distinct categories [11].

The first category includes two-stage algorithms, which separate the task of detection into region proposal and candidate region classification stages. Though they achieve high accuracy, these algorithms often sacrifice computational speed. One of the most well-known methods among various candidate region extraction approaches is Region with CNN feature (R-CNN) proposed by Girshick et al. [12]. Fast R-CNN [13] was introduced to address the laborious training steps as well as time and memory consumption in R-CNN, resulting in significant improvements in performance and computational efficiency. Faster R-CNN [14] shares the convolutional layers between the candidate region generation network and the R-CNN detector. It only requires one pass of a convolutional neural network (CNN) to generate candidate regions and their corresponding features, enabling the use of deeper networks for generating more accurate candidate regions. Yang et al. [15], Zhang et al. [16], and Ma et al. [17] adopt ideas from R-CNN for ship target detection, yet the proposal part of these methods lacks stability. Building upon Faster R-CNN, which utilizes a pre-training mechanism to increase the robustness when the number of annotated samples is limited, Han et al. [18] conduct detection and recognition of 10 classes of objects, including ships. Li et al. [19] propose a network based on deep features that effectively detects offshore and nearshore ships with different scales, obtaining 81.15% meanaverage precision (*mAP*) on the NWPU-2 dataset. A fast and powerful deep CNN-based ship detection framework is presented in [20], leveraging deep CNN features for feature extraction, distinguishing ship targets using Region Proposal Networks (RPN), and performing bounding box regression, thereby achieving high precision and efficiency in different complex backgrounds. Yang et al. [21] introduce a region-based deep forest (RDF) detection algorithm composed of a basic region proposal network and a collection of deep forests, allowing adaptive learning of features from remote sensing data and effective differentiation between real ships and region proposals. R2CNN (Rotational R-CNN), proposed by [22], has achieved outstanding results in text detection. These methods have also been employed in ship target detection and demonstrated favorable results.

The second category includes one-stage algorithms, which do not require the region proposal stage of the two-stage methods and can directly output the object classification and bounding boxes. One representative one-stage object detection method is YOLO (You Only Look Once) [23]. YOLO is widely applied in ship detection because of its fast detection speed, almost achieving real-time performance. However, its localization accuracy is slightly lower compared to two-stage algorithms. YOLOv2 [24] addresses the vanishing gradient and exploding gradient problems in the backpropagation process by adding batch normalization to all convolutional layers. It also adjusts the input size, leading to a 7% improvement in recall rate. YOLOv3 [25] consists of the Darknet53 Backbone network and the Feature Pyramid Network (FPN) for multi-scale feature extraction. It improves both the detection speed and the accuracy of detecting small objects. Xu et al. [26] propose an improved YOLOv3 model for arbitrary direction ship detection in SAR images. This improved model meets the real-time requirements of ship target detection but is only suitable for pure maritime backgrounds. The YOLO algorithm has been continuously improved throughout its development [27–29], gradually moving towards more efficient detection. Wang et al. [30] utilize an improved version of YOLO for ship detection in remote sensing images, achieving improved detection performance, while still facing challenges in the detection of small, blurry ship targets. Zhang et al. [31] propose an improved YOLOv4 for ship key part detection. This algorithm introduces the SA model and constructs a feature extraction module. The proposed method improves the mAP of the baseline algorithm for critical ship parts by 12%, and the model calculation speed is fast, which can meet the requirements of real-time performance. Li et al. [32] proposes an improved YOLOv5 ship target detection algorithm, achieving improvements in both speed and accuracy. Ge et al. [33] introduce YOLOX, an algorithm that enhances detection performance through clever integration solutions such as data augmentation, anchor-free detection, and label classification [34]. However, there is still limited research on multi-scale ship target detection for optical remote sensing images in complex backgrounds.

Though previous research has shown promising results, previous algorithms for detecting ship targets in optical remote sensing images face challenges such as complex backgrounds and diverse target variations, which result in missed detections, false alarms, and high computational costs. Building upon the prior research and experience in this field, this paper adopts the anchor-free YOLOX as the baseline network and incorporates the Efficient Channel Attention module (ECA-Net) [35] into the YOLOX Backbone network. Additionally, improvements are made to the localization and confidence loss functions in the YOLOX Head, resulting in the development of the E2YOLOX-VFL algorithm. The E2YOLOX-VFL method aims to overcome the challenges of low detection accuracy and limited robustness in ship target detection from optical remote sensing images with complex backgrounds. The enhanced algorithm demonstrates a 9.28% improvement mAP on the HRSC2016 dataset [36]. By harnessing these advancements, the proposed approach enhances ship target detection performance, offering superior accuracy and robustness while maintaining efficient processing speed.

The main contributions of this study are outlined as follows:

1. We propose the E2YOLOX-VFL algorithm for ship target detection, which combines the anchor-free YOLOX network and ECA-Net to overcome the limitations of previous methods.

2. We propose Efficient Force-IoU (EFIoU) Loss to modify Intersection over Union (IoU) Loss, which improves the regression performance of the algorithm by taking into account geometric dimensions of the predicted bounding box (*B*) and the ground truth box (*G*), suppressing the impact of low-quality samples.
3. The localization and confidence loss functions are improved in the YOLOX Head, leading to enhanced detection accuracy and robustness in complex backgrounds.
4. We propose the Balanced Gaussian NMS (BG-NMS), which is used instead of non-maximum suppression (NMS) to address the issue of missed detections caused by dense arrangements and occlusions.
5. The proposed algorithm was experimentally validated using the HRSC2016 dataset, demonstrating significant improvements in *mAP* and overall detection performance.

To provide a clear analysis, the paper is structured as follows. Section 2 presents an introduction to the related work. Section 3 analyzes the construction methods of the E2YOLOX-VFL. The HRSC2016 dataset and evaluation metrics are introduced, and the experiments are discussed in detail in Section 4, followed by a comprehensive discussions in Section 5. Finally, concluding remarks are presented in Section 6 to summarize the findings of our research.

2. Related Work

In this section, we first review the development stages of optical remote sensing image ship target detection and recognition. Second, we introduce the research progress in coarse- and fine-grained ship classification. Third, we investigate the form of loss functions. Fourth, we introduce the development of NMS.

2.1. Development Stages of Optical Remote Sensing Image Ship Target Detection and Recognition

SAR has been used for ship detection for a long time. A paper was published on SAR image ship target detection in 1999 [37]. In 2004, a paper on ship target detection in optical remote sensing images was published [38]. With improvements to the quantity and quality of acquired images, ship target detection research expanded from pure sea surfaces to complex environments, and the research methodology shifted from traditional methods to deep learning-based approaches. If we summarize the development of ship target detection methods in optical remote sensing images from the point of view of dimensions such as the time period, image resolution, detection scenarios, and processing workflows, it can be divided into the following three stages.

1. Around 2010, the data sources for ship target detection in optical remote sensing images were mainly 5 m resolution optical images. Object analysis based on low-level features was utilized for ship target detection. Statistical and discriminatory analysis was carried out using the grayscale, texture, and shape characteristics to distinguish ship targets from the water surface. This achieved ship target detection in medium- to high-resolution and simple scenarios.
2. Around 2013, the data sources for ship target detection in optical remote sensing images were mainly 2 m resolution images. The images contained a large number of medium- and small-sized ship targets, which were abundant and easy to identify. Visual attention mechanisms and mid-level feature encoding theory were widely used for ship target localization and modeling in complex backgrounds with diverse intra-class variations. This enabled ship detection in more complex regions of high-resolution images.
3. Around 2016, a large number of high-resolution optical remote sensing images facilitated ship target detection based on deep learning. Convolutional neural networks (CNNs) were widely applied, mainly in two detection frameworks—classification-based and regression-based approaches—corresponding to step-by-step detection and integrated localization detection, respectively.

2.2. Ship Target Classification

Deep learning methods can achieve both coarse-grained and fine-grained ship recognition. Coarse-grained ship recognition involves distinguishing between different types of ships such as civilian ships, naval vessels, and aircraft carriers. Fine-grained ship recognition aims to identify specific categories of ships, such as cruisers, frigates, and so on.

2.2.1. Coarse-Grained Ship Recognition

Bentes et al. [39] discussed the application of convolutional neural networks in high-resolution remote sensing image classification and employed them for the discrimination of ocean targets. The networks accurately classified cargo ships, tanker ships, oil platforms, and wind-park structures [40]. In [41], a CNN model was used to identify five classes of maritime targets with an average classification accuracy of over 93%. Liu et al. [42] proposed an FCN-based detection method, and experimental results showed that it outperformed traditional methods based on morphological and semantic information, with a classification precision of 90.5% [43]. In the general detection and recognition process, ship detection and angle prediction are usually treated separately. Due to the significant aspect ratio of ship targets, omitting angle information would result in including a large amount of background information in the detection box and the inability to predict the orientation of ships, leading to poor performance in densely distributed ship detection and hampering ship recognition. Yang et al. [44] presented a rotation angle prediction method based on a multi-layer pyramid network, which allowed the ship detection and recognition process to be completed end-to-end and achieved angle prediction. This algorithm designs a rotation anchor strategy to predict the minimum circumscribed rectangle of the object so as to reduce the redundant detection region and improve the precision by 2.2%. Polat et al. [45] applied rotation R-CNN to ship target detection, enabling accurate feature extraction and localization of ships. The method outperforms baselines in mAP by 2.2%. Liu et al. [46] proposed a detection framework for ships in arbitrary directions, accurately locating ship areas by incorporating angle information into bounding box regression and improving the recognition performance of small ship targets. The method outperforms baselines in mAP by 1.7% and has advantages in dealing with small ships. Nie et al. [47] presented a method based on Mask R-CNN for ship segmentation and orientation estimation by outputting the positions of the ship's bow and stern key points, achieving ship target recognition and orientation determination. The precision of the ship direction estimation is 80.4% on the KASDC2018 dataset.

2.2.2. Fine-Grained Ship Recognition

The purpose of fine-grained image recognition is to identify different subclasses within the same major category. Due to the characteristics of ship targets and optical remote sensing imaging, ship targets exhibit both inter-class similarity and intra-class variability, which pose challenges for fine-grained ship recognition tasks. Due to limitations in optical remote sensing image quality and datasets, the task of fine-grained ship recognition is still in the exploratory stage. Feng et al. [48] proposed a two-stage framework for ship detection and classification. In the detection stage, a sequential local context module was introduced based on the Faster R-CNN framework to improve ship detection performance. After obtaining the detection results, a classification network was used to perform fine-grained classification on ship sub-images. The average classification accuracy for 19 types of targets is 81.5%. Sun et al. [49] and others proposed an end-to-end detection and classification framework based on Cascade R-CNN [50], which converted horizontal bounding boxes to rotated bounding boxes, effectively addressing the issue of imprecise localization of horizontal ship bounding boxes, but only recognizing five specific ship classes. Wang et al. [51] presented a two-stage fine-grained ship recognition method using generative adversarial networks, achieving an average precision rate improvement of 2% in recognizing 25 different ship classes. Li et al. [52] proposed a dual-channel ship target recognition method that combines global and local features, enhancing the complementarity between different features and improving feature robustness, leading to fine-grained ship classification. Based on the FGSC-23 dataset, the accuracy of this model is 86.36%, which is better than classical models such as R-CNN. Zhou et al. [53] constructed an end-to-end framework for ship detection and fine-grained classification in marine environments, employing the ResNeSt model as the backbone of the detector to enhance the algorithm's fine-grained classification capability. The *mAP* of this model on the FGSC dataset is 86%.

2.3. Loss Functions

IoU measures the ratio of the intersection to the union of B and G. Equation (1) presents the formula for IoU, which is widely employed as a metric in the field of object detection [54].

$$\text{IoU} = \frac{|G \cap B|}{|G \cup B|} \tag{1}$$

Hence, the formula for calculating the loss using IoU is represented by Equation (2):

$$\text{IoU Loss} = -\ln(\text{IoU}) \tag{2}$$

The IoU Loss ranges from 0 to 1, where a smaller value indicates a closer distance between B and G. A value of 0 indicates a perfect overlap, and a value of 1 indicates no overlap. By minimizing the IoU Loss, B gradually approaches G.

To use IoU as a measure of distance, it must first satisfy the mathematical distance axioms: non-negativity, symmetry, and the triangle inequality. Additionally, IoU possesses the advantageous property of scale invariance, meaning it is insensitive to changes in object size. This scale invariance makes IoU suitable for calculating localization losses for predicted bounding boxes. However, using IoU to calculate losses also has some drawbacks:

1. Lack of distance information: When B and G do not have any intersection, the IoU calculation results in 0, which fails to reflect the true distance between the two boxes. Additionally, since the loss value is 0, the gradient cannot be backpropagated, preventing the network from updating its parameters.
2. Insufficient differentiation between regression outcomes: In cases where multiple regression outcomes yield the same IoU value, the quality of regression differs among these outcomes. In such cases, IoU fails to reflect the regression performance.

These limitations highlight the shortcomings of using IoU for loss calculation. Alternative methods, such as incorporating additional metrics and penalties, have been developed to overcome these challenges and provide more comprehensive measures of distance.

To address the issues with IoU calculation, Generalized-IoU (GIoU) Loss was proposed. It aims to resolve the problem of obtaining a loss value of 0 when B and G do not intersect [55]. GIoU takes into account the completeness of the bounding boxes, even when they do not intersect. By incorporating the smallest convex polygon, GIoU provides a more reliable distance measure between B and G, preventing the loss from being 0 when there is no intersection.

The GIoU Loss function measures the dissimilarity between B and G, and minimizing this loss function during training aims to make B and G as close as possible.

However, GIoU Loss also has its limitations. When B completely encompasses G or is completely contained within it, the GIoU Loss becomes a constant value as long as the areas of B and G remain the same. Regardless of the position of B or G, GIoU Loss remains unchanged. In such cases, GIoU Loss fails to reflect the regression performance. In these scenarios, GIoU Loss essentially degrades to IoU, and the issues associated with IoU resurface.

To address the limitations of GIoU, an improved version called Distance-IoU (DIoU) is proposed [56]. However, DIoU Loss also has limitations: it does not consider the aspect ratio of B and G. This would result in the same loss calculation for the three scenarios when the positions of the center points in B are the same. This fails to reflect the actual situation.

To address the limitations of DIoU, the concept of considering the aspect ratio is incorporated, resulting in Complete-IoU (CIoU) [57]. The CIoU considers the aspect ratio of the boxes, which helps address the issues of DIoU. However, for predicted boxes with the same aspect ratio, the model cannot reflect the actual situation.

To address the limitations of CIoU, the Efficient-IoU (EIoU) Loss is proposed. EIoU introduces a term that separately penalizes the differences in width and height between B and G. The EIoU Loss comprises three main components: overlap loss, center distance loss, and aspect ratio loss. The overlap loss and center distance loss are derived from the CIoU method, and the aspect ratio loss directly reduces the disparity between the width and length of B and G, thereby facilitating quicker convergence [57]. Dividing the aspect ratio loss into the disparities between the predicted dimensions and the dimensions of the minimal enclosing box accelerates the convergence speed and enhances the regression accuracy.

2.4. NMS Strategy

Target detection algorithms based on convolutional neural networks often generate a large number of candidate boxes with different positions and sizes during the prediction stage. These candidate boxes are mostly concentrated in regions that potentially contain the target of interest. It is necessary to perform retention and suppression operations on these candidate boxes. These candidate boxes only contain coordinate information and class confidence scores. The coordinates alone cannot determine the optimal bounding box, and the class confidence score represents the probability of the presence of a specific object in the candidate box. A higher class confidence score indicates a higher likelihood of the candidate box containing a specific object. The selection of the optimal bounding box affects subsequent candidate box suppression operations. If the candidate box with the highest class confidence score is not selected as the optimal bounding box, and other candidate boxes are chosen instead, this may lead to the removal of the candidate box with the highest class confidence score during the candidate box suppression stage. Repeating this operation results in lower localization accuracy for the majority of retained optimal bounding boxes and a decrease in detection accuracy. Therefore, traditional non-maximum suppression algorithms typically use class confidence scores as the evaluation criterion for the optimal bounding box. The candidate boxes are sorted in descending order based on class confidence scores, and the candidate box with the highest class confidence score is selected as the optimal bounding box. Then, the optimal bounding box is retained, and candidate boxes with an IoU greater than a certain threshold are suppressed using this box. If there are multiple objects among these candidate boxes, the candidate box with the highest class confidence score in the remaining candidate boxes is selected as the optimal bounding box for the next object, and the suppression process is repeated until all candidate boxes are removed.

Traditional NMS algorithms consider candidate boxes with a higher overlap with the optimal bounding box as redundant boxes. Therefore, a threshold is set to determine whether a candidate box is redundant. Liu et al. [58] discovered that setting different thresholds in different object detection scenarios can improve recall rate and proposed the Adaptive-NMS algorithm. The Adaptive-NMS algorithm adds a density prediction module in the network to learn the density of the boxes and determines the threshold based on the sparsity of object distribution in the image, thereby achieving a higher recall rate. The Soft-NMS algorithm [59] improves the recall rate, to some extent, by reducing instead of eliminating the class confidence scores of highly overlapping bounding boxes through decay. However, these algorithms do not address the issue of low localization accuracy of the selected optimal bounding boxes.

3. Methodology

In this section, we construct the E2YOLOX-VFL network by improving on the baseline YOLOX, which is an anchor-free detection algorithm. The E2YOLOX-VFL consists of three components: an improved Backbone, a Neck, and an improved Head. The details of the improvements in each component are described below. Please refer to Figure 1 for the network architecture diagram.

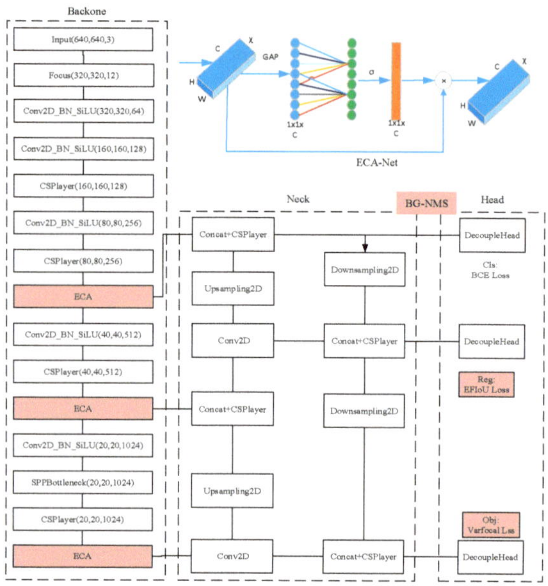

Figure 1. Overall network architecture of E2YOLOX-VFL.

3.1. *Adding ECA Attention Mechanism to Backbone*

In optical remote sensing images, the variable backgrounds, complex scenes, and small target proportions pose challenges for effectively transmitting relevant feature information to the deep network. This often results in missed or false detections due to the inability to accurately locate the targets. To enhance the detection accuracy of these targets, attention mechanisms were incorporated into the Backbone to preserve more detailed information.

Through the integration of the ECA attention mechanism into the network, E2YOLOX-VFL improves the preservation of detailed feature information from optical remote sensing images, thereby enhancing the overall detection accuracy of targets.

ECA-Net provides a novel approach to channel interaction by directly assigning attention weights to corresponding channels through group convolution. By adjusting the feature information based on the magnitude of attention weights, ECA-Net enhances the model's utilization of relevant information.

ECA-Net applies global average pooling to the feature maps and conducts local cross-channel interactions using convolution with a specific kernel size, effectively leveraging channel information. The sigmoid activation function is employed to obtain the weight for each channel, and the original feature map is then recalibrated. This recalibration process can either weaken or strengthen specific region features, which proves beneficial for the detection of smaller targets, as shown in the upper right corner of Figure 1.

Given the aggregated features without channel reduction $y \in R^C$, channel attention can be learned through Equation (3):

$$\omega = \sigma(Wy) \tag{3}$$

Here, W is a $C \times C$ parameter matrix.

ECA-Net explores an alternative method for capturing cross-channel interactions within a field, aiming to ensure efficiency and effectiveness. Specifically, it uses a band matrix W_k to learn channel attention. The banded convolution for cross-channel interaction within channel neurons W_k is as follows.

$$\begin{matrix} w^{1,1} & \cdots & w^{1,k} & 0 & 0 & \cdots & \cdots & 0 \\ 0 & w^{2,2} & \cdots & w^{2,k+1} & 0 & \cdots & \cdots & 0 \\ \vdots & \vdots & \vdots & \vdots & \ddots & \vdots & \vdots & \vdots \\ 0 & \cdots & 0 & 0 & \cdots & w^{C,C-k+1} & \cdots & w^{C,C} \end{matrix} \tag{4}$$

W_k involves $k \times C$ learnable parameters. The convolutional kernel, denoted by k, and the number of channels, represented as C, are crucial components within the network. The convolutional kernel and channel number mapping function is as follows.

$$k = \varphi(C) = \left| \frac{\log_2(C)}{r} + \frac{b}{r} \right|_{odd} \tag{5}$$

Among them, $\gamma = 2$, $b = 1$, and k takes the nearest odd result.

The strip convolution operation involves shifting each row of elements by one bit compared to the previous row in order to achieve channel interaction while also reducing the number of parameters. This process simplifies the convolution calculation, as illustrated in Figure 2. Different colors represent the upper and lower rows of elements.

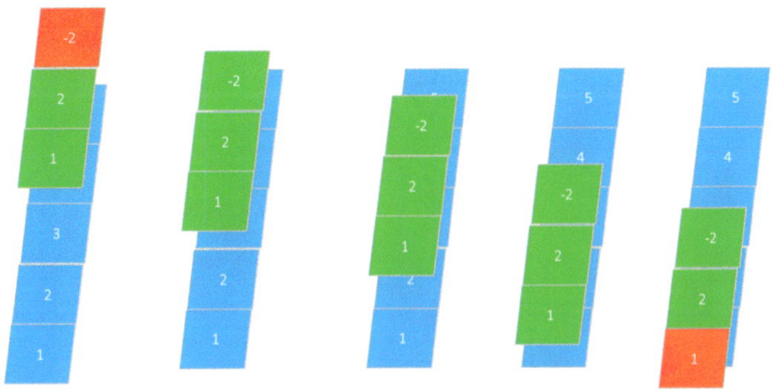

Figure 2. Simplified diagram of calculation process. Different colors represent the upper and lower rows of elements.

Overall, the ECA module uses non-reduced Global Average Pooling (GAP) to aggregate convolutional features, adaptively determines the kernel size k, performs one-dimensional convolution, and then learns channel attention using the sigmoid function.

In our network structure, ECA is added after three feature output layers, which correspond to target features of different scales. (80, 80) corresponds to large-scale targets, (40, 40) corresponds to mesoscale targets, and (20, 20) corresponds to small-scale targets. By incorporating attention mechanisms into the Backbone at three positions, the network becomes capable of capturing large-, small-, and medium-sized target information more accurately.

3.2. Proposing EFIoU Loss as the Localization Loss Function

The localization loss function, also known as the regression loss function, calculates the distance deviation between the final B and G. It computes the loss value as the discrepancy between the two through a specific function and then adjusts the weight parameters using error backpropagation to iteratively minimize the difference, bringing B closer to G. The YOLOX algorithm utilizes the IoU Loss as the localization loss function, but it does not consider overlap loss, center distance loss, aspect ratio loss, or loss distortion caused by low-quality samples.

To address these issues, we propose EFIoU_Loss. EFIoU evolves from EIoU, which is expressed as follows:

$$\text{EFIoU\_Loss} = 1 - \text{IoU} + (1 - \text{IoU})^{\gamma}\left(\frac{\rho^2(b, b^{gt})}{(c_w)^2 + (c_h)^2} + \frac{\rho^2(w, w^{gt})}{(c_w)^2} + \frac{\rho^2(h, h^{gt})}{(c_h)^2}\right) \quad (6)$$

where $(1 - \text{IoU})^{\gamma}$ is the modulation factor used to focus high-quality samples and suppress low-quality samples. γ is a modulation parameter, and when γ is 0, it is EIoU_Loss. In object detection, the predicted boxes obtained from anchor points with a small IoU compared to the ground truth are considered to be low-quality samples. The function of $(1 - \text{IoU})^{\gamma}$ is such that when the IoU is small, indicating a considerable deviation of the predicted box from the ground truth, signifying a low-quality sample, it becomes larger, thereby inhibiting EFIoU_Loss increase. However, when the IoU is large, signifying proximity between the predicted box and the ground truth, the sample is likely a high-quality sample, and the value of $(1 - \text{IoU})^{\gamma}$ basically remains unchanged. This highlights how EFIoU can inhibit the loss of low-quality samples to ensure the network focuses more on high-quality samples. Compared to other IoUs, EFIoU can more accurately describe the difference between the real box and the predicted box. From Section 4.4, it can improve the optimization efficiency and accuracy of target detection networks. Therefore, the original IoU Loss function in the YOLOX algorithm can be improved by replacing it with the EFIoU Loss.

3.3. Adopting Varifocal Loss as the Confidence Loss Function

The confidence value is pivotal in determining whether an object within the bounding box is a positive or negative sample. If the confidence value exceeds the set threshold, it is classified as a positive sample, whereas a value below the threshold indicates a negative sample or background. The confidence loss function employed in YOLOX originally utilized the Cross-Entropy (CE) loss function [60]. However, CE loss suffers from the problem of imbalanced positive and negative samples. This imbalance can substantially impact the accuracy of first-order algorithms when compared to second-order algorithms. To address this issue, our study proposes an improvement to the confidence loss function in YOLOX by replacing the CE function with Varifocal Loss [61]. Varifocal Loss helps balance positive and negative samples and classes, as well as challenging and easy samples.

Standard Cross-Entropy (SCE) Loss is used to assess the discrepancy between target values and predicted values. But its limitation is that all samples are assigned equal weights. When there is an imbalance between positive and negative samples in the dataset, the impact of simple negative samples becomes dominant. In cases with a small number of samples, difficult samples and positive samples cannot significantly contribute to the detection accuracy, leading to poor performance.

Varifocal Loss is a novel loss function designed to train a dense object detector in predicting the IoU-aware Classification Score (IACS), drawing inspiration from Focal Loss.

In contrast to Focal Loss, which treats positives and negatives equally, Varifocal Loss applies an asymmetrical treatment to them [61].

$$VFL(p,q) = \begin{cases} -q(q\log(p) + (1-q)\log(1-p)), if(q > 0) \\ -\alpha p^{\gamma}\log(1-p), if(q = 0) \end{cases} \quad (7)$$

where p is the predicted IACS, and q is the target IoU score.

For positive training examples, q is determined as the IoU between the generated bounding box and the corresponding ground truth box (gt_IoU). Conversely, for negative training examples, the training target q for all classes is set to 0.

As illustrated in Equation (7), the Varifocal Loss reduces the impact of negative examples (q = 0) by incorporating the p^{γ} factor. However, it does not diminish the weight of positive examples (q > 0) in the same manner. This is because positive samples are relatively infrequent compared to negatives, and preserving their learning information is crucial.

Moreover, positive samples are weighted based on the training target q. If a positive sample has a higher gt_IoU, it contributes more to the loss. This places emphasis on positive examples during training, as they are more crucial for achieving higher average precision (AP) compared to low-quality examples.

3.4. Proposing BG-NMS to Replace NMS

Another improvement introduced in the YOLOX algorithm is the replacement of the traditional NMS [62] by our proposed BG-NMS. BG-NMS provides a more flexible and adaptive way to handle overlapping bounding boxes during the post-processing stage of object detection. Instead of suppressing all overlapping boxes with a fixed threshold, BG-NMS assigns reduced confidence scores to the overlapping boxes based on their IoU values. This allows for better localization of objects and helps prevent the suppression of nearby objects, which are potentially valid detections. BG-NMS improves the overall detection performance by providing smoother and more accurate bounding box predictions. The results are shown in and Sections 4.4 and 5.

The NMS can be expressed as Equation (8).

$$S_i = \begin{cases} S_i, iou(M,b_i) < N_t \\ 0, iou(M,b_i) \geq N_t \end{cases} \quad (8)$$

When using NMS, consider M as the bounding box with the highest confidence score, and b_i represents other candidate bounding boxes. The IoU denotes the IoU between M and b_i. If the IoU between M and b_i is above a certain threshold, NMS directly sets the confidence score of b_i to zero. This means that NMS considers b_i as a redundant bounding box and removes it from the candidate boxes.

When multiple objects are densely packed together, traditional NMS can aggressively remove a significant number of tightly arranged bounding boxes, which may lead to missed detections. To overcome this challenge, a Balance Gaussian-weighted BG-NMS algorithm is introduced. The BG-NMS can be expressed as Equation (9).

$$S_i = \begin{cases} S_i, iou(M,b_i) < N_t \\ S_i\alpha e^{\frac{iou(M,b_i)^2}{\sigma}}, iou(M,b_i) \geq N_t \end{cases} \quad (9)$$

BG-NMS applies a Balanced Gaussian-weighted function to the confidence scores, which resembles the density function of a normal distribution with a penalty term. α is the balance factor, with a range of 0–1. The bounding box M serves as the center point of this distribution. As the IoU between M and b_i decreases, indicating a higher proximity of b_i to the center point, the function's value for the weighting on bi increases. α plays a role in balancing the speed of weighted value increase. This Balanced Gaussian weighting brings flexibility and allows BG-NMS to retain more information from closely arranged bounding boxes. This improves the detection performance for densely arranged ship targets by preserving the bounding box information in dense object arrangements.

4. Results

4.1. HRSC2016 Dataset

The HRSC2016 dataset is one of the few high-resolution optical remote sensing ship datasets that subdivides ship categories, with three levels of category labeling. The first level (L1) is classified as ships, the second level (L2) is roughly classified as aircraft carriers, warships, merchant ships, and submarines, and the third level (L3) is a detailed classification of each model. We are conducting ship detection for L3. The dataset consists of 1061 images, including 2976 samples. All images are from six famous ports on Google Earth, with a size of 300 × 300–1500 × 900 pixels, with a spatial resolution of 0.4–2 m [36].

The dataset is divided into three files: Train, Test, and ImageSets. The Train and Test directories are divided into AllImages that only contain ship images and Annotations that only contain annotation information. The images are named in port order and stored in BMP format. The annotation information of the images is stored in an XML file. In addition, the Segmentations file under the Test file also contains ship segmentation images, i.e., semantically segmented labels, stored in PNG format [36]. The distribution of various samples in the training, validation, and testing sets is as in Figure 3.

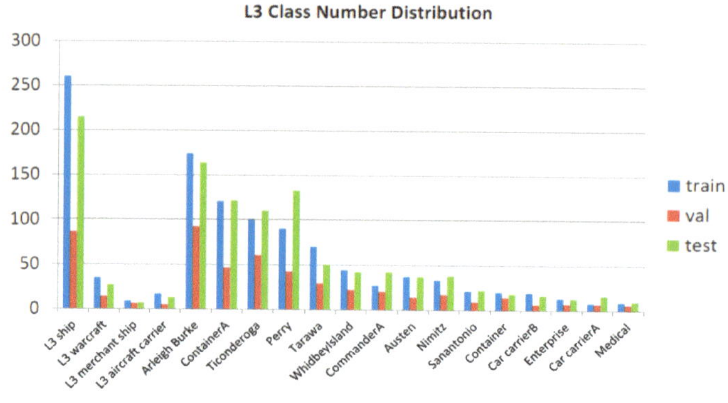

Figure 3. L3 class number distribution.

Given the characteristics of the dataset and the requirements of the model, this paper preprocesses and manipulates the data to make them suitable for analysis and modeling. Three data cleaning strategies are employed:

1. Removing unannotated data from the dataset to exclude it from training.
2. Converting BMP format images in the dataset to JPG format to align with the model used in this research.
3. Enhancing low-contrast images to highlight ship targets.

4.2. Evaluation Metrics

The evaluation metrics employed in this research encompass *Precision*, *Recall*, *F1 score*, *AP*, and *mAP* as selected performance measures [63]. In evaluating the algorithm, ships are considered positive examples, and non-ships are treated as negative examples. In this context, True Positive (*TP*) signifies the count of accurately identified ships, False Positive (*FP*) represents the count of non-ships inaccurately labeled as ships, and False Negative

(*FN*) corresponds to the count of ships that went undetected. The definitions for *Precision*, *Recall*, *F*1, and *AP* are outlined below.

$$Precision = \frac{TP}{TP + FP} \tag{10}$$

$$Recall = \frac{TP}{TP + FN} \tag{11}$$

$$F1 = \frac{2 \times Precision \times Recall}{Precision + Recall} \tag{12}$$

$$AP = \int_0^1 precision \text{ smooth}(recall)d(recall) \tag{13}$$

$$mAP = \frac{1}{N}\sum_{i=1}^{N} AP_i \tag{14}$$

Precision represents the proportion of correctly located ships out of all the location results. Recall, on the other hand, represents the proportion of correctly located ships out of all the ships included in the test set. The Precision rate reflects the false alarm rate (false alarm rate = 1 − Precision rate), and the Recall rate reflects the miss rate (miss rate = 1 − Recall rate). Only when both Precision and Recall are maintained at high levels can the effectiveness of the method be demonstrated. AP, on the other hand, refers to the Precision–Recall curve, with the area under the curve indicating the average accuracy of the test results. *mAP* is the mean of multiple categories of *APs*, representing the overall performance of the algorithm on the dataset. It is a commonly used performance metric.

4.3. Experimental Conditions

The hardware configuration was as follows: Windows 10, AMD Ryzen 7 4800 H, 16.0 GB RAM, GPU 4 GHz.

The software platform included Python 3.8, Torch 1.10.0, CUDA 11.0.3, and cuDNN 8.0.5.

Parameter settings: On top of the determined neural network structure, training the neural network requires the determination of several important parameters that greatly impact the experimental results. Some parameters need to be tested repeatedly to find the optimal values. For this experiment, 90% of the training set is allocated for training purposes, and the remaining 10% is reserved for validation. The batch size is configured as 8, a common practice to align with the capabilities of the graphics card and the number of images processed per training iteration. The term "epoch" denotes the number of complete passes through the entire training set by the neural network. In this particular study, the epoch count is set at 100. The Adam gradient descent optimizer is utilized, initialized with a learning rate of 0.01, and a cosine annealing schedule is employed to gradually decrease the learning rate. The momentum value is established as 0.937.

As the number of epochs increases, the loss gradually decreases until it stabilizes. The optimal epoch value is determined when the rate of change becomes slow. Generally, a slightly larger initial value is chosen, and training can be manually stopped when the loss reaches stability. Figure 4 depicts the change in loss values during the training. It shows a decreasing trend in loss until it stabilizes after approximately 85 epochs.

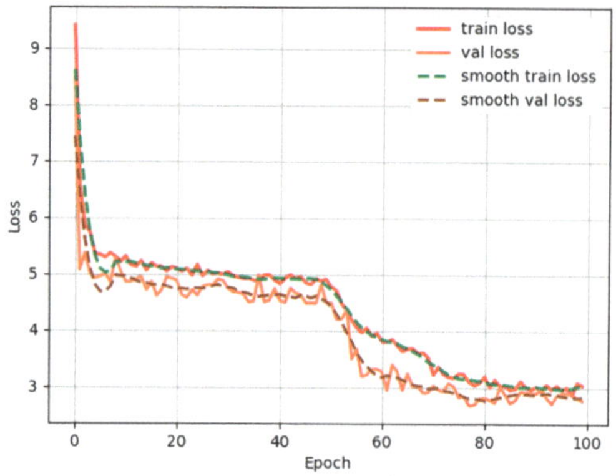

Figure 4. Epoch_loss change curve.

4.4. *Ablation Experiments*

To assess the efficacy of the proposed enhancements, ablation experiments are performed. These experiments aim to validate the impact and effectiveness of the introduced improvements in enhancing the ship detection performance. The compared models are listed in Table 1.

1. YOLOX: The original YOLOX model.
2. YOLOX-BGNMS: Change NMS to BG-NMS.
3. YOLOX-ECA: An improved model in which ECA-Net is added to the Backbone network of YOLOX.
4. YOLOX-VFL: An improved model in which the confidence loss function of YOLOX is replaced with Varifocal Loss.
5. YOLOX-EFIoU: An improved model in which the localization loss function of YOLOX is replaced with EFIoU Loss.
6. E2YOLOX: An improved model in which ECA-Net is added to the Backbone network of YOLOX, and the localization loss function is replaced with EFIoU Loss.
7. E2YOLOX-VFL: An improved model in which modifications are made to the Backbone network, confidence loss function, and localization loss function of YOLOX.

Table 1. Networks used for training.

Networks	Illustrate
YOLOX	Original network
YOLOX-BGNMS	Change NMS to BGNMS
YOLOX-ECA	Add ECA module to YOLOX Backbone
YOLOX-VFL	Change the confidence loss of YOLOX from CE to Varifocal Loss
YOLOX-EFIoU	Change the confidence loss of YOLOX from IoU Loss to EFIoU Loss
E2YOLOX	Add ECA module to YOLOX Backbone, and change the location loss from IoU Loss to EFIoU Loss
E2YOLOX-VFL	Add ECA module to YOLOX Backbone, change NMS to BG-NMS, change the confidence loss from CE to Varifocal Loss, and change the location loss from IoU Loss to EFIoU Loss

These models are evaluated through ablative experiments on the HRSC2016 dataset. The findings obtained from the conducted experiments are consolidated and presented in Table 2 for reference and analysis purposes.

Table 2. Results of ablation experiments.

Model \ Module	YOLOX	BG-NMS	ECA	Varifocal Loss	EFIoU	*mAP*/%
YOLOX [1]	√	×	×	×	×	64.54
YOLOX [2]	√	×	×	×	×	64.82
YOLOX-BGNMS	√	√	×	×	×	66.96
YOLOX-ECA	√	×	√	×	×	68.87
YOLOX-VFL	√	×	×	√	×	69.28
YOLOX-EFIoU	√	×	×	×	√	70.27
E2YOLOX	√	×	√	×	√	70.64
E2YOLOX-VFL	√	√	√	√	√	73.82

Footnote: YOLOX [2] is the result of image enhancement, and YOLOX [1] represents the result without image enhancement.

To illustrate the advantage of the ECA module in more accurately extracting large-, small-, and medium-sized targets, we sequentially add ECA after the three feature layers (80, 80), (40, 40), and (20, 20). The experimental results are shown in Table 3, which evaluates the ECA detection performance.

Table 3. The results of adding ECA after the three feature layers.

Model \ Module	ECA	ECA	ECA	*mAP*/%
YOLOX	×	×	×	64.54
(80, 80)	√	×	×	66.36
(40, 40)	√	√	×	67.13
(20, 20)	√	√	√	68.87

We conducted a comparative experiment between EIoU and EFIoU to illustrate the advantages of the modulation factor. The experimental results are shown in Table 4. EFIoU has increased by 2.7% in *mAP* compared to EIoU. This shows the effectiveness of EFIoU.

Table 4. The results of EIoU and EFIoU.

Model \ Module	EIoU	EFIoU	*mAP*/%
YOLOX	×	×	64.54
YOLOX-EIoU	×	√	67.57
YOLOX-EFIoU	×	√	70.27

5. Discussion

From the data in Table 2, it can be seen that replacing NMS with BG-NMS improves the *mAP* for ship detection by 2.42%. BG-NMS improves the accuracy of dense target detection, which can refer to Figure 5.

From Table 2, it becomes apparent that an enhancement is achieved by adding the ECA attention module to the Backbone network of the original model, which improves the *mAP* for ship detection by 4.33%. This enhancement improves the network's feature extraction capability, allowing for better capture of target information and alleviating cases of missing small targets.

Replacing the confidence loss function with Varifocal Loss results in a 4.74% improvement in *mAP* for ship detection. This addresses the issues of imbalanced positive and negative samples and class imbalance.

By replacing the localization loss function with EFIoU Loss, the *mAP* for ship detection improves by 5.73%. This resolves the problem of regression failure when the two bounding boxes are not intersecting and the inability to reflect regression for targets of different sizes with the same aspect ratio.

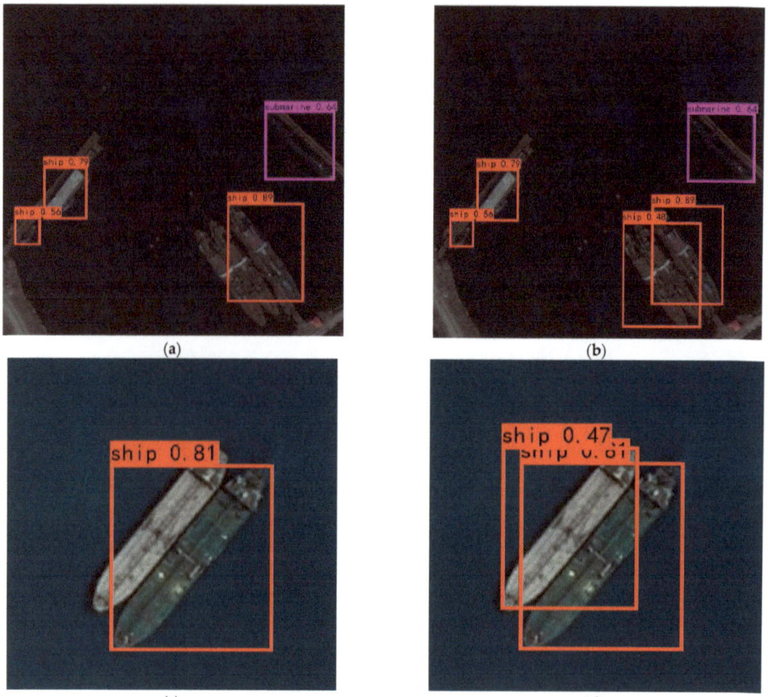

(a) (b) (c) (d)

Figure 5. Comparison of detection performance between NMS and BG-NMS approach: (**a**,**c**) NMS detection result, (**b**,**d**) BG-NMS detection result.

When both the ECA attention module and EFIoU Loss are incorporated into the original model, the *mAP* for ship detection improves by 6.10%. This indicates that the joint improvement approach is effective. When all improvements are added to the original model, the *mAP* for ship detection improves by 9.28%. The outcomes obtained clearly illustrate the substantial influence of the proposed improvements.

Furthermore, the study compares the results of incorporating the ECA module into the Backbone network versus incorporating it into the FPN network. Additionally, upon reviewing Table 5, it is evident that the utilization of the ECA in this research, compared to the CBAM, achieves a slightly lower *mAP* by 0.44%, but with a reduction in parameters by 0.4MB. Incorporating the ECA module into the Backbone network demonstrates better results, with higher *mAP* and lower parameter quantity compared to incorporating it into the FPN network.

Additionally, the detection results in L3 of the E2YOLOX-VFL model were compared with the results in L3 of the SRBBS-Fast-RCNN and SRBBS-Fast-RCNN-R models reported in the literature proposing the HRSC2016 [36] dataset and in Li et al. [64]. As observed from

Table 6, the E2YOLOX-VFL model outperforms the results reported in [36], and it performs just a little lower than SHDRP9 in terms of ship detection *mAP*.

Table 5. Improvement results when adding different modules and adding ECA in different positions.

Model	*mAP*/%	Parameter Quantity
Adding CBAM to the Backbone	69.31	34.7 MB
Adding ECA to the Backbone	68.87	34.3 MB
Adding ECA to the FPN	61.75	55.2 MB

Table 6. Detection results of different models in the HRSC2016 dataset (L3).

Model	*mAP*/%
E2YOLOX-VFL	73.82
SRBBS-Fast-RCNN	60.77
SRBBS-Fast-RCNN-R	45.23
RC1	51.0
RC1 + Multi-NMS	43.7
Faster R-CNN	40.5
R-DFPN	43.5
R-DFPN + PSG97	71.0
SHDRP9	74.2

Figure 5 shows a comparison of the detection performance between the NMS approach and the BG-NMS approach. Figure 5a,c shows the detection results using the NMS approach, and Figure 5b,d shows the detection results using the BG-NMS approach. Despite some missed detections (in cases where prediction boxes of two targets overlap completely), the BG-NMS approach reduces instances of missing densely packed targets to some extent. Therefore, the BG-NMS algorithm exhibits better detection performance for densely packed targets.

To gain a more visual understanding of the detection performance of the E2YOLOX-VFL algorithm, the same image was subjected to detection using both the YOLOX and E2YOLOX-VFL networks. The detection results are depicted in Figure 6, where Figure 6a shows the detection result using YOLOX, and Figure 6b shows the detection result using E2YOLOX-VFL. It can be observed that the E2YOLOX-VFL algorithm performs better in detecting small targets and reduces instances of missing densely packed targets.

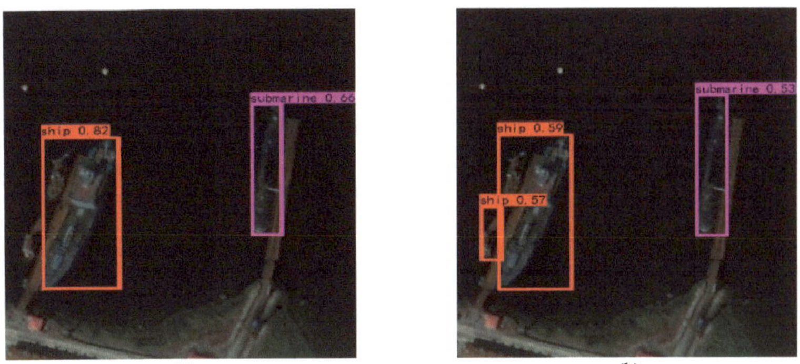

(a) (b)

Figure 6. Comparison of detection performance between the original and improved algorithms: (**a**)YOLOX detection result, (**b**) E2YOLOX-VFL detection result.

To illustrate how the new algorithm can improve the missed detection of densely packed targets, we compared the detection results of YOLOX and E2YOLOX-VFL in Figure 7. Figure 7a,c shows the detection result using YOLOX, and Figure 7b,d shows the detection result using E2YOLOX-VFL. It can be observed that the E2YOLOX-VFL algorithm can reduce instances of missing densely packed targets.

Applying the E2YOLOX-VFL algorithm to the location of a typhoon center, the results are shown in Figure 9. The *mAP* achieved using the E2YOLOX-VFL algorithm is 93.75%, outperforming YOLOX by 1.99%, which demonstrates the applicability of our method to other types of object detection and its overall effectiveness.

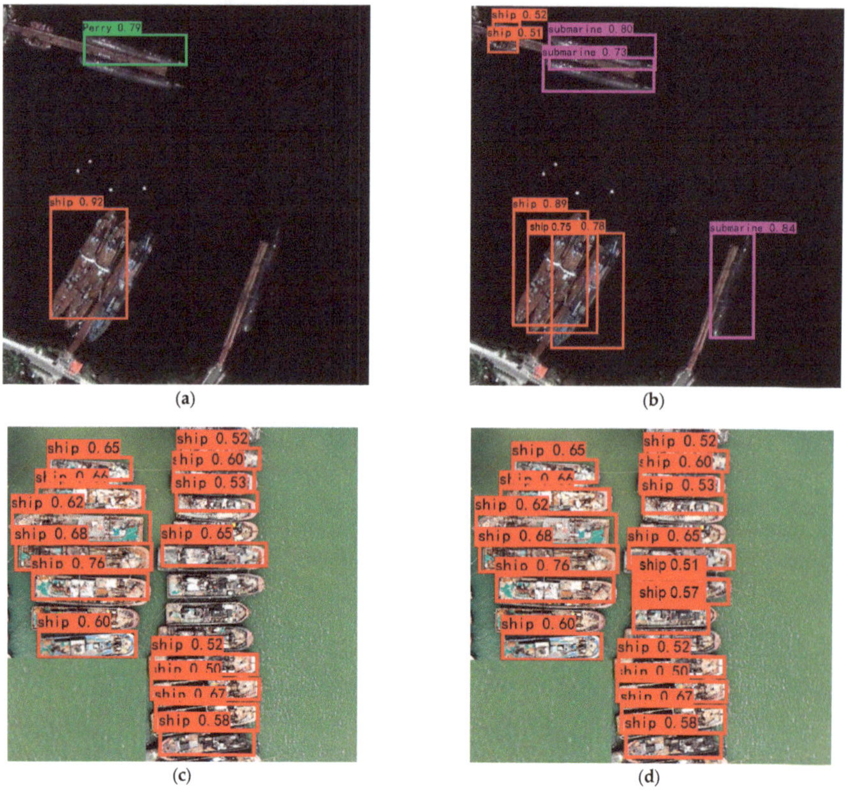

Figure 7. Comparison of detection performance of dense targets between the original and improved algorithms: (**a**,**c**) YOLOX detection result, (**b**,**d**) E2YOLOX-VFL detection result.

Figure 8 shows the effect of image enhancement. It can be seen that the target becomes clearer after image enhancement.

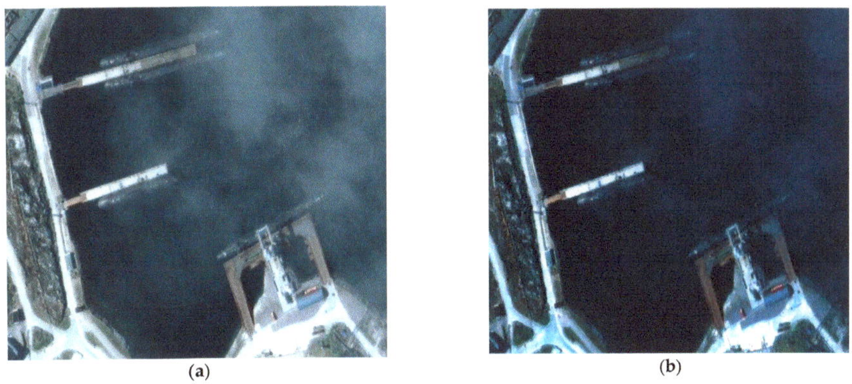

Figure 8. The effect of image enhancement: (**a**) without image enhancement, (**b**) with image enhancement.

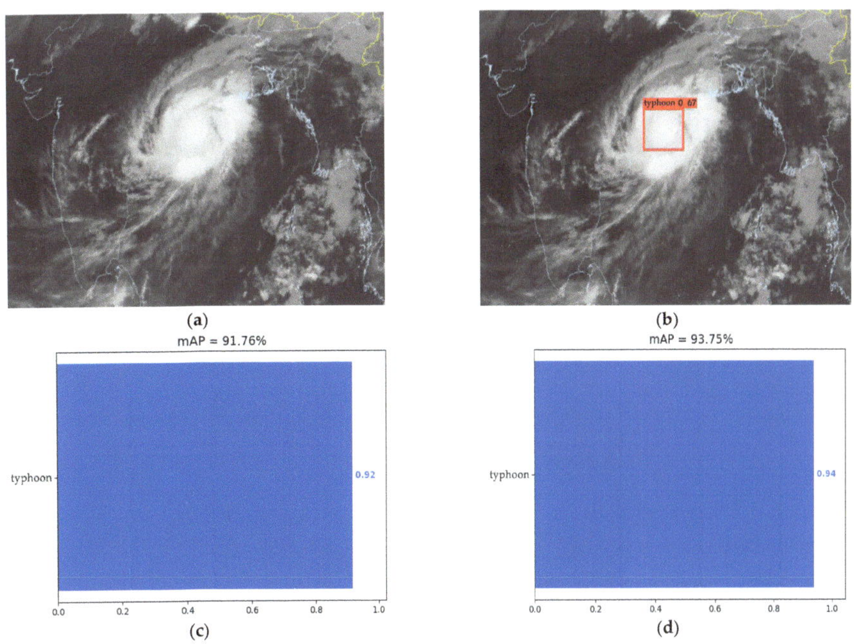

Figure 9. The performance of the algorithm in locating the center of a typhoon: (**a**) original typhoon image, (**b**) positioning typhoon center image, (**c**) YOLOX's *mAP*, (**d**) E2YOLOX-VFL's *mAP*.

6. Conclusions

In this research, we propose an improved algorithm based on the YOLOX model, E2YOLOX-VFL, to address the challenges faced in ship detection in optical remote sensing imagery, such as background interference, variations in target size and position, and class imbalance. By incorporating the ECA module into the Backbone network, modifying the localization and confidence loss functions in YOLO Head, and improving the non-maximum suppression strategy, we enhance the model's ability to detect occluded and multi-scale targets, address the issue of imbalanced positive and negative samples, and handle foreground–background imbalance. These improvements result in an enhanced robustness of the model. Experimental results demonstrate that the proposed E2YOLOX-VFL model achieves higher accuracy in ship detection compared to the original YOLOX model, and it performs favorably against other algorithms evaluated in this research. The model can be further applied to other types of object detection tasks.

Author Contributions: Conceptualization, Q.Z. and Y.W.; methodology, Q.Z.; software, Q.Z.; validation, Q.Z. and Y.Y.; formal analysis, Q.Z. and Y.Y.; investigation, Q.Z.; resources, Y.W.; data curation, Y.Y.; writing—original draft preparation, Q.Z.; writing—review and editing, Q.Z. and Y.W.; visualization, Q.Z.; supervision, Q.Z.; project administration, Y.W.; funding acquisition, Q.Z. All authors have read and agreed to the published version of the manuscript.

Funding: This research was funded by the National Natural Science Foundation of China, grant number 61573183, and Civil Aerospace during the 14th Five Year Plan period, guide number D010201.

Data Availability Statement: The data presented in this study are available on request from the corresponding author. The data are not publicly available due to copyright.

Acknowledgments: The authors are grateful for anonymous reviewers' critical comments and constructive suggestions.

Conflicts of Interest: The authors declare no conflicts of interest.

References

1. Bi, F.; Zhu, B.; Gao, L.; Bian, M. A Visual Search Inspired Computational Model for Ship Detection in Optical Satellite Images. *IEEE Geosci. Remote Sens. Lett.* **2012**, *9*, 749–753.
2. Zhao, M.; Zhang, X.; Kaup, A. Multitask Learning for SAR Ship Detection with Gaussian-Mask Joint Segmentation. *IEEE Trans. Geosci. Remote Sens.* **2021**, *14*, 5214516. [CrossRef]
3. Wan, H.; Huang, Z.; Xia, R.; Wu, B.; Sun, L.; Yao, B.; Liu, X.; Xing, M. AFSar: An Anchor-free SAR Target Detection Algorithm Based on Multiscale Enhancement Representation Learning. *IEEE Trans. Geosci. Remote Sens.* **2022**, *60*, 5219514. [CrossRef]
4. Tang, J.; Deng, C.; Huang, G.; Zhao, B. Compressed-Domain Ship Detection on Spaceborne Optical Image Using Deep Neural Network and Extreme Learning Machine. *IEEE Trans. Geosci. Remote Sens.* **2015**, *53*, 1174–1185. [CrossRef]
5. Zhu, C.; Zhou, H.; Wang, R.; Wang, R.; Guo, J. A Novel Hierarchical Method of Ship Detection from Spaceborne Optical Image Based on Shape and Texture Features. *IEEE Trans. Geosci. Remote Sens.* **2010**, *48*, 3446–3456. [CrossRef]
6. He, S.; Yang, S.; Shi, A.; Li, T. Application of Texture Higher-Order Classification Feature in Ship Target Detection on The Sea. *Opt. Electron. Technol.* **2008**, *6*, 79–82.
7. Zhang, D.; He, S.; Yang, S. A Multi-Scale Fractal Method for Ship Target Detection. *Laser. Infrared.* **2009**, *39*, 315–318.
8. Li, C.; Hu, Y.; Chen, X. Research on Marine Ship Detection Method of SAR Image Based on Fuzzy Theory. *Comput. App.* **2005**, *25*, 1954–1956.
9. Zhao, H.; Wang, P.; Dong, C.; Shang, Y. Ship Target Detection Using Multi-scale Visual Saliency. *Optisc. Precis. Eng.* **2020**, *28*, 1395–1403. [CrossRef]
10. Wang, H.; Zhu, M.; Lin, C.; Chen, D.; Yang, H. Ship Detection in Complex Sea Background in Optical Remote Sensing Images. *Optisc. Precis. Eng.* **2018**, *26*, 723–732. [CrossRef]
11. Zhang, C.; Xiong, B.; Kuang, G. Overview of Ship Target Detection in Optical Satellite Remote Sensing Images. *J. Rad. Sci.* **2020**, *35*, 637–647.
12. Girshick, R.; Donahue, J.; Darrell, T.; Malik, J. Rich Feature Hierarchies for Accurate Object Detection and Semantic Segmentation. In Proceedings of the 2014 IEEE Conference on Computer Vision and Pattern Recognition, Columbus, OH, USA, 23–28 June 2014.
13. Girshick, R. Fast R-CNN. In Proceedings of the 2015 IEEE International Conference on Computer Vision (ICCV), Santiago, Chile, 7–13 December 2015.
14. Ren, S.; He, K.; Girshick, R.; Sun, J. Faster R-CNN: Towards Real-Time Object Detection with Region Proposal Networks. *IEEE Trans. Pattern Anal. Mach. Intel.* **2017**, *39*, 1137–1149. [CrossRef] [PubMed]

15. Yang, L.; Su, J.; Huang, H.; Li, X. A Ship Target Detection Algorithm for SAR Images Based on Deep Multi-scale Feature Fusion CNN. *J. Opt.* **2020**, *40*, 0215002.
16. Zhang, R.; Yao, J.; Zhang, K.; Chen, F.; Zhang, J. S-CNN-Based Ship Detection from High-Resolution Remote Sensing Images. In Proceedings of the International Archives of the Photogrammetry Remote Sensing and Spatial Information Sciences, Prague, Czech Republic, 12–19 July 2016.
17. Ma, X.; Shao, L.; Jin, X. Ship Ttarget Detection Method in Visual Images Based on Improved Mask R-CNN. *J. B. Univ. Technol.* **2021**, *41*, 734–744.
18. Han, X.; Zhong, Y.; Zhang, L. An Efficient and Robust Integrated Geospatial Object Detection Framework for High Spatial Resolution Remote Sensing Imagery. *Remote Sens.* **2017**, *9*, 666. [CrossRef]
19. Li, Q.; Mou, L.; Liu, Q.; Wang, Y.; Zhu, X. HSF-Net: Multiscale Deep Feature Embedding for Ship Detection in Optical Remote Sensing Imagery. *IEEE Trans. Geosci. Remote Sens.* **2018**, *56*, 7147–7161. [CrossRef]
20. Yao, Y.; Jiang, Z.; Zhang, H.; Zhao, D.; Cai, B. Ship Detection in Optical Remote Sensing Images Based on Convolutional Neural Networks. *J. App. Remote Sens.* **2017**, *11*, 042611. [CrossRef]
21. Yang, F.; Xu, Q.; Li, B.; Ji, Y. Ship Detection from Thermal Remote Sensing Imagery through Region Based on Deep Forest. *IEEE Geosci. Remote Sens. Lett.* **2018**, *15*, 449–453. [CrossRef]
22. Jiang, Y.; Zhu, X.; Zhang, W. Ship Extraction Using Post CNN from High Resolution Optical Remotely Sensed Images. In Proceedings of the 3rd Information Networking, Electronics and Automation Control Conference (ITNEC), Chengdu, China, 15–17 March 2019.
23. Redmon, J.; Redmon, S.; Girshick, R.; Farhadi, A. You Only Look Once: Unified, Real-Time Object Detection. In Proceedings of the 2016 IEEE Conference on Computer Vision and Pattern Recognition (CVPR), Las Vegas, NV, USA, 27–30 June 2016.
24. Redmon, J.; Farhadi, A. YOLO9000: Better, Faster, Stronger. In Proceedings of the 2017 IEEE Conference on Computer Vision and Pattern Recognition (CVPR), Honolulu, HI, USA, 21–26 July 2017.
25. Chen, D.; Shao, H.; Zhang, J. Research on Improving YOLOv3′s Ship Detection Algorithm. *Mod. Electron. Technol.* **2023**, *46*, 101–106.
26. Xu, Y.; Gu, Y.; Peng, D.; Liu, J.; Chen, H. An Improved YOLOv3 Model for Arbitrary Direction Ship Detection in Synthetic Aperture Radar Images. *J. Mil. Eng.* **2021**, *42*, 1698–1707.
27. Zou, Z.; Shi, Z. Ship Detection in Spaceborne Optical Image with SVD Networks. *IEEE Trans. Geosci. Remote Sens.* **2016**, *54*, 5832–5845. [CrossRef]
28. Wang, R.; Li, J.; Duan, Y.; Cao, H.; Zhao, Y. Study on the Combined Application of CFAR and Deep Learning in Ship Detection. *J. Indian Soc. Remote Sens.* **2018**, *46*, 1413–1421. [CrossRef]
29. Zhu, X.; Lyu, S.; Wang, X. TPH-YOLOv5: Improved YOLOv5 Based on Transformer Prediction Head for Object Detection on Drone-Captured Scenarios. In Proceedings of the IEEE International Conference on Computer Vision Workshops(ICCVW), Montreal, BC, Canada, 11–17 October 2021.
30. Wang, X.; Jiang, H.; Lin, K. Ship Detection in Remote Sensing Images Based on Improved YOLO Algorithm. *J. B. Univ. Aeronaut. Astronaut.* **2020**, *46*, 1184–1191.
31. Zhang, D.; Wang, C.; Fu, Q. A Ship Critical Position Detection Algorithm Based on Improved YOLOv4 Tiny. *Radio Eng.* **2023**, *53*, 628–635.
32. Li, J.; Zhang, D.; Fan, Y.; Yang, J. Lightweight Ship Target Detection Algorithm Based on Improved YOLOv5. *Comput. App.* **2023**, *43*, 923–929.
33. Ge, Z.; Liu, S.; Wang, F.; Li, Z.; Ge, L. YOLOX: Exceeding YOLO Series in 2021. *arXiv* **2021**, arXiv:2107.08430.
34. Liu, F.; Chen, R.; Zhang, J.; Xing, K.; Liu, H.; Qin, J. R2YOLOX: A Lightweight Refined Anchor-Free Rotated Detector for Object Detection in Aerial Images. *IEEE Trans. Geosci. Remote Sens.* **2022**, *60*, 5632715. [CrossRef]
35. Wang, Q.; Wu, B.; Zhu, P.; Li, P.; Zuo, W.; Hu, Q. ECA-Net: Efficient Channel Attention for Deep Convolutional Neural Networks. In Proceedings of the 2020 IEEE/CVF Conference on Computer Vision and Pattern Recognition (CVPR), Seattle, WA, USA, 13–19 June 2020.
36. Liu, Z.; Yuan, L.; Weng, L.; Yang, Y. A High Resolution Optical Satellite Image Dataset for Ship Recognition and Some New Baselines. In Proceedings of the 6th International Conference on Pattern Recognition Applications and Methods (ICPRAM), Porto, Portugal, 24–26 February 2017.
37. Howard, D.; Roberts, S.; Brankin, R. Target Detection in SAR Imagery by Genetic Programming. *Adv. Eng. Softw.* **1999**, *5*, 303–311. [CrossRef]
38. Marre, F. Automatic vessel detection system on SPOT-5 optical imagery: A neuro-genetic approach. In Proceedings of the Fourth Meeting of the DECLIMS Project, Toulouse, France, 10–12 June 2004.
39. Bentes, C.; Velotto, D.; Lehner, S. Target Classification in Oceanographic SAR Images with Dep Neural Networks: Architecture and Initial Results. In Proceedings of the 2015 IEEE International Geoscience and Remote Sensing Symposium, Milan, Italy, 26–31 July 2015.
40. Bentes, C.; Frost, A.; Velotto, D.; Tings, B. Ship-Iceberg Discrimination with Convolutional Neural Networks in High Resolution SAR Images. In Proceedings of the 11th European Conference on Synthetic Aperture Radar Electronic Proceedings, Hamburg, Germany, 6–9 June 2016.

41. Fernandez, V.; Velotto, D.; Tings, B.; Van, H.; Bentes, C. Ship Classification in High and Very High Resolution Satellite SAR Imagery. In Proceedings of the Future Security 2016, Stuttgart, Germany, 29–30 September 2016.
42. Liu, G.; Zhang, Y.; Zheng, X.; Sun, X. A New Method on Inshore Ship Detection in High-Resolution Satellite Images Using Shape and Context Information. *IEEE Geosci. Remote Sens. Lett.* **2013**, *11*, 617–621. [CrossRef]
43. Abadi, M.; Barham, P.; Chen, J.; Chen, Z.; Zhang, X. TensorFlow: A System for Large-Scale Machine Learning. *arXiv* **2016**, arXiv:1605.08695.
44. Yang, X.; Sun, H.; Fu, K.; Yang, J.; Sun, X.; Yan, M.; Guo, Z. Automatic Ship Detection of Remote Sensing Images from Google Earth in Complex Scenes Based on Multi-Scale Rotation Dense Feature Pyramid Networks. *Remote Sens.* **2018**, *10*, 132. [CrossRef]
45. Liu, Z.; Hu, J.; Weng, L.; Yang, Y. Rotated region based CNN for ship detection. In Proceedings of the IEEE International Conference on Image Processing (ICIP), Beijing, China, 17–20 September 2017.
46. Liu, W.; Ma, L.; Chen, H. Arbitrary-Oriented Ship Detection Framework in Optical Remote-Sensing Images. *IEEE Geosci. Remote Sens. Lett.* **2018**, *15*, 937–941. [CrossRef]
47. Nie, M.; Zhang, J.; Zhang, X. Ship Segmentation and Orientation Estimation Using Keypoints Detection and Voting Mechanism in Remote Sensing Images. In Proceedings of the 16th International Symposium on Neural Networks (ISNN 2019), Moscow, Russia, 10–12 July 2019.
48. Feng, Y.; Diao, W.; Sun, X.; Yan, M. Towards Automated Ship Detection and Category Recognition from High-Resolution Aerial Images. *Remote Sens.* **2019**, *11*, 1901. [CrossRef]
49. Sun, J.; Zou, H.; Deng, Z.; Cao, X.; Li, M.; Ma, Q. Multiclass Oriented Ship Localization and Recognition in High Resolution Remote Sensing Images. In Proceedings of the 2019 IEEE International Geoscience and Remote Sensing Symposium, Yokohama, Japan, 28 July–2 August 2019.
50. Cai, Z.; Vasconcelos, N. Cascade R-CNN: Delving into High Quality Object Detection. In Proceedings of the 2018 IEEE/CVF Conference on Computer Vision and Pattern Recognition, Salt Lake City, UT, USA, 18–23 June 2018.
51. Wang, C.; Tian, J. Precise Recognition of Ship Target Based on Generative Adversarial Network Assisted Learning. *J. Intell. Syst.* **2020**, *15*, 296–301.
52. Li, M.; Sun, W.; Zhang, X.; Yao, L. Fine Grained Recognition of Ship Targets in Optical Satellite Remote Sensing Images Based on Global Local Feature Combination. *Spacecraft Rec. Remote Sens.* **2021**, *42*, 138–147.
53. Zhou, Q. Research on Ship Detection Technology in Ocean Optical Remote Sensing Images. Master's Thesis, Institute of Optoelectronics Technology, Chinese Academy of Sciences, Chengdu, China, 2021.
54. Yu, J.; Jiang, Y.; Wang, Z.; Huang, T. Unitbox: An Advanced Object Detection Network. In Proceedings of the 24th ACM International Conference on Multimedia, Amsterdam, The Netherlands, 1 October 2016.
55. Rezatofighi, H.; Tsoi, N.; Gwak, J.; Sadeghian, A.; Savarese, S. Generalized Intersection over Union: A Metric and A Loss for Bounding Box Regression. In Proceedings of the 2019 IEEE/CVF Conference on Computer Vision and Pattern Recognition (CVPR), Long Beach, CA, USA, 15–20 June 2019.
56. Zheng, Z.; Wang, P.; Liu, W.; Li, J.; Ye, R.; Ren, D. Distance-IoU Loss: Faster and Better Learning for Bounding Box Regression. *arXiv* **2019**, arXiv:1911.08287. [CrossRef]
57. Zhang, Y.; Ren, W.; Zhang, Z.; Jia, Z.; Wang, L.; Tan, T. Focal and Efficient IoU loss for Accurate Bounding Box Regression. *arXiv* **2022**, arXiv:2101.08158. [CrossRef]
58. Liu, S.; Huang, D.; Wang, Y. Adaptive NMS: Refining Pedestrian Detection in A Crowd. In Proceedings of the 2019 IEEE/CVF Conference on Computer Vision and Pattern Recognition (CVPR), Long Beach, CA, USA, 15–20 June 2019.
59. Navaneeth, B.; Bharat, S.; Rama, C.; Davis, L. Soft-NMS-Improving Object Detection with One Line of Code. In Proceedings of the 2017 IEEE International Conference on Computer Vision (ICCV), Venice, Italy, 24–27 October 2017.
60. Li, X.; Yu, L.; Chang, D.; Ma, Z.; Cao, J. Dual Cross-Entropy Loss for Small-Sample Fine-Grained Vehicle Classification. *IEEE Trans. Vehi. Tech.* **2019**, *68*, 4204–4212. [CrossRef]
61. Zhang, H.; Wang, Y.; Dayoub, F.; Sünderhauf, N. VarifocalNet: An IoU-Aware Dense Object Detector. In Proceedings of the 2021 IEEE/CVF Conference on Computer Vision and Pattern Recognition (CVPR), Nashville, TN, USA, 20–25 June 2021.
62. An, H.; Rodrigo, B.; Bernt, S. Learning Non-Maximum Suppression. In Proceedings of the 2017 IEEE Conference on Computer Vision and Pattern Recognition (CVPR), Honolulu, HI, USA, 21–26 July 2017.
63. Jiang, S. Research on Ship Detection Method of Optical Remote Sensing Image Based on Deep Learning. Master's Thesis, School of Electronic Information Engineering, Shanghai Jiao Tong University, Shanghai, China, 2019.
64. Li, B.; Xie, X.; Wei, X.; Tang, W. Ship detection and classification from optical remote sensing images: A survey. *Chin. J. Aeronaut.* **2021**, *343*, 145–163. [CrossRef]

remote sensing

MDPI

Article

DaliWS: A High-Resolution Dataset with Precise Annotations for Water Segmentation in Synthetic Aperture Radar Images

Shanshan Zhang [1], Weibin Li [1,2,*], Rongfang Wang [1], Chenbin Liang [2], Xihui Feng [3] and Yanhua Hu [4]

1 School of Artificial Intelligence, Xidian University, Xi'an 710071, China; 21171213865@stu.xidian.edu.cn (S.Z.); rfwang@xidian.edu.cn (R.W.); liangchenbin@xidian.edu.cn (C.L.)
2 Laboratory of Artificial Intelligence, Hangzhou Institute of Technology of Xidian University, Hangzhou 311231, China; liangchenbin@xidian.edu.cn
3 Key Laboratory of Coal Resources Exploration and Comprehensive Utilization, Ministry of Natural Resources, Xi'an 710021, China; fengxihui789@gmail.com
4 Department of Water Resources of Shaanxi Province, Xi'an 710004, China; sp.huyanhua@gmail.com
* Correspondence: weibinli@xidian.edu.cn

Abstract: The frequent occurrence of global flood disasters leads to millions of people falling into poverty each year, which poses immense pressure on governments and hinders social development. Therefore, providing more data support for flood disaster detection is of paramount importance. To facilitate the development of water body detection algorithms, we create the DaliWS dataset for water segmentation, which contains abundant pixel-level annotations, and consists of high spatial resolution SAR images collected from the GaoFen-3 (GF-3) satellite. For comprehensive analysis, extensive experiments are conducted on the DaliWS dataset to explore the performance of the state-of-the-art segmentation models, including FCN, SegNeXt, U-Net, and DeeplabV3+, and investigate the impact of different polarization modes on water segmentation. Additionally, to probe the generalization of our dataset, we further evaluate the models trained with the DaliWS dataset, on publicly available water segmentation datasets. Through detailed analysis of the experimental results, we establish a valuable benchmark and provide usage guidelines for future researchers working with the DaliWS dataset. The experimental results demonstrate the F1 scores of FCN, SegNeXt, U-Net, and DeeplabV3+ on the dual-polarization data of DaliWS dataset reach to 90.361%, 90.192%, 92.110%, and 91.199%, respectively, and these four models trained using the DaliWS dataset exhibit excellent generalization performance on the public dataset, which further confirms the research value of our dataset.

Keywords: dataset construction; water segmentation; synthetic aperture radar; deep learning; GF-3

Citation: Zhang, S.; Li, W.; Wang, R.; Liang, C.; Feng, X.; Hu, Y. DaliWS: A High-Resolution Dataset with Precise Annotations for Water Segmentation in Synthetic Aperture Radar Images. *Remote Sens.* **2024**, *16*, 720. https://doi.org/10.3390/rs16040720

Academic Editors: Yanni Dong, Xiaochen Yang and Qian Du

Received: 25 December 2023
Revised: 14 February 2024
Accepted: 16 February 2024
Published: 18 February 2024

1. Introduction

Frequent global flood disasters have become a pressing issue, resulting in millions of people falling into poverty each year. These events not only impose enormous pressure on governments, but also hinder social development significantly. In July 2021, catastrophic rainfall in Zhengzhou, Henan Province, China, led to 302 deaths or disappearances [1]. In February 2022, heavy rains in Queensland, Australia, caused eight deaths. In September 2022, Sudan experienced torrential rain and flooding that resulted in 146 deaths. The economic losses caused by flood disasters worldwide in the 21st century have exceeded USD 46 billion annually [2], and approximately 70% of the deaths caused by floods in China are attributed to flash flood disasters [3,4]. These devastating impacts of flood disasters highlight the urgency and importance of effective water resource management. During floods, the prolonged duration and immense destructive power of the floodwaters necessitate the timely and accurate acquisition of flood inundation extents [5], which is crucial for minimizing disaster losses and monitoring floods.

Traditional methods for water body extraction often require extensive manpower and material resources. These methods include direct comparison approaches such as the difference method, ratio method, regression analysis, as well as image transformation approaches such as the Normalized Difference Vegetation Index difference (NDVI), change vector analysis, principal component analysis, and texture-based analysis [6]. In the 1990s, with the emergence of machine learning, researchers began to apply more sophisticated approaches for water detection, including artificial neural networks [7], support vector machines (SVM), decision trees, random forests [7–9], multi-kernel learning, and various hybrid methods such as spectral mixture analysis, fuzzy clustering analysis, and bio-inspired evolutionary algorithms. In the early 21st century, researchers shifted their focus towards object-based image analysis, introducing techniques like Markov random fields, conditional random fields, object-level change vector analysis, and other technologies [6,10,11]. Simultaneously, object category comparison method emerged, including hybrid method at the pixel and object levels. In the 2010s, with the advent of remote sensing big data and artificial intelligence, deep learning such as autoencoders, neural networks [12], recurrent neural networks, and knowledge graphs gradually found application in water body identification.

Deep learning exhibits the ability to dynamically learn by assessing the differences between predicted outcomes and actual ground truth labels, wherein Convolutional Neural Networks (CNNs) stand out as a prominent branch of this domain. Currently, owing to ongoing innovations, CNNs have demonstrated remarkable success across a spectrum of computer vision tasks. Today, diverse standard CNN models have been widely employed for water body segmentation tasks. Several widely recognized and exceptional algorithms, such as SegNet [13], U-Net [14], RefineNet [15], PSPNet [16], Mask R-CNN [17], Deeplab series [18–21], DUPNet [22], CoANet [23], D-LinkNet [24], and PANet [25]. The advent of deep learning techniques has revolutionized the segmentation paradigm from traditional methods, significantly enhancing the accuracy and speed of water body segmentation.

As the data-driven algorithm, the deep learning model relies heavily on the quantity and quality of training data. For water body segmentation tasks, the primary data sources include optical imagery and Synthetic Aperture Radar (SAR) imagery [26]. Optical remote sensing sensors utilize the reflection of sunlight to generate images, constituting a passive imaging system. However, optical imagery is susceptible to factors such as weather conditions, lighting variations, and cloud cover, making it unable to provide all-weather remote sensing capabilities. On the other hand, SAR represents another remote sensing method that relies on active high-resolution microwave radar sensors for imaging. It actively emits electromagnetic waves towards the Earth's surface, and the surface generates echoes that are received by the radar sensor to generate SAR images. SAR images have notable advantages over optical images, since SAR satellites can perform all-weather and all-day data monitoring without being affected by factors such as clouds, fog, and illumination conditions [27]. Researchers and practitioners are more inclined to use SAR images as the preferred data source for water body segmentation tasks. Therefore, utilizing SAR images for water body segmentation is a better choice. In recent years, with the continuous development of SAR technology, the spatial resolution of SAR images has been progressively improving and has now reached sub-meter levels [12,28–30], which facilitates achieving high-precision water body extraction and provides feedback for promptly identifying disaster-stricken areas and mitigating the associated losses.

As is well known, publicly available datasets play a crucial role in driving advancements in the computer vision field and documenting its development process [31]. However, in the context of water body segmentation, there is a lack of accurately annotated high-resolution SAR image datasets. This shortage poses challenges such as unclear targets and inaccurate data, directly impacting the accuracy and generalization capability of models. To address this issue, this study constructs a precisely annotated high-resolution SAR image dataset for water body segmentation based on the GF-3 satellite imagery, termed the DaliWS dataset. Furthermore, several state-of-the-art semantic segmentation models are employed for experimental evaluation on our dataset. Through the analysis of experimental results, a valuable performance benchmark has been established.

Our contributions are summarized as follows:

1. To address the scarcity of publicly available accurately annotated high-resolution SAR image datasets for water body segmentation, this paper provides a manually annotated dataset using GF-3 satellite imagery. The dataset covers large rural water bodies in Dali County and provides pixel-level annotations for the training and validation of deep learning algorithms.
2. We explore the impact of different polarization modes, including HH polarization (single copolarization, horizontal transmit/horizontal receive), HV polarization (single copolarization, horizontal transmit/vertical receive), and HHHV (dual copolarization, horizontal transmit/horizontal receive) on water body segmentation tasks. on the water segmentation task, which provides reference for subsequent related research and facilitates the application of our dataset.
3. To further understand the characteristics of the dataset, this study extensively evaluates its performance using several state-of-the-art segmentation algorithms. The results demonstrate the dataset's inherent challenges and provide new opportunities for water segmentation research material.
4. This study conducts numerous experiments and establishes performance benchmarks specifically for our dataset, laying the foundation for future research. It is anticipated to provide valuable resources and references for the research and development of water body segmentation.

The structure of this paper is as follows: Section 2 presents the processing procedure for SAR images and the creation of the DaliWS dataset. Section 3 demonstrates the segmentation networks used for evaluation, data augmentation methods, and evaluation metrics employed in this study. Section 4 describes the experiments and presents the results. Finally, in Sections 5 and 6, we discuss and summarize the findings of this research.

2. Dataset and Material

2.1. Study Area

The GF-3 satellite captured the study area of the DaliWS dataset in Dali County, Weinan City, Shaanxi Province [32], China. Dali County is situated in the upstream region of the Yellow River, adjacent to Henan Province and Shanxi Province. It is known for its abundant water resources. In this area, multiple rivers converge, including the major Yellow River, the moderate-sized Wei River, and the smaller Beiluo River, forming a unique and rich water network. Additionally, there are numerous artificial water bodies, such as enclosed ponds and reservoirs, which provide ample irrigation and drinking water resources for the local population. Due to its special geographical location and abundant water resources, Dali County is considered a natural and typical research area for hydrological studies, and holds significant importance for water resource management and environmental protection. The satellite orbit coverage of the research area and data source is illustrated in Figure 1.

The study utilized the GF-3 satellite to acquire three sets of images on 24 September 2019, 28 February 2020, and 23 May 2020, respectively. The majority of the covered areas in these images overlap, with the geographical coordinates of the image center captured on 23 May 2020, being at 34°7′ N and 110°3′ E. The original dimensions of the three images are 24,338 × 13,872, 22,553 × 14,384, and 20,312 × 18,480 pixels, respectively, employing a dual-polarization (HH, HV) mode. The imaging mode is Fine Strip I mode of the FSI system, with a resolution of 5 m and a bandwidth of 50 km, providing high-resolution support for water body extraction. The high-quality data obtained from the satellite can be applied in various fields, such as land use classification, environmental monitoring, and disaster assessment. For detailed information, refer to Table 1.

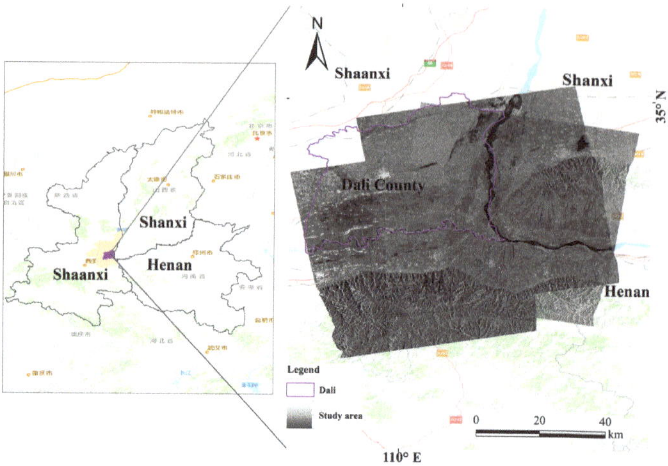

Figure 1. The DaliWS dataset. Geographic location of the study area in Shaanxi province, which is adjacent to Henan province, Shanxi province.

Table 1. Information used in the construction of DailiWS in this article, including sensor, satellite ground station, longitude, latitude, product level, incidence angle, etc.

Item	Parameter
Sensor	GF-3, China
Satellite Ground Station	Sanya Station, China
Longitude	110.3
Latitude	34.7
Imaging band	C bond
Imaging Mode	FSI
Resolution (m)	5
Product Level	L1A
Polarization mode	HH, HV
Incidence Angle (°)	46~49
Swath Width (km)	50

2.2. Data Sources

The data used in this study were obtained from the GF-3 satellite, which is China's first C-band high-resolution SAR satellite with significant importance in various fields such as global ocean observation, disaster reduction [33], glacier identification, land resource monitoring, and surface motion detection. GF-3 is also China's first low-earth orbit remote sensing satellite with a designed lifespan of 8 years, successfully launched in August 2016 [34,35]. To generate products adaptable to different environments, the SAR system of the GF-3 satellite incorporates advanced technologies such as adaptive terrain background classification, polarimetric SAR (PolSAR) [36], and cross-polarization ratio (XPOL), which significantly enhance the quality and application value of SAR data. Furthermore, the GF-3 satellite is designed with 12 different operating modes, achieving a spatial resolution of up to 1 m in spotlight mode.

SAR image products have different levels, and different processing of raw images generates products with varying levels. For different levels of products, the pixel values represent different physical quantities, including Digital Number (DN) values, amplitude values, intensity values, and backscatter coefficient values [37]. DN values are unsigned integer values commonly used for storing complex SAR image data, while amplitude values and intensity values correspond to the magnitude and squared magnitude of complex data, respectively. Intensity values can be further processed to obtain backscatter coefficient values after radiometric correction. The data products of the GF-3 satellite are classified into L0, L1, L2, and L3 standard products, as well as L4 industry application products. Each level of product undergoes different processing to generate subsequent levels of product.

Currently, one of the important research directions in remote sensing is water body extraction. In this study, we utilized GF-3 satellite imagery to create a corresponding dataset and trained multiple deep learning models, demonstrating through experiments that GF-3 satellites can provide high-precision data for water body extraction.

2.3. Data Preprocessing

This study is based on Single-Look Complex (SLC) images from the GF-3 satellite, which belong to the L1A level product, to create a SAR water body segmentation dataset. The raw data received by the SAR sensor is in the form of raw data (L0 level), which can be processed using the focusing algorithm to generate slant-range SLC images [38]. The pixel values of this level of product are 16-bit complex data composed of real and imaginary parts [39], where the real part reflects the intensity and the imaginary part reflects the phase [26]. To facilitate subsequent operations, SLC data undergoes a series of preprocessing steps aimed at reducing image noise, correcting terrain, and improving overall image quality for better extraction and analysis of target objects.

Intensity is one of the main characteristics of SAR images, and information about objects can be extracted based on intensity images. Therefore, SAR complex data needs to be converted to SAR intensity data. Also, because radar echo signals are coherently added together, SLC data products have a lot of speckle noise that needs to be removed using multi-look processing. Multi-look processing involves averaging the SLC data in the azimuth and range directions to weaken the impact of speckle noise on object imaging. The resulting data from multi-look processing are intensity data, completing the transformation from complex data to intensity data. As multi-look processing averages the noise, it improves radiometric resolution but reduces the spatial resolution of the product image.

Speckle noise is inherent in SAR images because the SAR system itself is a coherent system. Since speckle noise can significantly affect the interpretation of SAR image objects [40], in addition to multi-look processing, specialized speckle noise filtering is required during the preprocessing stage. Commonly used SAR filters include Lee [41], Frost [42], Kuan [43], and Gamma MAP filters [44], among others. The Frost filter is one of the most popular adaptive speckle filters and is an exponentially weighted adaptive filter that suppresses speckle noise in the image using the Minimum Mean Square Error (MMSE) estimation method. Within the kernel size of n-by-n, the center pixel value is replaced by a weighted sum of the values of the neighborhood in kernel [45]. In this study, a 5 × 5 Frost filter was chosen to smooth the obtained intensity data, as shown in Formula (1). Among the formulas, $f(i,j)$ represents the filtered pixel value, $g(i,j)$ represents the original pixel value, $m(i,j)$ indicates the average value of surrounding pixels, $w(i,j)$ indicates the weighting factor, $V(i,j)$ is the variance of pixel values in the neighborhood, $N(i,j)$ is the number of pixels in the neighborhood, and γ and β are adjustable parameters.

$$\begin{cases} f(i,j) = w(i,j) \times g(i,j) + (1 - w(i,j)) \times m(i,j) \\ \\ w(i,j) = e^{-\left(\frac{\gamma \times V(i,j)}{N(i,j)^{\beta}}\right)} \end{cases} \tag{1}$$

After multi-look processing and Frost filtering, SAR images need to be geocoded and radiometrically calibrated [46]. After the SAR system acquires reflection information from the Earth's surface, it is encoded in the radar coordinate system, known as the slant-range coordinate system. Generally, before applying SAR data, it needs to be transformed from the slant-range coordinate system to the geographic coordinate system. This process is known as SAR geocoding. The SAR system observes the ratio between the transmitted pulse and the received signal intensity, which is referred to as the backscattering [47]. Different SAR sensors or different receiving modes can affect the backscattering values. Radiometric calibration normalizes the backscattering values to a unified standard, mitigating differences between different SAR sensors. This process is known as SAR radiometric calibration.

The overall processing workflow of remote sensing imagery is shown in Figure 2. These steps complete the preprocessing of SAR images. The preprocessed SAR images are then input into image annotation software for the annotation task.

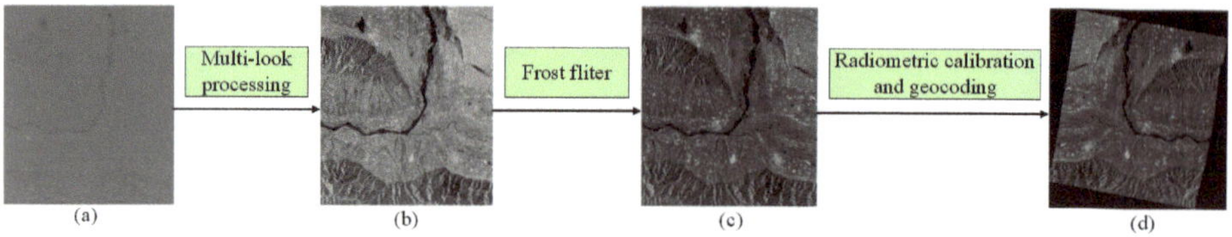

Figure 2. Processing workflow of GF-3 SLC imagery. (**a**) Original SLC imagery; (**b**) multi-view processing result imagery; (**c**) Frost filtering result imagery; (**d**) geocoding and radiometric calibration result imagery.

2.4. Label Generation

In this study, the DaliWS dataset is constructed from three scenes of GF-3 satellite images as described in Section 2.1. Due to significant overlap in most regions of the three images, we primarily use imagery from 23 May 2020 as the main reference, while the other two scenes are utilized to supplement areas not covered by the main image, ensuring comprehensive coverage of most water bodies within Dali County. The generation of dataset labels involved a purely manual annotation process, leveraging domain expertise and calibration with reference to Google Earth [39,48] and other historical images. This approach ensures more accurate annotation of the DaliWS dataset, providing crucial support for further scientific research and practical applications, particularly in the field of SAR image water body segmentation. Such a dataset not only enhances the reliability of the study, but also yields more precise results for practical applications [28].

2.4.1. Chip Creation and Sampling

Figure 3 shows the example used for dataset creation. From the figure, it can be observed that the main land features in the image are water bodies, mountains, and plains. The objective of this study is to create a high-resolution SAR image water segmentation dataset. Therefore, to reduce manual annotation effort and improve annotation accuracy, the image is divided into smaller image blocks before the formal annotation process. Additionally, in later stages of model training, large image blocks can lead to increased computational, memory, and GPU requirements, thereby reducing the parallel training capability of the models [49]. Considering these factors, in this study, the image is segmented into non-overlapping 256 $\times$ 256 image blocks. It is important to note that the width and height of the image are not divisible by 256, resulting in overlapping regions between the last column and last row image blocks with their preceding blocks. After completing the image segmentation process, image blocks containing water body areas are selected for annotation, while image blocks without water body areas are left unmarked. Algorithm 1 describes the process of image partitioning.

Algorithm 1 Image block segmentation

Input: Assigning the original size image as "Img", the image block size is "BlockSize".

Output: ImageBlocks

```
ImageBlocks = []
H,W = Img.shape
// Calculating the segmentation points for the dimensions "h" and "w".
h = Range(0,H,BlockSize)
w = Range(0,W,BlockSize)
h[−1] = (H-BlockSize)
w[−1] = (W-BlockSize)
for h → i do // Iterating through "h" and assigning each iterated element to "i".
    for w → j do
        ImageBlock = Img[i:i+BlockSize,j:j+BlockSize]
        ImageBlocks.add(ImageBlock)
    end for
end for
return ImageBlocks
```

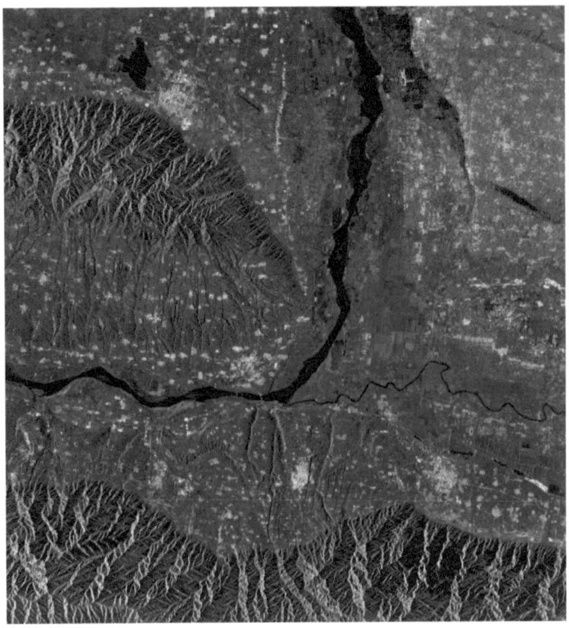

Figure 3. A GF-3 remote sensing image used to create the water body segmentation dataset.

In Algorithm 1, H and W represent the height and width of the original remote sensing image, respectively. BlockSize is the size of the image block. Range (a, b, c) is a sequence generation function that generates a sequence within the range [a, b] with a step size of c. After the sequence is generated, the last segmentation point needs to be updated to avoid boundary issues.

2.4.2. Hand Labeling

This section describes the process of water area annotation using the labeling tool Labelme. Labelme [50] is an offline annotation tool developed in Python that allows annotations to be made using polygons, rectangles, circles, points, and line segments.

The water segmentation dataset created in this study is a binary classification dataset. Non-water land features such as plains, farmland, and mountains are labeled as "background", while the remaining areas are labeled as "water". In addition to rivers, the water areas also include features such as rice fields and ponds, which pose challenges for this annotation task due to their small size. To improve the accuracy of the annotations, other data sources were used for auxiliary calibration. Specifically, Google Earth [48], expert judgment, and online maps from the National Geospatial Information Sharing Service Platform were used as reference data to calibrate features that were difficult to annotate.

In addition to small water bodies, accurately annotating the transition edges between water and background is also challenging. The suitability of annotating the edge regions will affect the effectiveness of subsequent model learning. During the annotation process, special attention was paid to carefully calibrating the transition edges between the background and water.

For annotated images, Labelme records the coordinates of the polygons in JSON format files. During model training, the labels are required to be binary images, where "0" represents the background region and "1" represents the water region. Therefore, after completing the annotation work for all image blocks, the JSON format labels are batch converted to binary PNG format images using the tools provided by Labelme. An example of the DaliWS dataset is shown in Figure 4, which includes five common types of water bodies (ponds, lakes, rivers, rice fields, and reservoirs) and their corresponding ground truth annotations. The first row displays the SAR image, and the second row displays the corresponding labels.

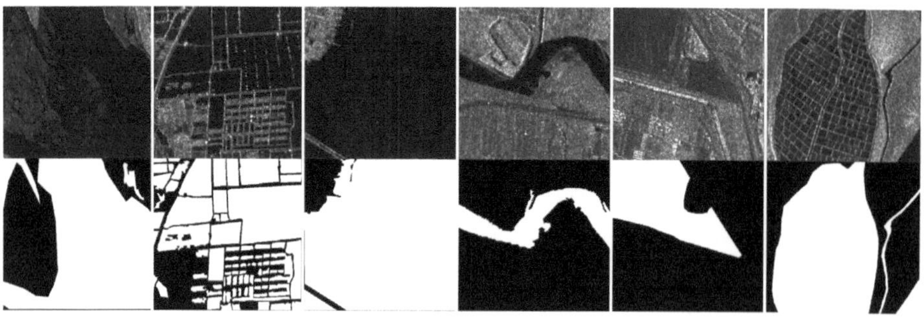

Figure 4. Examples of DaliWS dataset. The images in the top row are SAR images and those in the bottom row are corresponding annotation.

3. Experimental Study

3.1. Dataset Description

3.1.1. DaliWS Dataset

The DaliWS dataset (We provide the DaliWS dataset at: https://github.com/Dataset-RFGroup/DaliWS-Dataset) is a high-resolution SAR dataset for water body segmentation created in this paper based on GF-3 satellite imagery. The original SLC images undergo multi-view processing, frost filtering, geocoding, and radiometric calibration to generate three SAR images. Subsequently, the image is divided into blocks to obtain 2033 images of size 256 × 256 pixels. Finally, the water bodies are accurately annotated using the Labelme annotation tool to create a complete dataset.

Prior to training, the dataset is divided approximately in a 6:2:2 ratio. Data augmentation techniques such as flipping, contrast and brightness enhancement, translation, random noise, and rotation are applied to expand the training and validation sets.

3.1.2. HISEA-1 Dataset

To validate the generalization capability of the models trained on the DaliWS dataset, we selected a publicly available dataset for testing. This dataset, named HISEA1 flooding dataset [51] and abbreviated as HISEA-1 dataset in this paper, was created using HISEA-1 satellite images for water body segmentation. The HISEA-1 satellite carries a synthetic aperture radar (SAR), which is an active remote sensing system for observing the Earth. The HISEA-1 dataset comprises 2340 non-overlapping SAR images with VV polarization. Each image has a size of 256 × 256 pixels. The dataset covers an area of 20,000 square kilometers and includes various landforms such as rivers, tributaries, reservoirs, lakes, and rice fields.

3.2. *Evaluation Methods*

The workflow of this study on DaliWS is illustrated in Figure 5. Four models, namely FCN-8s, SegNeXt, U-Net, and DeeplabV3+, are selected for training on the DaliWS training set. After training, the models are evaluated on the DaliWS testing set by comparing the predicted results with the ground truth labels to obtain evaluation metrics. To test the generalization performance of the models trained on the DaliWS dataset, water extraction and evaluation metric calculations are performed on the publicly available HISEA dataset.

3.2.1. FCN-8s

In 2015, Jonathan Long first applied deep learning to the field of semantic segmentation and created Fully Convolutional Networks (FCN) [52]. By adopting the fully convolutional approach, FCN can accept input images of arbitrary sizes and perform pixel-level classification tasks in an end-to-end manner. The characteristics of FCN include replacing fully connected layers with convolutional layers for end-to-end pixel-level training. Additionally, the introduction of skip connections allows for the fusion of multi-scale features [53]. The FCN architecture consists of three series: 32s, 16s, and 8s, which employ different fusion strategies for prediction. Among them, FCN-8s has more parameters and computational complexity, and it achieves higher segmentation performance. Thus, in our experiments, we selected FCN-8s as the model for water segmentation.

3.2.2. SegNeXt

SegNeXt [54] network was proposed in 2022 and achieved the highest ranking in the Visual Object Classes (VOC) challenge at that time. Additionally, the network has demonstrated significant performance improvements on several mainstream semantic segmentation datasets. The major significance of SegNeXt lies in the search for a more cost-effective convolutional attention mechanism beyond transformer-based multi-head attention. SegNeXt introduces a novel Multi-Scale Convolutional Attention (MSCA) module. The MSCA module consists of three parts: deep convolution for aggregating local information, multi-branch deep stripe convolution for enlarging receptive fields, and 1 × 1 convolution for modeling relationships between different channels. SegNeXt network offers four different overall segmentation model sizes, and taking into account model size and performance, we have decided to use the SegNeXt-B model size in our dataset.

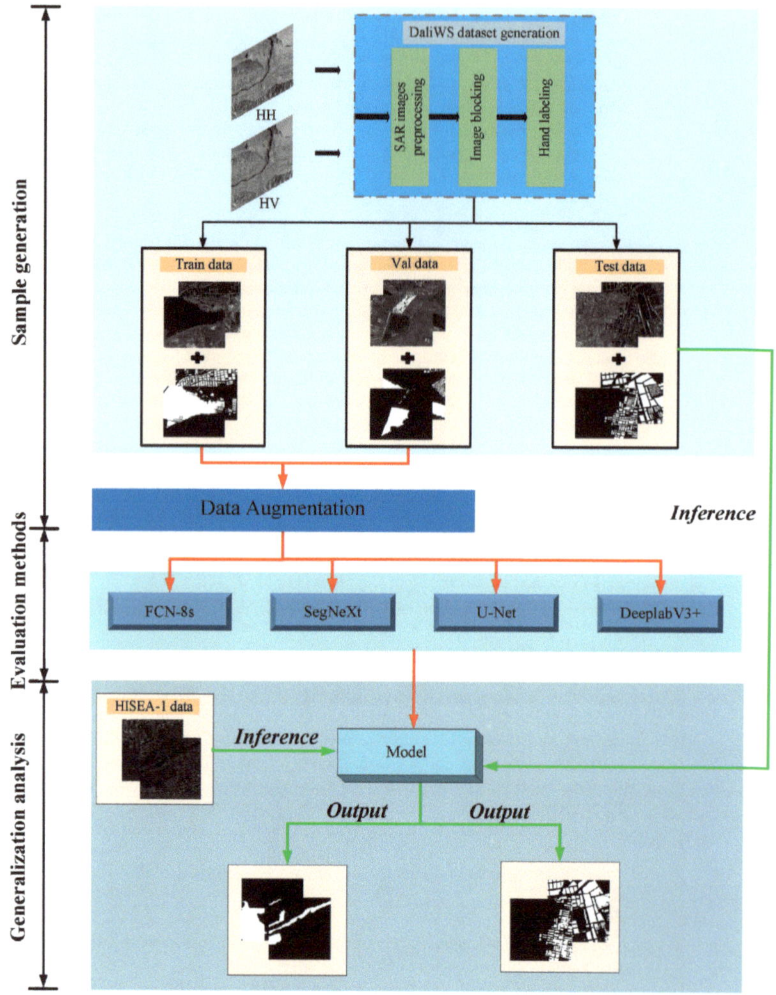

Figure 5. Model training, testing and generalization evaluation flow chart in this paper.

3.2.3. U-Net

As the name suggests, U-Net [14] is a network architecture that resembles the letter "U" and is commonly used for semantic segmentation tasks. It was initially proposed by Ronneberger for medical segmentation. The U-Net architecture consists of a contracting path and an expanding path, making it a typical encoder–decoder network. U-Net incorporates skip connections between the encoder and decoder to enable multi-scale information fusion. Due to its simplicity and effectiveness, U-Net can achieve high accuracy with limited training samples, making it widely applicable in various image segmentation tasks, including remote sensing image segmentation.

3.2.4. DeeplabV3+

Deeplab is a series of deep learning models for image semantic segmentation, initially proposed by the Google team and continuously improved. The final version of the Deeplab series, DeeplabV3+ [21], was introduced in 2018. It utilizes the Xception [55] architecture as the backbone network for extracting abstract and high-level semantic features and incorporates new decoders to achieve more precise object boundary segmentation. In this paper, we selected ResNet50 [56] as the backbone network for DeeplabV3+. One notable feature of DeeplabV3+ is the adoption of the Atrous Spatial Pyramid Pooling (ASPP) module. Additionally, the decoder module in DeeplabV3+ is simple yet effective, gradually recovering spatial features of the targets and achieving finer segmentation of object boundaries. These advantages make DeeplabV3+ stand out among many other methods.

3.3. *Experimental Settings*

3.3.1. Implementation Details

As shown in Table 2, all experiments were conducted on a workstation equipped with a 64-bit Intel(R) Xeon(R) Gold 6226R CPU @ 2.90 GHz processor, RECC DDR4 128 GB memory, and NVIDIA GeForce RTX 3090Ti graphics card, using the PyTorch framework. The network's input was images of size 256 $\times$ 256 pixels. Additionally, each deep learning model was trained for 100 epochs with the Adam optimizer [57], using an initial learning rate of 1×10^{-5} and a batch size of 16. The training time depends on the complexity of the model.

Table 2. The experimental environment used in this study.

Item	Configuration Version
Operating system	Linux CentiOS7.9
Processor	64 Intel(R) Xeon(R) Gold 6226R CPU @ 2.90 GHz
GPU	NVIDIA GeForce RTX 3090
CUDA	CUDA 11.6
Depth framework	PyTorch 1.13.1

3.3.2. Data Augmentation

Data augmentation plays a crucial role in data preprocessing before model training. As networks become deeper, the number of parameters to learn increases, which can lead to overfitting. To address this issue, many researchers employ data augmentation methods to increase the amount of data, enhance data diversity, alleviate overfitting, and improve the generalization and robustness of deep learning models [14,58].

To enhance data diversity and improve model performance, we applied the data augmentation methods listed in Table 3 to the original image patches obtained during the image segmentation stage. Specifically, these methods include: (1) horizontal/vertical/diagonal flipping; (2) contrast enhancement; (3) brightness enhancement; (4) random noise; (5) random translation; and (6) random rotation.

Table 3. Parameters of different data augmentation methods. / indicates that no parameter is set.

Augmentation Methods	Parameter
Horizontal flip	/
Vertical flip	/
Diagonal flip	/
Contrast enhancement	[0.5, 2.5]
Brightness enhancement	[0.5, 2.5]
Random noise	[0, 0.1]
Translation	[−1,1]
Random rotation	$[-\frac{\pi}{4}, \frac{\pi}{4}]$

Diagonal flipping flips the image along the top-left to bottom-right diagonal. Contrast enhancement is applied with a factor ranging from 0.5 to 2.5, while brightness enhancement is applied with a factor ranging from 0.5 to 2.5. Random noise enhancement randomly assigns 0–10% of pixels in each image to have a value of 255. Translation augmentation shifts the image along the top-left to bottom-right diagonal, where negative values indicate shifting from the bottom-right to top-left and positive values indicate shifting from the top-left to bottom-right. Random rotation randomly selects an angle within the range of $[-\frac{\pi}{4}, \frac{\pi}{4}]$ for rotation around the center. Figure 6 shows examples of the aforementioned data augmentation techniques.

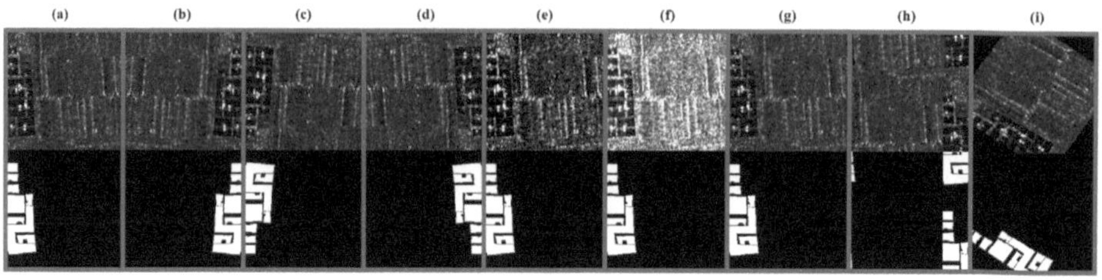

Figure 6. Diagrams of different data augmentation methods. (**a**) Original image. (**b**) Horizontal flip. (**c**) Vertical flip. (**d**) Diagonal flip. (**e**) Contrast enhancement. (**f**) Brightness enhancement. (**g**) Random noise. (**h**) Translation. (**i**) Random rotation.

3.3.3. Evaluation Metrics

The distribution of water and background in the DaliWS dataset is imbalanced, making it challenging to accurately segment the water regions. To comprehensively evaluate the performance of the selected four models on the DaliWS dataset, we employ multiple evaluation metrics. The specific evaluation metrics include accuracy, recall, precision, F1 score, mean Intersection over Union (mIoU), and segmentation time. Pixel accuracy (PA) refers to the ratio of correctly classified pixels to the total number of pixels. Recall (Rec) describes the ability to find all positive samples of a specific class. Precision (Pre) represents the ability of the classifier to label negative samples as positive samples [59]. The F1 score (F1) is a composite metric that balances precision and recall. mIoU is the average of Intersection over Union for each class. Since inference time is critical for disaster monitoring [51], the speed of water extraction is also included in the evaluation. Segmentation time refers to the time required for the model to complete the water segmentation task in an image. Higher image resolutions require more time.

Based on the confusion matrix shown in Table 4, the calculation formula is as follows:

$$PA = \frac{TP + TN}{TP + FN + FP + TN} \tag{2}$$

$$Rec = \frac{TP}{TP + FN} \tag{3}$$

$$Pre = \frac{TP}{TP + FP} \tag{4}$$

$$F1 = \frac{2 \times Pre \times Rec}{Pre + Rec} \tag{5}$$

$$mIoU = \frac{1}{k+1} \sum_{i=0}^{k} \frac{TP}{TP + FP + FN} \tag{6}$$

In the formulas, TP represents the number of pixels correctly predicted as water, FN represents the number of water pixels falsely predicted as background, FP describes the number of background pixels falsely predicted as water, and TN represents the number of background pixels correctly predicted as background.

Table 4. The confusion matrix.

		Prediction	
		Water	**Background**
Ground truth	Water	TP	FP
	Background	FN	TN

4. Results

4.1. Quantitative Analysis of Four Networks on DaliWS Dataset

To comprehensively explore the DaliWS dataset, we train four segmentation networks, e.g., FCN, SegNeXt, U-Net, and DeeplabV3+ on the training set. Subsequently, we evaluate the trained models using the test set. We then perform a quantitative comparison of the segmentation results of these four models on the DaliWS dataset and conduct an in-depth investigation into the impact of polarization modes on segmentation performance. To ensure experimental fairness, we employ consistent evaluation metrics, evaluation code, and maintain identical network parameter configurations as described in Section 3.3.2.

In the first experiment, we focus on examining the water body extraction performance of different models on the DaliWS dataset. Results from Table 5 indicate that all four models achieve F1 scores of over 85%, with U-Net exhibiting the best performance. In the dual-polarization mode, U-Net achieves F1, PA, and mIoU scores of 92.110%, 98.371%, and 85.374%, respectively. DeeplabV3+ closely followed, achieving F1, PA, and mIoU scores of 91.199%, 98.188%, and 83.822% in the dual-polarization mode. In comparison, FCN-8s and SegNeXt networks demonstrate relatively poorer performance in water area segmentation, with F1 scores of 90.361% and 90.192% in the dual-polarization mode. Table 6 displays the GFLOPs and Params for these four models with an input size of (1,1,256,256), revealing that the order of GFLOPs from highest to lowest is U-Net, DeeplabV3+, SegNeXt, and FCN. U-Net has the smallest parameter count while also having the highest computational workload.

In the second experiment, we delve into the impact of different polarization modes on segmentation performance. Firstly, FCN, SegNeXt, U-Net, and DeeplabV3+ achieve F1 scores of 90.361%, 90.192%, 92.110%, and 91.199%, respectively, on HH + HV dual-polarization data, demonstrating superior performance compared to HH and HV single-polarization data. This is understandable, as the fusion of HH and HV dual-polarization data provides more comprehensive and richer spatial features for water body segmentation models. On HH and HV single-polarization data, U-Net consistently performs the best, with F1 scores of 91.173% and 90.198%, respectively. Overall, dual-polarization outperforms single-polarization, with HH polarization being superior to HV polarization.

Table 5. Quantitative comparison of different models in terms of F1 score(%) , PA(%), Rec (%), Pre (%), and mIoU(%) on the DaliWS dataset.

Models	Polarization	F1(%)	PA(%)	Rec(%)	Pre(%)	mIoU(%)
FCN	HH	89.762	97.889	90.235	89.294	81.426
	HV	88.823	97.687	89.605	88.055	79.893
	HH+HV	**90.361**	**98.012**	**90.853**	**89.875**	**82.417**
SegNeXt	HH	90.090	97.967	**90.101**	90.080	81.967
	HV	87.894	97.605	84.779	91.246	78.403
	HH+HV	**90.192**	**98.023**	88.659	**91.780**	**82.137**
U-Net	HH	91.173	98.197	90.800	**91.549**	83.778
	HV	90.198	98.002	89.624	90.779	82.146
	HH+HV	**92.110**	**98.371**	**92.744**	91.484	**85.374**
DeeplabV3+	HH	90.898	98.170	89.126	**92.742**	83.315
	HV	89.994	97.906	**91.844**	88.216	81.808
	HH+HV	**91.199**	**98.188**	91.555	90.845	**83.822**

Bold indicates the maximum value of each evaluation index.

Table 6. The parameter sizes of each model are listed in the table below.

Model	Input-Size	GFLOPS	Params(M)	Time(s)
FCN	(1,1,256,256)	**7.78**	46.24	0.0100
SegNeXt	(1,1,256,256)	7.96	27.54	0.0324
U-Net	(1,1,256,256)	31.81	**16.69**	0.0093
DeeplabV3+	(1,1,256,256)	26.51	40.48	0.0138

"Times" refers to the inference time for a single image.

4.2. Qualitative Analysis of Four Networks on DaliWS Dataset

Figures 7 and 8 illustrate the partial water body prediction results of FCN-8s, SegNeXt, U-Net, and DeeplabV3+ on the DaliWS test dataset. From the figure, it can be observed that all models are able to predict large water bodies to some extent. However, they exhibit limitations in accurately segmenting boundaries, shadows, and small water bodies.

In the first row, it can be seen that SegNeXt and U-Net incorrectly identify a small tributary as water, while FCN and DeeplabV3+ correctly classify it as background. However, most methods erroneously label farmland as water. In the second row, U-Net misses a portion of the water body in the river. In the last row, none of the methods are able to effectively segment the water body and path boundaries in the shadow areas. Among these four models, FCN produces relatively coarse predictions due to the final feature map output not being of the same size as the original input. It requires 8-fold upsampling to restore the prediction map size. While this approach reduces computational complexity, it results in the loss of significant detail. For the DaliWS dataset, which has a spatial resolution of 5 m, such coarse boundaries severely impact the accuracy of water body extent calculations.

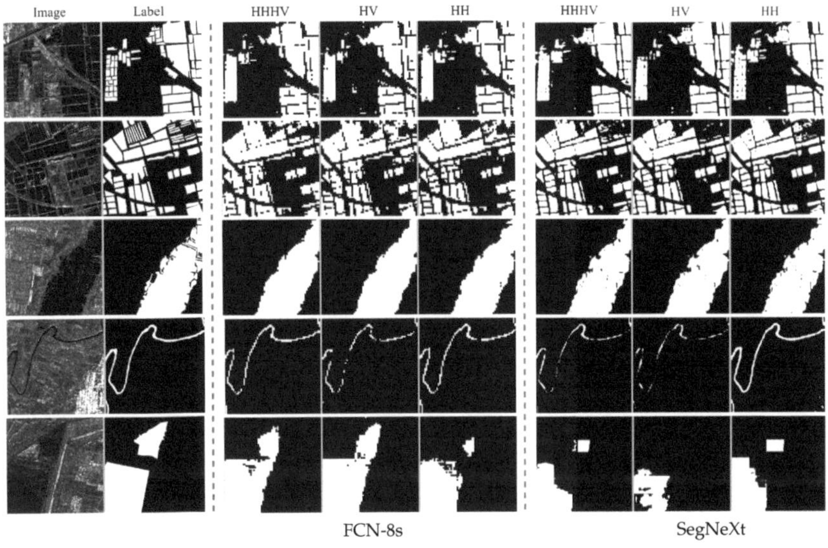

Figure 7. Comparison example images of FCN-8s and SegNeXt models trained on differently polarized data for water segmentation.

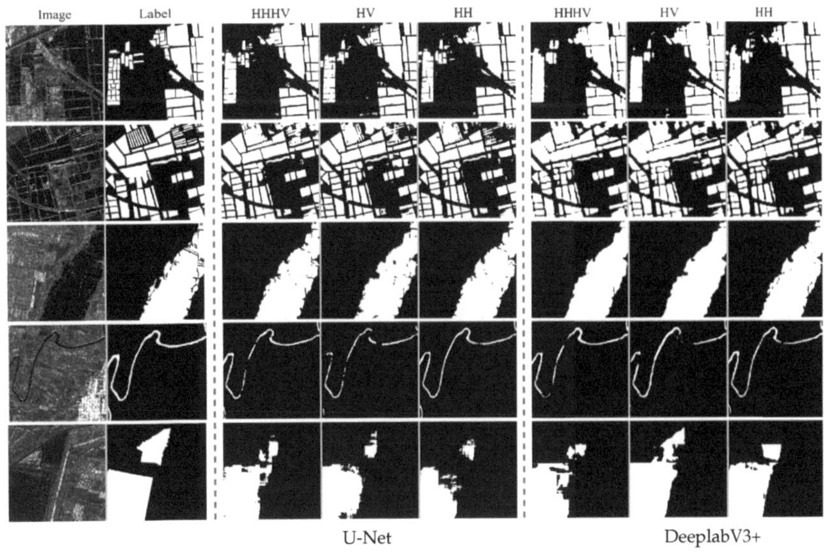

Figure 8. Comparison example images of U-Net and DeeplabV3+ models trained on differently polarized data for water segmentation.

The four models are trained using three different modes: HH, HV, and HH+HV. By observing the results in the first and second rows of Figures 7 and 8, it can be seen that U-Net achieves the best segmentation performance across all three modes. The models trained with HV single-polarization data performed the worst. In the fourth row of Figures 7 and 8, the models trained with HH single-polarization data show better segmentation performance compared to HH+HV dual-polarization data, particularly in accurately segmenting elongated rivers. In the fifth row of Figures 7 and 8, all four models struggled to segment water bodies with indistinct boundaries.

4.3. Generalization Analysis of Four Models Trained on the DaliWS Dataset

To assess the generalization capabilities of models trained on the DaliWS dataset, we conduct generalization experiments on the HISEA-1 dataset. To do this, different polarization data from DaliWS are used to train segmentation models like FCN-8s, SegNeXt, U-Net, and DeeplabV3+. These models are then tested on the HISEA-1 dataset to obtain segmentation metrics. The DaliWS dataset has two polarization modes, HH and HV, while the HISEA-1 dataset only has VV polarization. For each model, we perform three sets of experiments: (1) HH testing on VV, (2) HV testing on VV, and (3) HHHV testing on VV. It is noteworthy that, in experiment (3), we duplicate the VV data to create a dual-channel image.

The partial water body prediction results of FCN-8s, SegNeXt, U-Net, and DeeplabV3+ on the HISEA-1 dataset are shown in Figures 9 and 10. From the figures, it can be observed that HH polarization data overall outperformed HV and HH+HV, with the latter two polarization data showing instances of interrupted flow in predicting water bodies. The higher rate of false detections in the U-Net prediction map in Figure 10 demonstrates that for HH+HV polarization data; the models do not efficiently extract complementary features from both data types.

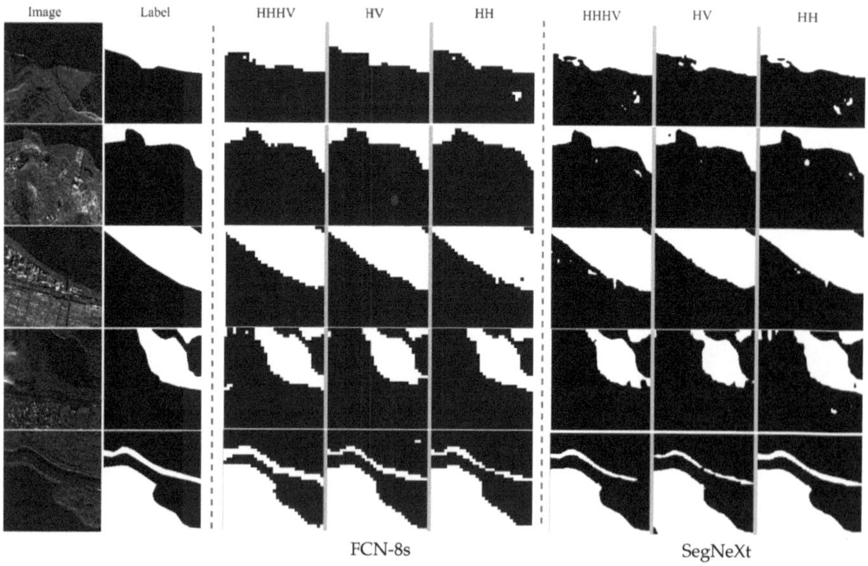

Figure 9. Comparison of example images for generalization predictions by FCN-8s and SegNeXt models.

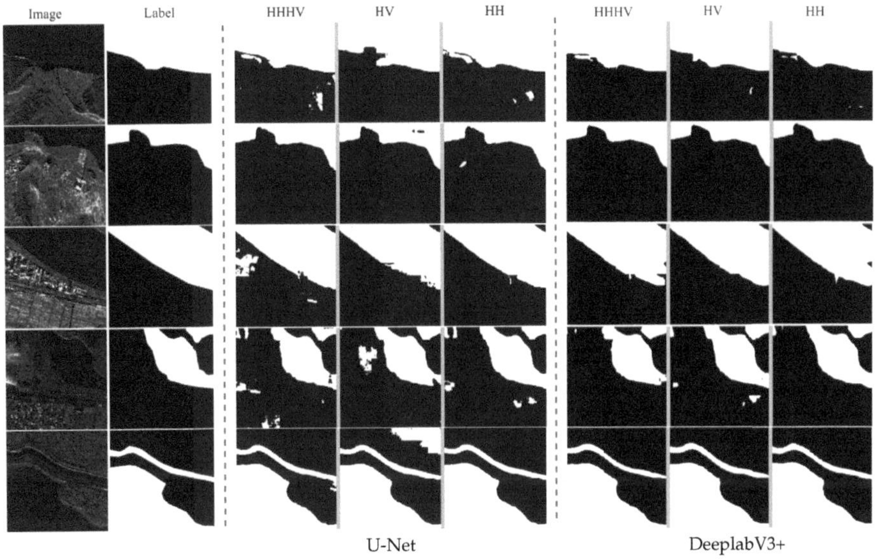

Figure 10. Comparison of example images for generalization predictions by U-Net and DeeplabV3+ models.

The generalization test results in Table 7 show that models trained with HH polarization data from the DaliWS dataset obtain F1 scores higher than 80% on the VV polarization data from the HISEA-1 dataset, showing better generalization performance. The models that are trained with HH+HV dual-polarization data come in second. Except for FCN, the HH+HV dual-polarization data exhibits commendable performance on the other three models. The generalization capability of HV polarization data are the weakest, significantly lower than that of HH and HH + HV. In terms of overall model generalization performance, SegNeXt exhibits the best results, with F1 scores of 85.276%, 77.307%, and 82.608% for HH, HV, and HH+HV, respectively. DeeplabV3+ follows closely, achieving F1 scores of 85.982%, 77.446%, and 81.650%. FCN demonstrates the poorest generalization performance. In summary, the four models trained on the DaliWS dataset in this study display excellent generalization capabilities on the public dataset, affirming the dataset's value for research purposes.

Table 7. Generalization results of four models trained on the DaliWS dataset.

Models	Polarization	F1(%)	PA(%)	Rec(%)	Pre(%)	mIoU(%)
FCN	HH	**83.890**	**93.543**	**82.770**	85.040	**72.250**
	HV	62.070	83.645	65.886	58.672	45.001
	HH+HV	75.867	91.597	65.028	**91.042**	61.117
SegNeXt	HH	**85.276**	**94.177**	**83.023**	87.654	**74.331**
	HV	77.307	92.064	66.553	**92.206**	63.008
	HH+HV	82.608	93.468	76.379	89.944	70.369

Table 7. *Cont.*

Models	Polarization	F1(%)	PA(%)	Rec(%)	Pre(%)	mIoU(%)
U-Net	HH	**85.391**	**94.313**	**81.827**	**89.280**	**74.506**
	HV	68.426	87.234	68.110	68.746	52.006
	HH+HV	80.365	91.937	81.240	79.508	67.175
DeeplabV3+	HH	**85.982**	**94.553**	**82.241**	90.080	**75.411**
	HV	77.446	90.715	78.487	76.433	63.194
	HH+HV	81.650	93.297	73.428	**91.945**	68.990

Bold indicates the maximum value of each evaluation index.

5. Discussion

In order to support research on Deep Neural Network (DEN)-based water body segmentation algorithms, this paper creates a high-resolution SAR image water body segmentation dataset. The dataset consists of L1A-level SLC images from the GF-3 satellite, captured on 24 September 2019, 28 February 2020, and 23 May 2020, respectively. The original images underwent multi-view processing, Frost filtering, geocoding, and radiometric calibration to generate the final images used for dataset creation. Subsequently, the images are partitioned and sampled, dividing the large image into non-overlapping 256 × 256 image blocks and selecting the parts containing water bodies for the dataset. Finally, the Labelme tool was used for image annotation.

After completing the construction of the DaliWS dataset, this study selects four segmentation networks, namely FCN, SegNeXt, U-Net, and DeeplabV3+, for training and evaluation on the dataset. Figure 11 shows a full comparison of how well these four models can segment and generalize across three polarization modes on the DaliWS dataset. All four models demonstrate outstanding segmentation performance on the DaliWS dataset. Among them, U-Net exhibits the best segmentation accuracy and inference time, achieving an F1 score of 92.110%. However, there is still room for improvement in the segmentation results, particularly in capturing details at the edges.

The performance of SegNeXt in Section 4.1 prompted deep reflection. Despite ranking highest in the VOC challenge, SegNeXt shows the poorest segmentation performance on the DaliWS dataset. In response, we provide a reasonable explanation by comparing the differences between the two datasets. Firstly, although the SegNeXt model excelled on the VOC dataset, it cannot be assumed that it would achieve the same performance on SAR images due to domain differences between natural images and SAR images, which may hinder effective knowledge transfer. Secondly, water body segmentation is a binary classification task, while SAR images are primarily composed of black and gray. A complex SegNeXt network may struggle to allow each parameter to learn sufficiently, whereas simpler segmentation networks like U-Net and Deeplabv3+ are more suitable for binary segmentation tasks on SAR images.

In Table 6, U-Net has the highest GFLOPS among the four models and the shortest inference time per image. In contrast, SegNeXt has significantly lower GFLOPS than U-Net, yet it has the longest inference time per image. Additionally, by comparing the GFLOPS and inference time of other models in Table 6, we conclude that GFLOPS is only a theoretical indicator for measuring model speed and cannot represent the actual inference speed of the model.

In the generalization experiments shown in Table 7, the HH polarization mode performs the best in generalization testing, while models trained using HV polarization data show significantly lower performance than HH polarization. We provide an explanation for this phenomenon: the HISEA-1 dataset used VV polarization, corresponding to HH, making the data distribution of these two modes more similar. On the other hand, although the HH+HV polarization mode integrates HH polarization information, the presence of HV polarization interferes with the overall segmentation results.

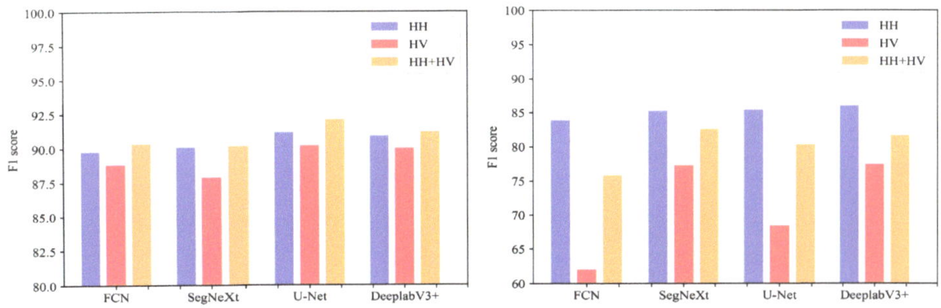

(**a**) DaliWS (**b**) HISEA-1

Figure 11. (**a**) F1 score histograms for four models trained on the DaliWS dataset with three different polarization modes; (**b**) F1 score histograms for four models trained on the DaliWS dataset with three polarization modes, evaluated on the HISEAR-1 dataset.

In summary, the experimental results of this study validate the effectiveness of the DaliWS dataset and compare and analyze the water body segmentation performance under different segmentation networks and polarization modes. This research holds significant implications for water remote sensing image analysis and water resource management, providing valuable insights for related studies and applications.

6. Conclusions

In conclusion, despite the rapid advancements in computer vision technology, its high-precision performance still relies heavily on precisely annotated datasets. This study aims to contribute significantly to SAR image water body segmentation tasks and offer valuable insights for future research endeavors. We meticulously create and extensively explore the DaliWS dataset, assessing it through a range of network models. While achieving satisfactory segmentation results across diverse network models, there remains room for enhancement in capturing finer details at water body edges.

Our investigation also delves deeply into the impact of different polarization modes on segmentation performance. And the experimental results demonstrate a certain correlation between datasets sharing similar polarization modes, thereby contributing to superior generalization performance. Notably, models trained on HH+HV datasets perform worse on the VV dataset. This shows that more research is needed to figure out how to combine multi-polarization data effectively.

DaliWS provides multi-polarized information and precise annotations, facilitating the extraction of practical disaster-related information. However, the dataset's sources are not yet comprehensive enough. In the future, we will continue to upgrade DaliWS by collecting multi-source remote sensing data and conducting pixel-level annotations, thereby offering greater convenience for flood disaster emergency responses. To sum up, our work establishes a high-resolution SAR image water body segmentation dataset, analyzes the performance of various segmentation networks and polarization modes, and provides crucial practical insights and references for water body remote sensing image analysis and water resource management.

Author Contributions: Conceptualization, S.Z. and C.L.; methodology, S.Z.; validation, S.Z. and R.W.; investigation, S.Z. and R.W.; resources, W.L., X.F. and Y.H.; data curation, S.Z. and R.W.; writing—original draft preparation, S.Z., C.L., W.L. and R.W.; writing—review and editing, S.Z. and C.L.; visualization, S.Z., C.L., W.L. and R.W.; supervision, W.L., X.F. and Y.H.; project administration,

S.Z., R.W. and W.L.; funding acquisition, W.L., R.W., X.F. and Y.H. All authors have read and agreed to the published version of the manuscript.

Funding: This research was funded by the National Natural Science Foundation of China (No. 62176196), Shaanxi Provincial Water Conservancy Fund Project (No. SLKJ2024-06), Research Project of Shaanxi Coal Geology Group (No. SMDZ-2023CX-14), the China Postdoctoral Science Foundation (No. 2023M742729) and the Key Research and Development Program of Shaanxi (No. 2023-YBNY-218).

Institutional Review Board Statement: Not applicable.

Informed Consent Statement: Not applicable.

Data Availability Statement: The data presented in this study are available on request from the corresponding author.

Acknowledgments: The authors would like to thank three anonymous reviewers for their helpful comments and suggestions, which significantly improved the quality of our paper.

Conflicts of Interest: The authors declare no conflicts of interest.

References

1. Liu, S.; Wang, J.H. Analysis of the monitoring ability of high-resolution satellites for the "21·7" heavy rain in Henan. *Acta Meteorol. Sin.* **2022**, *80*, 765–776.
2. Li, C.; Xie, J.G.L. A Review of the Study on Flash Flood Early Warning in China. *Pearl River* **2017**, *38*, 29–35.
3. Xu, J.; Wang, Y. Water Problems and Their Countermeasures in the Protection and Developing of the Yangtze River. *J. Yangtze River Sci. Res. Inst.* **2020**, *37*, 1–6.
4. Liu, C.;Nie, R. Research Conception and Achievement Prospect of Key Technologies for Forecast and Early Warning of Flash Flood and Sediment Disasters in Mountainous Rainstorm. *Adv. Eng. Sci.* **2020**, *52*, 1–8.
5. Wu, Y.; Ma, M. Urban flood prevention countermeasures in Nanjing of Jiangsu Province. *China Flood Drought Manag.* **2023**, *33*, 53–56.
6. Qi, X. Overview of Object-oriented Change Detection in Remote Sensing Images. *Beijing Surv. Mapping.* **2021**, *35*, 427–431.
7. Sarp, G.; Ozcelik, M. Water body extraction and change detection using time series: A case study of Lake Burdur, Turkey. *J. Taibah Univ. Sci.* **2017**, *11*, 381–391. [CrossRef]
8. Guo, Z.; Wu, L.; Huang, Y.; Guo, Z.; Zhao, J.; Li, N. Water-Body Segmentation for SAR Images: Past, Current, and Future. *Remote Sens.* **2022**, *14*, 1752. [CrossRef]
9. Rokni, K.; Ahmad, A.; Selamat, A.; Hazini, S. Water feature extraction and change detection using multitemporal Landsat imagery. *Remote Sens.* **2014**, *6*, 4173–4189. [CrossRef]
10. Zhang, X.; Zhang, H.; Wang, C. Water-change detection with Chinese Gaofen-3 simulated compact polarimetric SAR images. In Proceedings of the 2017 SAR in Big Data Era: Models, Methods and Applications (BIGSARDATA), Beijing, China, 13–14 November 2017; pp. 1–4.
11. Fan, K.; Wang, Z. Change detection of remote sensing images through DT-CWT and MRF. *Int. Arch. Photogramm. Remote Sens. Spat. Inf. Sci.* **2017**, *21*, 375–385. [CrossRef]
12. Xu, Z.; Wang, R.L.N. A novel approach to change detection in SAR images with CNN classification. *J. Radars* **2017**, *6*, 483–491.
13. Badrinarayanan, V.; Kendall, A.; Cipolla, R. Segnet: A deep convolutional encoder-decoder architecture for image segmentation. *IEEE Trans. Pattern Anal. Mach. Intell.* **2017**, *39*, 2481–2495. [CrossRef] [PubMed]
14. Ronneberger, O.; Fischer, P.; Brox, T. U-net: Convolutional networks for biomedical image segmentation. In Proceedings of the Medical Image Computing and Computer-Assisted Intervention–MICCAI 2015: 18th International Conference, Munich, Germany, 5–9 October 2015; Proceedings, Part III 18; Springer: Berlin, Germany, 2015; pp. 234–241.
15. Lin, G.; Milan, A.; Shen, C.; Reid, I. Refinenet: Multi-path refinement networks for high-resolution semantic segmentation. In Proceedings of the IEEE Conference on Computer Vision and Pattern Recognition, Honolulu, HI, USA, 21–26 July 2017; pp. 1925–1934.
16. Zhao, H.; Shi, J.; Qi, X.; Wang, X.; Jia, J. Pyramid Scene Parsing Network. In Proceedings of the IEEE Conference on Computer Vision and Pattern Recognition, Honolulu, HI, USA, 21–26 July 2017; pp. 2881–2890.
17. He, K.; Gkioxari, G.; Dollár, P.; Girshick, R. Mask R-CNN. In Proceedings of the IEEE International Conference on Computer Vision, Venice, Italy, 22–29 October 2017; pp. 2961–2969.
18. Chen, L.C.; Papandreou, G.; Kokkinos, I.; Murphy, K.; Yuille, A.L. Semantic image segmentation with deep convolutional nets and fully connected crfs. *arXiv* **2014**, arXiv:1412.7062.
19. Chen, L.C.; Papandreou, G.; Kokkinos, I.; Murphy, K.; Yuille, A.L. Deeplab: Semantic image segmentation with deep convolutional nets, atrous convolution, and fully connected crfs. *IEEE Trans. Pattern Anal. Mach. Intell.* **2017**, *40*, 834–848. [CrossRef] [PubMed]
20. Chen, L.C.; Papandreou, G.; Schroff, F.; Adam, H. Rethinking atrous convolution for semantic image segmentation. *arXiv* **2017**, arXiv:1706.05587.

21. Chen, L.C.; Zhu, Y.; Papandreou, G.; Schroff, F.; Adam, H. Encoder-decoder with atrous separable convolution for semantic image segmentation. In Proceedings of the European Conference on Computer Vision (ECCV), Munich, Germany, 8–14 September 2018; pp. 801–818.
22. Liu, Z.; Chen, X.; Zhou, S.; Yu, H.; Guo, J.; Liu, Y. DUPnet: Water Body Segmentation with Dense Block and Multi-Scale Spatial Pyramid Pooling for Remote Sensing Images. *Remote Sens.* **2022**, *14*, 5567. [CrossRef]
23. Mei, J.; Li, R.J.; Gao, W.; Cheng, M.M. CoANet: Connectivity attention network for road extraction from satellite imagery. *IEEE Trans. Image Process.* **2021**, *30*, 8540–8552. [CrossRef]
24. Zhou, L.; Zhang, C.; Wu, M. D-LinkNet: LinkNet with pretrained encoder and dilated convolution for high resolution satellite imagery road extraction. In Proceedings of the IEEE Conference on Computer Vision and Pattern Recognition Workshops, Salt Lake City, UT, USA, 18–23 June 2018; pp. 182–186.
25. Liu, S.; Qi, L.; Qin, H.; Shi, J.; Jia, J. Path aggregation network for instance segmentation. In Proceedings of the IEEE Conference on Computer Vision and Pattern Recognition, Salt Lake City, UT, USA, 18–23 June 2018; pp. 8759–8768.
26. Lei, Y.; Leng, X.Z. Construction and recognition performance analysis of wide-swath SAR maritime large moving ships dataset. *J. Radars* **2022**, *11*, 347–362.
27. Huang, Q.; Zhu, W.Y. Summary of research on construction of SAR image ship target detection dataset. *Telecommun. Eng.* **2021**, *61*, 1451–1458.
28. Zhu, D.; Geng, Z.X. SAR Database Construction for Ground Targets at Multiple Angles and Target Recognition Method. *J. Nanjing Univ. Aeronaut. Astronaut.* **2022**, *54*, 985–994.
29. Sun, X.; Wang, Z. AIR-SARShip-1.0: High-resolution SAR Ship Detection Dataset. *Leidaxuebao* **2019**, *8*, 852–863.
30. Wang, Y.; Wang, C.; Zhang, H.; Dong, Y.; Wei, S. A SAR dataset of ship detection for deep learning under complex backgrounds. *Remote Sens.* **2019**, *11*, 765. [CrossRef]
31. Russakovsky, O.; Deng, J.; Su, H.; Krause, J.; Satheesh, S.; Ma, S.; Huang, Z.; Karpathy, A.; Khosla, A.; Bernstein, M.; et al. Imagenet large scale visual recognition challenge. *Int. J. Comput. Vis.* **2015**, *115*, 211–252. [CrossRef]
32. Ke, H.L.H. Present state and change of phreatic water quality in Dali County, Shaanxi, China. *Geol. Bull. China* **2008**, *27*, 1196–1204.
33. Wen, N.; Zeng, F.; Dai, K.; Li, T.; Zhang, X.; Pirasteh, S.; Liu, C.; Xu, Q. Evaluating and Analyzing the Potential of the Gaofen-3 SAR Satellite for Landslide Monitoring. *Remote Sens.* **2022**, *14*, 4425. [CrossRef]
34. Zhao, L.; Zhang, Q.; Li, Y.; Qi, Y.; Yuan, X.; Liu, J.; Li, H. China's Gaofen-3 satellite system and its application and prospect. *IEEE J. Sel. Top. Appl. Earth Obs. Remote Sens.* **2021**, *14*, 11019–11028. [CrossRef]
35. Kang, W.; Xiang, Y.; Wang, F.; Wan, L.; You, H. Flood detection in gaofen-3 SAR images via fully convolutional networks. *Sensors* **2018**, *18*, 2915. [CrossRef]
36. Gao, H.; Wang, C.; Wang, G.; Zhu, J.; Tang, Y.; Shen, P.; Zhu, Z. A crop classification method integrating GF-3 PolSAR and Sentinel-2A optical data in the Dongting Lake Basin. *Sensors* **2018**, *18*, 3139. [CrossRef] [PubMed]
37. Freeman, A. SAR calibration: An overview. *IEEE Trans. Geosci. Remote Sens.* **1992**, *30*, 1107–1121. [CrossRef]
38. Choi, H.; Jeong, J. Despeckling images using a preprocessing filter and discrete wavelet transform-based noise reduction techniques. *IEEE Sens. J.* **2018**, *18*, 3131–3139. [CrossRef]
39. Xu, C.; Su, H. RSDD-SAR: Rotated Ship Detection Dataset in SAR Images. *J. Radars* **2022**, *11*, 581–599.
40. Jaybhay, J.; Shastri, R. A study of speckle noise reduction filters. *Signal Image Process. Int. J. (SIPIJ)* **2015**, *6*, 71–80. [CrossRef]
41. Wang, X. Lee filter for multiscale image denoising. In Proceedings of the 2006 8th International Conference on Signal Processing, Guilin, China, 16–20 November 2006; Volume 1.
42. Saxena, N.; Rathore, N. A review on speckle noise filtering techniques for SAR images. *Int. J. Adv. Res. Comput. Sci. Electron. Eng. (IJARCSEE)* **2013**, *2*, 243–247.
43. Deepa, N.; Nagarajan, N. Kuan noise filter with Hough transformation based reweighted linear program boost classification for plant leaf disease detection. *J. Ambient. Intell. Humaniz. Comput.* **2021**, *12*, 5979–5992. [CrossRef]
44. Mullissa, A.; Vollrath, A.; Odongo-Braun, C.; Slagter, B.; Balling, J.; Gou, Y.; Gorelick, N.; Reiche, J. Sentinel-1 sar backscatter analysis ready data preparation in google earth engine. *Remote Sens.* **2021**, *13*, 1954. [CrossRef]
45. Kulkarni, S.; Kedar, M.; Rege, P.P. Comparison of Different Speckle Noise Reduction Filters for RISAT-1 SAR Imagery. In Proceedings of the 2018 International Conference on Communication and Signal Processing (ICCSP), Tamilnadu, India, 3–5 April 2018; pp. 0537–0541.
46. Frulla, L.; Milovich, J.; Karszenbaum, H.; Gagliardini, D. Radiometric corrections and calibration of SAR images. In Proceedings of the IGARSS'98. Sensing and Managing the Environment. 1998 IEEE International Geoscience and Remote Sensing. Symposium Proceedings (Cat. No. 98CH36174), Seattle, WA, USA, 6–10 July 1998; Volume 2, pp. 1147–1149.
47. Manjusree, P.; Prasanna Kumar, L.; Bhatt, C.M.; Rao, G.S.; Bhanumurthy, V. Optimization of threshold ranges for rapid flood inundation mapping by evaluating backscatter profiles of high incidence angle SAR images. *Int. J. Disaster Risk Sci.* **2012**, *3*, 113–122. [CrossRef]
48. Gorelick, N.; Hancher, M.; Dixon, M.; Ilyushchenko, S.; Thau, D.; Moore, R. Google Earth Engine: Planetary-scale geospatial analysis for everyone. *Remote Sens. Environ.* **2017**, *202*, 18–27. [CrossRef]
49. Yuan, K.; Zhuang, X.; Schaefer, G.; Feng, J.; Guan, L.; Fang, H. Deep-learning-based multispectral satellite image segmentation for water body detection. *IEEE J. Sel. Top. Appl. Earth Obs. Remote Sens.* **2021**, *14*, 7422–7434. [CrossRef]
50. Torralba, A.; Russell, B.C.; Yuen, J. Labelme: Online image annotation and applications. *Proc. IEEE* **2010**, *98*, 1467–1484. [CrossRef]

51. Lv, S.; Meng, L.; Edwing, D.; Xue, S.; Geng, X.; Yan, X.H. High-Performance Segmentation for Flood Mapping of HISEA-1 SAR Remote Sensing Images. *Remote Sens.* **2022**, *14*, 5504. [CrossRef]
52. Long, J.; Shelhamer, E.; Darrell, T. Fully convolutional networks for semantic segmentation. In Proceedings of the IEEE Conference on Computer Vision and Pattern Recognition, Boston, MA, USA, 7–12 June 2015; pp. 3431–3440.
53. Bishop, C.M. *Pattern Recognition and Machine Learning*; Springer: Berlin, Germany, 2006; Volume 4.
54. Guo, M.H.; Lu, C.Z.; Hou, Q.; Liu, Z.; Cheng, M.M.; Hu, S.M. Segnext: Rethinking convolutional attention design for semantic segmentation. *arXiv* **2022**, arXiv:2209.08575.
55. Chollet, F. Xception: Deep learning with depthwise separable convolutions. In Proceedings of the IEEE Conference on Computer Vision and Pattern Recognition, Honolulu, HI, USA, 21–26 July 2017; pp. 1251–1258.
56. He, K.; Zhang, X.; Ren, S.; Sun, J. Deep residual learning for image recognition. In Proceedings of the IEEE Conference on Computer Vision and Pattern Recognition, Las Vegas, NV, USA, 26 June–1 July 2016; pp. 770–778.
57. Kingma, D.P.; Ba, J. Adam: A method for stochastic optimization. *arXiv* **2014**, arXiv:1412.6980.
58. Fan, Z.; Hou, J.; Zang, Q.; Chen, Y.; Yan, F. River Segmentation of Remote Sensing Images Based on Composite Attention Network. *Complexity* **2022**, *2022*, 7750281. [CrossRef]
59. Wu, X.; Zhan, C.; Lai, Y.K.; Cheng, M.M.; Yang, J. Ip102: A large-scale benchmark dataset for insect pest recognition. In Proceedings of the IEEE/CVF Conference on Computer Vision and Pattern Recognition, Long Beach, CA, USA, 16–17 June 2019; pp. 8787–8796.

remote sensing

Article

Multi-Scale Object Detection in Remote Sensing Images Based on Feature Interaction and Gaussian Distribution

Ruixing Yu [1,*], Haixing Cai [1], Boyu Zhang [1] and Tao Feng [1,2]

[1] School of Astronautics, Northwestern Polytechnical University, Xi'an 710072, China; caihaixing129@mail.nwpu.edu.cn (H.C.); boyuzhang@mail.nwpu.edu.cn (B.Z.); nestleblack@163.com (T.F.)
[2] Xi'an Institute of Electronic Engineering, China North Industries Group Corporation, Xi'an 710199, China
* Correspondence: yrxgigi@nwpu.edu.cn

Abstract: Remote sensing images are usually obtained from high-altitude observation. The spatial resolution of the images varies greatly and there are scale differences both between and within object classes, resulting in a diversified distribution of object scales. In order to solve these problems, we propose a novel object detection algorithm that maintains adaptability to multi-scale object detection based on feature interaction and Gaussian distribution in remote sensing images. The proposed multi-scale feature interaction model constructs feature interaction modules in the feature layer and spatial domain and combines them to fully utilize the spatial and semantic information of multi-level features. The proposed regression loss algorithm based on Gaussian distribution takes the normalized generalized Jensen–Shannon divergence with Gaussian angle loss as the regression loss function to ensure the scale invariance of the model. The experimental results demonstrate that our method achieves 77.29% mAP on the DOTA-v1.0 dataset and 97.95% mAP on the HRSC2016 dataset, which are, respectively, 1.12% and 1.41% higher than that of the baseline. These experimental results indicate the effectiveness of our method for object detection in remote sensing images.

Keywords: remote sensing images; object detection; feature interaction; Gaussian distribution

Citation: Yu, R.; Cai, H.; Zhang, B.; Feng, T. Multi-Scale Object Detection in Remote Sensing Images Based on Feature Interaction and Gaussian Distribution. *Remote Sens.* **2024**, *16*, 1988. https://doi.org/10.3390/rs16111988

Academic Editors: Yanni Dong, Xiaochen Yang and Qian Du

Received: 24 March 2024
Revised: 11 May 2024
Accepted: 21 May 2024
Published: 31 May 2024

1. Introduction

With the rapid development of remote sensing technology, the quantity and quality of remote sensing images have substantially improved. The objects of interest in remote sensing image object detection typically include harbors, bridges, airplanes, ships, vehicles, and other multi-type objects. The main tasks of object detection in remote sensing images are to localize and classify the objects of interest [1]. this is important in intelligence reconnaissance, object surveillance, disaster rescue, industrial applications, and daily life. It also serves as the basis for the subsequent work of object tracking, scene classification, and image understanding.

Most of the object detection algorithms in remote sensing images based on deep learning are derived from general object detection algorithms [2]. Due to the outstanding performance of deep learning algorithms in the field of natural image processing, object detection technology in remote sensing images has also developed from the traditional manual feature extraction to the current deep learning-based approach, and its performance has been greatly improved. Compared with natural images, remote sensing images are usually obtained from high-altitude observations. There are situations such as drastic object scale changes, diverse object orientation, small and densely distributed objects, and complex backgrounds. Therefore, there are still a lot of challenges for object detection in remote sensing image scenes.

Compared to natural images, remote sensing images feature complex object characteristics and a significant amount of background interference, the object visual information is not sufficient, and there is a lack of discriminative object features [3]. It is difficult to achieve good detection results by directly adopting the detection scheme of the natural scene. Most current detection methods are researched based on the intrinsic characteristics of remote sensing images. As shown in Figure 1, in remote sensing scenes, it is common to see objects of different classes and scales or objects of the same class but different sizes in the same field of view [4]. The scale difference among the objects is significant, which makes it difficult for the features of different objects to be synchronously transmitted to deep networks. As a result, the deep convolutional network, which relies on feature map representations, is unable to effectively capture the features of multi-scale objects. The detection model lacks good scale invariance, which in turn affects the detection accuracy of multi-scale objects [5]. Therefore, object detection in remote sensing images with large-scale variations is still a challenging problem.

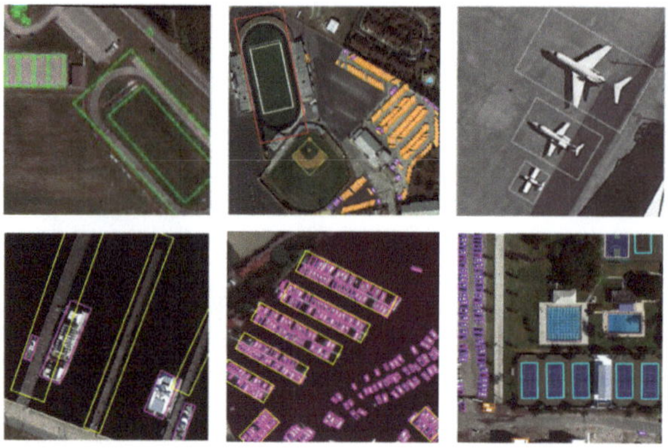

Figure 1. Large-scale variation in objects within remote sensing images.

To address the aforementioned challenges, we propose a multi-scale feature interaction (MSFI) network. At the feature layer, a cross-level feature interaction (CFI) module is constructed to facilitate interactions between feature layers at different levels of the feature pyramid. This module is based on the focal modulation network [6], which can efficiently process multi-scale features while avoiding redundant and dense connections and mitigating the loss of small object features caused by multiple down-samplings. In the spatial domain, a spatial feature interaction (SFI) network, based on context aggregation [3] with a residual connection, is adopted to enhance features and suppress background noise. The proposed multi-scale feature interaction algorithm enhances the information recognition and expression capabilities of the detector, enabling multi-scale detection. Meanwhile, commonly used loss functions in object detection, such as L-norm loss, lack scale invariance [7]. To adapt the model for the multi-scale detection task, we propose a regression loss based on Gaussian distribution (RLGD). We transform the regression problem in object detection into a problem of measuring the similarity between Gaussian distributions. Then, the generalized Jensen–Shannon divergence [8] that possesses the scale invariance is used to measure the similarity between Gaussian distributions and it is normalized as the regression loss function. On this basis, the Gaussian angular loss [9] is introduced to overcome the angle confusion problem of square bounding boxes in the rotation box representation method based on Gaussian distribution.

In this paper, we follow the Oriented R-CNN [10] object detector framework and apply the proposed algorithm to the object detection task in optical remote sensing images. The results are evaluated on the datasets of DOTA-v1.0 [4] and HRSC2016 [11]. Compared with other detectors, the proposed method achieved better detection results.

The main contributions in this paper are as follows:

(1) We propose a new multi-scale feature interaction (MSFI) network that constructs a cross-level feature interaction (CFI) module at the feature layer and introduces a spatial feature interaction (SFI) module in the spatial domain. By combining the feature interaction in the feature layer and spatial domain, the network makes full use of the spatial and semantic information of multi-level features to effectively localize and identify objects with drastic scale changes.

(2) We propose a new regression loss based on Gaussian distribution (RLGD), which uses the normalized generalized Jensen–Shannon divergence between Gaussian distributions as the regression loss function to make the model scale-invariant. The Gaussian angle loss is introduced to overcome the angle confusion problem in Gaussian modeling.

(3) The experimental results show that the proposed method has superior detection capabilities compared to the baseline and other state-of-the-art detectors. Our method obtains mAP of 77.29% on the DOTA V.1.0 test set and mAP of 97.95% on the HRSC2016 test set.

2. Related Work

2.1. Object Detection in Remote Sensing Images

Advanced remote sensing image object detection algorithms usually rely on the two-stage RCNN [12] framework, which consists of a region proposal network and a region CNN detection head. In recent years, a large amount of research has focused on oriented bounding box representations for remote sensing image object detection, and several variations of the RCNN framework have been proposed. The RoI Transformer [13] applies multi-layer fully connected layers from the horizontal anchor box to learn the rotated anchor box and extracts features in the boxes for further regression and classification. ReDet [14] provides an explicit encoding for equivariance and rotation invariance and proposes a rotation-invariant RoI alignment. Oriented RCNN [10] introduces a new bounding box encoding system and proposes an oriented region proposal network (oriented RPN) that directly generates high-quality oriented proposals. It also introduces the Oriented R-CNN head for refining and recognizing oriented regions of interest (oriented RoIs). Gliding Vertex [15] adds four sliding offset variables to the classical horizontal bounding box representation to address the instability in training loss caused by the periodicity of the rotation angle.

Different from the two-stage detection framework, the one-stage detection framework directly employs a network to transform the object detection problem into a regression problem. For the remote sensing detection tasks, S2ANet [16] extracts robust object features by means of a feature alignment module and orientation detection module. Some approaches [17–20] consider remote sensing detection as a point detection task. In addition, DRN [21] uses the attention mechanism to dynamically refine the features extracted by the backbone for more accurate prediction. RSDet [22] effectively mitigates the discontinuity of regression loss by weakening the loss value of the training samples in the bounded case from the perspective of improving the loss function. R3Det [23] detects objects quickly and accurately by using a stepwise regression approach from coarse-grained to fine-grained. AOPG [24] uses a progressive regression method to refine the bounding box from coarse-grained to fine-grained. LSKNet [25] can dynamically adjust its large spatial receptive field to better model the ranging context of various objects in remote sensing scenarios. ARC [26] proposes a convolution kernel that rotates adaptively to extract object features with varying orientations in different images. In addition to CNN-based frameworks, the O2DETR [27] and AO2-DETR [28] also introduced the Transformer-based detection framework DETR [29] to the remote sensing detection tasks, which provides another way to solve the object detection problem in remote sensing images.

In recent years, the Transformer architecture has also promoted the advancement of remote sensing image object detection. Such algorithms are not directly equivalent to single-stage or two-stage algorithms. However, when they are applied to object detection, single-stage or two-stage detection can be achieved in different ways. The core idea of the Transformer is to focus on different parts of the image through attention mechanisms, but it does not directly define whether the detection process is single-stage or two-stage. Attention mechanism-based Transformers [30], initially designed for natural language processing, have made their impactful presence felt in multiple vision tasks, including object detection. The core principle behind Transformers is attention—determining a specific part of the image where the model should focus at a given time [31]. This attention allocation can be immensely beneficial for object detection, where spatial information and an understanding of the object hierarchy in the image are often crucial. Furthermore, several studies have highlighted the feasibility and efficiency of hybrid models that combine the spatial dominance of CNNs with the dynamic attention capabilities of Transformers, referred to herein as CNN–Transformer hybrids. Diverse experiments, ranging from DETR [29], a pioneering model applying Transformers to object detection, to Swin Transformer [32], a CNN–Transformer hybrid, have illustrated the potency of this amalgamation.

2.2. *Small Infrared Object Detection*

This section delves into the recent progress made in small infrared object detection, with a particular focus on the contributions of several groundbreaking studies. Firstly, the ISNet model [33] emphasizes the significance of 'shape' in detecting infrared small targets and integrating a shape-aware feature extraction module within a convolutional neural network framework. This innovative approach has proven effective in enhancing the recognition of small targets, which often exhibit irregular and amorphous shapes against complex backgrounds.

Following this, Rkformer [34] introduces a novel application of the Runge–Kutta Transformer coupled with random-connection attention mechanisms. This model adeptly captures the dynamic and transient characteristics of small targets, facilitating robust detection even within highly intricate scenes. Further exploration into feature analysis is presented in [35], where a method for feature compensation and cross-level correlation is proposed. This technique aims to offset the loss of critical target details and leverages the intrinsic correlation between different feature levels, showing marked effectiveness in distinguishing small targets from background noise.

Similarly, Chfnet [36] proposes a curvature half-level fusion network that merges curvature-driven contour features with conventional intensity features, significantly boosting the discriminative capability for single-frame infrared small target detection. Drawing inspiration from thermodynamic principles, [37] introduces a multi-feature network designed to effectively segregate small targets from their surroundings. This thermodynamics-inspired approach has shown exceptional proficiency, especially in scenarios where traditional intensity-based methods fall short. DMEF-Net [38] offers a lightweight solution tailored to limited sample scenarios, further underscoring the diverse approaches being developed to tackle the challenges of infrared small target detection. Addressing the computational efficiency of infrared small target detection, IRPruneDet [39] incorporates wavelet structure-regularized soft channel pruning. This method significantly reduces computational overhead without compromising detection accuracy, presenting an ideal solution for real-time applications.

2.3. *Multi-Scale Object Detection*

2.3.1. Multi-Scale Object Detection in Remote Sensing Images

Remote sensing images are usually obtained from high-altitude observation, the spatial resolution of the images varies widely, and there are scale differences between and within object classes. The object scale in remote sensing images changes drastically, so the designed model must have good scale adaptability. This means the model should retain high recognition ability across various scales, even with drastic changes in multiple classes of objects. Multi-scale feature fusion can effectively alleviate the challenging problem of large object scale changes.

The most typical module of multi-scale feature fusion is the feature pyramid [40], which integrates deep semantic information with shallow contour and position information and uses different feature maps to uniformly locate the object to be detected at multiple scales, to achieve accurate object detection [41]. CATNet [3] proposes a dense feature pyramid network with spatial context pyramids to improve the multi-scale object feature extraction process, thus improving object detection accuracy. Liu et al. [42] propose a hybrid network named TransConvNet. The model pays more attention to the aggregation of global and local information to improve the information representation ability of feature maps with different resolutions, while an adaptive feature fusion network is designed to mitigate the problem of drastic changes in object scales. Li et al. [43] aggregate the features of global spatial locations at multiple scales on the Transformer and model the interaction between pairwise instances to improve the detection performance of objects at different scales.

In addition to multi-scale feature fusion, Yang et al. [44] propose a measurement method based on Kullback–Leibler divergence (KLD), which is scale-invariant and can localize the object to be detected well. Zhu et al. [45] propose Transformer prediction head YOLOv5 (TPH-YOLOv5), which replaces some convolutional and cross-stage partial (CSP) structures in the original YOLOv5 with Transformer encoder and Transformer prediction head to mitigate the problem of drastic scale changes of the object. Xu et al. [46] propose the adaptive zoom network (AdaZoom), which freely scales the object-focusing region with flexible aspect ratios and scales for multi-scale object detection. And they use co-training to promote the coordination between AdaZoom and the object detector to further improve the detection performance. Chalavadi et al. [47] propose a multi-scale object detection network with hierarchical extended convolution. Their model employs parallel extended convolution to learn contextual information about different classes of objects across multiple scales and fields of view, effectively enhancing its ability to detect objects at various scales.

2.3.2. Multi-Scale Features

Multi-scale features have been widely used in the field of object detection. Higher-level feature maps emphasize advanced semantic information, while lower-level feature maps focus on texture and localization information. The fusion of multi-scale information can effectively combine high-resolution and high-semantic features, enhancing the recognition and expression of object information. Effectively fusing them is the key to improving the detection model [42]. Some current studies have achieved certain results [41]. However, in the process of multi-scale feature fusion, in order to obtain higher-level semantic information, the feature map is downsampled several times, which leads to the disappearance of small object features and is very unfriendly to small objects. Additionally, various noises are also amplified during the process of feature fusion. How to deal with such a situation is a critical problem that needs to be solved at present [48].

The multiscale feature propagation strategy [49] propagates features along a fixed path, while flexible information flow reduces information confusion and better aggregates multiscale features to avoid the problem of small object features disappearing due to multiple down-sampling. Cross-scale feature fusion pre-fuses features at a uniform scale and subsequently generates feature maps of different specifications for subsequent detection. This approach samples a larger span on feature layers that deviate from the center layer, whereas feature fusion at the corresponding scales separately avoids loss of information [50]. Furthermore, the attention mechanism introduced by Transformer [30] has demonstrated effectiveness in multi-scale object detection. The attention mechanism focuses on regions of interest within a large amount of input information, reducing focus on other details and filtering out irrelevant data. This can enhance both the efficiency and accuracy of object detection tasks. TPH-YOLOv5 [45] leverages the Transformer architecture to enhance YOLOv5 and incorporates a self-attention mechanism to effectively detect multi-scale objects.

2.3.3. Scale Invariance of Regression Loss

In the field of object detection, commonly used loss functions such as L1-norm do not have scale invariance, while regression loss based on intersection over union (IoU) [51,52] can be applied to various object detection tasks with different scales and shapes. This is achieved by optimizing the overlap of the bounding boxes to train the network parameters, which is suitable for object detection tasks with different scales and shapes. However, the method of IOU loss is non-differentiable in the oriented bounding box and cannot be directly applied to rotating object detection in remote sensing images. Additionally, the existing deep learning framework operators do not easily accommodate the skew IoU loss, so the IoU loss is not widely used in the field of object detection based on oriented bounding boxes.

The Gaussian Wasserstein distance model (GWD) [7] transforms a rectangular box representation with arbitrary orientation into the form of a two-dimensional Gaussian distribution. This transformation reframes the bounding box parameter regression problem as a similarity measurement between two Gaussian distributions. It replaces the traditional method of calculating differences between bounding box parameters at various scales and units, used in L1 loss calculations. The KLD model [44] computes the Kullback–Leibler divergence (KLD) between Gaussian distributions, which can approximate the degree of overlap between the predicted bounding boxes and the true bounding boxes and is suitable for object detection tasks with various scales and shapes.

3. Materials and Methods

3.1. Overall Architecture

In order to improve the multi-scale detection capability in remote sensing images, we propose a multi-scale feature interaction (MSFI) network to take full advantage of the spatial and semantic information of multi-level features, and the proposed regression loss based on the Gaussian distribution (RLGD) algorithm adopts regression loss with scale invariance. In this paper, we follow the Oriented R-CNN [10] object detector framework and apply the proposed method to the object detection task in remote sensing images. Oriented R-CNN is a two-stage object detection framework based on a five-parameter rotated bounding box with good accuracy and efficiency, which mainly consists of an oriented region proposal network (oriented RPN) and an Oriented R-CNN head. Specifically, oriented RPN is employed in the first stage to directly generate high-quality-oriented proposals in a nearly cost-free manner. The Oriented R-CNN head, equipped with rotated RoIAlign and a fully connected (FC) layer, is used to classify and regress from the oriented RPN. The overall architecture of the proposed multi-scale remote sensing image object detection model presented in this paper is shown in Figure 2. The structural details of each part of our method are described in the following.

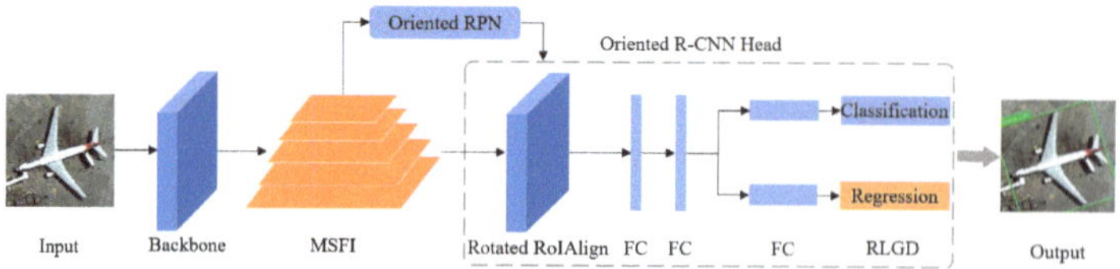

Figure 2. Overall framework of the proposed algorithm.

3.2. Multi-Scale Feature Interaction Network

This section proposes a multi-scale feature interaction (MSFI) network, including a cross-level feature interaction (CLFI) module composed of a top-down grounding feature interaction (GFI) module and a bottom-up rendering feature interaction (RFI) module, as well as a spatial feature interaction (SFI) module. The overall architecture is shown in Figure 3a. The MSFI network proposed in this section carries out feature interactions between different layers of features and within the same scale. This multi-scale feature interaction strategy can help the model integrate deep and shallow features more efficiently, thus capturing more comprehensive information about the object.

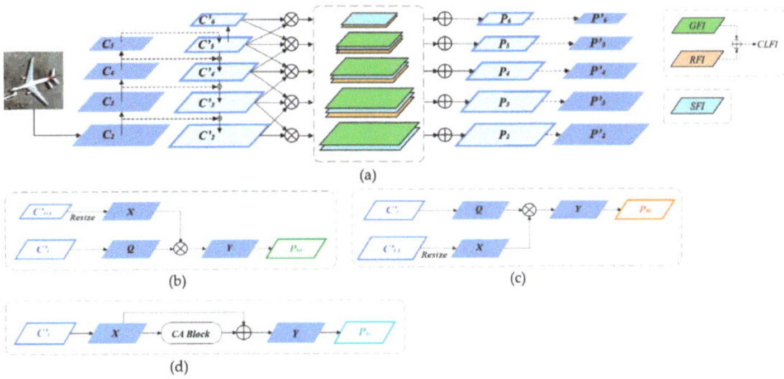

Figure 3. Overall architecture of the multi-scale feature interaction network. (**a**) Multi-scale feature interaction network; (**b**) grounding feature interaction module; (**c**) rendering feature interaction module; (**d**) spatial feature interaction module.

3.2.1. Cross-Level Feature Interaction Network

In this section, we propose a cross-level feature interaction network (CLFI) that interacts with the features of adjacent layers in the feature pyramid. The feature pyramid obtained after feature interaction is constant in size and has richer contextual information. In the CLFI network, adjacent feature maps are scaled to perform grounding feature interaction (GFI) and rendering feature interaction (RFI), respectively. Among these, the GFI module facilitates the interaction of lower-level feature maps with higher-level feature maps, whereas the RFI module enables the interaction of higher-level feature maps with lower-level ones. These feature interactions occur between adjacent feature layers and are based on the focal modulation network.

As shown in Figure 3b, the GFI module is a kind of top-down feature interaction [53], which grounds the "concept" in the higher-level feature map to the "pixels" in the lower-level feature map. And the size of the output is the same as that of the lower-level feature map. Specifically, in the GFI module, the lower-level feature map is $C_i^{'} \in \mathbb{R}^{H \times W \times C}$, the higher-level feature map is $C_{i+1}^{'} \in \mathbb{R}^{H/2 \times W/2 \times C}$, and the feature map obtained after feature interaction is $P_{Gi} \in \mathbb{R}^{H \times W \times C}$.

In the GFI module, we set $Q = C_i^{'}$. The bilinear interpolation algorithm is used to upsample $C_{i+1}^{'}$ to obtain X, keeping its resolution consistent with Q_i. We adopt the focal modulation network [6] instead of the attention mechanism [54], feature interaction is performed on Q and X to obtain Y:

$$y_i = f(q_i) \odot m(i, X) \tag{1}$$

where $f(q_i)$ is a query projection function, $\odot$ is multiplied by elements, and $m(\cdot)$ is a context aggregation function whose output is a modulator. The focal modulation network [6] is shown in Figure 4. The feature map $P_{Gi} = Y$ obtained after feature interaction is the output of the GFI module.

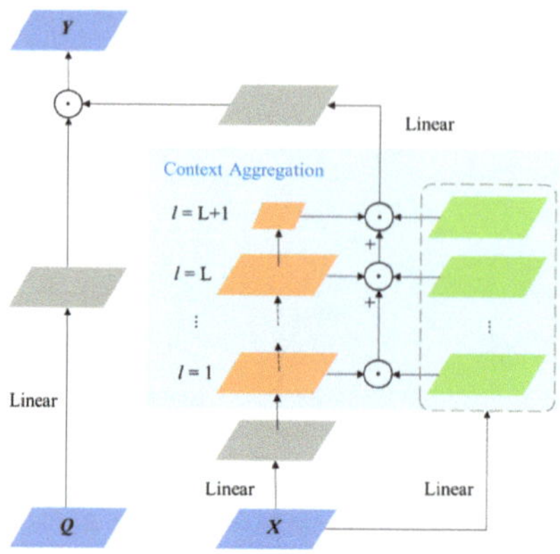

Figure 4. The foundation for feature interaction in the CLFI.

As shown in Figure 3c, the RFI module adopts a bottom-up approach [53] and uses the visual attributes of "pixels" in lower-level feature maps to render "concepts" in higher-level feature maps. The size of the output is the same as that of a higher-level feature map. Specifically, in the RFI module, given that higher-level feature map is $C_i^{'} \in \mathbb{R}^{H\times W\times C}$, the lower-level feature map is $C_{i-1}^{'} \in \mathbb{R}^{2H\times 2W\times C}$, and the feature map obtained after feature interaction is $P_{Ri} \in \mathbb{R}^{H\times W\times C}$.

In the RFI module, similar to the GFI module, let $Q = C_i^{'}$. The maximum pool algorithm is applied to reduce the resolution of the feature map $C_{i-1}^{'}$ to obtain X, whose resolution is consistent with $Q \in \mathbb{R}^{H\times W\times C}$. Like the GFI module, feature map Y is obtained by combining the features of Q and X based on the focal modulation network. The feature map $P_{Ri} = Y$ obtained after feature interaction is the output of the RFI module.

3.2.2. Spatial Feature Interaction Network

The cross-level feature interaction network performs feature interactions for different levels of feature maps. But the feature pyramid still contains spatial information. Therefore, this section introduces the spatial feature interaction (SFI) network to further enhance features by learning the global spatial context within each level. The structure of the SFI network is shown in Figure 3d, which is composed of residual connection and context aggregation block (CAB) [3]. Given the input feature map $C_i^{'} \in \mathbb{R}^{H\times W\times C}$, the output feature map $P_{Si} \in \mathbb{R}^{H\times W\times C}$ is obtained after SFI, which has the same size as the input feature map $C_i^{'}$.

In the SFI module, let $X = C_i^{'}$ perform the pixel-by-pixel spatial feature interaction:

$$Y = X + CAB(X) \tag{2}$$

where Y is the output feature map after the spatial feature interaction. $CAB(\cdot)$ [3] denotes the context aggregation block, as shown in Figure 5. Specifically, the context aggregation block is denoted as follows:

$$CAB(X_j) = Sigmoid(W_{v1}X_j) \odot \left[W_{v2}\left(\sum_{k=1}^{N_i}\right) \frac{e^{W_{v3}X_k}}{\sum_{m=1}^{N_i} e^{W_{v3}X_m}} W_{v4}X_k \right] \tag{3}$$

where X is the feature map containing N pixels, $j, m, k \in \{1, \ldots, N_i\}$ denotes the index of each pixel, $\odot$ is multiplied element-wise, and $W_{v1}, W_{v2}, W_{v3}, W_{v4}$ is the linear transformation matrix used to project the feature map, which is mapped using a 1×1 convolution. Let the feature map, $P_{Si} = Y$, obtained after spatial feature interaction be the output of the SFI module.

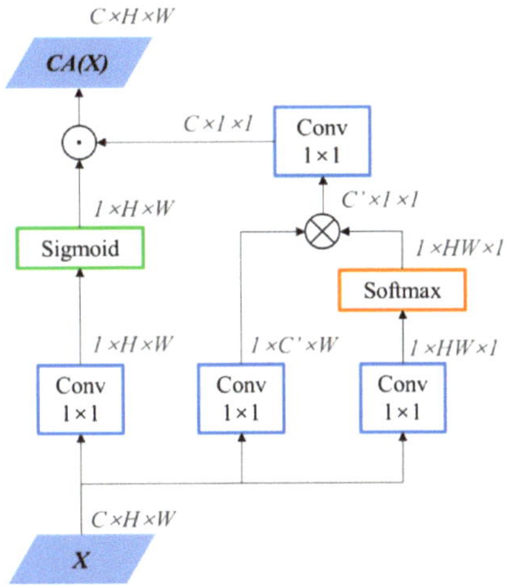

Figure 5. The foundation for feature interaction in the SFI.

3.2.3. Multi-Scale Feature Interaction Network

The overall architecture of the multi-scale feature interaction network is shown in Figure 3. Firstly, the input image is a feature extracted by the backbone network and denoted as $C = \{C_2, \ldots, C_5\}$, where C_i represents the feature map at level i. Then, a pyramid-shaped feature map $C' = \left\{C_2', \ldots, C_5'\right\}$ with different layers is obtained by enhancing the features through the lateral connection and top-down pathways [40], as shown in Figure 6a. Specifically, a 1×1 convolution is used on each feature layer for lateral connection to reduce and unify its channel dimensions to 256. A bilinear interpolation algorithm is used to upsample the spatial resolution of higher-level feature maps by a factor of 2, which is then summed element-by-element with the lower-level feature maps. And C_5' is convoluted by 3 × 3 with a step size of 2 to obtain C_6'. $C_2' \sim C_6'$ have the same number of channels but different resolutions.

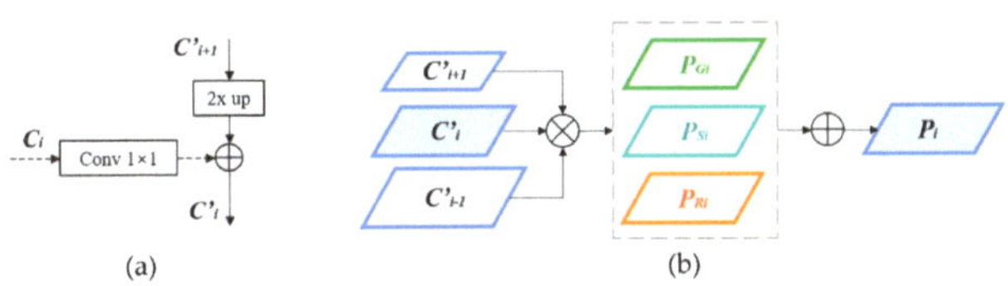

Figure 6. Multi-scale feature interaction process. (**a**) The lateral connection and top-down pathways; (**b**) multi-scale feature interaction from C_i' to P_i at the i-th layer.

The multi-scale feature interaction is performed on the feature map pyramid $C' = \left\{C_2', \ldots, C_6'\right\}$ to obtain $P = \{P_2, \ldots, P_6\}$, where P_i denotes the feature map at the i-th level.

The resolution of the feature map $C_i^{'}$ and P_i is $1/i^2$ of that of the input image. The multi-scale feature interaction from $C_i^{'}$ to P_i at the i-th layer is shown in Figure 6b.

For the multi-scale feature interaction network, P_{Gi} and P_{Ri} are obtained from cross-layer feature interactions, and P_{Si} is obtained from spatial feature interactions. Let $P_i = P_{Gi} + P_{Ri} + P_{Si}$, where $P_i \in \{P_2, \ldots, P_6\}$ is the feature map obtained by combining feature interactions in the feature layer and the spatial domain. Then $P_i = \left\{P_2^{'}, \ldots, P_6^{'}\right\}$ is obtained by a 3 × 3 convolution. $P_2^{'} \sim P_6^{'}$ have the same number of channels but different resolutions.

3.3. *Regression Loss Based on Gaussian Distribution*

3.3.1. Oriented Bounding Box Representation based on Gaussian Distribution

In this section, we propose a regression loss algorithm based on Gaussian distribution. Firstly, the rectangular box representation with arbitrary orientation is transformed into the form of a two-dimensional Gaussian distribution. Replacing the oriented box representation with Gaussian distribution can solve the bounding problem, improve the square-like problem, and avoid the inconsistency between the metrics and the loss in the detection of rotating objects [7]. Specifically, the representation of a five-parameter-oriented bounding box $B(x, y, w, h, \theta)$ is transformed into the form of two-dimensional Gaussian distribution $N(\mu, \Sigma)$ [7]:

$$\begin{cases} \mu = (x, y)^T \\ \Sigma^{1/2} = R\Lambda R^T & = \begin{pmatrix} \cos\theta & -\sin\theta \\ \sin\theta & \cos\theta \end{pmatrix} \begin{pmatrix} \frac{w}{2} & 0 \\ 0 & \frac{h}{2} \end{pmatrix} \begin{pmatrix} \cos\theta & \sin\theta \\ -\sin\theta & \cos\theta \end{pmatrix} \\ & = \begin{pmatrix} \frac{w}{2}\cos^2 + \frac{h}{2}\sin^2\theta & \frac{w-h}{2}\cos\theta\sin\theta \\ \frac{w-h}{2}\cos\theta\sin\theta & \frac{w}{2}\sin^2\theta + \frac{h}{2}\cos^2\theta \end{pmatrix} \end{cases} \quad (4)$$

where R denotes the rotation matrix and Λ denotes the diagonal matrix of eigenvalues. According to the above equation, the prediction bounding box and the truth bounding box are transformed into two-dimensional Gaussian distributions $X_p \sim N_p(\mu_p, \Sigma_p)$ and $X_t \sim N_t(\mu_t, \Sigma_t)$, respectively. This transformation allows for the regression problem in object detection to be converted into a problem of measuring the similarity between two Gaussian distributions.

3.3.2. Regression Loss of Normalized GJSD

Due to the significant size differences among various categories of objects, or even within the same category at different resolution acquisitions in remote sensing image analysis, it is crucial for the detection algorithm to possess excellent scale adaptability. To address these challenges, we propose a regression loss model based on the generalized Jensen–Shannon divergence (GJSD) [8]. Specifically, GJSD is utilized to measure the similarity between two Gaussian distributions, $X_p \sim N_p(\mu_p, \Sigma_p)$ and $X_t \sim N_t(\mu_t, \Sigma_t)$. The calculations are detailed below:

$$D_{GJS}(N_p \parallel N_t) = (1-\alpha)D_{KL}(N_\alpha \parallel N_p) + \alpha D_{KL}(N_\alpha \parallel N_t) \quad (5)$$

where D_{KL} represents the Kullback–Leibler divergence (KLD) [44] and $D_{KL}(N_\alpha \parallel N_p)$ is denoted as follows:

$$D_{KL}(N_\alpha \parallel N_p) = \frac{1}{2}(\mu_\alpha - \mu_p)^T \Sigma_p^{-1}(\mu_\alpha - \mu_p) + \frac{1}{2}Tr\left(\Sigma_p^{-1}\Sigma_\alpha\right) + \frac{1}{2}\ln\frac{|\Sigma_p|}{|\Sigma_\alpha|} - 1 \quad (6)$$

The definition of $D_{KL}(N_\alpha \parallel N_t)$ is the same as above, which gives $D_{GJS}(N_p \parallel N_t)$, denoted by the following:

$$D_{GJS}(N_p \parallel N_t) = \frac{1-\alpha}{2}(\mu_\alpha - \mu_p)^T \Sigma_p^{-1}(\mu_\alpha - \mu_p) + \frac{1-\alpha}{2} Tr\left(\Sigma_p^{-1}\Sigma_\alpha\right) + \frac{1-\alpha}{2}\ln\frac{|\Sigma_p|}{|\Sigma_\alpha|} - 1$$
$$+ \frac{\alpha}{2}(\mu_\alpha - \mu_t)^T \Sigma_t^{-1}(\mu_\alpha - \mu_t) + \frac{\alpha}{2} Tr\left(\Sigma_t^{-1}\Sigma_\alpha\right) + \frac{\alpha}{2}\ln\frac{|\Sigma_t|}{|\Sigma_\alpha|} \quad (7)$$

$N_\alpha(\mu_\alpha, \Sigma_\alpha)$ is given by the following formula:

$$\begin{cases} \Sigma_\alpha = \left[(1-\alpha)\Sigma_p^{-1} + \alpha\Sigma_t^{-1}\right]^{-1} \\ \mu_\alpha = \Sigma_\alpha\left[(1-\alpha)\Sigma_p^{-1}\mu_p + \alpha\Sigma_t^{-1}\mu_t\right] \end{cases} \quad (8)$$

The parameter α controls the weights of the two distributions, and its value is set to 0.5. Then, normalizing the distance function $D_{GJS}(N_p \parallel N_t)$ yields a regression loss of the following:

$$L_{GJSD} = 1 - \frac{1}{\tau + f(D_{GJS})}, \quad \tau \geq 1 \quad (9)$$

The symbol $f(\cdot)$ represents a nonlinear function, typically including functions such as $sqrt(\cdot)$ and $log(\cdot)$. The symbol τ is a hyperparameter that modulates the entire loss function. In this case, $f(\cdot)$ is set to $sqrt(\cdot)$ and τ is set to 2 [55].

From Equations (7) and (8), it can be seen that each term in $D_{GJS}(N_p \parallel N_t)$ consists of a partial parameter coupling, which makes all parameters form a chain coupling relationship. During the optimization process of GJSD-based regression loss, the parameters interact and co-optimize with each other, resulting in a self-modulating optimization mechanism. GJSD is capable of measuring the similarity between the Gaussian distributions of the truth bounding box and the prediction bounding box. It inherits the scale invariance of KLD while overcoming its asymmetric drawbacks. The scale invariance of GJSD is analyzed as follows:

If there is a full-rank matrix M, then $X_p' = MX_p \sim N_p'(M\mu_p, M\Sigma_p M^T)$ and $X_t' = MX_t \sim N_t'(M\mu_t, M\Sigma_t M^T)$. From formulas (5) to (8), the following can be obtained:

$$D_{GJS}\left(N_p' \parallel N_t'\right) = D_{GJS}(N_p \parallel N_t) \quad (10)$$

When $M = kI$ (I is a unit matrix and $I \neq 0$), the scale invariance of GJSD is proved (see the detailed proof in Appendix A). Therefore, using the GJSD-based regression loss model, which employs the oriented bounding box representation of the Gaussian distribution, can mitigate the issue of large object scale variation in remote sensing images and enhance the effectiveness of object detection.

3.3.3. Gaussian Angle Loss for the Problem of Angle Confusion

Although the regression loss based on GJSD has scale invariance, when the bounding box $B(x, y, w, h, \theta)$ is square, its two-dimensional Gaussian distribution $N(\mu, \Sigma)$ is a circle that cannot accurately represent the direction. Specifically, when $w = h$, the following can be obtained from Equations (4):

$$\begin{cases} \mu = (x, y)^T \\ \Sigma = \begin{pmatrix} w^2/4 & 0 \\ 0 & h^2/4 \end{pmatrix} \end{cases} \quad (11)$$

It can be seen that the angular information in it is lost, which can lead to angular confusion for square objects, thus affecting the performance of detection.

To address this issue, the Gaussian angle loss function is introduced as a solution to the angle confusion problem of square bounding boxes in the rotation box representation method based on Gaussian distribution [9]. The Gaussian angle loss function is as follows:

$$L_{GA} = \beta e^{4\lambda - \lambda\left(\frac{w_t}{h_t} + \frac{h_t}{w_t}\right)^2} \sin^2(2\Delta\theta) \tag{12}$$

where $\Delta\theta = \theta_p - \theta_t$. λ and β are hyperparameters. To reduce the effect on non-square objects, the values of λ and β are set to 3 and 0.25, respectively. For further analysis, we have the following:

$$\frac{\partial L_{GA}}{\partial \theta_p} = 2e^{4\lambda - \lambda\left(\frac{w_t}{h_t} + \frac{h_t}{w_t}\right)^2} \sin(4\Delta\theta) \tag{13}$$

When $w_t = h_t$, $\partial L_{GA}/\partial\theta_p = 2\sin(4\Delta\theta)$. As the aspect ratio approaches 1, L_{GA} exhibits a periodic variation of 90° as the angle deviation $\Delta\theta$ increases, which is consistent with the angular period of a square. As the aspect ratio increases, $\partial L_{GA}/\partial\theta_p$ approaches 0 and L_{GA} becomes invalid.

The relationship between the regression loss based on GJSD (L_{GJSD}) and Gaussian angle loss (L_{GA}) with the bounding box aspect ratio and angle deviation, respectively, is shown in Figure 7, which visualizes the above properties.

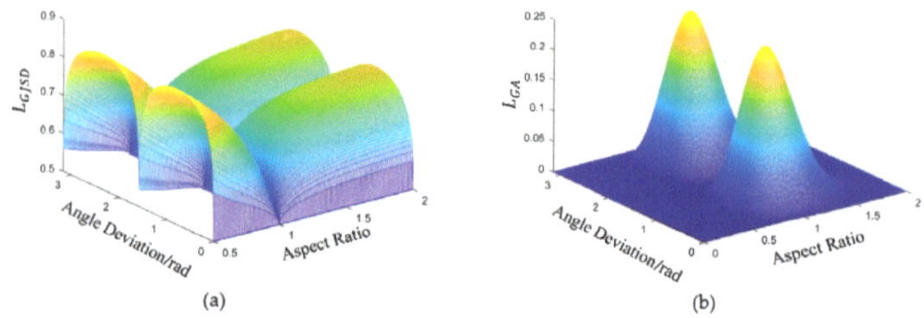

Figure 7. Diagram of aspect ratio (X-axis), angle deviation (Y-axis), and loss value (Z-axis). (**a**) Z-axis for L_{GJSD}; (**b**) Z-axis for L_{GA}.

3.3.4. Regression Loss Based on Gaussian Distribution

Based on the discussion of the above issues, we propose the regression loss based on Gaussian distribution (RLGD), which can be expressed as follows:

$$L_{reg} = L_{GJSD} + L_{GA} \tag{14}$$

Among them, L_{GJSD} is the regression loss based on GJSD, and L_{GA} is the Gaussian angle loss, which is used to address the issue of angle confusion in square bounding boxes caused by the rotating box representation method based on Gaussian distribution. Combining L_{GJSD} and L_{GA} can enhance the accuracy and performance of oriented object detection. When the aspect ratio of the bounding box approaches 1, L_{GA} can be used as compensation for L_{GJSD} to improve the detection accuracy of objects with a square bounding box. Moreover, when the aspect ratio of the bounding box is not 1, L_{GA} approaches 0, which does not affect the value of the original loss function.

4. Experiments and Results

4.1. Datasets

In recent years, with the widespread application of remote sensing image object detection techniques across various fields, several institutions have collected and organized a wide variety of remote sensing datasets. In this paper, we conduct extensive experimental validation of the proposed method on two publicly available remote sensing image datasets, namely, the DOTA-v1.0 dataset [4] and HRSC2016 dataset [11].

4.1.1. DOTA-v1.0

DOTA-v1.0 [4] is a dataset specifically designed for object detection in remote sensing images. The images in this dataset originate from a diverse range of sources, including aerial images collected by various sensors and platforms, which contribute to the rich scene variations and practical application value of the dataset. Furthermore, the objects within the images exhibit extensive variations in scale, orientation, and shape. It comprises 2806 images with a total of 188,282 instance objects labeled using quadrilateral bounding boxes. The dataset contains 15 categories: plane (PL), baseball diamond (BD), bridge (BR), ground track field (GTF), small vehicle (SV), large vehicle (LV), ship (SH), tennis court (TC), basketball court (BC), storage tank (ST), soccer ball field (SBF), roundabout (RA), harbor (HA), swimming pool (SP) and helicopter (HC). The images in the DOTA-v1.0 dataset have varying pixel sizes, ranging from 800 × 800 to 4000 × 4000. The dataset is divided into three sets: training, validation, and test sets (1411, 458, and 937 images, respectively). The training and validation sets provide labels of the object annotation boxes in the images, while the test set does not provide true labels. After testing, the predicted bounding box labels are submitted to the official DOTA-v1.0 dataset test website to obtain the final detection results.

4.1.2. HRSC2016

HRSC2016 [11] is a dataset containing high-resolution remotely sensed ship images, known as the High-Resolution Ship Collection 2016. Due to the complex backgrounds of remote sensing images in this dataset, the significant similarity between object ships and coastal textures, as well as the diverse scale variations of ships of various sizes present in the same image, we chose this dataset to test our model. It consists of 1070 images containing 2976 ship objects in the rectangular box labeled format. The ships vary in size, appearance, and orientation, with over 20 different types represented. The HRSC2016 dataset comprises images with pixel sizes ranging from 300 × 300 to 1500 × 900 pixels, with the majority exceeding 1000 × 600 pixels. These images feature a spatial resolution between 0.3 and 3 m. The dataset includes 626 images in the training set and 444 images in the test set, all of which are accompanied by accurate real-world labels.

4.2. Implementation Details

Our experiment was conducted on the Ubuntu 18.04 operating system with a single NVIDIA TESLA T4. We utilized PyTorch 1.9.0 as a deep learning framework and Python 3.8 was used as the development language. Oriented R-CNN [10] was used as the object detection framework. ResNet50 and ResNet101 [56] were used as our backbones. They were pre-trained on ImageNet [57]. Horizontal and vertical flips were used as data augmentation techniques during training. The batch size for the experiment was set to 2. We optimized the overall network using the stochastic gradient descent (SGD) algorithm, with a momentum of 0.9 and a weight decay of 0.0001. The effectiveness of the remote sensing image object detection algorithms was evaluated using the mean average precision (mAP), the number of parameters (Params), and the floating-point operations (Flops).

On the DOTA-v1.0 dataset, the original images were cropped into 1024 × 1024 patches with a cropping stride set to 824, resulting in a 200-pixel overlap between two adjacent patches. The number of epochs was set to 12 for the DOTA dataset. The initial learning rate was set at 0.005 and reduced by a factor of 10 at epochs 8 and 11.

On the HRSC2016 dataset, the short side of the original images was resized to 800 pixels, while the long side was kept to a maximum of 1333 pixels, preserving the original aspect ratio of the images. The number of epochs was set to 36 for the HRSC2016 dataset. The initial learning rate was set to 0.005 and divided by 10 at epochs 24 and 33.

4.3. Results on the DOTA-v1.0 Dataset

We compare the proposed model with other advanced remote sensing image object detection algorithms on the DOTA-v1.0 dataset. Table 1 displays the experimental results, including the average precision (AP) of the 15 classes, respectively, on the DOTA-v1.0 dataset and the mean average precision (mAP) of all the classes. The backbone networks, R-50 and R-101, are denoted as ResNet-50 and ResNet-101, respectively. Compared to other algorithms, our method achieves the highest detection accuracy of 77.29% mAP. Compared to the baseline, our method improves the detection accuracy (mAP)by 1.15% and 1.12% when using ResNet-50 and ResNet-101 as the backbones, respectively. The experimental results presented above demonstrate the effectiveness of our method in addressing the problem of large-scale variation in remote sensing image object detection and improving detection accuracy. The comparison results of various algorithms for object detection on the DOTA-v1.0 dataset are visualized in Figure 8. It can be seen that our method outperforms other models in terms of AP for the eight classes (PL, BR, SV, LV, SH, TC, BC, and ST). And the AP values of the proposed model are generally higher than those of other models. This suggests that our approach is effective at enhancing object detection in remote sensing images.

Table 1. Detection results of different algorithms on the DOTA-v1.0 dataset(%). The bold represents the highest detection accuracy among a class or all classes.

Method	Backbone	PL	BD	BR	GTF	SV	LV	SH	TC	BC	ST	SBF	RA	HA	SP	HC	mAP
RetinaNet-O [58]	R-50	88.67	77.62	41.81	58.17	74.58	71.64	79.11	90.29	82.18	74.32	54.75	60.60	62.57	69.67	60.64	68.43
Faster R-CNN-O [12]	R-50	88.44	73.06	44.86	59.09	73.25	71.49	77.11	90.84	78.94	83.90	48.59	62.95	62.18	64.91	56.18	69.05
RoI Transformer [13]	R-101	88.64	78.52	43.44	75.92	68.81	73.68	83.59	90.74	77.27	81.46	58.39	53.54	62.83	58.93	47.67	69.56
S2ANet [16]	R-50	89.30	80.11	50.97	73.91	78.59	77.34	86.38	90.91	85.14	84.84	60.45	66.94	66.78	68.55	51.65	74.13
SASM [59]	R-50	86.42	78.97	52.47	69.84	77.30	75.99	86.72	90.89	82.63	85.66	60.13	68.25	73.98	**72.22**	62.37	74.92
Gliding Vertex [15]	R-101	89.64	85.00	52.26	**77.34**	73.01	73.14	86.82	90.74	79.02	86.81	59.55	**70.91**	72.94	70.86	57.32	75.02
AOPG [24]	R-101	89.14	82.74	51.87	69.28	77.65	82.42	88.08	90.89	86.26	85.13	60.60	66.30	74.05	67.76	58.77	75.39
ARC [26]	R-50	89.28	78.77	53.00	72.44	79.81	77.84	86.81	90.88	84.27	86.20	60.74	68.97	66.35	71.25	**65.77**	75.49
Oriented R-CNN [10]	Swin-T	90.29	81.83	52.51	73.96	79.13	84.47	88.86	91.88	86.37	86.02	58.44	64.21	71.16	69.15	57.55	75.72
Oriented R-CNN [10]	R-50	89.52	84.24	53.19	70.50	79.20	83.30	88.21	90.90	86.03	84.95	61.69	67.06	74.40	69.80	54.51	75.83
Oriented R-CNN [10]	R-101	89.63	84.10	54.87	72.57	78.89	83.21	88.37	91.01	86.32	85.08	61.99	67.09	**74.84**	69.10	55.54	76.17
Ours	Swin-T	90.03	84.72	52.92	74.00	78.82	84.21	88.92	91.60	87.11	86.01	61.00	66.62	74.51	69.96	57.67	76.54
Ours	R-50	91.43	**85.62**	55.05	72.24	80.52	84.94	90.06	92.85	87.79	86.94	**62.29**	66.80	74.47	70.26	53.41	76.98
Ours	R-101	**92.39**	79.57	**56.14**	74.07	**82.04**	**86.16**	**90.98**	**93.97**	**88.42**	**87.74**	61.72	66.18	70.54	70.90	58.55	**77.29**

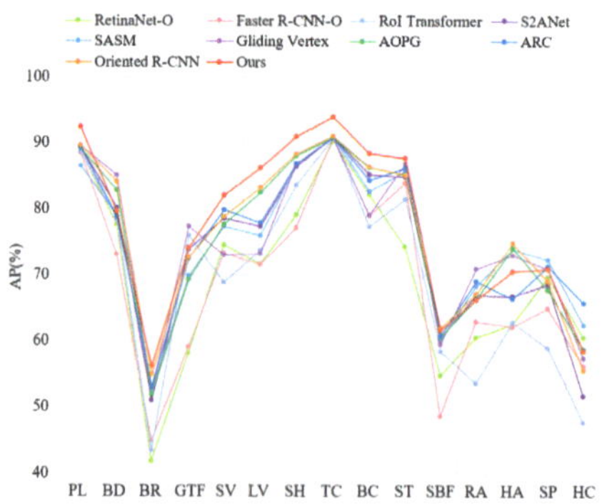

Figure 8. Detection results of different algorithms on the DOTA-v1.0 dataset.

4.4. Results on the HRSC2016 Dataset

For the HRSC2016 dataset, we compare the proposed model with other advanced remote sensing image object detection models. Table 2 shows the experimental results, including the mAP values of different methods under PASCAL VOC 2007 [60] and VOC 2012 metrics. Compared to other algorithms, the proposed model achieves better detection results, with mAP(07) and mAP(12) reaching 90.59% and 97.95%, respectively. Compared to the baseline, our method improves the detection accuracies by 1.56% mAP(12) and 1.41% mAP(12) based on the backbone of ResNet-50 and ResNet-101, respectively, which proves the advancement of our method. As demonstrated in Figure 9, the comparison results of mAP(07) and mAP(12) obtained by different algorithms for object detection on the HRSC2016 dataset are intuitively reflected. It can be seen that the detection accuracy metrics mAP(07) and mAP(12) of our method are higher compared to other algorithms. This indicates that our algorithm demonstrates superior performance in remote sensing image object detection.

Table 2. Detection results of different algorithms on the HRSC2016 dataset (%). The bold represents the highest detection accuracy.

Method	Backbone	mAP(07)	mAP(12)
RoI Transformer [13]	ResNet-101	86.20	\
Gliding Vertex [15]	ResNet-101	88.20	\
RetinaNet-O [58]	ResNet-101	89.18	95.21
R3Det [23]	ResNet-101	89.26	96.01
DCL [61]	ResNet-101	89.46	96.41
CSL [62]	ResNet-101	89.62	96.10
S2ANet [16]	ResNet-101	90.17	95.01
SASM [59]	ResNet-101	90.27	\
AOPG [24]	ResNet-101	90.34	96.22
ARC [26]	ResNet-50	90.41	\
Oriented R-CNN [10]	Swin-T	89.70	\
Oriented R-CNN [10]	ResNet-50	90.23	96.11
Oriented R-CNN [10]	ResNet-101	90.43	96.54
Ours	Swin-T	90.42	\
Ours	ResNet-50	90.55	97.67
Ours	ResNet-101	**90.59**	**97.95**

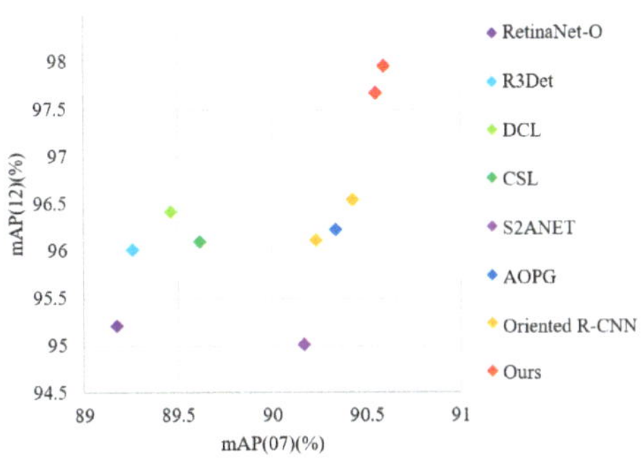

Figure 9. Detection results of different algorithms on the HRSC2016 dataset.

4.5. Visualization Results

The remote sensing image object detection results of the model proposed in this paper are visualized as shown in Figure 10. As can be seen in Figure 10, there are significant size differences between different categories or within the same category of objects in remote sensing images. The detection algorithm must be highly adaptable to handle objects with significant scale changes. And the model should be capable of detecting both tiny and oversized objects. Figure 10 illustrates the detection results of our method in various multi-scale scenarios. The proposed approach can effectively detect various classes of objects with significant differences in scale, including objects of the same classes that vary in size, such as planes and ships, as well as objects of different classes that vary significantly in size, such as vehicles and baseball fields, ships, and harbors, and so on. The experimental results demonstrate that the model proposed in this paper has a strong ability to detect multi-scale objects.

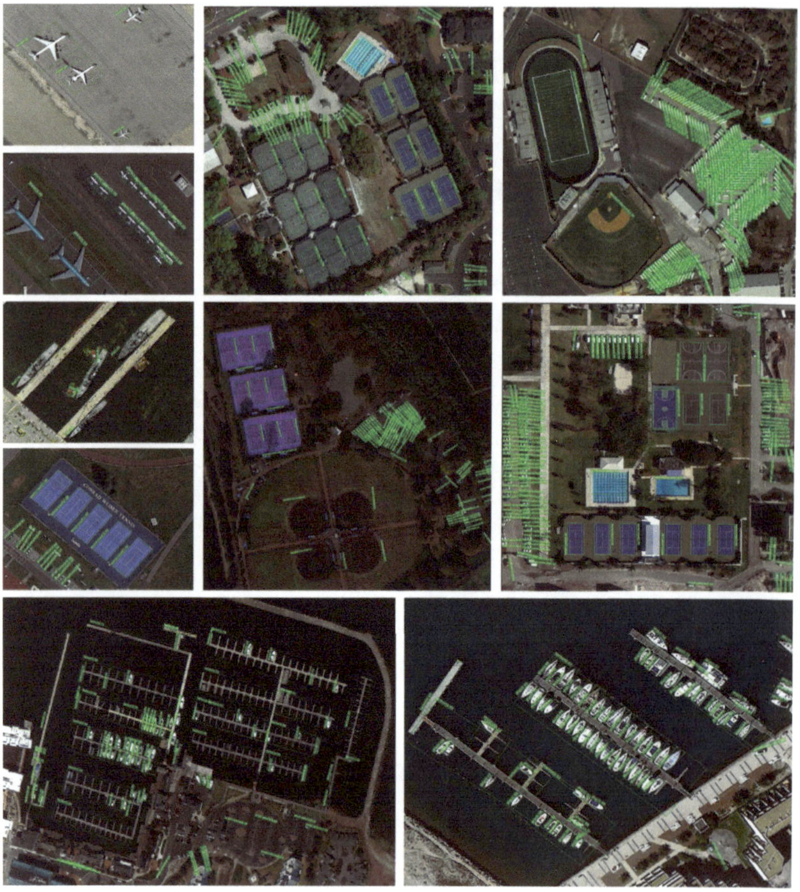

Figure 10. Visualization of remote sensing image object detection.

5. Discussion

5.1. Ablation Study

In order to verify the effectiveness of each component of the proposed algorithm, ablation experiments were conducted on the DOTA-v1.0 dataset and HRSC2016 dataset. The baseline used was Oriented R-CNN with ResNet-50 serving as the backbone for feature extraction. The results of the ablation experiments are presented in Table 3.

The proposed multi-scale feature interaction (MSFI) method in this paper achieves a detection accuracy of 76.81% mAP on the DOTA-v1.0 dataset, which is an improvement of 0.98% compared to the baseline. In addition, on the HRSC2016 dataset, the detection accuracy metrics mAP(07) and mAP(12) reached 90.50% and 97.52%, respectively, which are superior to the baseline. This indicates that our method enables the network to effectively locate and identify objects with drastic scale changes by utilizing the spatial and semantic information of multilevel features.

The regression loss based on the Gaussian distribution (RLGD) algorithm achieves a detection accuracy of 76.70% mAP, representing a 0.87% improvement over the baseline. Moreover, the model outperforms the baseline with detection accuracy metrics of 90.35% mAP(07) and 96.62% mAP(12) on the HRSC2016 dataset. The experimental results demonstrate that the proposed algorithm—by adopting the normalized generalized Jensen–Shannon divergence with Gaussian angle loss as the regression loss function—can render the model scale-invariant, improving the effectiveness of object detection in remote sensing images across multiple scales.

By combining the proposed models of MSFI and RLGD, the detection accuracy, measured by mAP on the DOTA-v1.0 dataset, improved to 76.98%, which is a relative increase of 1.15% compared to the baseline. Meanwhile, the detection accuracy metrics mAP(07) and mAP(12) on the HRSC2016 dataset are 90.55% and 97.67%, representing a significant improvement of 0.32% and 1.56% from the baseline. The experimental results show that the improvements made by the proposed algorithm effectively enhance object detection capabilities in remote sensing images. In order to further explore the benefits of each component of the proposed algorithm, more detailed experiments and discussions were carried out.

Table 3. Ablation experimental results of our method on the DOTA-v1.0 dataset and HRSC2016 dataset (%). The bold represents the highest detection accuracy. ✓ indicates that the algorithm adopts this module.

Method	MSFI	RLGD	mAP	mAP(07)	mAP(12)
Baseline			75.83	90.23	96.11
	✓		76.81	90.50	97.52
		✓	76.70	90.35	96.62
Ours	✓	✓	**76.98**	**90.55**	**97.67**

5.2. Effect of the MSFI

In this paper, we propose a multi-scale feature interaction (MSFI) network that includes a cross-layer feature interaction (CLFI) module consisting of a grounding feature interaction (GFI) module and a rendering feature interaction (RFI) module, as well as a spatial feature interaction (SFI) module. The computational input of FLOPs is 1024 × 1024. Table 4 presents the results of the ablation experiments.

Table 4. Ablation experimental results of the MSFI network on the DOTA-v1.0 dataset. The bold represents the highest detection accuracy. ✓ indicates that the algorithm adopts this module.

Method	GFI	RFI	SFI	mAP(%)	Params(M)	FLOPs(G)
Baseline				75.83	41.14	211.43
GFI	✓			76.42	41.44	222.36
RFI		✓		76.07	41.44	217.81
CLFI	✓	✓		76.51	41.44	228.73
SFI			✓	76.39	41.27	217.20
MSFI	✓	✓	✓	**76.81**	41.57	234.51

From the experimental results, it can be seen that the detection accuracies improve with the addition of the GFI, RFI, and CLFI modules, achieving mAP values of 76.42%, 76.07%, and 76.51%, respectively, all of which exceed those in the baseline. This shows that the cross-layer feature interaction method effectively integrates deep and shallow features, capturing comprehensive information about the object. Additionally, the SFI module achieves a detection accuracy of 76.39% mAP, which is higher than the baseline.

This indicates that the method effectively utilizes spatial domain information to enhance features and suppress background noise.

Therefore, each sub-module of the MSFI algorithm enhances the object detection effect in remote sensing images. The MSFI effectively combines these sub-modules and improves the object detection effect with only a slight increase in computation. As a result, the detection accuracy, measured by mAP, is significantly improved.

5.3. Effect of the RLGD

This paper proposes a regression loss algorithm based on Gaussian distribution (RLGD), which includes an oriented bounding box representation based on the Gaussian distribution (GD), a regression loss function based on generalized Jensen–Shannon divergence (GJSD), and Gaussian angle loss for the angle confusion problem (GA). The ablation experiment results are presented in Table 5. The baseline model utilizes the smooth-L1 regression loss function [10]. The GWD denotes the oriented bounding box representation of the Gaussian distribution and employs the normalized Wasserstein distance as the regression loss function [7].

Table 5. Ablation experimental results of the RLGD algorithm on the DOTA-v1.0 dataset and HRSC2016 dataset (%). The bold represents the highest detection accuracy. ✓ indicates that the algorithm adopts this module.

Method	GD	GJSD	GA	mAP	mAP(07)	mAP(12)
Baseline				75.83	90.23	96.11
GWD	✓			75.91	90.17	96.19
GJSD	✓	✓		76.12	90.30	96.58
RLGD	✓	✓	✓	**76.70**	**90.35**	**96.62**

The experimental results show that the detection accuracy of the GWD method achieves 75.91% mAP, which is higher than that of the baseline model. This transformation of the regression problem into a measure of similarity between Gaussian distributions helps improve detection accuracy by mitigating the negative effects of scale and unit differences between parameters. Moreover, the detection accuracy metric mAP for the GJSD method is 76.12%, which is an improvement compared to the baseline model. This suggests that the GJSD-based regression loss method with scale invariance can enhance the scale adaptation ability of the model. In addition, when the RLGD model is enhanced with the introduction of GA, the detection accuracy, measured by mAP, improves to 76.70%, which is 0.87% higher than the baseline model. This indicates that the integration of Gaussian angle loss effectively addresses the angle confusion issue associated with the rotated bounding box representation of the Gaussian distribution for square bounding boxes, further enhancing object detection.

As a consequence, all components of the RLGD algorithm enhance the object detection effect of remote sensing images. The synergistic effects of these components significantly improve the detection accuracy.

5.4. Visualization Analysis

In this section, we compare the detection results of our method with the baseline Oriented R-CNN, as shown in Figure 11. Figure 11 displays three groups of detection results, each containing several ship objects. The objects within the same category vary significantly in size, reflecting large-scale changes in remote sensing image objects. Figure 11 illustrates the comparison of detection results between the baseline and our method. The first row displays the baseline results, where the objects of small ships in the selected and enlarged boxes are not detected, resulting in missed detections. In contrast, the second row shows the improved results of our method, accurately detecting the small ships in the selected and enlarged boxes and reducing the missed detection rate. The model presented in this paper significantly improves the detection effect compared to the baseline. Our method accurately detects several small-sized ship objects in these figures while also ensuring the detection of large-sized ships, making it more effective in multi-scale object detection than the baseline.

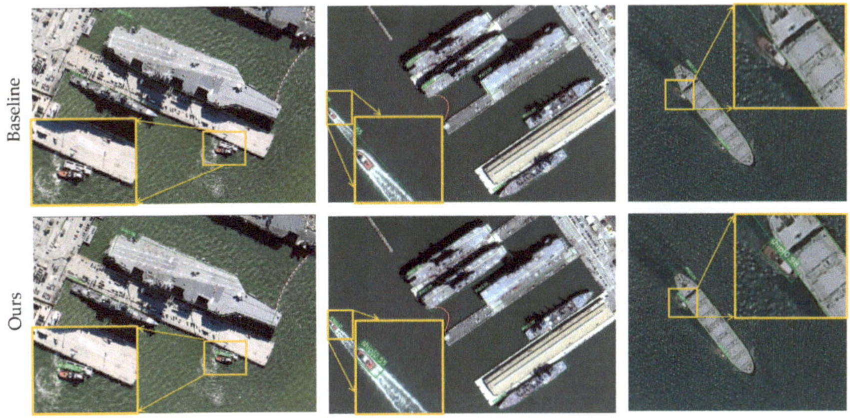

Figure 11. Comparison of detection results between the baseline and our method.

5.5. Limitations

This paper explores the issue of multi-scale object detection in remote sensing images. While some progress has been made in improving the performance of detection algorithms, there are still open issues.

One such issue is that the multi-scale feature interaction model achieves feature interaction through the attention mechanism. This enhances detection accuracy but has a large impact on model speed due to the large number of parameters and high computational complexity of the mechanism. In future work, the attention mechanism can be improved to optimize the computation while maintaining detection accuracy.

Additionally, the algorithm for regression loss based on Gaussian distribution uses Gaussian representation, which requires additional matrix transformations and may increase the computational load. Moreover, this method is unsuitable for detecting objects with large aspect ratios. Therefore, further optimization is necessary to reduce the computational effort caused by matrix transformations and improve the effect of object detection with large aspect ratios.

6. Conclusions

In this paper, we propose a multi-scale object detection algorithm based on feature interaction and Gaussian distribution to address the problem of scale variation in objects within remote sensing images. In the proposed method, the multi-scale feature interaction (MSFI) network combines feature interactions at the feature layer and in the spatial domain so that the network can make full use of the spatial and semantic information of multi-level features. The proposed regression loss (RLGD) algorithm adopts the method based on Gaussian distribution instead of rotating bounding box representation and improves it by proposing a scale-invariant regression loss. Experimental results demonstrate that our proposed approach can effectively improve the ability of multi-scale object detection. Compared to other advanced algorithms, our method shows superior detection results on remote sensing image datasets.

Author Contributions: Conceptualization: R.Y. and H.C.; methodology: R.Y. and H.C.; writing—original draft preparation: R.Y., H.C. and B.Z.; writing—review and editing: H.C., B.Z. and T.F. All authors have read and agreed to the published version of the manuscript.

Funding: This research received no external funding.

Data Availability Statement: The experiments are evaluated on publicly open datasets. The datasets can be accessed in their corresponding published papers [4,11].

Conflicts of Interest: Author Tao Feng was employed by the company China North Industries Group Corporation. The remaining authors declare that the research was conducted in the absence of any commercial or financial relationships that could be construed as a potential conflict of interest.

Appendix A

Proof of Scale Invariance of GJSD

Suppose that M is a full-rank matrix and there are two Gaussian distributions, denoted as $X_p' = MX_p \sim N_p'\left(\mu_p', \Sigma_p'\right)$ and $X_t' = MX_t \sim N_t'\left(\mu_t', \Sigma_t'\right)$, where:

$$\begin{cases} \mu_p' = M\mu_p, & \Sigma_p' = M\Sigma_p M^T \\ \mu_t' = M\mu_t, & \Sigma_t' = M\Sigma_t M^T \end{cases} \tag{A1}$$

According to Equation (5), the GJSD between these two Gaussian distributions is expressed as:

$$D_{GJS}(N_p' \parallel N_t') = (1-\alpha)D_{KL}(N_\alpha' \parallel N_p') + \alpha D_{KL}(N_\alpha' \parallel N_t') \tag{A2}$$

where $N_\alpha'\left(\mu_\alpha', \Sigma_\alpha'\right) = N_\alpha\left(M\mu_\alpha, M\Sigma_\alpha M^T\right)$. According to Equation (8) there is:

$$\begin{aligned} \Sigma_\alpha' &= \left[(1-\alpha)\left(M^T\right)^{-1}\Sigma_p^{-1}M^{-1} + \alpha\left(M^T\right)^{-1}\Sigma_t^{-1}M^{-1}\right]^{-1} \\ &= \left\{\left(M^T\right)^{-1}\left[(1-\alpha)\Sigma_p^{-1} + \alpha\Sigma_t^{-1}\right]M^{-1}\right\}^{-1} \\ &= M\left[(1-\alpha)\Sigma_p^{-1} + \alpha\Sigma_t^{-1}\right]^{-1}M^T \\ &= M\Sigma_\alpha M^T \end{aligned} \tag{A3}$$

and

$$\begin{aligned} \mu_\alpha' &= M\Sigma_\alpha M^T\left[(1-\alpha)\left(M^T\right)^{-1}\Sigma_p^{-1}M^{-1}M\mu_p + \alpha\left(M^T\right)^{-1}\Sigma_t^{-1}M^{-1}M\mu_t\right] \\ &= M\Sigma_\alpha M^T\left(M^T\right)^{-1}\left[(1-\alpha)\Sigma_p^{-1}\mu_p + \alpha\Sigma_t^{-1}\mu_t\right] \\ &= M\Sigma_\alpha\left[(1-\alpha)\Sigma_p^{-1}\mu_p + \alpha\Sigma_t^{-1}\mu_t\right] \\ &= M\mu_\alpha \end{aligned} \tag{A4}$$

It can be obtained from Equation (6):

$$\begin{aligned} D_{KL}\left(N_\alpha' \parallel N_p'\right) &= \frac{1}{2}(\mu_\alpha - \mu_p)^T M^T\left(M^T\right)^{-1}\Sigma_p^{-1}M^{-1}M(\mu_\alpha - \mu_p) \\ &\quad + \frac{1}{2}Tr\left(\left(M^T\right)^{-1}\Sigma_p^{-1}M^{-1}M\Sigma_\alpha M^T\right) + \frac{1}{2}\ln\frac{|M\Sigma_p M^T|}{|M\Sigma_\alpha M^T|} - 1 \\ &= \frac{1}{2}(\mu_\alpha - \mu_p)^T\Sigma_p^{-1}(\mu_\alpha - \mu_t) + \frac{1}{2}Tr\left(M^T\left(M^T\right)^{-1}\Sigma_p^{-1}\Sigma_\alpha\right) + \frac{1}{2}\ln\frac{|\Sigma_p|}{|\Sigma_\alpha|} - 1 \\ &= D_{KL}(N_\alpha \parallel N_p) \end{aligned} \tag{A5}$$

Similarly,

$$D_{KL}\left(N_{\alpha}^{'} \parallel N_{t}^{'}\right) = D_{KL}(N_{\alpha} \parallel N_{t}) \tag{A6}$$

It can be concluded that:

$$\begin{aligned} D_{GJS}\left(N_{p}^{'} \parallel N_{t}^{'}\right) &= (1-\alpha)D_{KL}\left(N_{\alpha}^{'} \parallel N_{p}^{'}\right) + \alpha D_{KL}\left(N_{\alpha}^{'} \parallel N_{t}^{'}\right) \\ &= (1-\alpha)D_{KL}(N_{\alpha} \parallel N_{p}) + \alpha D_{KL}(N_{\alpha} \parallel N_{t}) \\ &= D_{GJS}(N_{p} \parallel N_{t}) \end{aligned} \tag{A7}$$

Therefore, GJSD has affine invariance. The scale invariance of GJSD is proved when $M = kI$ (I is an identity matrix and $I \neq 0$).

References

1. Cheng, G.; Han, J. A survey on object detection in optical remote sensing images. *Isprs J. Photogramm. Remote Sens.* **2016**, *117*, 11–28. [CrossRef]
2. Li, K.; Wan, G.; Cheng, G.; Meng, L.; Han, J. Object detection in optical remote sensing images: A survey and a new benchmark. *ISPRS J. Photogramm. Remote Sens.* **2020**, *159*, 296–307. [CrossRef]
3. Liu, Y.; Li, H.; Hu, C.; Luo, S.; Luo, Y.; Chen, C.W. Learning to aggregate multi-scale context for instance segmentation in remote sensing images. *IEEE Trans. Neural Netw. Learn. Syst.* 2024, *early access*. [CrossRef]
4. Xia, G.S.; Bai, X.; Ding, J.; Zhu, Z.; Belongie, S.; Luo, J.; Datcu, M.; Pelillo, M.; Zhang, L. DOTA: A large-scale dataset for object detection in aerial images. In Proceedings of the IEEE Conference on Computer Vision and Pattern Recognition, Salt Lake City, UT, USA, 18–23 June 2018; pp. 3974–3983.
5. Liao, Y.; Wang, H.; Lin, C.; Li, Y.; Fang, Y.; Ni, S. Research progress of deep learning-based object detection of optical remote sensing image. *J. Commun.* **2022**, *43*, 190–203.
6. Yang, J.; Li, C.; Dai, X.; Gao, J. Focal modulation networks. *Adv. Neural Inf. Process. Syst.* **2022**, *35*, 4203–4217.
7. Yang, X.; Yan, J.; Ming, Q.; Wang, W.; Zhang, X.; Tian, Q. Rethinking rotated object detection with gaussian wasserstein distance loss. In Proceedings of the International Conference on Machine Learning, PMLR, Virtual, 18–24 July 2021; pp. 11830–11841.
8. Nielsen, F. On a generalization of the Jensen–Shannon divergence and the Jensen–Shannon centroid. *Entropy* **2020**, *22*, 221. [CrossRef]
9. Yu, X.; Lu, J.; Lin, M.; Zhou, L.; Ou, L. MKIoU loss: Toward accurate oriented object detection in aerial images. *J. Electron. Imaging* **2023**, *32*, 033030. [CrossRef]
10. Xie, X.; Cheng, G.; Wang, J.; Yao, X.; Han, J. Oriented R-CNN for object detection. In Proceedings of the IEEE/CVF International Conference on Computer Vision, Montreal, BC, Canada, 11–17 October 2021; pp. 3520–3529.
11. Liu, Z.; Yuan, L.; Weng, L.; Yang, Y. A high resolution optical satellite image dataset for ship recognition and some new baselines. In Proceedings of the International Conference on Pattern Recognition Applications and Methods, Porto, Portugal, 24–26 February 2017; Volume 2, pp. 324–331.
12. Ren, S.; He, K.; Girshick, R.; Sun, J. Faster r-cnn: Towards real-time object detection with region proposal networks. *Adv. Neural Inf. Process. Syst.* **2015**, *28* , 91–99. [CrossRef]
13. Ding, J.; Xue, N.; Long, Y.; Xia, G.S.; Lu, Q. Learning RoI transformer for oriented object detection in aerial images. In Proceedings of the IEEE/CVF Conference on Computer Vision and Pattern Recognition 2019, Long Beach, CA, USA, 16–17 June 2019; pp. 2849–2858.
14. Han, J.; Ding, J.; Xue, N.; Xia, G.S. Redet: A rotation-equivariant detector for aerial object detection. In Proceedings of the IEEE/CVF Conference on Computer Vision and Pattern Recognition 2021, Virtual, 19–25 June 2021; pp. 2786–2795.
15. Xu, Y.; Fu, M.; Wang, Q.; Wang, Y.; Chen, K.; Xia, G.S.; Bai, X. Gliding vertex on the horizontal bounding box for multi-oriented object detection. *IEEE Trans. Pattern Anal. Mach. Intell.* **2020**, *43*, 1452–1459. [CrossRef]
16. Han, J.; Ding, J.; Li, J.; Xia, G.S. Align deep features for oriented object detection. *IEEE Trans. Geosci. Remote Sens.* **2021**, *60*, 1–11. [CrossRef]
17. Law, H.; Deng, J. Cornernet: Detecting objects as paired keypoints. In Proceedings of the European Conference on Computer Vision (ECCV) 2018, Munich, Germany, 8–14 September 2018; pp. 734–750.
18. Wang, J.; Yang, W.; Li, H.C.; Zhang, H.; Xia, G.S. Learning center probability map for detecting objects in aerial images. *IEEE Trans. Geosci. Remote Sens.* **2020**, *59*, 4307–4323. [CrossRef]
19. Zhou, X.; Wang, D.; Krähenbühl, P. Objects as points. *arXiv* **2019**, arXiv:1904.07850.
20. Li, W.; Chen, Y.; Hu, K.; Zhu, J. Oriented reppoints for aerial object detection. In Proceedings of the IEEE/CVF Conference on Computer Vision and Pattern Recognition 2022, New Orleans, LA, USA, 18–24 June 2022; pp. 1829–1838.
21. Pan, X.; Ren, Y.; Sheng, K.; Dong, W.; Yuan, H.; Guo, X.; Ma, C.; Xu, C. Dynamic refinement network for oriented and densely packed object detection. In Proceedings of the IEEE/CVF Conference on Computer Vision and Pattern Recognition 2020, Seattle, WA, USA, 13–19 June 2020; pp. 11207–11216.

22. Qian, W.; Yang, X.; Peng, S.; Yan, J.; Guo, Y. Learning modulated loss for rotated object detection. In Proceedings of the AAAI Conference on Artificial Intelligence 2021, Vancouver, BC, Canada, 2–9 February 2021; Volume 35, pp. 2458–2466.
23. Yang, X.; Yan, J.; Feng, Z.; He, T. R3det: Refined single-stage detector with feature refinement for rotating object. In Proceedings of the AAAI Conference on Artificial Intelligence 2021, Vancouver, BC, Canada, 2–9 February 2021; Volume 35, pp. 3163–3171.
24. Cheng, G.; Wang, J.; Li, K.; Xie, X.; Lang, C.; Yao, Y.; Han, J. Anchor-free oriented proposal generator for object detection. *IEEE Trans. Geosci. Remote Sens.* **2022**, *60*, 1–11. [CrossRef]
25. Li, Y.; Hou, Q.; Zheng, Z.; Cheng, M.M.; Yang, J.; Li, X. Large selective kernel network for remote sensing object detection. In Proceedings of the IEEE/CVF International Conference on Computer Vision 2023, Paris, France, 4–6 October 2023 pp. 16794–16805.
26. Pu, Y.; Wang, Y.; Xia, Z.; Han, Y.; Wang, Y.; Gan, W.; Wang, Z.; Song, S.; Huang, G. Adaptive rotated convolution for rotated object detection. In Proceedings of the IEEE/CVF International Conference on Computer Vision 2023, Paris, France, 4–6 October; pp. 6589–6600.
27. Ma, T.; Mao, M.; Zheng, H.; Gao, P.; Wang, X.; Han, S.; Ding, E.; Zhang, B.; Doermann, D. Oriented object detection with transformer. *arXiv* **2021**, arXiv:2106.03146.
28. Dai, L.; Liu, H.; Tang, H.; Wu, Z.; Song, P. Ao2-detr: Arbitrary-oriented object detection transformer. *IEEE Trans. Circuits Syst. Video Technol.* **2022**, *33*, 2342–2356. [CrossRef]
29. Carion, N.; Massa, F.; Synnaeve, G.; Usunier, N.; Kirillov, A.; Zagoruyko, S. End-to-end object detection with transformers. In Proceedings of the European Conference on Computer Vision, Glasgow, UK, 23-28 August 2020; pp. 213–229.
30. Vaswani, A.; Shazeer, N.; Parmar, N.; Uszkoreit, J.; Jones, L.; Gomez, A.N.; Kaiser, Ł.; Polosukhin, I. Attention is all you need. *Adv. Neural Inf. Process. Syst.* **2017**, *30*, 5998–6008.
31. Bello, I.; Zoph, B.; Vaswani, A.; Shlens, J.; Le, Q.V. Attention Augmented Convolutional Networks. In Proceedings of the IEEE/CVF International Conference on Computer Vision (ICCV) 2019, Seoul, Republic of Korea, 27 October–2 November 2019.
32. Liu, Z.; Lin, Y.; Cao, Y.; Hu, H.; Wei, Y.; Zhang, Z.; Lin, S.; Guo, B. Swin transformer: Hierarchical vision transformer using shifted windows. In Proceedings of the IEEE/CVF International Conference on Computer Vision 2021, Montreal, QC, Canada, 10–17 October 2021; pp. 10012–10022.
33. Zhang, M.; Zhang, R.; Yang, Y.; Bai, H.; Zhang, J.; Guo, J. ISNet: Shape matters for infrared small target detection. In Proceedings of the IEEE/CVF Conference on Computer Vision and Pattern Recognition, New Orleans, LA, USA, 18–24 June 2022; pp. 877–886.
34. Zhang, M.; Bai, H.; Zhang, J.; Zhang, R.; Wang, C.; Guo, J.; Gao, X. Rkformer: Runge-kutta transformer with random-connection attention for infrared small target detection. In Proceedings of the 30th ACM International Conference on Multimedia, Lisbon, Portugal, 10–14 October 2022; pp. 1730–1738.
35. Zhang, M.; Yue, K.; Zhang, J.; Li, Y.; Gao, X. Exploring feature compensation and cross-level correlation for infrared small target detection. In Proceedings of the 30th ACM International Conference on Multimedia, Lisbon, Portugal, 10–14 October 2022; pp. 1857–1865.
36. Zhang, M.; Li, B.; Wang, T.; Bai, H.; Yue, K.; Li, Y. Chfnet: Curvature half-level fusion network for single-frame infrared small target detection. *Remote Sens.* **2023**, *15*, 1573. [CrossRef]
37. Zhang, M.; Yang, H.; Yue, K.; Zhang, X.; Zhu, Y.; Li, Y. Thermodynamics-Inspired Multi-Feature Network for Infrared Small Target Detection. *Remote Sens.* **2023**, *15*, 4716. [CrossRef]
38. Ma, T.; Yang, Z.; Song, Y.F.; Liang, J.; Wang, H. DMEF-Net: Lightweight Infrared Dim Small Target Detection Network for Limited Samples. *IEEE Trans. Geosci. Remote Sens.* **2023**, *61*, 1–15. [CrossRef]
39. Zhang, M.; Yang, H.; Guo, J.; Li, Y.; Gao, X.; Zhang, J. IRPruneDet: Efficient Infrared Small Target Detection via Wavelet Structure-Regularized Soft Channel Pruning. In Proceedings of the AAAI Conference on Artificial Intelligence 2024, Vancouver, BC, Canada, 22–25 February 2024; Volume 38, pp. 7224–7232.
40. Lin, T.Y.; Dollár, P.; Girshick, R.; He, K.; Hariharan, B.; Belongie, S. Feature pyramid networks for object detection. In Proceedings of the IEEE Conference on Computer Vision and Pattern Recognition 2017, Honolulu, HI, USA, 21–26 July 2017; pp. 2117–2125.
41. Li, Q.; Mou, L.; Liu, Q.; Wang, Y.; Zhu, X.X. HSF-Net: Multiscale deep feature embedding for ship detection in optical remote sensing imagery. *IEEE Trans. Geosci. Remote Sens.* **2018**, *56*, 7147–7161. [CrossRef]
42. Liu, X.; Ma, S.; He, L.; Wang, C.; Chen, Z. Hybrid network model: TransConvNet for oriented object detection in remote sensing images. *Remote Sens.* **2022**, *14*, 2090. [CrossRef]
43. Li, Q.; Chen, Y.; Zeng, Y. Transformer with transfer CNN for remote-sensing-image object detection. *Remote Sens.* **2022**, *14*, 984. [CrossRef]
44. Yang, X.; Yang, X.; Yang, J.; Ming, Q.; Wang, W.; Tian, Q.; Yan, J. Learning high-precision bounding box for rotated object detection via kullback-leibler divergence. *Adv. Neural Inf. Process. Syst.* **2021**, *34*, 18381–18394.
45. Zhu, X.; Lyu, S.; Wang, X.; Zhao, Q. TPH-YOLOv5: Improved YOLOv5 based on transformer prediction head for object detection on drone-captured scenarios. In Proceedings of the IEEE/CVF International Conference on Computer Vision, Montreal, BC, Canada, 11–17 October 2021; pp. 2778–2788.
46. Xu, C.; Wang, J.; Yang, W.; Yu, L. Dot distance for tiny object detection in aerial images. In Proceedings of the IEEE/CVF Conference on Computer Vision and Pattern Recognition 2021, Nashville, TN, USA, 20–25 June 2021; pp. 1192–1201.
47. Chalavadi, V.; Jeripothula, P.; Datla, R.; Ch, S.B.; Mohan, K.C. mSODANet: A network for multi-scale object detection in aerial images using hierarchical dilated convolutions. *Pattern Recognit.* **2022**, *126*, 108548. [CrossRef]

48. Shi, Z.; Wu, C.; Li, C.; You, Z.; Wang, Q.; Ma, C. Object detection techniques based on deep learning for aerial remote sensing images: A survey. *Chin. J. Image Graph.* **2023**, *28*, 2616–2643.
49. Liu, S.; Qi, L.; Qin, H.; Shi, J.; Jia, J. Path aggregation network for instance segmentation. In Proceedings of the IEEE Conference on Computer Vision and Pattern Recognition 2018, Salt Lake City, UT, USA, 18–23 June 2018; pp. 8759–8768.
50. Zhang, D.; Zhang, H.; Tang, J.; Wang, M.; Hua, X.; Sun, Q. Feature pyramid transformer. In Proceedings of the Computer Vision–ECCV 2020: 16th European Conference, Glasgow, UK, 23–28 August 2020, Proceedings, Part XXVIII 16; Springer: Berlin/Heidelberg, Germany, 2020; pp. 323–339.
51. Zheng, Z.; Wang, P.; Liu, W.; Li, J.; Ye, R.; Ren, D. Distance-IoU loss: Faster and better learning for bounding box regression. In Proceedings of the AAAI Conference on Artificial Intelligence 2020, New York, NY, USA; 7–12 February 2020; Volume 34, pp. 12993–13000.
52. Rezatofighi, H.; Tsoi, N.; Gwak, J.; Sadeghian, A.; Reid, I.; Savarese, S. Generalized intersection over union: A metric and a loss for bounding box regression. In Proceedings of the IEEE/CVF Conference on Computer Vision and Pattern Recognition 2019, Long Beach, CA, USA, 15–20 June 2019; pp. 658–666.
53. Wang, X.; Girshick, R.; Gupta, A.; He, K. Non-local neural networks. In Proceedings of the IEEE Conference on Computer Vision and Pattern Recognition 2018, Salt Lake City, UT, USA, 18–23 June 2018; pp. 7794–7803.
54. Khan, S.; Naseer, M.; Hayat, M.; Zamir, S.W.; Khan, F.S.; Shah, M. Transformers in vision: A survey. *ACM Comput. Surv. (CSUR)* **2022**, *54*, 1–41. [CrossRef]
55. Hou, L.; Lu, K.; Yang, X.; Li, Y.; Xue, J. G-rep: Gaussian representation for arbitrary-oriented object detection. *Remote Sens.* **2023**, *15*, 757. [CrossRef]
56. He, K.; Zhang, X.; Ren, S.; Sun, J. Deep residual learning for image recognition. In Proceedings of the IEEE Conference on Computer Vision and Pattern Recognition 2016, Las Vegas, NV, USA; 27–30 June 2016; pp. 770–778.
57. Deng, J.; Dong, W.; Socher, R.; Li, L.J.; Li, K.; Fei-Fei, L. Imagenet: A large-scale hierarchical image database. In Proceedings of the 2009 IEEE Conference on Computer Vision and Pattern Recognition 2009, Miami, FL, USA, 20–25 June 2009; pp. 248–255.
58. Lin, T.Y.; Goyal, P.; Girshick, R.; He, K.; Dollár, P. Focal loss for dense object detection. In Proceedings of the IEEE International Conference on Computer Vision 2017, Venice, Italy, 22–29 October 2017; pp. 2980–2988.
59. Hou, L.; Lu, K.; Xue, J.; Li, Y. Shape-adaptive selection and measurement for oriented object detection. In Proceedings of the AAAI Conference on Artificial Intelligence 2022, Virtual, 22 February–1 March 2022; Volume 36, pp. 923–932.
60. Everingham, M.; Van Gool, L.; Williams, C.K.; Winn, J.; Zisserman, A. The pascal visual object classes (voc) challenge. *Int. J. Comput. Vis.* **2010**, *88*, 303–338. [CrossRef]
61. Yang, X.; Hou, L.; Zhou, Y.; Wang, W.; Yan, J. Dense label encoding for boundary discontinuity free rotation detection. In Proceedings of the IEEE/CVF Conference on Computer Vision and Pattern Recognition, Nashville, TN, USA, 20–25 June 2021; pp. 15819–15829.
62. Yang, X.; Yan, J. Arbitrary-oriented object detection with circular smooth label. In Proceedings of the Computer Vision—ECCV 2020: 16th European Conference, Glasgow, UK, 23–28 August 2020, Proceedings, Part VIII 16; Springer: Berlin/Heidelberg, Germany, 2020, pp. 677–694.

remote sensing

Article

L1RR: Model Pruning Using Dynamic and Self-Adaptive Sparsity for Remote-Sensing Target Detection to Prevent Target Feature Loss

Qiong Ran [1], Mengwei Li [1], Boya Zhao [2], Zhipeng He [2,3] and Yuanfeng Wu [2,*]

1 College of Information Science and Technology, Beijing University of Chemical Technology, Beijing 100029, China; ranqiong@mail.buct.edu.cn (Q.R.); 2021200819@mail.buct.edu.cn (M.L.)
2 Key Laboratory of Computational Optical Imaging Technology, Aerospace Information Research Institute, Chinese Academy of Sciences, Beijing 100094, China; zhaoby@aircas.ac.cn (B.Z.); hezhipeng23@mails.ucas.ac.cn (Z.H.)
3 School of Electronic, Electrical and Communication Engineering, University of Chinese Academy of Sciences, Beijing 100049, China
* Correspondence: wuyf@aircas.ac.cn

Abstract: Limited resources for edge computing platforms in airborne and spaceborne imaging payloads prevent using complex image processing models. Model pruning can eliminate redundant parameters and reduce the computational load, enhancing processing efficiency on edge computing platforms. Current challenges in model pruning for remote-sensing object detection include the risk of losing target features, particularly during sparse training and pruning, and difficulties in maintaining channel correspondence for residual structures, often resulting in retaining redundant features that compromise the balance between model size and accuracy. To address these challenges, we propose the L1 reweighted regularization (L1RR) pruning method. Leveraging dynamic and self-adaptive sparse modules, we optimize L1 sparsity regularization, preserving the model's target feature information using a feature attention loss mechanism to determine appropriate pruning ratios. Additionally, we propose a residual reconstruction procedure, which removes redundant feature channels from residual structures while maintaining the residual inference structure through output channel recombination and input channel recombination, achieving a balance between model size and accuracy. Validation on two remote-sensing datasets demonstrates significant reductions in parameters and floating point operations (FLOPs) of 77.54% and 65%, respectively, and a 48.5% increase in the inference speed on the Jetson TX2 platform. This framework optimally maintains target features and effectively distinguishes feature channel importance compared to other methods, significantly enhancing feature channel robustness for difficult targets and expanding pruning applicability to less difficult targets.

Keywords: remote sensing imagery; target detection; model pruning; sparse regularization

Citation: Ran, Q.; Li, M.; Zhao, B.; He, Z.; Wu, Y. L1RR: Model Pruning Using Dynamic and Self-Adaptive Sparsity for Remote-Sensing Target Detection to Prevent Target Feature Loss. *Remote Sens.* **2024**, *16*, 2026. https://doi.org/10.3390/rs16112026

Academic Editors: Yanni Dong and Gemine Vivone

Received: 3 April 2024
Revised: 30 May 2024
Accepted: 31 May 2024
Published: 5 June 2024

1. Introduction

Real-time target detection is a crucial task of intelligent remote-sensing satellite systems [1]. It requires imaging, data transmission, and real-time data processing capabilities on the imaging payload side. The requirement of high computing power and the resource-constrained conditions of the edge computing platform pose significant challenges for real-time processing on the imaging payload side due to the over-parameterization of remote-sensing image-processing models. Therefore, compressing the model is the most effective solution.

Model pruning is an essential model compression algorithm that removes redundant weights from deep neural networks. It maintains a high detection performance while reducing the number of parameters and computational complexity, improving the model's detection efficiency.

Few studies on pruning methods have focused on target detection models for remote-sensing images. The network slimming (NS) method [2], which uses a scaling factor in the batch normalization (BN) [3] layer to evaluate the importance of the feature channels, is a preferred method. NS focuses on the classification of natural scenes; thus, it may not be applicable for other target detection tasks. The target detection model must perform regression and classification; it contains many parameters. Therefore, large-scale pruning while maintaining model accuracy is often not possible.

Sparse regularization techniques are essential for striking a balance between the model's size and accuracy. Common methods for achieving sparsity include L0, L1, and L2 norms. The L0 norm represents the number of non-zero elements and can achieve a highly sparse model solution. However, due to the non-convexity and computational complexity of the L0 norm, it is challenging to apply directly. The L1 norm, which represents the sum of the absolute values of the elements, can induce sparsity by driving the model parameters toward zero with a fixed stride, resulting in a sparse solution with many zero parameters. The L2 norm, which represents the square root of the sum of the squared elements, influences sparsity in relation to the magnitude of the elements. The larger the element, the greater the stride, causing most parameters to approach zero without reaching zero exactly. This process, which helps in smoothing the model parameters, cannot directly produce sparse solutions.

The L1 norm is easier to solve compared to the L0 norm and also promotes sparsity more readily than the L2 norm. However, remote-sensing images differ significantly from natural-scene images, with most targets being small and background features being complex; the background information overwhelms the target information. As a result, achieving sparse optimal solutions with L1 regularization for remote-sensing images is challenging.

As shown in Figure 1, the feature activation map displays regions of high attention in dark red and regions of low attention in dark blue. The background image in the feature areas of interest is the input optical remote-sensing image, with the shaded areas representing regions of high attention in the feature activation map. At the beginning of sparse training, the feature areas of interest indicate that the region of interest for this feature channel is concentrated on the airplane target. After sparse training with L1 regularization, the feature areas of interest show that the region of interest for this feature channel shifts mainly to the background, with the attention on the airplane target decreasing. This indicates that directly applying L1 regularization to sparsity in remote-sensing image object-detection models tends to focus on a significant amount of background features, leading to the loss of target features and confusion between background and target feature channels.

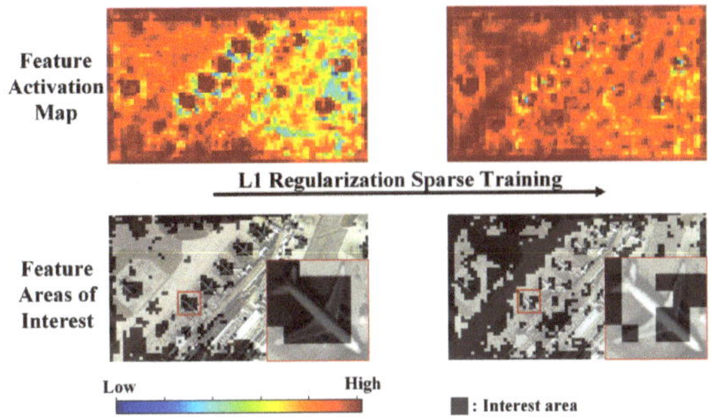

Figure 1. Comparison of changes in the feature activation map and the feature areas of interest map of the same feature channel during L1 regularization.

References [4–6] used fine-tuning and knowledge distillation to improve the feature extraction performance of the model and prevent feature loss. However, the model lost features during sparse training and pruning. Therefore, this paper uses the feature attention loss (L_{FA}) in the detection layer to represent the degree of loss of the target features during model pruning. It is defined as follows:

$$FA(D) = \frac{1}{k}\sum_{i=1}^{k} sum(F \times M) \quad (1)$$

$$L_{FA} = \frac{1}{c}\sum_{i=1}^{c}(1 - \frac{FA^{(i)}_{pruned}}{FA^{(i)}_{sparse}}) \quad (2)$$

where D represents the input dataset; k represents the number of images in the dataset; $sum(\cdot)$ represents the summation function of the tensor; F represents the feature activation value of the detection layer; M represents the target region mask; and c represents the number of model detection layers. Equation (1) calculates the feature attention of the detection layer on the target by computing the mean activation in the target region, reflecting the model's attention to target features. Equation (2) then calculates the change in feature attention between the pruned model and the sparse model at each detection layer, measuring the effectiveness of the target features in the pruned model.

As shown in Figure 2, the relationship between various pruning ratios and target feature information is illustrated. In the feature activation map, regions of high attention are depicted in dark red, while regions of low attention are depicted in dark blue. Each column represents the same feature channel under different feature attention losses (L_{FA}) for the two columns of feature activation maps. The percentage in red text represents the L_{FA} at that pruning rate. The zoomed-in areas in the feature activation map correspond to the zoomed-in airplane target in the input image. When the L_{FA} is only 0.36%, the two feature activation maps shown maintain a high level of attention on the airplane target. When the L_{FA} is 28.6%, the attention to the airplane target in the two feature activation maps is relatively reduced but still retains some target feature information. When the L_{FA} reaches 45.75%, the attention to the airplane target in the two feature activation maps is almost entirely lost.

Therefore, the feature attention loss can measure the loss of target features. When the feature attention loss is around 1%, it can be considered that the pruned model essentially retains the target feature information of the sparse model. Conversely, if the feature attention loss is higher, it indicates a loss of target features in the pruned model.

Residual learning can enhance the model's feature extraction ability [7], learning ability, and generalization performance. However, pruning affects the summation of feature channels in residual models. References [8–11] used the OR method in a residual model, and Reference [6] did not prune the model. Neither approach provides a balance between model size and accuracy.

To address the above issues, the L1RR was proposed to retain the feature information of remote sensing targets and achieve a balance between model size and accuracy. The implementation of L1RR has the following key steps. First, the scaling factors in the BN layer are the criterion for feature channel importance. No other parameters are required. The magnitude of these scaling factors reflects the significance of the respective feature channels and indicates the quantity of target feature information encapsulated within each channel. Second, the dynamic and self-adaptive sparsity method based on L1 is used for the remote-sensing target-detection model. This is achieved through the dynamic sparse module (DSM), which dynamically tunes the sparsity stride of each feature channel based on the amount of remote-sensing target information. Furthermore, the self-adaptive sparse module (SASM) adapts the overall sparsity stride of each layer according to variations in the number of channels containing remote-sensing target information. Third, pruning masks are constructed using scaling factors, with feature channel pruning for the remote-sensing target-detection model. The residual reconstruction procedure (RRP) is proposed to eliminate redundant features from the residual structure. It achieves forward propagation inference of pruned residual structures by reorganizing feature maps within the residual structure, utilizing both output channel recombination (OCR) and input channel recombination (ICR) methodologies. Finally, the proposal of feature attention loss as the cut-off condition in model pruning ensures the effectiveness of target features in the pruned model. This is verified by comparing the differences in target attention between the pruned model and the sparse model.

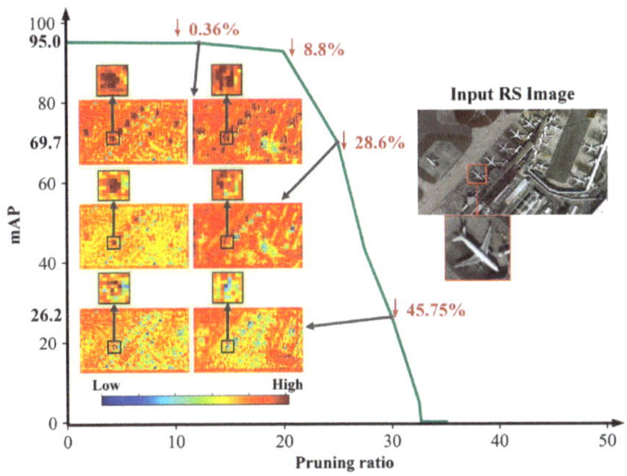

Figure 2. Feature activation maps of the pruned model in remote sensing scenarios based on L1 regularization. In the feature activation map, regions of high attention are depicted in dark red, while regions with low attention are in dark blue. The percentage in red text represents the feature attention loss (L_{FA}) at that pruning rate. The zoomed-in areas in the feature activation map correspond to the zoomed-in airplane target in the input image.

The proposed method uses the YOLOv5s model to compare the latest target detector and pruning methods on NWPU VHR-10 [12] and RSOD [13] datasets. The experimental results show that the proposed pruning framework outperforms the other methods in terms of the pruning ratio and detection accuracy.

The contributions of this paper are as follows:

- The proposed dynamic and self-adaptive sparsity method enhances model efficiency by controlling sparse ratios with DSM for the channel aspect and SASM for the layer aspect, respectively. In this way, the effective target features are given lower pruning factors and keep the training procedure away from over-sparsity.
- A new model pruning strategy is proposed, using feature attention loss as the criterion to find the balance point between the pruning ratio and accuracy. Other than enumerating the pruned ratio to find a trade-off model, the L_{FA} supplies a new direct metric for measuring target attention differences among original and pruned models, thereby ensuring the pruned model's performance and reliability.
- A residual reconstruction procedure method is proposed, which only alters forward propagation with effective logical rules and pruning factors, resulting in compact and practical residual features.

The rest of this paper is organized as follows. Section 2 describes the related work on deep learning-based pruning of target detection models for remote sensing applications. The proposed method for target detection model pruning is described in Section 3. Section 4 describes the experiment and results. Section 5 describes the ablation study. Section 6 concludes the paper.

2. Related Work

2.1. *Remote-Sensing Target Detection*

One-stage target-detection algorithms perform target location and classification simultaneously. Two-stage target-detection algorithms first detect the candidate region where the target may exist and then classify the candidate region and target position [14].

The former algorithms have better real-time performance and are more suitable for deployment on edge computing platforms. You only look once (YOLO) [15] was the first one-stage target detection model, achieving end-to-end optimization. YOLOv2 [16] improved the original model by introducing anchor and batch normalization. YOLOv3 [17] was an improved model through a more efficient backbone network and spatial pyramid pooling. YOLOv4 [18] and YOLOv5 [19] inherit the advantages of YOLOv3 and have an optimized model structure, improving the detection performance. YOLOX [20] is an anchor-free one-stage target-detection model that improves upon YOLOv3 by adding optimization algorithms such as data augmentation, decoupled head, and Similarity Overlap Threshold Assigner (SimOTA), providing a lightweight model suitable for multi-scale analysis. PP-YOLOE [21] is an anchor-free one-stage target detector with a more efficient dynamic matching strategy and task-aligned head. A unified model structure is used to enable multi-scale analysis. YOLOv7 [22] is an anchor-based model optimized for real-time target detection. The model has a coarse-to-fine label assignment strategy. It includes a new model reparameterization module and uses a concatenation-based scaling method to improve efficiency and enable multi-scale target detection. YOLOv8 [23] is a lightweight model for multi-scale target detection based on an anchor-free architecture. The model replaces the C3 module in the YOLOv5 model with a more lightweight C2f module and uses a decoupled head and task-aligned assigner for sample matching.

Most target-detection networks for remote-sensing images have improved detection networks to achieve better performance. R-YOLO [24] contains an MS transformer module based on YOLOv5 and a convolutional block attention module (CBAM) to enhance the network's feature extraction capability. The priori frame design of YOLOv5 was improved to make it suitable for detecting targets in rotated images. SPH-YOLOv5 [25] uses Swin-transformer prediction heads (SPHs) and normalization-based attention modules (NAMs) to improve the detection accuracy for small targets in satellite images. ECAP-YOLO [26] has a high detection accuracy for small targets and low computational complexity because it reduces model depth by leveraging shallow features for detecting small targets. CSD-YOLO [27] combines spatial pyramid pooling and a shuffle attention mechanism with a feature pyramid network (SAS-FPN) based on YOLOv7, improving the loss function and the model's performance for detecting ships in synthetic aperture radar (SAR) images.

Although existing one-stage target-detection models have a high accuracy for remote-sensing target detection, many parameters and FLOPs limit the inference time in edge computing platforms. The coarse-grained lightweight implementations do not balance the model's size and accuracy.

2.2. *Pruning of Remote-Sensing Target-Detection Models*

Unstructured and structured pruning methods exist. Unstructured pruning, also known as fine-grained pruning, refers to the pruning of unstructured modules such as weights. This strategy significantly affects model inference speed while maintaining accuracy. Structured pruning, also known as coarse-grained pruning, refers to the pruning of structured modules such as feature channels. This method retains some of the redundant weight parameters and significantly improves model inference and hardware acceleration.

Several methods have been used recently for target detection model pruning, including NS [2], soft filter pruning (SFP) [28], and performance-aware global channel pruning (PAGCP) [29]. NS utilizes the scaling factor of the BN layer to perform L1 sparse training and differentiate the importance of the feature channels. SFP uses a criterion and a layer pruning ratio to perform soft pruning and model training to differentiate the importance of the feature channels. PAGCP uses sequentially greedy channel pruning based on a performance-aware oracle criterion. It can prune multi-task models without sparse regularization.

The NS method has advantages in flexibility and simplicity of implementation because it uses feature-channel-level pruning. Additionally, the method does not require a sparse regularization of model weights, making it widely applicable for pruning target-detection models.

Most remote-sensing target-detection models use NS to increase the inference speed and reduce model complexity to facilitate the deployment on hardware platforms. Light-YOLOv4 [11] prunes unimportant channels and layers using NS pruning, reducing the network size, improving accuracy, and increasing inference speed. Ghost-YOLOS [30] prunes the YOLOv3 model using NS pruning to facilitate field-programmable gate array (FPGA) hardware acceleration. The model utilizes knowledge distillation to compensate for accuracy degradation due to pruning and is deployed on the FPGA after quantization. Lite-YOLOv5 [31] reduces the number of parameters and computational complexity of the YOLOv5 model. It prunes the YOLOv5 model using NS and uses fine-tuning to compensate for accuracy degradation caused by pruning.

NS pruning is widely used in remote-sensing target detection to reduce the number of model parameters. However, this method does not consider the feature loss that may occur during sparse training and pruning. This paper aims to balance model size and accuracy in target detection. Therefore, we optimize the L1 sparse regularization of the feature channel and layer dimension to increase the ability to distinguish meaningful channels. Additionally, we use L_{FA} to determine the effective pruning ratio.

3. Methodology

Figure 3 shows the L1RR framework. It consists of three modules: DSM, SASM, and RRP. It is implemented as follows:

Firstly, the DSM is used to dynamically adjust the model sparsity in the feature channel dimension through sparse strides, allocating larger sparse strides to background feature channels and smaller sparse strides to target feature channels. By dynamically adjusting sparse strides in feature channels, it enhances the distinguishability of feature channels and minimizes the likelihood of target feature loss. The SASM is employed to adjust the model sparsity in the layer dimension adaptively through balance factors. It decreases the sparsity rate for layers with higher sparsity levels and increases the sparsity rate for layers with lower sparsity levels. This adaptive adjustment of layer sparsity rates prevents over-sparsity of layers, which could lead to the loss of target features. Sparse training of the DSM and SASM are used. The target-detection model assigns a high scaling factor to the target feature channel and a low scaling factor to the background feature channel. The sparse model is obtained after N epochs of training, resulting in the ability to distinguish target features from background features.

Secondly, feature channels with high scaling factors (black rectangles) are retained while those with low scaling factors (white rectangles) are pruned by constructing a pruning mask for the sparse model to extract the scaling factors.

Thirdly, the sparse model is pruned using the pruning mask. Convolutional pruning is applied to prune the feature channels, except for the residual structure. RRP is used to prune the feature channels for the residual structure. It utilizes OCR and ICR to remove redundant channels through model pruning. Convolutional pruning and RRP prune the sparse model, resulting in the final pruned target-detection model. Finally, feature attention loss is used to test the effectiveness of the target features in the pruned model.

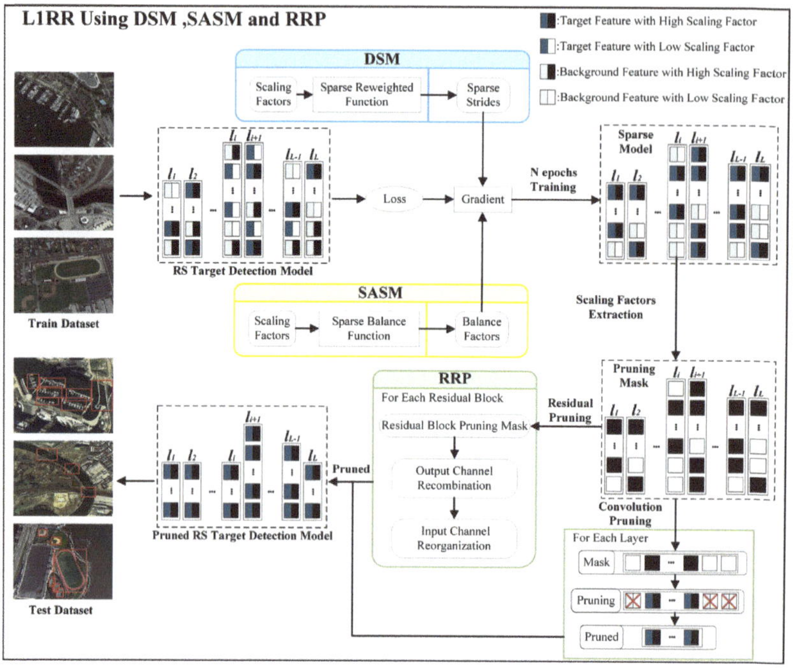

Figure 3. Block Diagram of L1RR.

3.1. Dynamic Sparse Module and Self-Adaptive Sparse Module for Sparse Training

3.1.1. Dynamic Sparse Module

Due to small targets and a large proportion of background information in remote-sensing images, $L1$ sparse regularization with a fixed stride can result in target feature loss. Thus, reweighted $L1$ regularization [32] is utilized to obtain dynamically adjusted weights in addition to $L1$ sparse regularization, ensuring that each element has a sparsity ratio. The sparse reweighted regularization is formulated as follows:

$$Reweighted_{L1}(x) = \alpha^t |x|_{L1} \tag{3}$$

The sparse reweighted function is formulated as follows:

$$\alpha^t = \frac{1}{|x| + \epsilon}, t \in T = [t, 2t, 3t, \ldots nt], n \leq 10 \tag{4}$$

where α is a positive weight; α^t denotes the α of the t-th update, initialized with $\alpha^{(0)} = 0$; and $|\cdot|_{L1}$ denotes $L1$ regularization.

The sparse training is divided into n phases, each consisting of t epochs. At the start of training, α is initialized to 0, which is equivalent to normal model training for t epochs. α is updated at the beginning of the 2nd phase to perform sparse training. Subsequently, a new round of updates occurs at the beginning of the 3rd phase, and so on, until the end of training.

3.1.2. Self-Adaptive Sparse Module

Since remote-sensing images are dominated by small targets, shallow rather than deep feature maps are commonly used because they contain more information on small targets. However, the traditional method for achieving sparsity is to apply a uniform sparsity level to the model. This approach can result in excessive sparsity of shallow features, causing feature loss and a low model accuracy.

Therefore, we propose an adaptive sparse module based on the exponential function. This module adjusts the sparsity degree layer-by-layer by considering the difference in the layer sparsity degree. The objective is to prevent feature loss caused by over-sparsity. Additionally, experiments have shown that a scaling factor of less than 0.01 has a negligible impact on the model's detection performance. The model's sparsity degree is the ratio of the number of scaling factors of less than 0.01 to the total number of scaling factors.The sparsity of the current layer is the ratio of the number of filters with a scaling factor of less than 0.01 to the total number of filters. The sparse balance function is formulated as follows:

$$\lambda_t^{(i)} = 2^{a(\rho - \frac{N_p^{(i)}}{N^{(i)}})}, i = 1, \ldots, L \tag{5}$$

where $\lambda_t^{(i)}$ is an exponential function with a base of two, denoting the sparse balance factor of the i-th layer; t denotes the balance factor of the t-th update; initialization occurs with $\lambda_0 = 1$; and t does not exceed 10; a is a hyperparameter to adjust the intensity of the change in the balance factor; ρ denotes the current sparsity of the model; N_p and N denote the number of pruned filters and the total number of filters in the current layer, respectively.

Since filters are more important in shallow than deep networks, the feature extraction ability of shallow networks is affected by the sparsity constraint. Thus, we propose a λ-decay strategy. It is used when the sparsity of a layer remains unchanged for a long time during sparse training:

$$\lambda_t^{(i)} = 2^{(a(\rho - \frac{N_p^{(i)}}{N^{(i)}} - s)}, i = 1, \ldots, L \tag{6}$$

where s denotes the number of times the layer is very sparse.

3.1.3. Dynamic and Self-Adaptive Sparsity

Figure 4 illustrates the implementation of the DSM and SASM to optimize model sparsity. For instance, in layer i, the sparsity optimization for the target detection model is as follows:

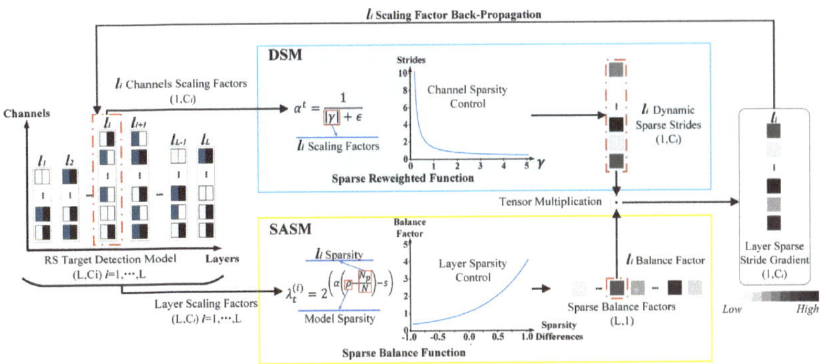

Figure 4. The single-layer sparse stride gradient construction. The color shades of the squares represent the weights. Darker colors indicate larger weights, while lighter colors indicate smaller weights.

In the first step, the scaling factor of the layer i's channel dimension is input into the sparse reweighted function in the DSM. Dynamic sparse strides are generated from the feature channel dimension. DSM assigns larger sparse strides to background feature channels and smaller sparse strides to target feature channels, increasing the distinguishability between feature channels and preventing the loss of target features. The scaling factor in the layers is input into the sparse balance function in the SASM, generating the balance factor from the layer dimension. SASM reduces the sparse stride for layers with sparsity higher than the model's overall sparsity and for layers where sparsity is saturated, while increasing the sparse stride for layers with sparsity lower than the model's overall sparsity, preventing over-sparsity from causing the loss of target features.

In the second step, the layer's sparse stride gradient (LSSG) of layer i is obtained by tensor-multiplying the dynamic strides and the balance factor of layer i. The LSSG and the training gradient are used to adjust the scaling factor of layer i through back-propagation. This approach enables cooperative sparsity control of feature channels and layers, increasing the distinguishability of feature channels and decreasing the likelihood of losing target features.

Sparse training is performed using DSM and SASM with the LSSG to achieve model sparsity. The sparse gradient is defined as follows:

$$\nabla_{sparse} = \nabla_{\gamma^{(i)}} + \theta \lambda_t^{(i)} \nabla_{Reweighted_{L1}(\gamma^{(i)})}, i = 1, \ldots, L, t \in T = [t, 2t, 3t, \ldots, nt] \quad (7)$$

where ∇ denotes the gradient; $\gamma^{(i)}$ denotes the i-th layer's scaling factors; $Reweighted_{L1}(\cdot)$ denotes the DSM; $\lambda_t^{(i)}$ denotes the SASM of the i-th layer; and θ denotes the sparsity ratio of the model.

Algorithm 1 describes the sparse training

Algorithm 1 Sparse Training Process

input: $update\ epoch\ T, sparse\ ratio\ \theta, current\ epoch\ e$
output: $Sparse\ model$
for $each\ layer\ i$ **do**
    $initialize\ the\ \alpha_i^{(0)} = 0,$
    $initialize\ the\ \lambda_{(0)}^i = 1,$
end for
repeat
    if $e\ is\ in\ T$ **then**
        $Sort\ all\ scaling\ factors\ to\ get\ model\ sparsity\ \rho$
        for $each\ layer\ i$ **do**
            $Update\ \alpha_i^{(t)}\ via\ Equation\ (4),$
            $Update\ \lambda_{(t)}^i\ via\ Equation\ (5),$
            if $the\ sparsity\ is\ high$ **then**
                $Update\ \lambda_{(t)}^i\ via\ Equation\ (6),$
            end if
        end for
    end if
    $Use\ Equation\ (7)\ to\ perform\ sparse\ training,$
until down
$return\ the\ sparse\ model,$

3.2. Model Convolution Pruning and Residual Reconstruction Procedure

The resulting sparse model undergoes feature channel pruning to obtain a pruned model. The sparse model is defined as follows:

$$[W^{(i)}]_{sparse} = [W^{(i)} \in R^{c^{(i)} \times r^{(i)} \times k_1^{(i)} \times k_2^{(i)}}]i = 1, \ldots, L \quad (8)$$

The pruned model is expressed as follows:

$$[W_p^{(i)}]_{pruned} = [W_p^{(i)} \in R^{c_p^{(i)} \times r_p^{(i)} \times k_1^{(i)} \times k_2^{(i)}}]i = 1, \ldots, L, c_p^{(i)} = c^{(i)} \times M^{(i)}, r_p^{(i)} = r^{(i)} \times M^{(i-1)} \quad (9)$$

where $[W^{(i)}]_{sparse}$ and $[W_p^{(i)}]_{pruned}$ denote the sparse and pruned models, respectively; $W^{(i)}$ and $W_p^{(i)}$ denote the weight tensor and the pruned weight tensor of the i-th layer, respectively; $c^{(i)}$ denotes the number of output channels for the i-th layer; $r^{(i)}$ denotes the number of input channels for the i-th layer; $c_p^{(i)}$ denotes the number of output channels after pruning for the i-th layer; $r_p^{(i)}$ denotes the number of input channels after pruning for the i-th layer; $k_1^{(i)} \times k_2^{(i)}$ denotes the size of the two-dimensional spatial kernel for the i-th layer; $M^{(i)}$ denotes the pruning mask for the i-th layer. $M^{(i-1)}$ does not exist when the i-th layer is the first layer; thus, only the output channels are pruned in the first layer.

There is a correspondence between the output channels of one layer and the input channels of the next layer. Therefore, the pruning process for the model's feature channels can be described as pruning the model in both the output channel dimension and the input channel dimension. The pruned output channels can be represented as the tensor multiplication of the sparse model's weights with the pruning mask from the output channel dimension ($c_p^{(i)} = c^{(i)} \times M^{(i)}$), while the pruned input channels can be represented as the tensor multiplication of the sparse model's weights with the pruning mask from the previous layer's input channel dimension ($r_p^{(i)} = r^{(i)} \times M^{(i-1)}$).

During model pruning, threshold screening is conducted to obtain the scaling factor of the BN layer in each network layer. The threshold value of the pruning model is evaluated using L_{FA}. If the L_{FA} is around 1%, the pruning is considered effective. However, if it exceeds 1%, the feature information is significantly affected, and the pruning is invalid.

Furthermore, the residual structure of the target detection model can be difficult to prune. The original residual structure is depicted in Figure 5. The outputs of $Conv1$ and $Conv2$ must be summed in the same dimension and require the same pruning mask. However, $Conv1$ and $Conv2$ typically have different pruning masks.

Figure 5 shows the proposed RRP. It consists of OCR and ICR. $Conv1$, $Conv2$, and $Conv3$ represent different convolutional layers, with $Conv1$ and $Conv2$ forming a residual structure. M denotes the mask, where black squares indicate a mask value of 1, white squares indicate a mask value of 0, and the numbers within the squares represent channel indices. The main steps of RRP are as follows:

First, OCR uses the pruning mask M_{Conv1} for $Conv1$ and the pruning mask M_{Conv2} for $Conv2$ to obtain M_{And} using an intersection operation. M_{And} is used to prune the residual structure while preserving the original additive channels I_{Conv1} and I_{Conv2}. M_{Sub1} is obtained by subtracting $Conv1$ from $Conv2$. It is used to compute the channels in $Conv2$ that cannot be added, resulting in S_{Conv1}. M_{Sub2} is derived by subtracting $Conv2$ from $Conv1$. It is used to compute the channels in $Conv2$ that cannot be added, resulting in S_{Conv2}. The resulting M_{And} and M_{Sub1} are used to prune $Conv1$, producing the pruned channels I_{Conv1} and S_{Conv1} for $Conv1$. M_{And} and M_{Sub2} are used to prune $Conv2$, producing the pruned channels I_{Conv2} and S_{Conv2} for $Conv2$. I_{Conv1} and I_{Conv2} are summed, and S_{Conv1} and S_{Conv2} are concatenated. The results of the summing and concatenation are the input for $Conv3$.

ICR utilizes three pruning masks (M_{And}, M_{Sub1}, and M_{Sub2}) obtained from OCR to determine the output channel order of the residual structure's output feature maps. This channel order is then used as the pruning mask for the input channels of the next layer's $Conv3$, resulting in a new $Conv3$.

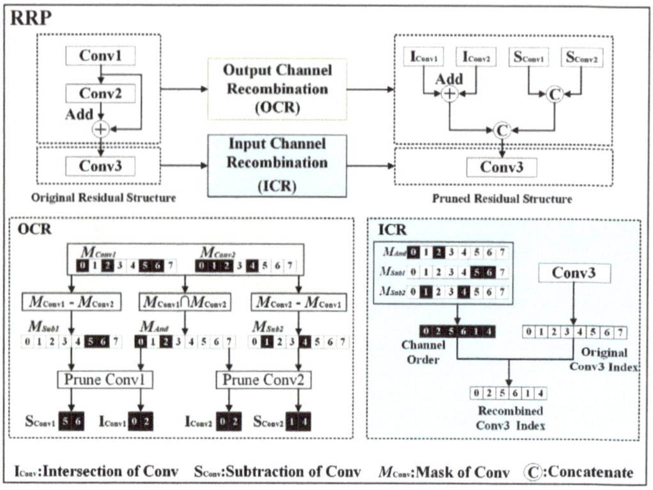

Figure 5. Schematic diagram of residual reconstruction procedure.

The RRP does not consider convolutional layers with the same pruning mask; thus, the pruned model has the same accuracy but fewer parameters. The output of the $Conv1$ is expressed as follows:

$$(W^{(Conv1)})^{2D}_{out} \in R^{r^{(Conv1)} \times c^{(Conv1)} \times k_1^{(Conv1)} \times k_2^{(Conv1)}} \tag{10}$$

The output of the $Conv2$ is expressed as follows:

$$(W^{(Conv2)})^{2D}_{out} \in R^{r^{(Conv2)} \times c^{(Conv2)} \times k_1^{(Conv2)} \times k_2^{(Conv2)}} \tag{11}$$

The output of the residual structure is expressed as follows:

$$(W^{residual})^{2D}_{out} = (W^{(Conv1)})^{2D}_{out} + (W^{(Conv2)})^{2D}_{out} \tag{12}$$

The correspondence between the output feature maps is no longer one-to-one due to differences in pruning masks. We use the pruning mask to address this. We obtain the pruning mask M_{Conv1} for the $Conv1$'s feature map $(W^{(Conv1)})^{2D}_{out}$ and the pruning mask M_{Conv2} for the $Conv2$'s feature map $(W^{(Conv2)})^{2D}_{out}$ of the residual module. Intersection M_{And} and subtraction (M_{Sub1} and M_{Sub2}) operations are performed. The output feature map of the residuals is expressed as follows:

$$(W_p^{residual})^{2D}_{out} = \left\{[(W_p^{(Conv1)})^{2D}_{out} \times M_{And}] + [(W_p^{(Conv2)})^{2D}_{out} \times M_{And}], (W_p^{(Conv1)})^{2D}_{out} \times M_{Sub1}, (W_p^{(Conv2)})^{2D}_{out} \times M_{Sub2}\right\} \tag{13}$$

After the output feature maps of the residuals have been combined, it is necessary to reorder the inputs of the kernels/filters in the next layer using the channel order. The pruning structure of the next layer's kernels/filters of the residual structure is expressed as follows:

$$K_j^{(Conv3)} \in R^{r^{(Conv3)} \times \Phi \times k_1^{(Conv3)} \times k_2^{(Conv3)}}, j = 1, \ldots, c^{(Conv3)} \tag{14}$$

where Φ denotes the mask of the channel order after reconstruction.

The pruning process is defined in Algorithm 2.

Algorithm 2 Pruning Process

input: *sparse model* $[W^{(i)}]_{sparse}$, *threhold* δ,
output: *pruned model*
for *each layer i* **do**
    $mask\ M_j^{(i)} = \begin{cases} 0, \gamma_j^{(i)} \leq \delta, \\ 1, \gamma_j^{(i)} > \delta, \end{cases}$
end for
use M initialize the pruned model $[W_p^{(i)}]_{pruned}$,
for *layer i in the sparse model* **do**
    Use Equation (9) *to obtain the parameters of* $W_p^{(i)}$;
    if *layer i is residual* **then**
        Use Equation (13) *to construct the output channels*
    else if *layer* $i-1$ *is residual* **then**
        Use Equation (14) *to adjust the input channels*
    end if
end for
Use Equations (1) *and* (2) *to compute* L_{FA},
if L_{FA} *is not around* 1% **then**
    Adjust the threshold to repruning.
end if
return the pruned model

4. Experiment

Experiments were conducted using the proposed pruning framework to verify its effectiveness. Two remote-sensing datasets and the YOLOv5s model were used. This section describes the two remote-sensing datasets, the NWPU VHR-10 and the RSOD datasets. The implementation details, including the training strategy and evaluation metrics, are presented. The accuracy, detection speed, and model size of the proposed and other lightweight networks are compared. The accuracy, parameter pruning ratio, and FLOP pruning ratio of the proposed and other pruning methods are contrasted.

4.1. Datasets

The NWPU VHR-10 dataset [12] is a geospatial target-detection dataset containing ten categories with image sizes of 600 × 600 to 1100 × 1100. The target classes include airplane (PL), ship (SH), storage tank (ST), baseball diamond (BD), tennis court (TC), basketball court (BC), ground track field (GT), harbor (HA), bridge (BR), and vehicle (VH). The dataset has 650 labeled images. We used 80% of the images for training and 20% for testing.

The RSOD dataset [13] is a geospatial target-detection dataset containing four categories: aircraft, oiltank, overpass, and playground. The dataset has 936 images. Similarly, an 80%/20% training/testing ratio was used.

4.2. Implementation Details

4.2.1. Training Strategy

Nearly identical training strategies were used for the two datasets. The anchor-based object detector YOLOv5s model with version 6.1 was utilized. Table 1 shows the anchor settings of the YOLOv5s model, customized to align with the sizes and distributions of targets in the remote-sensing datasets using the k-means clustering method. A stochastic gradient descent (SGD) was employed to optimize the loss function during training. The training epoch was 300, with a sparsity rate of 0.001, a learning rate of 0.01, a warm-up learning rate of 0.001, and 1000 iterations. The dynamic adaptive updating epoch was 40. The non-maximum suppression intersection over union (IoU) threshold during testing was 0.6, and the confidence threshold was 0.001.

Table 1. The anchor settings of the YOLOv5s model.

Datasets	Anchor Setting
NWPU VHR-10	[23,23, 31,31, 33,48], [48,35, 52,53, 92,42], [72,66, 199,97, 165,196]
RSOD	[14,12, 21,19, 30,26], [37,35, 49,43, 62,58], [78,77, 180,223, 284,291]

4.2.2. Evaluation Metrics

Several metrics were used to assess the target detection model's performance: mean average precision (mAP), number of parameters (*Params*), number of FLOPs, frames per second (*FPS*), Parameter Prune Ratio (PR_{Param}), and FLOP Prune Ratio (PR_{FLOP}).

The mAP is based on recall and precision. The true positive (*TP*), false positive (*FP*), and false negative (*FN*) are calculated. A *TP* occurs when the overlap ratio of the predicted box and ground truth exceeds a given threshold. Conversely, an *FP* occurs when the overlap ratio is lower. An *FN* occurs if there is no match between the prediction and ground truth. The equations for recall and precision are as follows:

$$precision = \frac{TP}{TP + FP} \tag{15}$$

$$recall = \frac{TP}{TP + FN} \tag{16}$$

The area below the precision–recall (P-R) curve is the average precision (*AP*), and the mean of the *APs* of all categories is the mAP:

$$AP = \int_0^1 P(r)dr \tag{17}$$

$$mAP = \frac{1}{m}\sum_1^m AP_i \tag{18}$$

The equations for PR_{Param} and PR_{FLOP} are as follows:

$$PR_{Param/FLOP} = \frac{\#PrunedParams/FLOPs}{\#OriginalParams/FLOPs} \tag{19}$$

The mAP is a measure of the detection accuracy of the target-detection model. *Params*, *FLOPs*, and *FPS* are used to measure the complexity and detection speed of the target-detection model. PR_{Param} and PR_{FLOP} are used to measure the pruning performance.

4.3. Results

4.3.1. Target-Detection Model Results

We pruned the YOLOv5s model and compared it with other models and the proposed model on two remote-sensing datasets.

Table 2 depicts the results of the proposed model and the latest lightweight target-detection model on the NWPU VHR-10 dataset. The comparison focuses on the target category accuracy, mAP, model size, and FPS. Optimal values in each category are indicated in bold.

Table 2. Detection results on the NWPU VHR-10 dataset.

Method	PL	SH	ST	BD	TC	BC	GT	HA	BR	VH	mAP	Model Size	FPS	
													Gtx1660	TX2
Baseline	99.5	93.9	99.4	98.1	94.9	95.7	99.5	89.7	89.6	90.8	95.1	28.1 MB	86.2	15.31
YOLOv7-tiny [22]	99.5	88.2	99.4	97.7	94.6	93.4	99.6	88.9	88.3	72.5	92.2	23.8 MB	54.3	17.99
SSD-VGG16 [33]	90.7	77.3	79.9	94.4	82.6	77.7	90.1	84.3	61.8	62.8	80.2	99.8 MB	14.8	3.26
YOLOv5-lite-c [34]	99.5	90.4	98.5	98.3	92.9	**96.8**	99.5	91.2	91.5	75.7	93.4	17.3 MB	74.1	19.37
Nanodetplus-m [35]	**100**	90.7	99.6	**99.3**	93.9	96.4	99.5	93.3	95.9	75.3	94.4	33.4 MB	-	-
YOLOX-s [20]	99.9	90.4	**99.9**	99.0	97.4	90.9	**100**	89.2	89.4	86.2	94.2	35.2 MB	72.9	9.96
YOLOv3 [17]	90.9	90.9	90.8	99.1	90.9	90.9	99.5	90.1	90.9	90.4	92.4	246 MB	25.8	4.52
LAR-YOLOv8 [36]	99.3	90.7	89.0	96.6	87.8	85.8	99.4	86.1	84.8	88.6	90.8	-	-	-
Faster R-CNN [37]	99.4	**95.0**	67.7	93.9	91.7	88.6	98.6	82.3	68.7	81.8	86.8	330.7 MB	9.1	1.37
CANet [38]	100	86.0	99.3	97.3	97.8	84.8	98.4	90.4	89.2	90.3	93.3	-	-	-
CFIBSS [39]	100	86.7	92.5	96.3	91.4	95.8	100	**97.1**	94.9	**93.2**	94.8	-	-	-
YOLOv8-n [23]	99.4	88.0	97.5	98.7	**98.3**	93.8	99.4	88.7	92.8	86.4	94.3	12.0 MB	76.9	13.95
Ours	99.5	88.7	99.4	97.8	95.1	93.4	99.3	93.9	**97.5**	85.7	**95.0**	**6.5 MB**	**105.5**	**22.73**

The pruned model achieved an mAP of 95.0%, significantly higher than the other models. Compared to object-detection models designed for natural scenes, it achieved a 2.6% and a 8.2% higher mAP than that of the large target detection models YOLOv3 [17] and Faster R-CNN [37], respectively. Additionally, the model size was reduced by 97.3% and 98%. The pruned model achieved 2.8% and 0.7% higher mAPs than the latest lightweight models YOLOv7-tiny [22] and YOLOv8n [23], respectively. Additionally, the model size was reduced by 64.5% and 24.8% for single-precision floating-point storage. It also exhibited a significantly higher FPS than both. Compared to the remote-sensing object-detection models, the pruned model achieved 4.2% and 1.7% higher mAPs than the latest optical remote-sensing models LAR-YOLOv8 [36] and CANet [38], respectively. Furthermore, the pruned model was only 23% the size of the baseline model. On NVIDIA Jetson TX2, the pruned model had a 48.5% higher model inference speed than the baseline model.

Although the overall detection accuracy of the pruned model was 0.1% that of the baseline model, the accuracy for ship and vehicle targets was 5.2% and 5.1% lower, respectively. However, the accuracy of detecting harbor and bridge targets was 4.2% and 7.9% higher, respectively, while that of the remaining targets remained unchanged. The model's detection accuracy fluctuated due to the removal of redundant features in the baseline model during sparse modeling. The pruned model had the same target-detection accuracy as the baseline model for small targets and a significantly higher accuracy for large targets, such as bridges and harbors.

The proposed pruning framework does not require pre-training or fine-tuning; thus, the regularization constraints are implemented during sparse training. This can cause differences in the detection accuracy of certain targets; however, the detection accuracy was comparable to that of the baseline model.

Figure 6 compares the detection results of the baseline (a,c,e,g,i,k) and pruned (b,d,f,h,j,l) models on the NWPU VHR-10 dataset. In Figure 6a,b, the baseline model missed the bridge target, whereas the pruned model correctly identified it. In Figure 6c,d, the baseline model missed the harbor target, while the pruned model correctly detected it. Figure 6e,f show that the baseline model had a false alarm for the harbor target, whereas the pruned model correctly detected all targets without false alarms. Similarly, the baseline model had a missed detection for the baseball diamond target, while the pruned model correctly detected it (Figure 6i,j). Figure 6k,l indicate that the baseline model incorrectly identified a ship (false alarm), whereas the pruned model correctly identified this area as the background.

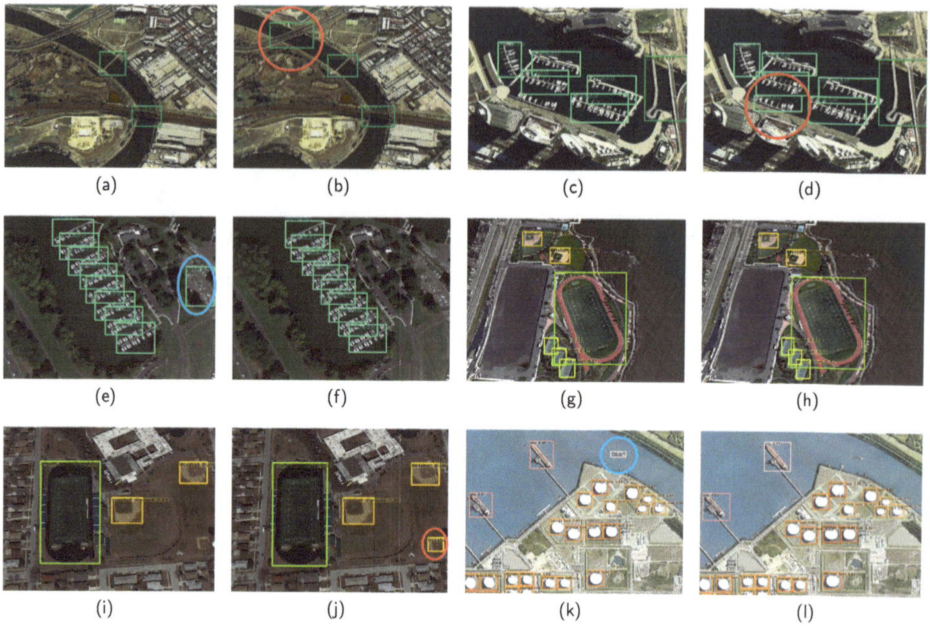

Figure 6. The detection results for the baseline (**a**,**c**,**e**,**g**,**i**,**k**) and pruned (**b**,**d**,**f**,**h**,**j**,**l**) models on the NWPU VHR-10 dataset. The red circles represent targets detected by the pruned model but not by the baseline model, and the blue circles represent false alarms for the baseline model.

The pruned and baseline models provided similar detection results in various complex scenarios. However, the pruned model had a higher detection accuracy for bridges ((a) and (b)) and harbors ((c) and (d), (e) and (f)) in some detection scenarios.

Figure 7 shows the heatmaps derived from Grad-CAM++ [40] for the baseline (a,c,e,g,i,k) and pruned (b,d,f,h,j,l) models on the NWPU VHR-10 dataset.

The baseline model focused on the target and background features, resulting in false alarms and missed detection. Figure 7a,b show that the baseline model detected bridge targets by focusing on the features of the target area. However, due to low attention to the target features, it classified the target as the background in some cases and missed some targets. In contrast, the pruned model detected all three bridge targets correctly. As shown in Figure 7c,d, the baseline model focused more on the background, resulting in missed detection. In contrast, the pruned model focused on the harbor and bridge targets and was not influenced by the background information, resulting in the correct detection of all targets. Figure 7e,f show that the baseline model's attention was primarily focused on the background, leading to false alarms. In contrast, the pruned model focused more on the harbor target and was not influenced by the background information, correctly detecting all targets. Figure 7i,j show that the baseline model's attention was on the target and its edges, resulting in missed detection in complex backgrounds. In contrast, the pruned model focused primarily on the target and was unaffected by the background features of the target edges. In the remaining detection scenarios (Figure 7g,h,k,l), the baseline model focused on the target and the background, resulting in correct target detection, although a significant amount of irrelevant background information was considered. In contrast, the pruned model focused solely on the target.

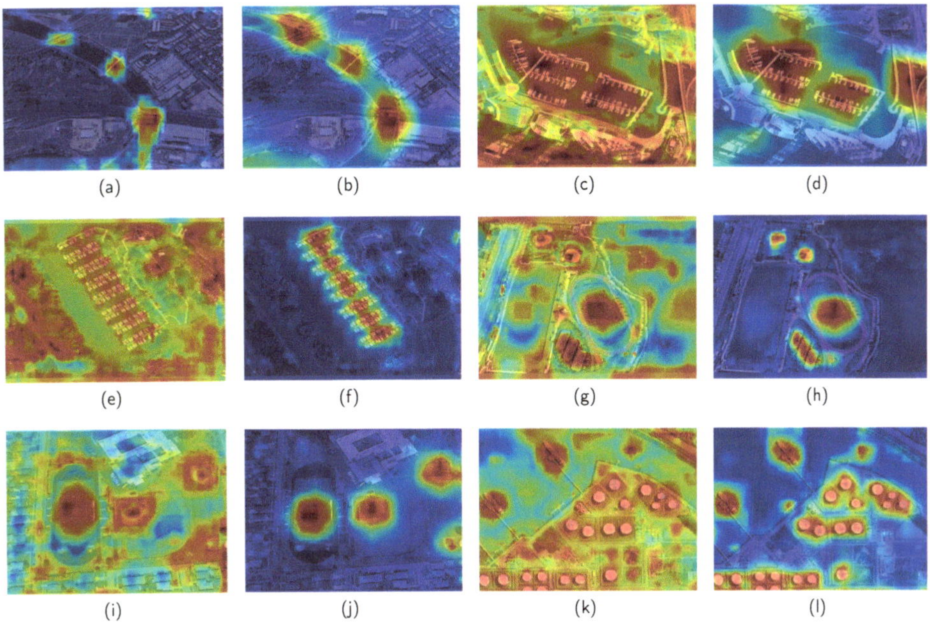

Figure 7. The heatmap derived from Grad-CAM++ for the baseline (**a**,**c**,**e**,**g**,**i**,**k**) model and pruned (**b**,**d**,**f**,**h**,**j**,**l**) model on the NWPU VHR-10 dataset.

Table 3 compares the accuracy, mAP, model size, and FPS of the latest lightweight target-detection models on the RSOD dataset. The optimal values in each category are indicated in bold.

The pruned model achieved a 97.3% mAP with a model size of only 6.6 MB. Compared to object-detection models designed for natural scenes, the mAP was 20.1% and 4.4% higher than that of the large target detection models YOLOv3 [17] and Faster R-CNN [37], respectively. Additionally, the model size was 97.3% and 98% smaller for single-precision floating-point storage. The mAP was also 2.3% higher than that of the lightweight model YOLOv7-tiny [22]. Compared to optical remote-sensing object-detection models, the pruned model achieved 3.0% and 2.1% higher mAP than the latest optical remote-sensing models LAR-YOLOv8 [36] and CANet [38], respectively. The model size was only 23.5% of the baseline model. The inference speed on the NVIDIA Jetson TX2 was 38.2% higher than that of the baseline model.

The pruning model's overall detection accuracy was 0.5% lower than that of the baseline model. However, it had the same accuracy as the baseline model for different complex scenes. The target accuracy was 0.9% lower for the aircraft class, 0.4% lower for the oiltank class, and 0.5% lower for the overpass class compared to the baseline model. The target accuracy of the playground class was the same. Redundant features were eliminated from the baseline model while preserving its ability to detect small targets like airplanes and large targets like playgrounds.

Table 3. Detection results on the RSOD dataset.

Method	Aircraft	Oiltank	Overpass	Playground	mAP	Model Size	FPS	
							Gtx1660	TX2
Baseline	96.1	98.8	96.7	99.5	97.8	28.1 MB	83.8	15.08
YOLOv3 [17]	74.3	73.8	75.1	85.2	77.1	246 MB	23.2	3.74
OYOLO [41]	**96.6**	**98.8**	77.1	98.8	92.8	162 MB	-	-
EDCNet [42]	96.1	94.1	45.9	94.1	82.5	12.2 MB	-	-
YOLOv5-lite-c [34]	93.4	96.8	91.4	99.3	95.2	17.3 MB	61.0	16.75
YOLOv7-tiny [22]	95.9	97.8	86.9	99.0	94.9	12 MB	43.3	15.17
Faster R-CNN [37]	70.8	90.2	78.7	98.1	84.5	330.5 MB	8.7	1.12
LAR-YOLOv8 [36]	94.8	98.2	85.4	98.9	94.3	-	-	-
CANet [38]	91.8	97.1	94.1	97.9	95.2	-	-	-
Nanodetplus-m [35]	91.0	95.8	92.6	**100**	94.8	33.4 MB	-	-
Ours	95.2	98.4	**96.2**	99.5	**97.3**	**6.6 MB**	**97.1**	**20.83**

Figure 8 shows the detection results for the baseline (a,c,e,g) and pruned (b,d,f,h) models on the RSOD dataset.

In Figure 8, comparing Figure 8a,b, The baseline model missed two small aircraft targets, whereas the pruned model correctly detected them. In Figure 8c,d, the baseline model missed three oiltank targets, while the pruned model correctly detected two. In the remaining detection scenarios (Figure 8e–h), the pruned and baseline models had the same detection results for large targets, such as overpasses and playgrounds.

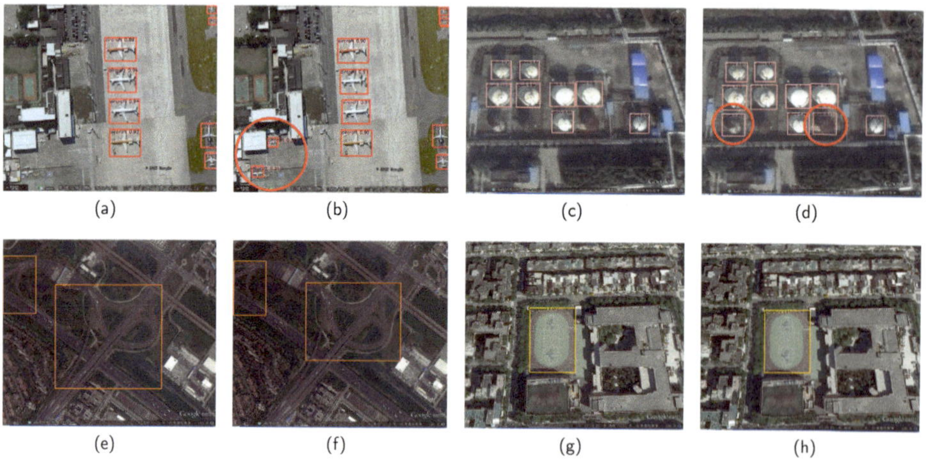

Figure 8. Comparison of detection results for the baseline (**a**,**c**,**e**,**g**) and pruned (**b**,**d**,**f**,**h**) models in complex scenes of the RSOD dataset. The red circles indicate targets detected by the pruned model but not by the baseline model.

Figure 9 compares the feature heatmaps derived from Grad-CAM++ [40] for the baseline (a,c,e,g) and pruned (b,d,f,h) models on the RSOD dataset.

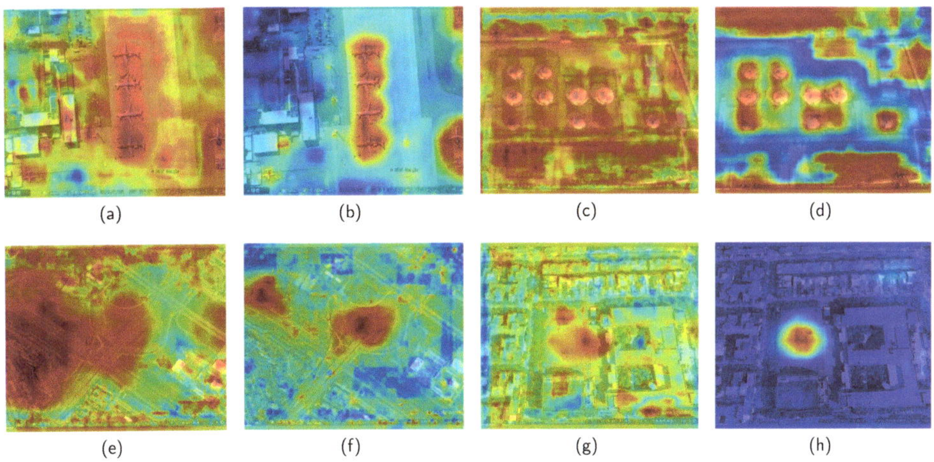

Figure 9. The heatmap derived from Grad-CAM++ for the baseline (**a**,**c**,**e**,**g**) and pruned (**b**,**d**,**f**,**h**) models on the RSOD dataset.

The baseline model focused on target and background features in all detection scenarios. In Figure 9g,h, the baseline model had a large focus area when detecting the playground target due to the complexity of the background information. In contrast, the pruned model focused solely on the target. In the remaining detection scenarios (Figure 9a–f), the baseline model's attention covered almost the entire image. Although it detected the target, its focus was also on the background features, resulting in missed detection. In contrast, the pruned model's attention was primarily on the target; it detected all targets correctly.

4.3.2. Model Pruning Results

The comparison of the proposed pruning framework and other pruning methods and the performances of the YOLOv5s model and the NS [2], SFP [28], and PAGCP [29] methods on two remote-sensing datasets are listed in Table 4. The optimal values in each category are indicated in bold.

On the NWPU VHR-10 dataset, a comparison of the proposed high pruning ratio (>80%) model (Ours-HP) and the NS method showed that our model was 0.8% more accurate than the NS method, with a 5.83% higher parameter pruning ratio and 3.75% higher FLOP pruning ratio. To compare our model with the SFP and PAGCP methods, we used the highest-accuracy pruning model (Ours-BA) due to the low pruning ratio of SFP with PAGCP. Our model had a 3.1% higher accuracy than the SFP method, with a 25.4% higher parameter pruning ratio and a 20% higher FLOP pruning ratio. Our pruning method had a 2.58-times higher parameter pruning ratio and a 2.71 times higher FLOP pruning ratio than the PAGCP method, with almost the same accuracy.

On the RSOD dataset, our model had a 3.4% higher accuracy, a 4.84% higher parameter pruning ratio, and an 11.25% higher FLOP pruning ratio than the NS method. We compared Ours-BA with the SFP and PAGCP methods. Our model achieved 7.2% higher accuracy, a 26.25% higher parameter pruning ratio, and a 20% higher FLOP pruning ratio than the SFP method. The parameter pruning ratio was 35.83% higher, the FLOP pruning ratio was 27.5% higher, and the accuracy was 1.6% higher than for the PAGCP method. The experimental results demonstrate that our pruning method outperformed other pruning methods for the target-detection models.

Table 4. Comparison of L1RR and other pruning methods on remote-sensing datasets.

Dataset	Method	Training Settings		mAP	PR	
		Pre-Defind?	Fine-Tuning?		Params	FLOPs
NWPU VHR-10	Baseline	N/A	N/A	95.1	0	0
	NS [2]	✗	✓	91.2	81.87%	72.5%
	Ours-HP	✗	✗	92.0	**87.7%**	**76.25%**
	SFP [28]	✓	✗	92.5	37.60%	32.5%
	Ours-BA	✗	✗	**95.6**	63%	52.5%
	PAGCP [29]	✗	✓	95.5	24.45%	19.4%
	Ours-BA	✗	✗	95.6	63%	52.5%
	Ours	✗	✗	95.0	77.54%	65%
RSOD	Baseline	N/A	N/A	97.8	0	0
	NS [2]	✗	✓	93.9	72.30%	48.75%
	Ours	✗	✗	97.3	77.16%	60%
	SFP [28]	✓	✗	90.7	34.12%	30%
	Ours-BA	✗	✗	97.9	60.37%	50%
	PAGCP [29]	✗	✓	96.3	24.54%	22.5%
	Ours-BA	✗	✗	**97.9**	60.37%	50%
	Ours	✗	✗	97.3	**77.16%**	**60%**

Figure 10 shows the mAP of different pruning methods for different pruning ratios. The proposed pruning framework had the highest detection accuracy and pruning ratio on the NWPU VHR-10 and RSOD datasets. The mAP of several algorithms decreased significantly and fell below 90% after the pruning ratio reached 80%. The SFP method exhibited a significant decrease in the mAP at a pruning ratio of 40%. The NS method maintained a higher accuracy, but the accuracy was below 90% when the pruning ratio exceeded 70%. In contrast, the proposed pruning framework maintained 90% and 95%, respectively.

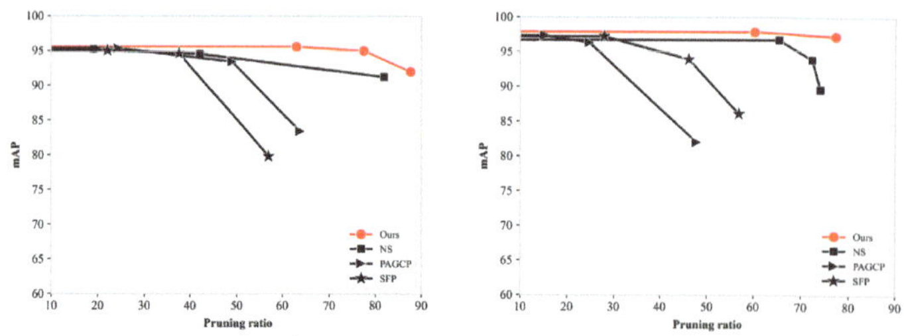

(a) (b)

Figure 10. Mean average precision of different pruning methods for different pruning ratios: (**a**) NWPU VHR-10 dataset, and (**b**) RSOD dataset.

Figure 11 shows the mAP for the targets for different pruning ratios and pruning methods on the NWPU VHR-10 dataset. The mAP of several targets (ship, storage tank, and others) for the SFP method declined significantly after the pruning ratio reached 50%. The PAGCP method exhibited a significant accuracy degradation (lower than 60%) for basketball court, harbor, and other targets. Although the NS method showed a similar detection accuracy for the ship and tennis court classes compared to our method, its detection accuracy for the harbor, bridge, and vehicle classes was significantly lower than ours. Therefore, the proposed pruning framework optimally maintains target features for more targets across a wider pruning ratio range and accurately distinguishes the importance of feature channels.

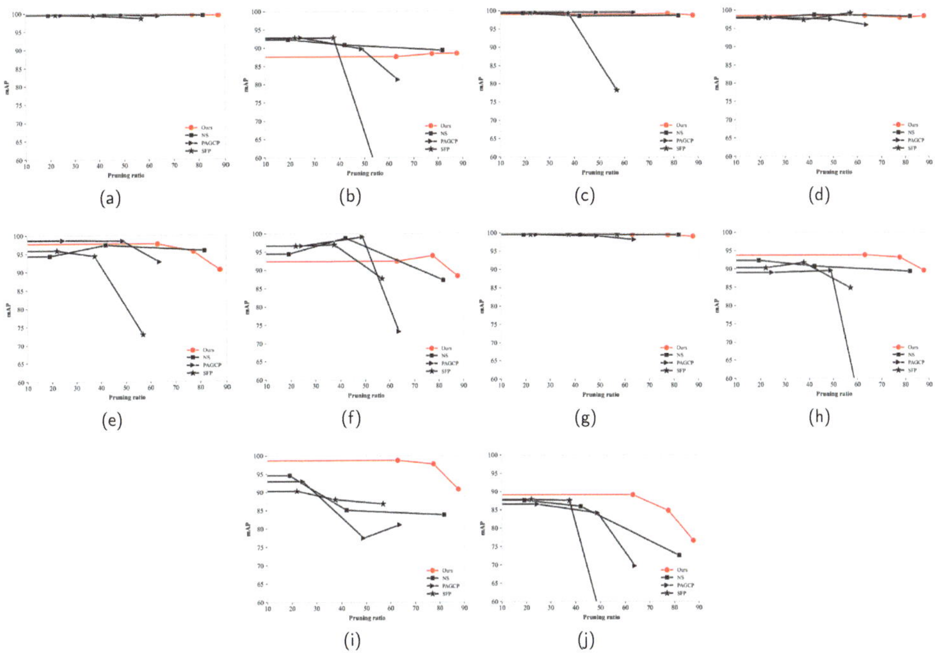

Figure 11. Mean average precision for various targets for different pruning methods and pruning ratios on the NWPU VHR10 dataset: (**a**) airplane; (**b**) ship; (**c**) storge tank; (**d**) baseball diamond; (**e**) tennis court; (**f**) basketball court; (**g**) ground track field; (**h**) harbor; (**i**) bridge and (**j**) vehicle.

Figure 12 displays the mAP for various targets for different pruning ratios and methods on the RSOD dataset. The SFP and PAGCP methods showed significant accuracy degradation for the aircraft and overpass targets (below 60%) after the pruning ratio reached 50%. The NS method exhibited substantial accuracy degradation for all targets when the pruning ratio exceeded 70%. In contrast, the proposed pruning framework only showed a slight decrease in accuracy for the overpass target, while the accuracy remained above 95% for a wide pruning ratio range.

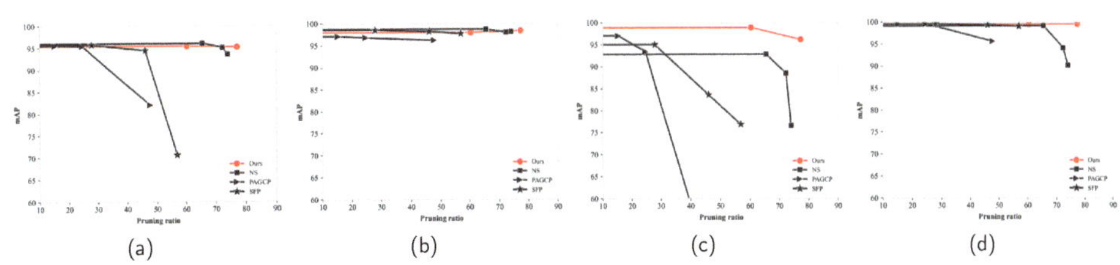

Figure 12. Mean average precision for various targets for different pruning ratios and methods on the RSOD dataset: (**a**) aircraft; (**b**) oiltank; (**c**) overpass and (**d**) playground.

5. Ablation Study

The proposed pruning approach comprises three modules: DSM, SASM, and RRP. We conducted ablation experiments to compare the influence of different modules on the accuracy and model parameters. The following analysis was conducted.

- Sparse training of the model using only the DSM.
- Sparse training of the model using the DSM and RRP.
- Sparse training of the model using the DSM and SASM.
- The DSM and SASM are used to simplify the model and to implement RPP.

The results in Table 5 indicate that the Yolov5s model had significantly fewer parameters and the highest accuracy. The optimal values in each category are indicated in bold. Specifically, the number of parameters was reduced by 55.23% with virtually no impact on the mAP when only the DSM was used on the NWPU VHR-10 dataset. The combination of the DSM and SASM resulted in a 72.02% reduction in the number of parameters without sacrificing accuracy. The use of the DSM, SASM, and RRP resulted in a 77.54% reduction in the number of parameters while maintaining accuracy. Similar results were observed on the RSOD dataset. Using the DSM reduced the parameter count by 57.29%, with a minor decrease in the mAP. The combination of the DSM and SASM reduced the parameter count by 75.29%. Using all three modules reduced the parameters by 76.97% without compromising accuracy.

Table 5. Results of ablation study.

Dataset		RRP	DSM	SASM	mAP	Params	PR_{Param}	Model Size	L_{FA}
NWPU VHR-10	Baseline	N/A	N/A	N/A	95.1	7046599	0%	28.1 MB	-
			✓		95.0	3154552	55.23%	12.9 MB	0.71%
		✓	✓		95.0	2364345	66.45%	9.7 MB	1.04%
			✓	✓	95.0	1971722	72.02%	8.1 MB	0.93%
		✓	✓	✓	95.0	**1582936**	**77.54%**	**6.5 MB**	1.14%
RSOD	Baseline	N/A	N/A	N/A	97.8	7030417	0%	28.1 MB	-
			✓		96.8	3002861	57.29%	12.0 MB	0.46%
		✓	✓		97.1	2677029	61.92%	10.7 MB	0.88%
			✓	✓	97.1	1731560	75.29%	7.0 MB	0.74%
		✓	✓	✓	97.3	**1619017**	**76.97%**	**6.6 MB**	0.6%

The accuracy of the pruned models with different pruning ratios trained on the NWPU VHR-10 dataset and the RSOD dataset was compared. We assessed four pruning ratios, labeled as "Prune-pruning ratio": "Prune-40" refers to the model where the feature channels were pruned by 40%. "Baseline" denotes the base training result for the YOLOv5s model. "Sparse" denotes the sparse model at the end of sparse training, with an optimum pruning ratio of 66.5% on the NWPU VHR-10 dataset and 73.2% on the RSOD dataset.

Table 6 presents the experimental results of the six models on the NWPU VHR-10 dataset. The sparse model differed slightly from the baseline model, resulting in 0.001 lower mAP@0.5. When the pruning ratio was lower than the optimal one, Prune-20 and Prune-40 exhibited no difference in mAP compared to the sparse model. Additionally, they had fewer parameters and FLOPs, and the L_{FA} values were low (0.25% and 0.24%, respectively). Prune-66.5 had the same mAP as the sparse model and significantly fewer parameters and FLOPs when the pruning ratio was optimal. The L_{FA} value was only 1.14%, and no noticeable difference in the feature extraction ability was observed compared to the sparse model. The mAP@0.5 of Prune-70 was 0.017 lower than that of the sparse model when the pruning ratio exceeded the optimal values. Although the number of parameters and FLOPs were significantly lower, the L_{FA} value was 6.79%, affecting the model's feature extraction ability.

Table 7 compares the performances of the six models on the RSOD dataset. The mAP@0.5 of the sparse model was 0.005 lower than that of the baseline model. Prune-20, Prune-60, and the sparse model had the same mAP when the pruning ratio was lower than the optimal one. Additionally, they had fewer parameters and FLOPs. The L_{FA} values were only 0.5% and 0.57%, respectively. Prune-73.2 had the same mAP as the sparse model when the pruning ratio was optimal. Additionally, the number of parameters and FLOPs were significantly lower, and the L_{FA} value was only 0.6%. The feature extraction abilities of the Prune-73.2 and the sparse model were similar. Prune-77 had a 0.479 lower mAP@0.5 than the sparse model when the pruning ratio exceeded the optimum value. It had fewer parameters and FLOPs. However, the high L_{FA} value (19.4%) indicated a low feature extraction ability.

Table 6. YOLOv5s pruning results for sparse models with different pruning ratios on the NWPU VHR-10 dataset.

	Precision	**Recall**	**mAP@0.5**	**mAP@0.5:0.95**	**Params**	**GFLOPs**	L_{FA}
Baseline	96.6	90.2	95.1	62.6	7046599	16.0	-
Sparse	93.2	90.2	95.0	59.7	7046599	16.0	0%
Prune-20	93.2	90.1	95.0	59.6	4586338	10.5	0.25%
Prune-40	93.1	90.1	95.0	59.6	2985918	7.6	0.24%
Prune-65.5	94.1	89.3	95.0	59.8	1582936	5.6	1.14%
Prune-70	94.7	85.8	93.3	58.7	1380728	5.1	6.79%

Table 7. YOLOv5s pruning results for sparse models with different pruning ratios on the RSOD dataset.

	Precision	**Recall**	**mAP@0.5**	**mAP@0.5:0.95**	**Params**	**GFLOPs**	L_{FA}
Baseline	96.6	90.2	97.8	67.2	7030417	16.0	-
Sparse	87.0	96.3	97.3	63.1	7030417	16.0	0%
Prune-20	87.7	95.3	97.3	63.0	4571508	11.4	0.5%
Prune-40	87.9	95.0	97.3	63.0	2051935	7.1	0.57%
Prune-73.2	94.1	89.3	97.3	63.0	1619017	6.4	0.6%
Prune-77	88.5	41.2	49.4	27.6	1501089	6.0	19.4%

To validate the effectiveness of L1RR in microwave remote-sensing images, the YOLOv5 model was pruned for synthetic aperture radar (SAR) image ship detection using the high-resolution SAR images dataset (HRSID) [43]. The primary comparisons were made in terms of mAP, parameters, model size, and GFLOPs.

Table 8 shows the model pruning results for microwave remote-sensing images on the HRSID dataset. The optimal values in each category are indicated in bold. The pruned model achieved a 92.4% mAP with a model size of only 7.2 MB, resulting in a parameter pruning rate and FLOP pruning rate of 77.8% and 61.25%, respectively. Compared to object-detection models designed for natural scenes, it outperformed the large target detection model Faster R-CNN [37] by 8.4% in mAP, while reducing the model size by 98% when stored in single-precision floating-point format. Additionally, it surpassed the lightweight model YOLOX-tiny [22] by 3.7% in mAP. When compared to models designed for microwave remote-sensing scenes, the pruned model achieved a higher mAP than SAR-Net [44] and ZeroSARNas [45] by 7.7% and 1.1%, respectively. The parameters were reduced by 96.3% and 10.8%, respectively. Furthermore, compared to pruning methods NS [2] and SFP [28], the pruned model achieved a higher mAP than YOLOv5s + NS and YOLOv5s + SFP by 2.5% and 1.1%, respectively, while achieving a higher parameter-pruning rate of 13.57% and 42.9% and a higher FLOP-pruning rate of 16.25% and 30%, respectively.The experimental results demonstrate that our proposed method achieves competitive detection performance even in microwave remote-sensing scenarios.

Table 8. Detection results on the HRSID dataset.

Method	mAP@0.5	Params	Model Size	GFLOPs
Baseline	93.2	7.02 M	28 MB	16.0
YOLOX-tiny [22]	88.7	5.0 M	19.5 MB	23.7
Faster R-CNN [37]	74.0	41.5 M	330 MB	91.4
SAR-Net [44]	84.7	42.6 M	-	104.2
ZeroSARNas [45]	91.3	1.75 M	-	**2.67**
YOLOv5s + NS	89.9	2.26 M	11 MB	8.4
YOLOv5s + SFP	91.3	4.57 M	18.2 MB	11.0
Ours	**92.4**	**1.56 M**	**7.2 MB**	6.2

Figure 13 shows the detection results on the HRSID dataset. In the comparison of the target detection model results, the pruned model (Figure 13b,d) demonstrates detection performance that is essentially consistent with the baseline model (Figure 13e,g). However, in the comparison of heatmaps, attentions of the baseline model (Figure 13e) primarily focus on background areas, whereas attentions of the L1RR model (Figure 13f) mainly concentrate on target areas. In the baseline (Figure 13g), attention is dispersed across the entire image. Although it correctly detects the target, its attention relies on the target's edge features. In contrast, the L1RR model (Figure 13h) focuses on the target. This demonstrates that L1RR can effectively distinguish between target and background features in the SAR images, thereby preserving the integrity of the target feature information.

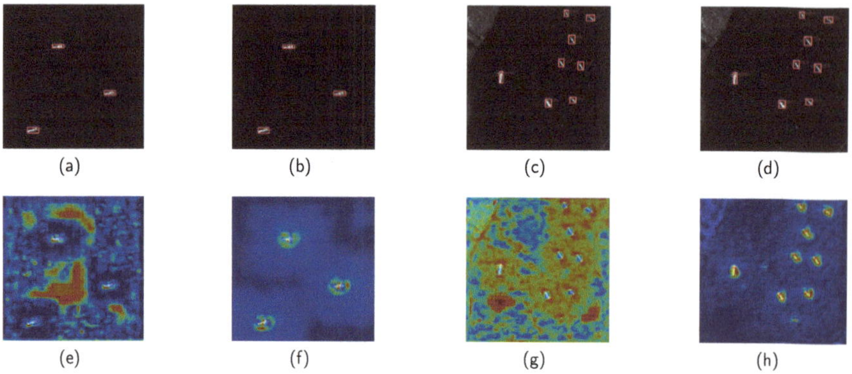

Figure 13. Comparison of detection results and heatmaps for the baseline (**a**,**c**,**e**,**g**) and pruned (**b**,**d**,**f**,**h**) models on the HRSID dataset.

6. Conclusions

This paper proposes a pruning method to ensure a balance in the size and accuracy of remote-sensing target-detection models. The L1RR framework prevents target feature loss during sparse training and pruning. An evaluation of the framework on the NWPU VHR-10 and RSOD datasets demonstrates that the proposed method results in a smaller model size, higher accuracy, and faster detection speed compared to existing lightweight target-detection methods. Moreover, it optimally maintains target features and effectively distinguishes feature channel importance compared to conventional pruning techniques.

Importantly, it achieves substantially higher performance than alternate pruning methods for remote-sensing targets with higher detection difficulty. Future studies will investigate the combination of model pruning and model quantization to enhance the efficiency of remote-sensing target detection.

Author Contributions: Conceptualization, Q.R., Y.W. and B.Z.; methodology, M.L. and B.Z.; software, Z.H.; validation, Q.R., Y.W. and B.Z.; formal analysis, M.L. and B.Z.; investigation, Y.W.; resources, Q.R.; data curation, Q.R. and M.L.; writing—original draft preparation, Q.R., Y.W., B.Z. and M.L.; writing—review and editing, Q.R., Y.W. and B.Z.; visualization, Q.R.; supervision, Q.R. and Y.W.; project administration, Y.W.; funding acquisition, Y.W. and B.Z. All authors have read and agreed to the published version of the manuscript.

Funding: This work was supported by the National Key R&D Program of China under Grant 2021YFA0715203.

Data Availability Statement: All data used in this paper are sourced from publicly available data.

Acknowledgments: The author would like to thank the Northwestern Polytechnical University for providing the NWPU VHR-10 data set, as well as Wuhan University for providing the RSOD data set.

Conflicts of Interest: The authors declare no conflicts of interest.

References

1. Zhang, B.; Wu, Y.; Zhao, B.; Chanussot, J.; Hong, D.; Yao, J.; Gao, L. Progress and challenges in intelligent remote sensing satellite systems. *IEEE J. Sel. Top. Appl. Earth Obs. Remote Sens.* **2022**, *15*, 1814–1822. [CrossRef]
2. Liu, Z.; Li, J.; Shen, Z.; Huang, G.; Yan, S.; Zhang, C. Learning efficient convolutional networks through network slimming. In Proceedings of the IEEE Conference on Computer Vision and Pattern Recognition (CVPR), Venice, Italy, 22–29 October 2017; pp. 2736–2744.
3. Ioffe, S.; Szegedy, C. Batch normalization: Accelerating deep network training by reducing internal covariate shift. In Proceedings of the International Conference on Machine Learning (ICML), Lille, France, 6–11 July 2015; pp. 448–456.
4. Wang, J.; Cui, Z.; Zang, Z.; Meng, X.; Cao, Z. Absorption Pruning of Deep Neural Network for Object Detection in Remote Sensing Imagery. *Remote Sens.* **2022**, *14*, 6245. [CrossRef]
5. Fu, Y.; Zhou, Y.; Yuan, X.; Wei, L.; Bing, H.; Zhang, Y. Efficient Esophageal Lesion Detection using Polarization Regularized Network Slimming. In Proceedings of the 2022 IEEE 8th International Conference on Cloud Computing and Intelligent Systems (CCIS), Chengdu, China, 26–28 November 2022; pp. 46–52.
6. Xu, Y.; Bai, Y. Compressed YOLOv5 for Oriented Object Detection with Integrated Network Slimming and Knowledge Distillation. In Proceedings of the 2022 3rd International Conference on Information Science, Parallel and Distributed Systems (ISPDS), Guangzhou, China, 22–24 July 2022; pp. 394–403.
7. He, K.; Zhang, X.; Ren, S.; Sun, J. Deep residual learning for image recognition. In Proceedings of the IEEE Conference on Computer Vision and Pattern Recognition (CVPR), Las Vegas, NV, USA, 27–30 June 2016; pp. 770–778.
8. Liu, R.; Cao, J.; Li, P.; Sun, W.; Zhang, Y.; Wang, Y. NFP: A No Fine-tuning Pruning Approach for Convolutional Neural Network Compression. In Proceedings of the 2020 3rd International Conference on Artificial Intelligence and Big Data (ICAIBD), Chengdu, China, 28–31 May 2020; pp. 74–77.
9. Choi, K.; Wi, S.M.; Jung, H.G.; Suhr, J.K. Simplification of Deep Neural Network-Based Object Detector for Real-Time Edge Computing. *Sensors* **2023**, *23*, 3777. [CrossRef] [PubMed]
10. Zhang, P.; Zhong, Y.; Li, X. SlimYOLOv3: Narrower, faster and better for real-time UAV applications. In Proceedings of the IEEE International Conference on Computer Vision Workshops (ICCV), Seoul, Republic of Korea, 27–28 October 2019; pp. 37–45.
11. Ma, X.; Ji, K.; Xiong, B.; Zhang, L.; Feng, S.; Kuang, G. Light-YOLOv4: An Edge-Device Oriented Target Detection Method for Remote Sensing Images. *IEEE J. Sel. Top. Appl. Earth Obs. Remote Sens.* **2021**, *14*, 10808–10820. [CrossRef]
12. Cheng, G.; Han, J.; Zhou, P.; Guo, L. Multi-class geospatial object detection and geographic image classification based on collection of part detectors. *Isprs J. Photogramm. Remote Sens.* **2014**, *98*, 119–132. [CrossRef]
13. Long, Y.; Gong, Y.; Xiao, Z.; Liu, Q. Accurate object localization in remote sensing images based on convolutional neural networks. *IEEE Trans. Geosci. Remote Sens.* **2017**, *55*, 2486–2498. [CrossRef]
14. Zou, Z.; Shi, Z.; Guo, Y.; Ye, J. Object detection in 20 years: A survey. *Proc. IEEE* **2023**, *111*, 257–276. [CrossRef]
15. Redmon, J.; Divvala, S.; Girshick, R.; Farhadi, A. You only look once: Unified, real-time object detection. In Proceedings of the IEEE Conference on Computer Vision and Pattern Recognition (CVPR), Las Vegas, NV, USA, 27–30 June 2016; pp. 779–788.
16. Redmon, J.; Farhadi, A. YOLO9000: Better, faster, stronger. In Proceedings of the IEEE Conference on Computer Vision and Pattern Recognition (CVPR), Honolulu, HI, USA, 21–26 July 2017; pp. 7263–7271.
17. Redmon, J.; Ali, F. Yolov3: An Incremental Improvement. *arXiv* **2018**, arXiv:1804.02767.

18. Bochkovskiy, A.; Wang, C.Y.; Liao, H.Y.M. YOLOv4: Optimal Speed and Accuracy of Object Detection. *arXiv* **2020**, arXiv:2004.10934.
19. Glenn, J.; Ayush, C.; Jing, Q. YOLO by Ultralytics. 2021. Available online: https://github.com/ultralytics/yolov5 (accessed on 10 April 2023).
20. Ge, Z.; Liu, S.; Wang, F.; Li, Z.; Sun, J. Yolox: Exceeding yolo series in 2021. *arXiv* **2021**, arXiv:2107.08430.
21. Xu, S.; Wang, X.; Lv, W.; Chang, Q.; Cui, C.; Deng, K.; Wang, G.; Dang, Q.; Wei, S.; Du, Y.; et al. PP-YOLOE: An evolved version of YOLO. *arXiv* **2022**, arXiv:2203.16250.
22. Wang, C.Y.; Bochkovskiy, A.; Liao, H.Y.M. YOLOv7: Trainable bag-of-freebies sets new state-of-the-art for real-time object detectors. In Proceedings of the IEEE Conference on Computer Vision and Pattern Recognition (CVPR), Vancouver, BC, Canada, 17–24 June 2023; pp. 7464–7475.
23. Glenn, J.; Ayush, C.; Jing, Q. YOLO by Ultralytics. 2023. Available online: https://github.com/ultralytics/ultralytics (accessed on 15 July 2023).
24. Hou, Y.; Shi, G.; Zhao, Y.; Wang, F.; Jiang, X.; Zhuang, R.; Mei, Y.; Ma, X. R-YOLO: A YOLO-Based Method for Arbitrary-Oriented Target Detection in High-Resolution Remote Sensing Images. *Sensors* **2022**, *22*, 5716. [CrossRef] [PubMed]
25. Gong, H.; Mu, T.; Li, Q.; Dai, H.; Li, C.; He, Z.; Wang, W.; Han, F.; Tuniyazi, A.; Li, H.; et al. Swin-Transformer-Enabled YOLOv5 with Attention Mechanism for Small Object Detection on Satellite Images. *Remote Sens.* **2022**, *14*, 2861. [CrossRef]
26. Kim, M.; Jeong, J.; Kim, S. ECAP-YOLO: Efficient Channel Attention Pyramid YOLO for Small Object Detection in Aerial Image. *Remote Sens.* **2021**, *13*, 4851. [CrossRef]
27. Chen, Z.; Liu, C.; Filaretov, V.F.; Yukhimets, D.A. Multi-scale ship detection algorithm based on YOLOv7 for complex scene SAR images. *Remote Sens.* **2023**, *15*, 2071. [CrossRef]
28. He, Y.; Kang, G.; Dong, X.; Fu, Y.; Yang, Y. Soft filter pruning for accelerating deep convolutional neural networks. *Sensors* **2018**, *18*, 1–15.
29. Ye, H.; Zhang, B.; Chen, T.; Fan, J.; Wang, B. Performance-aware approximation of global channel pruning for multitask cnns. *IEEE Trans. Pattern Anal. Mach. Intell.* **2023**, *45*, 10267–10284. [CrossRef] [PubMed]
30. Yang, R.; Chen, Z.; Wang, B.A.; Guo, Y.; Hu, L. A Lightweight Detection Method for Remote Sensing Images and Its Energy-Efficient Accelerator on Edge Devices. *Sensors* **2023**, *23*, 6497. [CrossRef]
31. Xu, X.; Zhang, X.; Zhang, T. Lite-YOLOv5: A Lightweight Deep Learning Detector for On-Board Ship Detection in Large-Scene Sentinel-1 SAR Images. *Remote Sens.* **2022**, *14*, 1018. [CrossRef]
32. Candes, E.J.; Wakin, M.B.; Boyd, S.P. Enhancing sparsity by reweighted L1 minimization. *J. Fourier Anal. Appl.* **2008**, *14*, 877–905. [CrossRef]
33. Liu, W.; Anguelov, D.; Erhan, D.; Szegedy, C.; Reed, S.; Fu, C.Y.; Berg, A.C. SSD: Single Shot Multibox Detector. In Proceedings of the 14th European Conference on Computer Vision(ECCV), Amsterdam, The Netherlands, 11–14 October 2016; Volume 14, pp. 21–37.
34. Chen, X.; Gong, Z. YOLOv5-Lite: Lighter, Faster, and Easier to Deploy. 2021. Available online: https://github.com/ppogg/YOLOv5-Lite (accessed on 10 June 2023).
35. RangiLyu. Nanodet-Plus: Super Fast and Lightweight Anchor-Free Object Detection Model. 2021. Available online: https://github.com/RangiLyu/nanodet (accessed on 10 June 2023).
36. Yi, H.; Liu, B.; Zhao, B.; Liu, E. Small Object Detection Algorithm Based on Improved YOLOv8 for Remote Sensing. *IEEE J. Sel. Top. Appl. Earth Obs. Remote Sens.* **2024**, *17*, 1734–1747. [CrossRef]
37. Ren, S.; He, K.; Girshick, R.; Sun, J. Faster r-cnn: Towards real-time object detection with region proposal networks. In Proceedings of the Advances in Neural Information Processing Systems 2015 (NIPS), Montreal, QC, Canada, 7–12 December 2015; Volume 28.
38. Shi, L.; Kuang, L.; Xu, X.; Pan, B.; Shi, Z. CANet: Centerness-aware network for object detection in remote sensing images. *IEEE Trans. Geosci. Remote Sens.* **2022**, *60*, 1–13. [CrossRef]
39. Li, Z.; Wang, Y.; Zhang, Y.; Gao, Y.; Zhao, Z.; Feng, H.; Zhao, T. Context Feature Integration and Balanced Sampling Strategy for Small Weak Object Detection in Remote-Sensing Imagery. *IEEE Geosci. Remote Sens. Lett.* **2024**, *21*, 1. [CrossRef]
40. Chattopadhay, A.; Sarkar, A.; Howlader, P.; Balasubramanian, V.N. Grad-CAM++: Generalized gradient-based visual explanations for deep convolutional networks. In Proceedings of the 2018 IEEE Winter Conference on Applications of Computer Vision (WACV), Lake, Tahoe, NV, USA, 12–15 March 2018; pp. 839–847.
41. Xie, T.; Han, W.; Xu, S. OYOLO: An Optimized YOLO Method for Complex Objects in Remote Sensing Image Detection. *IEEE Geosci. Remote Sens. Lett.* **2023**, 1. [CrossRef]
42. Kang, H.; Liu, Y. Efficient Object Detection with Deformable Convolution for Optical Remote Sensing Imagery. In Proceedings of the 2022 5th International Conference on Pattern Recognition and Artificial Intelligence (PRAI), Chengdu, China, 19–21 August 2022; pp. 1108–1114.
43. Wei, S.; Zeng, X.; Qu, Q.; Wang, M.; Su, H.; Shi, J. HRSID: A high-resolution SAR images dataset for ship detection and instance segmentation. *IEEE Access* **2020**, *8*, 120234–120254. [CrossRef]

44. Gao, S.; Liu, J.; Miao, Y.; He, Z. A high-effective implementation of ship detector for SAR images. *IEEE Geosci. Remote Sens. Lett.* **2021**, *19*, 1–5. [CrossRef]
45. Wei, H.; Wang, Z.; Hua, G.; Ni, Y. A Zero-Shot NAS Method for SAR Ship Detection Under Polynomial Search Complexity. *IEEE Signal Process. Lett.* **2024**, *31*, 1329–1333. [CrossRef]

remote sensing

MDPI

Article

A Reparameterization Feature Redundancy Extract Network for Unmanned Aerial Vehicles Detection

Shijie Zhang [1,2,3], Xu Yang [1,2,3], Chao Geng [1,2,3,*] and Xinyang Li [1,2,3]

1 National Laboratory on Adaptive Optics, Chengdu 610209, China; zhangshijie22@mails.ucas.ac.cn (S.Z.); yangxu@ioe.ac.cn (X.Y.); xyli@ioe.ac.cn (X.L.)
2 Institute of Optics and Electronics, Chinese Academy of Sciences, Chengdu 610209, China
3 University of Chinese Academy of Sciences, Beijing 100049, China
* Correspondence: gengchao@ioe.ac.cn

Abstract: In unmanned aerial vehicles (UAVs) detection, challenges such as occlusion, complex backgrounds, motion blur, and inference time often lead to false detections and missed detections. General object detection frameworks encounter difficulties in adequately tackling these challenges, leading to substantial information loss during network downsampling, inadequate feature fusion, and being unable to meet real-time requirements. In this paper, we propose a Real-Time Small Object Detection YOLO (RTSOD-YOLO) model to tackle the various challenges faced in UAVs detection. We further enhance the adaptive nature of the Adown module by incorporating an adaptive spatial attention mechanism. This mechanism processes the downsampled feature maps, enabling the model to better focus on key regions. Secondly, to address the issue of insufficient feature fusion, we employ combined serial and parallel triple feature encoding (TFE). This approach fuses scale-sequence features from both shallow features and twice-encoded features, resulting in a new small-scale object detection layer. While enhancing the global context awareness of the existing detection layers, this also enriches the small-scale object detection layer with detailed information. Since rich redundant features often ensure a comprehensive understanding of the input, which is a key characteristic of deep neural networks, we propose a more efficient redundant feature generation module. This module generates more feature maps with fewer parameters. Additionally, we introduce reparameterization techniques to compensate for potential feature loss while further improving the model's inference speed. Experimental results demonstrate that our proposed RTSOD-YOLO achieves superior detection performance, with $mAP_{50}/mAP_{50:95}$ reaching 97.3%/51.7%, which represents improvement of 3%/3.5% over YOLOv8, and 2.6%/0.1% higher than YOLOv10. Additionally, it has the lowest parameter count and FLOPs, making it highly efficient in terms of computational resources.

Keywords: drone detection; occlusion awareness; sequence feature fusion; attention mechanism

Citation: Zhang, S.; Yang, X.; Geng, C.; Li, X. A Reparameterization Feature Redundancy Extract Network for Unmanned Aerial Vehicles Detection. *Remote Sens.* **2024**, *16*, 4226. https://doi.org/10.3390/rs16224226

Academic Editors: Yanni Dong, Xiaochen Yang and Qian Du

Received: 21 October 2024
Revised: 31 October 2024
Accepted: 1 November 2024
Published: 13 November 2024

1. Introduction

In recent years, the rapid development of UAV technology has brought about changes in many industries, ranging from agricultural monitoring, disaster assessment to city management, and been used increasingly widely [1–5]. As a flexible and efficient aerial platform, UAVs can quickly cover large areas and provide high-resolution images and video data. Due to their small size and flexibility, UAVs play an important role in a variety of fields. Therefore, most studies focus on the detection and processing of images acquired by UAVs, but neglect the detection and regulation of UAVs themselves. Due to the small scale, complex background, target occlusion, and motion blurring [6], it is difficult to identify and localize UAVs in images or videos.

Previous UAV detection methods mainly used audio, radar, radio frequency (RF) signals, and computer vision techniques [7]. Hauzenberger et al. applied linear predictive coding (LPC) to detect UAVs [8] through their unique sounds, though this is easily disturbed by noise. Mohajerin et al. used radar trajectories for detection [9], but radar signals are limited in bad weather. Al-Emadi et al. employed a CNN to analyze RF signals between UAVs and controllers for detection [10].

In recent years, the widespread application of neural networks in the field of computer vision has provided new ideas for UAV detection. Target detection algorithms are mainly divided into traditional manual feature-based detection methods and deep neural network-based detection techniques. Traditional target detection methods rely on manual feature extraction and rule-based classifiers, which have limited performance when dealing with complex scenes and varying environments. The accuracy and robustness of target detection have significantly improved with the advent of deep learning, especially with the development of convolutional neural networks (CNNs).

General object detectors are primarily categorized into two types: single-stage detectors and two-stage detectors. Two-stage detectors typically involve two phases: region proposal for generating candidate boxes and bounding box regression, such as R-CNN [11], Fast R-CNN [12], Faster R-CNN [13], VFNet [14], and CenterNet2 [15]. These detectors tend to perform well in terms of accuracy but often suffer from lower real-time performance. Although many improvements have been made to the R-CNN family, they still fail to completely overcome the speed limitations of two-stage detectors.

In response, researchers combined the two phases into a single step, leading to the development of single-stage detectors such as SSD [16], RetinaNet [17], and YOLO [18]. However, these single-stage detectors often compromise detection accuracy, and directly applying them to UAV detection may lead to suboptimal results. Moreover, most previous research aimed at improving detection accuracy has often overlooked the limited computational power of practical deployment devices.

Our motivation stems from the fact that current general-purpose object detectors cannot be directly applied to drone detection scenarios, particularly due to challenges such as small scale, occlusion, and complex backgrounds. These general detectors often struggle to effectively address these issues. Therefore, we have made improvements to existing advanced detectors YOLOv8 and proposed the RTSOD-YOLO model, aiming to enhance the model's detection performance and real-time capabilities. A new downsampling module was introduced to retain more global and detailed information while reducing the number of parameters.

Additionally, to fully utilize the relationships between pyramid model features, we incorporated a dual TFE module to achieve comprehensive fusion of features at different scales in the pyramid model. We also introduced a scale-sequence feature fusion (SSFF) module to further extract and merge the results of dual encodings into a new small-object detection layer, enriching it with both global and detailed information.

Given that the widespread redundancy in current mainstream CNNs offers a comprehensive understanding of features but generates significant resource overhead, we designed a new efficient feature redundancy generation module. This module maintains redundant information while reducing the model's parameters. We also introduced reparameterization techniques, further improving inference speed. Finally, to address the issue of small-scale objects being easily occluded, we incorporated an occlusion-aware attention mechanism, which enhances the model's ability to detect occlusions.

In order to detect UAVs targets accurately and efficiently, this paper proposes the RTSOD-YOLO model, which has the following five main improvements compared with YOLOv8s:

1. We add an adaptive spatial attention mechanism to the Adown module. After downsampling, the feature maps undergo spatial enhancement, enabling the model to focus on more important regions and reduce attention to irrelevant features.
2. We introduce a new detection layer that draws inspiration from the TFE and SSFF modules. Given that the feature maps from the P2 layer contain critical information for small object detection, we perform scale-sequence feature fusion based on the P2

feature maps and the outputs from the two TFE modules, resulting in a detection layer specifically focused on small objects.
3. We design an efficient redundant feature generation module, replacing the original convolution module. This module ensures a comprehensive understanding of features while reducing the model's parameter count and computational cost. By incorporating reparameterization techniques, the model's inference speed is further enhanced.

The remainder of this paper is organized as follows: In Section 2, we provide an overview of related work on drone detection techniques, focusing on traditional image processing methods and deep learning approaches. Section 3 details the proposed method, including the enhancements to the TFE and SSFF modules. In Section 4, we present the experimental setup and results, evaluating the performance of our model against state-of-the-art methods. Section 5 concludes the paper and outlines future research directions.

2. Related Work

Unlike counter-UAV detection tasks, current research primarily focuses on target detection and tracking from the perspective of the UAV itself. UAV-based detection and tracking often involve a top-down, bird's-eye view, which provides a wide field of view but also introduces new challenges such as high-density targets, small objects, and complex backgrounds.

In recent years, significant research has been conducted on UAV detection. The main UAV detection techniques include radar sensors [19–25], RF sensors [10,26–30], audio sensors [31–36], and vision-based detector technologies [37–49]. A comparison of the characteristics of these detection technologies is presented in Table 1.

Table 1. Comparison of detection methods.

Method	Principle	Advantages	Disadvantages
Radar-based	Utilizes radio waves for detection and localization of nearby objects.	Long-range detection; resilient to adverse weather; can analyze communication spectra.	Limited detection due to low radar cross section; performance drops in low altitudes; high cost and deployment complexity.
RF-based	Captures wireless signals from UAVs to detect their radio frequency emissions.	Long-range detection; resilient to adverse weather; can analyze communication spectra.	Cannot identify autonomous drones; susceptible to RF interference; vulnerable to hacking.
Acoustic-based	Detects UAVs based on sound signatures produced during operation.	Cost-effective; no line-of-sight required; quick deployment.	Affected by background noise; limited detection range; sensitive to wind conditions.
Vision-based	Uses visual data captured by cameras to detect UAVs.	Provides visual confirmation; nonintrusive; cost-effective.	Limited range; requires line-of-sight; affected by weather and lighting conditions.

(1) Radar is an electromagnetic technology that uses radio waves to detect and locate nearby objects. It can calculate important target features such as distance, speed, azimuth, and elevation [19]. Some moving parts of UAVs produce unique radar echoes, and a method was proposed based on the analysis of these echo signals [23], focusing on developing patterns and features to identify UAVs based on rotor blade types. In [24], the authors analyzed the micro-Doppler characteristics of different UAVs and bird species, confirming that K-band (24 GHz) or millimeter-wave radar systems are effective in detecting UAVs.

(2) RF detection is mainly due to the presence of electronic components on UAVs, such as radio transmitters and GPS receivers, which emit energy detectable by RF sensors. Detection systems based on this technology capture the communication signals between the controller and the UAVs, and analyze these signals to detect the UAV [26–28]. Nemer I. et al. proposed a machine-learning-based UAV detection and recognition system, introducing a new layered learning approach to recognize different types of UAV RF signals through a

four-layer classifier [26]. In [27], the authors compared various machine learning techniques for RF detection and verified the good classification performance of the XGBoost algorithm. Some researchers have applied deep learning to this problem; for example, in [10], a simple CNN achieved better detection accuracy. In the study [29], VGG-16 was used to train compressed RF signals for UAV detection and recognition.

(3) Audio detection is based on sound information. During flight, UAVs produce distinct acoustic features due to their engines, propellers, and aerodynamic properties, which can be used to detect UAV targets. The most common sound feature is the noise generated by the propeller blades, which typically has a larger amplitude. By analyzing features such as frequency, amplitude, modulation, and duration, UAVs can be identified. This research area includes both machine learning and deep learning approaches. In [32], the authors developed a simple and effective UAV detection system using machine learning, classifying through balanced random forests and multilayer perceptrons. Tejera-Berengue et al. conducted research on the dependency of various machine-learning-based UAV detection systems on distance, demonstrating the superior performance of random forests in detection and the excellent distance performance of linear classifiers.

(4)Vision-based detection is mainly divided into traditional handcrafted feature extraction and deep learning methods. In recent years, deep learning has demonstrated superior performance in various fields, and most recent studies have adopted deep learning techniques [37–49]. These methods are generally classified into two types: two-stage detectors and single-stage detectors. In [37], a two-stage detector was used, employing a Mask R-CNN with two backbones, and it was shown that a ResNet-50 network outperforms MobileNet in terms of performance. However, while this method provides good detection accuracy, its detection speed is relatively low. Single-stage detectors, on the other hand, offer better speed performance.

In [38], the authors proposed the AD-YOLOv5s algorithm to detect low-altitude UAVs. Firstly, the detection of small targets is addressed by feature enhancement. Secondly, the Ghost module and depth-separable convolution (DSConv) are used to reduce a large number of redundant parameters in YOLOv5s, which enables the network to be used in embedded devices and perform inference at a high speed. Comparatively, some researchers have discarded the requirement of real time and pursued extreme detection accuracy. The author of [39] compared the performance of YOLOv4, YOLOv5, and DETR in terms of accuracy and speed. In [40], the author created a low-visibility UAVs flight dataset and proposed an improved YOLOv5 algorithm. It obtains more texture and contour information in small targets by adding a new scale to the model, and performs multiscale feature fusion to reduce the loss of information. For the problem of balancing detection speed and accuracy, Selvis S.S. et al. verified the higher real-time performance of YOLOv5 compared to YOLOv4 and overcame the problem of high-precision detection [41].

With the continuous development of deep learning technologies, multiple variants based on the YOLO architecture have emerged to address issues such as detection accuracy, speed, and complexity across different application scenarios. PP-YOLOE [42], proposed by PaddlePaddle, is one such variant that significantly enhances detection efficiency through multiprecision training and mixed-precision inference. It offers various configurations from small to large, catering to different computational resource environments. PP-YOLOE maintains high detection accuracy while achieving rapid inference speeds, making it highly suitable for real-world deployment. On the other hand, DAMO-YOLO [43], developed by Alibaba's DAMO Academy, employs neural architecture search (NAS) technology and innovative modules like RepGFPN and AlignedOTA to boost detection efficiency and precision. This model, available in different configurations such as tiny, small, and medium, reduces parameter count and inference time, making it ideal for edge devices and resource-constrained environments. YOLO-MS [44] focuses on addressing multiscale object detection challenges, particularly optimizing small object detection by fusing information from multiple feature layers, thus improving performance when detecting small targets. These model variants leverage innovative network designs, feature fusion strategies, and optimization techniques to enhance the overall performance of YOLO, further extending its applicability across various use cases.

There are also many studies that have conducted different research on datasets for UAVs detection [45,46]. In [47], the authors collected a total of 2395 images of drones and birds from publicly available online resources for training, and Zhao et al. proposed a visible light mode dataset called Dalian University of Technology Anti-UAV dataset (DUT Anti-UAV). It contains a detection dataset with a total of 10,000 images and a tracking dataset with 20 videos that include short-term and long-term sequences [48].

3. Methodology

YOLOv8 [50] is one of the most advanced real-time detectors; it consists of three parts: backbone, neck, and head. This paper proposes an improved network RTSOD-YOLO based on YOLOv8 network for UAVs detection. We have made several improvements to address the issues in the network, such as downsampling, feature fusion, and model speed. First, we replaced the original convolution module with an adaptive Adown module for downsampling, allowing the model to retain rich information from higher layers. Next, we designed an efficient redundant feature generation module to preserve the comprehensive understanding of input features while reducing computational resources. Additionally, we employed a dual TFE module in a serial-parallel structure to perform two encodings, merging global information from deeper features and texture information from shallower features. This is followed by a new SSFF module that fuses these and P2's features, forming a new detection layer specifically designed for small object detection, thereby improving the model's capability in detecting small targets. Finally, we introduced an occlusion-aware attention module in each detection head, enhancing the network's ability to detect objects under occlusion; the architecture of the network is shown in Figure 1.

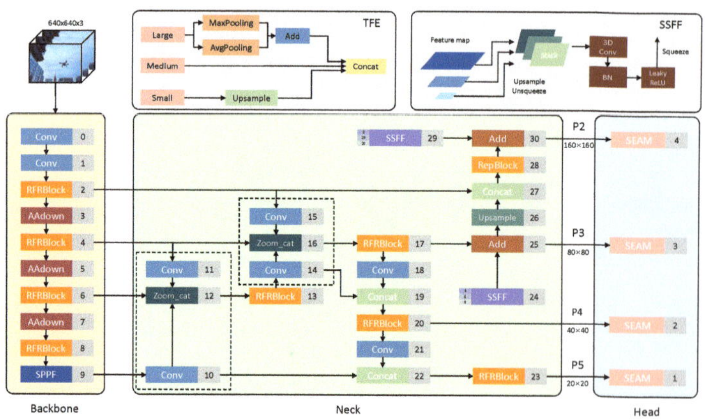

Figure 1. The architecture of RTSOD-YOLO.

3.1. *Adaptive Adown Downsampling*

YOLOv8 extracts five layers of feature maps with different scales, P1–P5, corresponding to 320×320, 160×160, 80×80, 40×40, and 20×20. However, P3, P4, and P5 are mainly used for feature fusion, which is followed by detection. These feature maps of different scales have different receptive fields and, therefore, contain different semantic and location information. Larger feature maps contain sufficient location information, but not enough semantic information. In YOLOv8, downsampling is achieved solely through convolution to extract deeper features, but this method results in the loss of many important feature information during the downsampling, and with the increase of downsampling, the lost feature information will further increase. Therefore, this paper introduces the Adown [51] downsampling module, whose structure is shown in Figure 2.

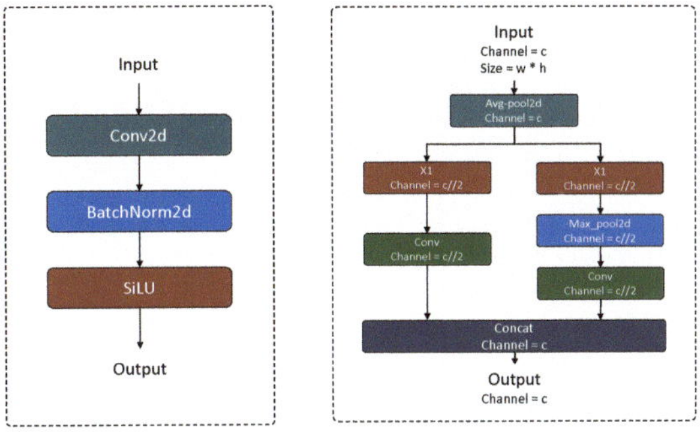

(a) (b)

Figure 2. Convolution and Adown downsampling. (**a**) Convolution. (**b**) Adown.

The approach employs average pooling initially to reduce the size of the feature map. Subsequently, it applies convolution and maximum pooling to the bisected channels. Finally, it seamlessly stitches the two components together, forming the ultimate downsampling feature map. As we all know, max pooling is used to retain the most prominent local activation, such as strong edges, textures, and key features that are important for fine-grained recognition, and average pooling provides a more generalized and smooth representation by capturing broader patterns and reducing the influence of noise. Thus, we use convolutions to help integrate the sharp, high-activation features from max pooling with the smooth, global context captured by average pooling.

To avoid potential inconsistencies between the outputs of the two pooling methods, the Adown module employs a learnable feature weighting mechanism. This allows the network to dynamically adjust the contributions of max pooling and average pooling based on the specific task or input, ensuring that the most relevant information is retained in the downsampled feature maps.

We represent the features obtained by the Adown module as $F \in \mathbb{R}^{C\times H\times W}$, where C is the number of channels, and H and W are the height and width of the feature map. We newly apply both global average pooling and global max pooling along the channel dimension to capture different aspects of the feature map:

$$F_{\text{avg}}^{\text{spatial}} = \text{AvgPool}(F) \in \mathbb{R}^{1\times H\times W} \tag{1}$$

$$F_{\text{max}}^{\text{spatial}} = \text{MaxPool}(F) \in \mathbb{R}^{1\times H\times W} \tag{2}$$

The results of average pooling and max pooling are concatenated along the channel dimension to form a feature map:

$$F_{\text{concat}} = \left[F_{\text{avg}}^{\text{spatial}}, F_{\text{max}}^{\text{spatial}}\right] \in \mathbb{R}^{2\times H\times W} \tag{3}$$

We apply a 7×7 convolution to the concatenated feature map to generate a spatial attention weight map:

$$M_{\text{spatial}}(F) = \sigma(\text{Conv}_{7\times 7}(F_{\text{concat}})) \in \mathbb{R}^{1\times H\times W} \tag{4}$$

where σ is the sigmoid activation function to normalize the attention map to the range $[0,1]$. The original input feature map F is multiplied element-wise with the spatial attention map $M_{\text{spatial}}(F)$:

$$F' = M_{\text{spatial}}(F) \odot F \tag{5}$$

where $\odot$ denotes the element-wise multiplication.

By leveraging the method above, the Adaptive Adown module ensures that downsampling not only reduces the spatial dimensions but also retains more information focused on key areas. This leads to more robust feature maps that maintain a balance between local details and global context.

3.2. Reparameter Redundant Feature Generation Module

In a GhostNet study [52], it is noted that CNNs often compute intermediate feature maps that exhibit significant redundancy which refers to the presence of highly similar feature maps. An ample amount of redundant information in the feature maps of well-trained deep neural networks often ensures a thorough understanding of the input data. The effective utilization of feature map redundancy plays a crucial role in the performance of CNNs. However, in mainstream CNNs, there is widespread redundancy in the intermediate feature maps. Therefore, we aim to reduce the resources required, such as the convolution filters used to generate them. The Ghost Module is designed based on this concept, aiming to generate a portion of redundant feature maps using computationally "cheap" operations, such as 3×3 and 5×5 convolutions. By doing so, the Ghost Module reduces computational overhead and the number of model parameters.

From this perspective, we propose the Reparameterization Feature Redundancy Generation Block (RFR-Block), shown in Figure 3b, as a replacement for the C2f module in the network. This block generates redundant features using more cost-efficient operations, thereby reducing both the number of parameters and the FLOPs of the model.

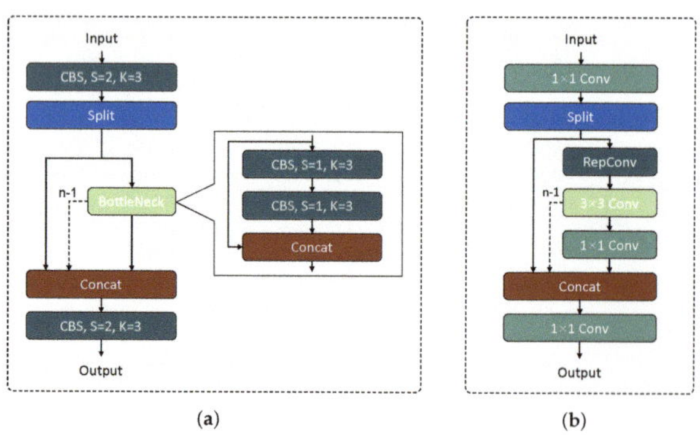

Figure 3. C2f and RFR-Block module. (**a**) Structure of C2f. (**b**) Structure of RFR-Block.

To compensate for the accuracy loss caused by discarding the original BottleNeck module, we employed the reparameterization technique RepConv. RepConv offers a way to convert the multibranch structure used during training into a more efficient single-branch structure during inference, which allows for faster inference speed and helps to improve the real-time performance of our model. The structure of RepConv is illustrated in Figure 4, and its key concept involves merging the convolution and normalization layers into a single convolution. The normalization process can be expressed by Equation (6).

$$\begin{aligned}\hat{x}_i &= \gamma \cdot \frac{x_i - \mu}{\sqrt{\sigma^2 + \varepsilon}} + \beta \\ &= \frac{\gamma}{\sqrt{\sigma^2 + \varepsilon}} \cdot x_i + \left(\beta - \frac{\gamma \cdot \mu}{\sqrt{\sigma^2 + \varepsilon}}\right)\end{aligned} \tag{6}$$

Think of this as being of the form $y = \omega x + b$, where $\omega_{BN} = \frac{\gamma}{\sqrt{\sigma^2+\varepsilon}}$, and $b_{BN} = \beta - \frac{\gamma \cdot \mu}{\sqrt{\sigma^2+\varepsilon}}$, so the process of convolution and normalization can be represented as Equation (7):

$$\begin{aligned}\hat{x} &= w_{BN} \cdot (w_{conv} \cdot x + b_{conv}) + b_{BN} \\ &= (w_{BN} \cdot w_{conv}) \cdot x + (w_{BN} \cdot b_{conv} + b_{BN})\end{aligned} \tag{7}$$

The conversion of the 1×1 convolution and the two unprocessed branches into a 3×3 convolution for fusion can be expressed as follows:

$$\begin{aligned}\hat{x}_i &= \sum_{k=0,1,3} (w_{fuse}^{k*k} x_i + b_{fuse}^{k*k}) + (w_{BN}^{0*0} x_i + b_{BN}^{0*0}) \\ &= (w_{fuse}^{3*3} + w_{fuse}^{1*1} + w_{BN}^{0*0}) \cdot x_i + (b_{fuse}^{3*3} + b_{fuse}^{1*1} + b_{BN}^{0*0})\end{aligned} \tag{8}$$

where fuse denotes the convolutional parameter after fusion of convolution and normalization.

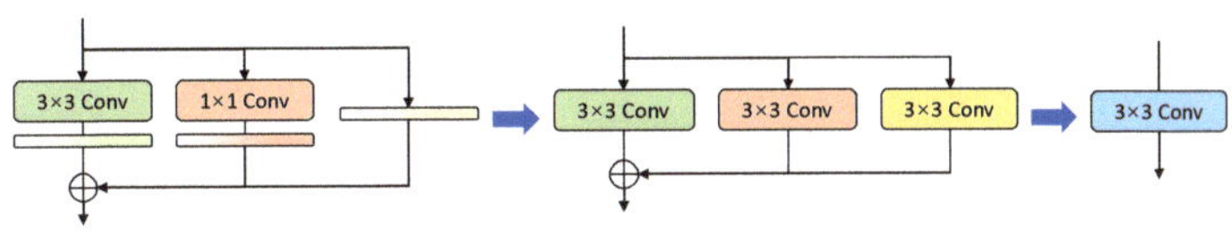

Figure 4. RepConv schematic diagram.

In the subsequent steps, a cost-effective operation of 3×3 convolution is employed to generate redundant feature maps. Following this, the feature maps undergo downsampling using a 1×1 convolution. Multiple feature maps are then combined or stitched together to form a new RFR-Block module.

We constructed this efficient redundant feature generation module using cost-effective operations and reparameterization techniques. This module retains the comprehensive understanding provided by redundant features while reducing the computational resources required by the model. Additionally, the reparameterization technique not only compensates for some of the lost accuracy but also allows the model to achieve faster inference speed.

The proposed RFR-Block module consists of the following components:

1. A 1×1 convolution is used to change the channels (where the change is influenced by different scale coefficients according to the size of the model), and then divide it into two parts along the channel dimension.
2. One part remains unprocessed, while the other part undergoes feature extraction through RepConv to compensate for the loss of accuracy. Additionally, cost-effective 3×3 and 1×1 convolutions are used to generate the necessary redundant features.
3. All intermediate feature maps are concatenated and passed through final convolution layers (1×1) which changes the channel to output channel, yielding the final output.

3.3. Small Target Detection with SSFF and TFE

In YOLOv8, the classical feature pyramid model is employed for feature fusion. However, the network utilizes simple concatenation and summation for fusing the pyramid features, without fully exploiting the connections between them. To address this limitation, we introduce the SSFF module and TFE module [53].

The TFE module aims to improve feature fusion by segmenting features from three different scales: large, medium, and small. It includes the addition of large-scale feature maps and feature zooming to refine the detailed feature information. Figure 5 illustrates the structure of the TFE module.

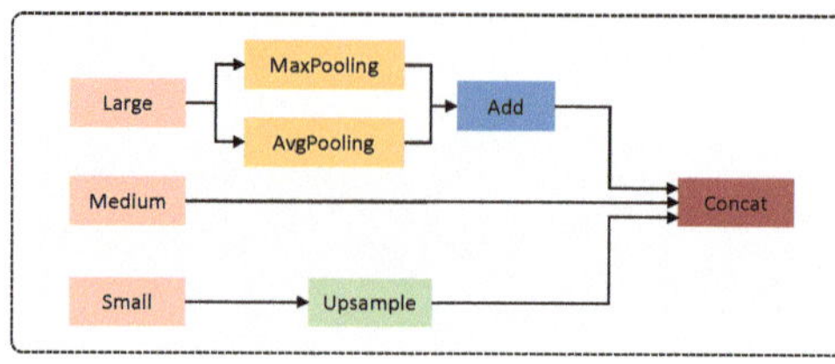

Figure 5. Triple feature encoding module.

Prior to feature encoding, the number of feature channels at each scale is adjusted to match the main feature map by the convolution module. Subsequently, the small-scale feature map undergoes upsampling using nearest-neighbor interpolation, which preserves the local strong semantic feature information from the low-resolution image. Conversely, the large-scale feature maps are downsampled through a combination of maximum pooling and average pooling. This downsampling approach aims to retain the global location information from the high-resolution image, along with the diverse feature information of the target image. Finally, the three scales of features are concatenated in the channel dimension.

In the SSFF module, which is illustrated in Figure 6, a 1×1 convolution is first applied to the P4 and P5 feature levels to reduce their channels to 256. Then, nearest neighbor interpolation adjusts their spatial dimensions to match P3. Next, the unsqueeze method is used to expand the feature maps along the first axis, changing them from 3D (height, width, channels) to 4D (depth, height, width, channels). After that, the 4D feature maps are concatenated along the first dimension depth to form a combined 3D feature map. Finally, 3D convolutions, followed by 3D batch normalization and SiLU activation, are applied to extract and refine the scale-sequence features. This approach effectively combines the high-dimensional information from deep feature maps with the detailed information from shallow feature maps.

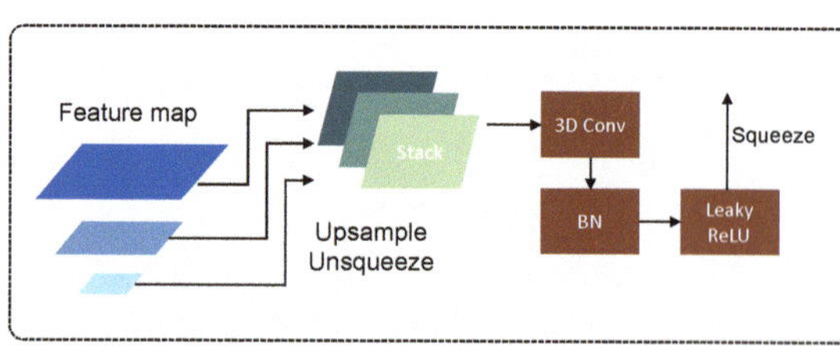

Figure 6. Scale-sequence feature fusion module.

The SSFF module is primarily designed based on the P3 feature layer, but since our network is aimed at detecting UAVs, which are typically small-scale targets, we applied scale-sequence fusion to the P2 feature layer. By integrating the output from the P2 feature layer with two TFE modules, we created a scale space that fuses shallower features—critical for small object detection. This fusion output serves as a new detection layer, enhancing the model's capability to detect small-scale targets.

3.4. Separated and Enhancement Attention Module

To address issues caused by occlusion between targets, this paper introduces a novel occlusion-aware attention mechanism called the Separated and Enhancement Attention Module (SEAM) [54]. The SEAM, depicted in Figure 7, seeks to mitigate problems such as alignment errors and feature loss resulting from occlusion.

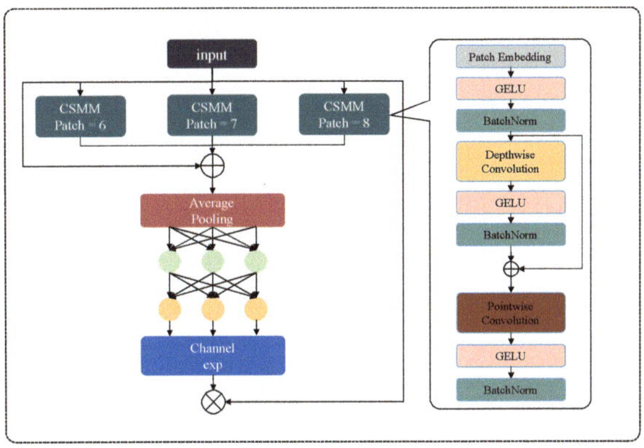

Figure 7. Separated and Enhancement Attention Module.

The SEAM incorporates a multistep approach to tackle the challenges posed by occlusion between targets. Firstly, it applies depthwise separable convolution with residuals to process the inputs. Depthwise convolutions apply a single filter to each input channel independently, which significantly reduces the number of parameters and computations compared to traditional convolutions. This parameter efficiency makes the model lighter and faster, which is particularly beneficial for real-time applications. Furthermore, by processing each channel separately, depthwise convolutions enable the model to learn specific features tailored to individual channels, enhancing the effectiveness of feature extraction without introducing excessive computational complexity.

However, it neglects the relationships between the channels. To compensate for this loss, the outputs of the depthwise separable convolutions are subsequently combined through pointwise (1×1) convolution. Pointwise convolutions serve to combine the outputs of depthwise convolutions across different channels, allowing for effective mixing of information. This capability is crucial for capturing interactions between various feature maps, leading to richer and more complex feature representations. Additionally, pointwise convolutions can adjust the number of channels, providing flexibility in the architecture. This characteristic is particularly useful for adapting the model to different input sizes or for reducing dimensionality after depthwise operations, ultimately contributing to improved model performance.

Then, a two-layer fully connected network is used to fuse the information from each channel, enabling the network to enhance the connections between all channels. The outputs derived from the fully connected layer are then subjected to an exponential function, mapping them from the range [0, 1] to [1, e]. This mapping facilitates improved tolerance towards positional errors. Finally, the module's output is multiplied with the original input features, enabling the model to effectively handle target occlusion.

4. Experiments

Based on the improvements before, we proposed the RTSOD-YOLO model and validated its performance through several experiments, including ablation studies and comparative experiments. In this section, we present and discuss our experimental setup, datasets, evaluation metrics, and the experimental results.

4.1. Evaluation Metrics

In deep learning, particularly in tasks like classification, object detection, and segmentation, evaluation metrics play a crucial role in measuring the performance of models. Below are some commonly used evaluation metrics:

(1) Accuracy is the ratio of correctly predicted instances to the total instances, used to evaluate the overall performance of classification models.

$$\text{Accuracy} = \frac{TP + TN}{TP + TN + FP + FN}$$

where true positive (TP): The number of instances correctly predicted as positive; true negative (TN): The number of instances correctly predicted as negative; false positive (FP): The number of instances incorrectly predicted as positive (Type I error); false negative (FN): The number of instances incorrectly predicted as negative (Type II error).

(2) Precision is the ratio of correctly predicted positive observations to the total predicted positive observations.

$$\text{Precision} = \frac{TP}{TP + FP}$$

(3) Recall is the ratio of correctly predicted positive observations to all actual positives.

$$\text{Recall} = \frac{TP}{TP + FN}$$

(3) Average precision (AP) In object detection, AP is the area under the precision–recall (PR) curve, measuring performance at different confidence thresholds.

$$AP = \int_0^1 P(R)\,dR$$

(4) Mean average precision (mAP) is the mean of average precision scores across all object classes in a multiclass detection task. mAP is computed by averaging the APs for each class of the objects, and getting the ultimate mAP score.

$$\text{mAP} = \frac{1}{C}\sum_{i=1}^{C} AP_i$$

where C is the number of classes and AP_i is the average precision for class i.

(5) Intersection over union (IoU) measures the overlap between the predicted bounding box and the ground truth bounding box.

$$\text{IoU} = \frac{\text{Area of Overlap}}{\text{Area of Union}}$$

(6) Frames per second (FPS) measures how many frames the model can process per second, indicating its real-time performance. FPS is computed by dividing the number of frames processed by the algorithm by the overall time taken by the algorithm, and obtaining the ultimate FPS score.

(7) Confusion matrix is a table that summarizes the performance of a classification algorithm by comparing the predicted labels with the actual labels. It consists of four components:

$$\begin{bmatrix} TN & FP \\ FN & TP \end{bmatrix}$$

(8) Number of parameters: The total number of learnable parameters in a model, indicating its complexity.

(9) Floating point operations (FLOPs) measure the number of floating-point operations required to perform one forward pass of the model.

Accuracy, precision, recall, and F1-dcore are basic metrics for classification models, while AP and mAP are typically used in object detection tasks. IoU evaluates how well bounding boxes match in object detection. FPS measures the real-time performance of a model, and the confusion matrix visually displays true and false positives/negatives. The number of parameters along with FLOPs are used to evaluate model complexity and computational efficiency.

4.2. Dataset

To ensure the model's inference and detection feasibility in real-world scenarios, this paper constructs a dataset using UAV images captured under natural conditions with complex backgrounds. The data collection process involves gathering images from both network sources and actual scene photography. Public dataset websites such as Kaggle and Paddle Paddle are utilized to obtain a diverse range of image data. From the collected images, a total of 14,600 samples are carefully selected. These samples encompass color images, infrared images, targets of varying scales, and diverse complex detection scenarios which are shown in Figure 8.

Figure 8. The unmanned aerial vehicle dataset.

To facilitate comprehensive evaluation, the dataset is divided into training, validation, and test sets in an 8:1:1 ratio. This partitioning ensures a balanced representation of the data while enabling robust training, validation, and assessment of the model's performance.

4.3. Experimental Setup

In this paper, the YOLOv8 model is adopted as the baseline network and implemented using the PyTorch framework (version 2.0.1). The experiments are conducted on a system equipped with an Intel Core™ i5-13490 CPU with 32 GB of RAM, running the Windows 11 operating system. To leverage parallel acceleration for the network, an NVIDIA RTX4060Ti GPU with 16 GB of video memory is utilized, along with CUDA version 11.8. The hyperparameter settings are shown in Table 2.

Table 2. Hyperparameter settings.

Hyperparameter	Value
Epochs	50
Warmup epochs	3
Image size	640
Batch size	32
Learning rate	0.01
Weight decay	0.0005
Momentum	0.937
Optimizer	SGD

4.4. Comparison Experiments

To track the evaluation metrics, we plotted the values during the iterations, which can be found in Appendix A. Figure A1 presents an overall training summary of the model. The loss curves exhibit a downward trend, indicating that both training and validation losses are minimized during the training process. The metric curves show an upward trend, suggesting that the model's performance improves throughout the iterations of training.

Table 3 presents a comparison of the proposed RTSOD-YOLO model with other classical advanced real-time detection methods on our drone dataset, including YOLOv5, AD-YOLOv5s, YOLOv6 [55], Damo-YOLO, Gold-YOLO [56], YOLO-MS , YOLOv8, YOLOv9 (where GELAN is a simplified version that does not include the PGI module), and YOLOv10 [57]. Our model has only 5.2 million parameters, yet it achieves the highest accuracy of 97.3% while maintaining a relatively high inference speed of 241.2 frames per second.

Table 3. Comparisons with state-of-the-art real-time object detectors.

Model	mAP_{50} (%)	$mAP_{50:95}$ (%)	FLOPs (G)	Params (M)	FPS
YOLOv5-S	93.3	47.4	23.8	9.1	144.1
AD-YOLOv5s	94.3	48.5	26.7	12.8	-
YOLOv6-S	94.1	48.5	44.0	16.3	243.9
YOLOv8-S	94.3	48.2	28.4	11.1	238.6
Damo-YOLO-S	94.4	49.1	37.8	12.3	187.5
Gold-YOLO-S	94.6	48.6	46.0	21.5	223.7
YOLO-MS	95.1	50.7	31.1	8.1	173.5
YOLOv9-S	94.4	**52.0**	38.7	9.6	196.1
GELAN-S	94.2	48.8	26.2	7.1	252.4
YOLOv10-S	94.7	51.6	24.4	8.0	**270.6**
OURS	**97.3**	51.7	**21.1**	**5.2**	241.2

We also analyzed the confusion matrices for each model across different scenarios, as illustrated in Figure A2 in Appendix B. This analysis allows us to compare the actual performance of the models under various conditions.

The evaluation focuses on three key scenarios for drone object detection: occlusion, intense light, and dim light conditions. Each scenario consists of 200 instances, corresponding to the performance of YOLOv5, YOLOv6, YOLOv8, YOLOv9, YOLOv10, and our proposed model. This comprehensive assessment provides valuable insights into how each model responds to challenging environmental factors, enabling a clearer understanding of their strengths and weaknesses in practical applications. For further details, please refer to Appendix B.

Additionally, in Figure 9, we provide visual comparisons of each model across different scenarios. It can be observed that in complex conditions such as occlusion and intense lighting, our model maintains excellent detection performance, while the other detectors exhibit varying degrees of false positives and false negatives.

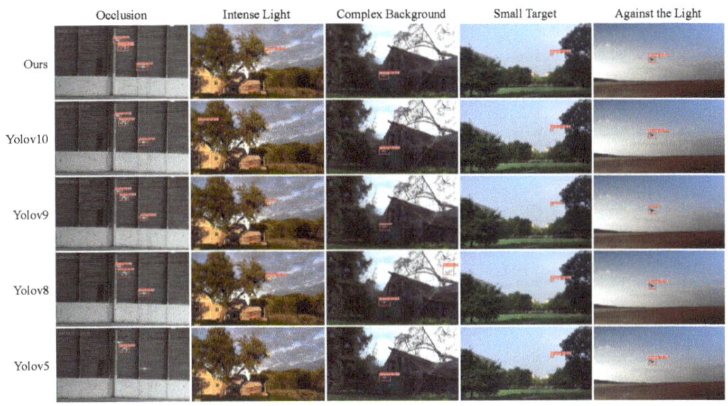

Figure 9. Detection performance in different scenarios.

To further validate the performance of our proposed model, we conducted comparative experiments on several datasets. These include commonly used UAV detection datasets such as Anti-UAV300 [58], Drone Detection [59], and Drone vs. Bird [60], which represent various real-world conditions, including different weather conditions, occlusions, and lighting variations.

Among the datasets we used, Anti-UAV includes both infrared and RGB images of UAVs, covering six types of UAVs (such as DJI and Parrot) captured under two different lighting conditions (daytime and nighttime). The images span various backgrounds, such as buildings, clouds, and trees, providing diverse scenarios for detection tasks. The Drone Detection dataset, on the other hand, focuses on UAV images in various environments with significant scale variations, making it ideal for evaluating how well models perform under changing target sizes. Lastly, the Drone vs. Bird dataset presents a challenging task of differentiating between UAVs and birds, both of which are small-scale aerial targets with similar backgrounds. The close similarity in appearance between drones and birds, combined with the complex background, makes detection and recognition particularly difficult in this dataset.

We conducted further experiments on the datasets mentioned above, and the results are shown in Table 4. Our model ranked first in the Anti-UAV dataset with an accuracy of 98.4%/65.1%. Additionally, our model also achieved the best performance in the Drone vs. Bird dataset. Although the $mAP_{50:95}$ in the Drone Detection dataset did not reach the highest, it was only second to YOLOv9. This validates that our model's generalization performance across multiple datasets surpasses that of currently advanced real-time detectors.

Table 4. Generalization performance comparison across different datasets.

Method	Anti-UAV300		Drone Detection		Drone vs. Bird	
	mAP_{50}	$mAP_{50:95}$	mAP_{50}	$mAP_{50:95}$	mAP_{50}	$mAP_{50:95}$
YOLOv5	94.6	62.3	86.8	50.4	89.8	47.3
AD-YOLOv5s	95.6	62.6	87.2	51.3	90.5	47.8
YOLOv6	95.4	62.5	86.9	51.1	88.6	47.1
YOLOv8	96.8	64.5	88.6	53.5	91.4	50.4
Damo-YOLO	97.3	64.3	88.7	52.8	91.5	50.3

Table 4. *Cont.*

Method	Anti-UAV300		Drone Detection		Drone vs. Bird	
	mAP_{50}	$mAP_{50:95}$	mAP_{50}	$mAP_{50:95}$	mAP_{50}	$mAP_{50:95}$
Gold-YOLO	96.9	63.9	87.4	52.7	91.9	50.2
YOLO-MS	97.4	64.7	90.8	52.9	92.8	50.7
YOLOv9	97.1	64.5	91.1	**54.2**	92.6	50.5
YOLOv10	96.7	64.1	90.4	53.2	92.4	50.4
Ours	**98.4**	**65.1**	**91.3**	53.6	**94.0**	**51.6**

4.5. Ablation Study

We also conducted ablation experiments on each module to validate the effectiveness of our improvements and proposed modules. First, we validated the contributions of each module to the network's performance. Next, we compared the proposed RFR-Block with the C3, C2f, and ELAN modules. Finally, we assessed the performance of the SEAM in enhancing occlusion awareness.

4.5.1. Effect of the Improved Methods

Table 5 illustrates the contribution of each of our improved modules to enhancing detection performance. By integrating the small object detection layer that fuses spatial and scale features, as well as our occlusion-aware attention mechanism, we achieved improvements in the network's detection accuracy, resulting in an increase of 2.0% and 1.7% in mAP_{50} and $mAP_{50:95}$, respectively. Additionally, the proposed RFR-Block maintains detection accuracy while reducing the computational resources required, with FLOPs decreased by 8.1 G and parameters reduced by 3.24 M.

Table 5. Ablation experiments on the UAVs dataset.

Model	S&T-P2	SEAM	Adown	RFR	mAP_{50} (%)	$mAP_{50:95}$ (%)	FLOPs (G)	Params (M)
YOLOv8s					94.3	48.2	28.4	11.13
a	√				96.3	49.9	35.8	9.05
b		√			95.4	49.8	25.7	10.65
c			√		95.4	51.3	25.7	9.48
d				√	94.3	48.0	20.3	7.89
Ours-A	√				96.3	49.9	35.8	9.05
Ours-B	√	√			96.3	50.2	31.7	8.80
Ours-C	√	√	√		97.1	51.4	29.7	7.68
Ours-D	√	√	√	√	**97.3**	**51.7**	**21.1**	**5.18**

4.5.2. Effect of the RFR-Block

Although the C2f module is powerful in feature extraction, it requires numerous parameters and FLOPs. In order to further reduce the complexity of our network model, we replaced the original C2f module with the newly designed RFR-Block module. We compared this module with the previous C3, ELAN, and C2f modules in v5, v7, and v8 of the YOLO series, and the results are presented in Table 6. Our RFR-Block achieves the lowest number of parameters and computation, with the only trade-off being a slightly lower FPS compared to the C3 module in YOLOv5.

Table 6. Comparison of C3, ELAN, C2f, and RFR-Block modules.

Name	All-Time	Mean-Time	FPS	FLOPs (G)	Params (K)
C3	2.10088	0.00105	951.983364	4.832	295.680
ELAN	2.97037	0.00149	673.317590	8.053	492.288
C2f	2.58631	0.00129	773.303101	7.516	459.520
RFR-Block	2.49479	0.00125	801.669814	3.691	226.176

4.5.3. Effect of the SEAM

We conducted a comparison of occlusion capabilities by simulating occluded targets using random erasing (RE) in the original test set, as shown in Figure 10. RE generates a black occlusion in the target region, thus simulating a situation where the target is partially occluded, and the occluded portion accounts for 10% to 50% of the target size.

Figure 10. Test dataset augmented with random erasing.

We evaluated the performance of the baseline model against the model with the introduced occlusion-aware attention mechanism. The results in Table 7 indicate that the improved model showed an enhancement in detection accuracy on the occluded test set, with mAP_{50} increasing by 3.2% and $mAP_{50:95}$ improving by 2.9%, thereby validating the effectiveness of the modifications.

Additionally, we compared the performance of the two models on both the occluded and nonoccluded test sets. Due to the impact of occlusion, both models exhibited a decrease in accuracy on the occluded test set; however, the network with the attention mechanism experienced a significantly smaller drop in performance compared to the baseline model. This indicates that the SEAM possesses a certain degree of resilience against occlusion interference.

Table 7. Results for the network introducing SEAM on the original and masked dataset.

Model	mAP_{50} (%)	$mAP_{50:95}$ (%)
YOLOv8s (Origin)	94.3	48.2
YOLOv8s (RE)	86.9 (7.4 ↓)	43.4 (4.8 ↓)
+SEAM-head (Origin)	95.4	49.8
+SEAM-head (RE)	90.1(5.3 ↓)	46.3 (3.5 ↓)

5. Conclusions

This paper presents an accurate and efficient real-time drone detection model, RTSOD-YOLO, which reintegrates pyramid features and optimizes the generation of redundant features, allowing the model to focus more on small object detection. We implemented various improvements within the YOLOv8 framework. The small object detection layer, which integrates spatial and scale features, enhances the model's capability to detect small targets. The RFR-Block module reduces the computational resources required by the model, while the SEAM attention mechanism improves the network's ability to perceive occlusions. Extensive experimental results demonstrate that this model can handle drone detection tasks under various conditions, significantly improving accuracy while maintaining inference speed. Our approach clearly outperforms currently popular real-time object detectors. Compared to the baseline network, our model achieved an accuracy improvement of 3.0%/3.5%, while reducing the number of parameters and computational cost by 25.7% and

53.1%, respectively. Also, there was a slight enhancement in inference speed. Additionally, in the ablation studies, we further validated the effectiveness of each module, providing a research basis for future improvements. However, due to the limited variety of drones in the dataset used in this study, the model's generalization ability to different drone shapes needs further enhancement. In the future, we will collect more drone data to enrich and improve the dataset.

Author Contributions: Conceptualization, S.Z., C.G. and X.L.; methodology, S.Z. and X.Y.; software, S.Z.; validation, S.Z., X.Y., C.G. and X.L.; investigation, X.L.; resources, C.G., X.Y. and X.L.; original draft preparation, S.Z.; review and editing, S.Z., C.G. and X.L.; visualization, S.Z. All authors have read and agreed to the published version of the manuscript.

Funding: This work was funded by the National Natural Science Foundation of China grant number 62175241, U2141255 and 62305344.

Data Availability Statement: The datasets analyzed during the current study are not publicly available but are available from the corresponding author on reasonable request.

Conflicts of Interest: The authors declare no conflicts of interest.

Appendix A

The loss curves demonstrate a consistent downward trajectory, reflecting a reduction in both training and validation losses throughout the training process. This trend indicates that the model is effectively learning and optimizing its parameters, leading to improved performance over time.

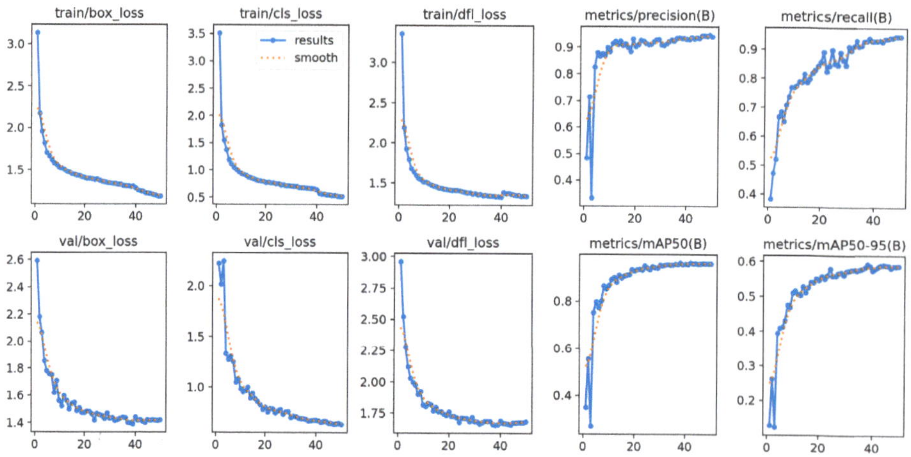

Figure A1. Summary of training.

Conversely, the metric curves display a steady upward trend, which signifies that the model's performance enhances with each training iteration. This improvement suggests that the model is not only fitting the training data well but is also generalizing effectively to the validation set. Such behavior is indicative of a well-optimized training process, where the model successfully captures the underlying patterns in the data, ultimately leading to robust performance in real-world applications.

Overall, the convergence of the loss curves alongside the improvement in metric curves illustrates the model's effectiveness and its potential for reliable predictions.

Appendix B

In different scenarios, the models exhibit varying degrees of missed detections and false detections. In the case of occlusion, the main issue is that the texture feature information of the target is obscured, leading to missed detections. In strong light conditions, the texture information of the target may blend into the background, resulting in missed detections. Additionally, the complex background can create textures that resemble the target, leading to missed detections and false detections. Under dim lighting, significant loss of detail information in the target primarily results in missed detections.

Based on the confusion matrix, our model exhibits the lowest false detection and missed detection rates under occlusion, strong lighting, and low light conditions compared to other models shown in Figure A2 in Appendix B.

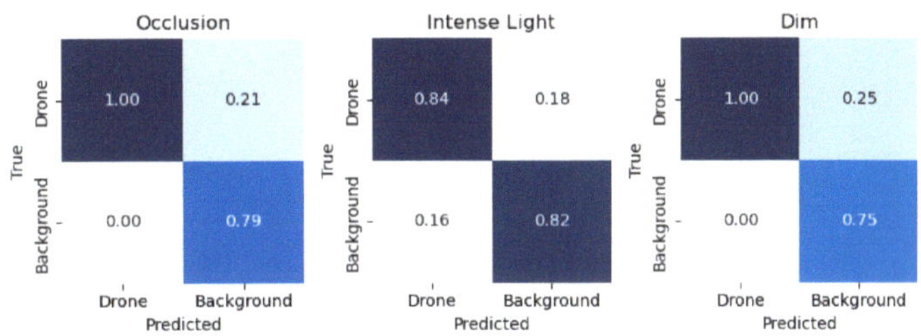

(a)

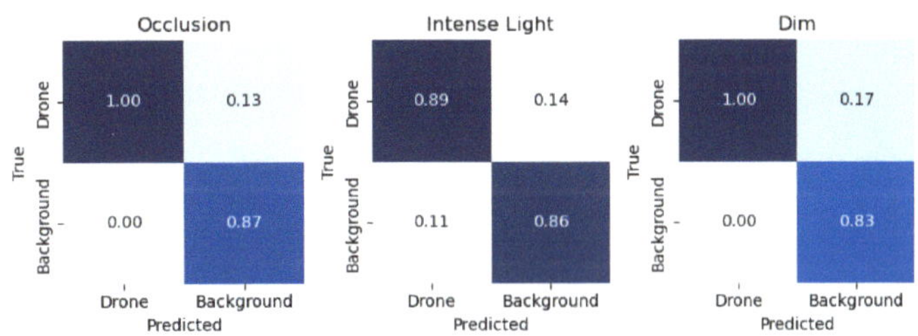

(b)

Figure A2. *Cont.*

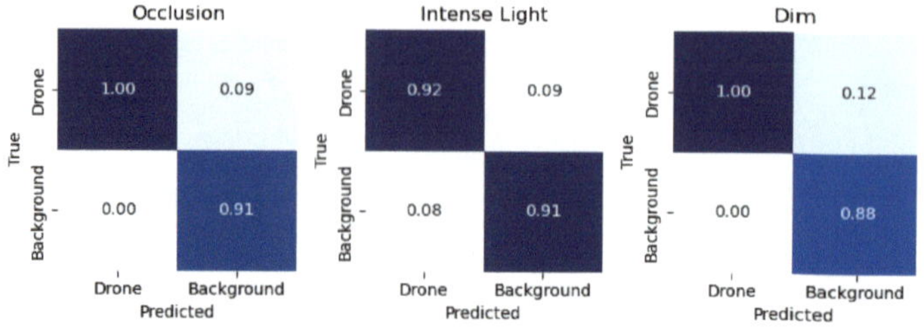

(**c**)

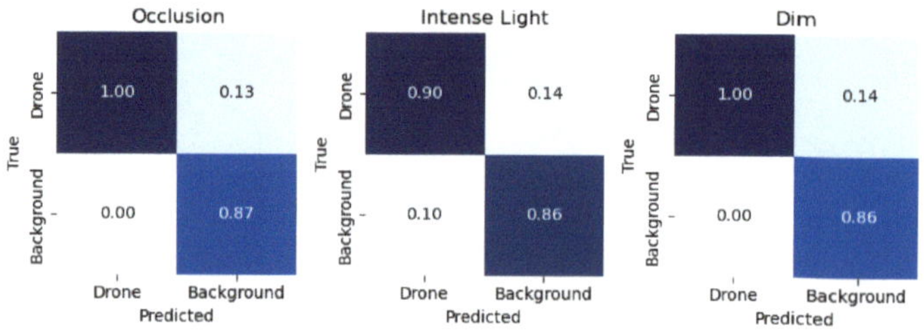

(**d**)

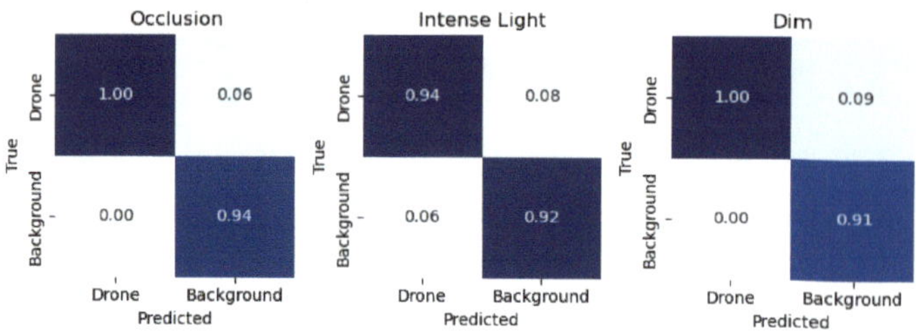

(**e**)

Figure A2. Confusion matrices of different models in different scenarios (occlusion, strong light irradiation, and dim scenes). (**a**) Confusion matrix of YOLOv5. (**b**) Confusion matrix of YOLOv8. (**c**) Confusion matrix of YOLOv9. (**d**) Confusion matrix of YOLOv10. (**e**) Confusion matrix of RTSOD-YOLO.

References

1. Azad, H.; Mehta, V.; Dadboud, F.; Bolic, M.; Mantegh, I. Air-to-Air Simulated Drone Dataset for AI-powered problems. In Proceedings of the 2023 IEEE/AIAA 42nd Digital Avionics Systems Conference (DASC), Barcelona, Spain, 1–5 October 2023; pp. 1–7.
2. Huttner, J.P.; Friedrich, M. Current Challenges in Mission Planning Systems for UAVs: A Systematic Review. In Proceedings of the 2023 Integrated Communication, Navigation and Surveillance Conference (ICNS), Herndon, VA, USA, 18–20 April 2023; pp. 1–7.
3. Mohsan, S.A.H.; Khan, M.A.; Noor, F.; Ullah, I.; Alsharif, M.H. Towards the Unmanned Aerial Vehicles (UAVs): A Comprehensive Review. *Drones* **2022**, *6*, 147. [CrossRef]

4. Farlík, J.; Gacho, L. Researching UAV Threat—New Challenges. In Proceedings of the 2021 International Conference on Military Technologies (ICMT), Brno, Czech Republic, 8–11 June 2021; pp. 1–6.
5. Mohsan, S.A.H.; Othman, N.Q.H.; Li, Y.; Alsharif, M.H.; Khan, M.A. Unmanned aerial vehicles (UAVs): Practical aspects, applications, open challenges, security issues, and future trends. *Intell. Serv. Robot.* **2023**, *16*, 109–137. [CrossRef] [PubMed]
6. Marvasti-Zadeh, S.M.; Cheng, L.; Ghanei-Yakhdan, H.; Kasaei, S. Deep Learning for Visual Tracking: A Comprehensive Survey. *IEEE Trans. Intell. Transp. Syst.* **2022**, *23*, 3943–3968. [CrossRef]
7. Tao, L.; Hong, T.; Chao, X. Drone identification and location tracking based on YOLOv3. *Chin. J. Eng.* **2020**, *42*, 463–468.
8. Hauzenberger, L.; Holmberg Ohlsson, E. Drone Detection Using Audio Analysis. Master's Thesis, Lund University, Lund, Sweden, 2015.
9. Mohajerin, N.; Histon, J.; Dizaji, R.; Waslander, S.L. Feature extraction and radar track classification for detecting UAVs in civillian airspace. In Proceedings of the 2014 IEEE Radar Conference, Cincinnati, OH, USA , 19–23 May 2014; pp. 0674–0679.
10. Al-Emadi, S.; Al-Senaid, F. Drone detection approach based on radio-frequency using convolutional neural network. In Proceedings of the 2020 IEEE International Conference on Informatics, IoT, and Enabling Technologies (ICIoT), Doha, Qatar, 2–5 February 2020; pp. 29–34.
11. Girshick, R.; Donahue, J.; Darrell, T.; Malik, J. Rich Feature Hierarchies for Accurate Object Detection and Semantic Segmentation. In Proceedings of the 2014 IEEE Conference on Computer Vision and Pattern Recognition, Columbus, OH, USA, 23–28 June 2014; pp. 580–587.
12. Girshick, R. Fast r-cnn. *arXiv* **2015**, arXiv:1504.08083.
13. Ren, S.; He, K.; Girshick, R.; Sun, J. Faster R-CNN: Towards Real-Time Object Detection with Region Proposal Networks. *IEEE Trans. Pattern Anal. Mach. Intell.* **2017**, *39*, 1137–1149. [CrossRef]
14. Zhang, H.; Wang, Y.; Dayoub, F.; Sünderhauf, N. VarifocalNet: An IoU-aware Dense Object Detector. In Proceedings of the 2021 IEEE/CVF Conference on Computer Vision and Pattern Recognition (CVPR), Nashville, TN, USA, 20–25 June 2021; pp. 8510–8519.
15. Zhou, X.; Koltun, V.; Krähenbühl, P. Probabilistic two-stage detection. *arXiv* **2021**, arXiv:2103.07461.
16. Liu, W.; Anguelov, D.; Erhan, D.; Szegedy, C.; Reed, S.; Fu, C.Y.; Berg, A.C. SSD: Single Shot MultiBox Detector. In Computer Vision–ECCV 2016, Proceedings of the 14th European Conference, Amsterdam, The Netherlands, 11–14 October 2016; Leibe, B., Matas, J., Sebe, N., Welling, M., Eds.; Springer International Publishing: Cham, Switzerland, 2016; pp. 21–37.
17. Lin, T.Y.; Goyal, P.; Girshick, R.; He, K.; Dollár, P. Focal Loss for Dense Object Detection. In Proceedings of the 2017 IEEE International Conference on Computer Vision (ICCV), Venice, Italy, 22–29 October 2017; pp. 2999–3007.
18. Redmon, J.; Divvala, S.; Girshick, R.; Farhadi, A. You Only Look Once: Unified, Real-Time Object Detection. In Proceedings of the Computer Vision & Pattern Recognition, Las Vegas, NV, USA, 27–30 June 2016.
19. Taha, B.; Shoufan, A. Machine learning-based drone detection and classification: State-of-the-art in research. *IEEE Access* **2019**, *7*, 138669–138682. [CrossRef]
20. Batool, S.; Frezza, F.; Mangini, F.; Simeoni, P. Introduction to radar scattering application in remote sensing and diagnostics. *Atmosphere* **2020**, *11*, 517. [CrossRef]
21. Liu, J.; Xu, Q.Y.; Chen, W.S. Classification of bird and drone targets based on motion characteristics and random forest model using surveillance radar data. *IEEE Access* **2021**, *9*, 160135–160144. [CrossRef]
22. Wang, C.; Tian, J.; Cao, J.; Wang, X. Deep learning-based UAV detection in pulse-Doppler radar. *IEEE Trans. Geosci. Remote Sens.* **2021**, *60*, 1–12. [CrossRef]
23. Yan, J.; Hu, H.; Gong, J.; Kong, D.; Li, D. Exploring Radar Micro-Doppler Signatures for Recognition of Drone Types. *Drones* **2023**, *7*, 280. [CrossRef]
24. Rahman, S.; Robertson, D.A. Radar micro-Doppler signatures of drones and birds at K-band and W-band. *Sci. Rep.* **2018**, *8*, 17396. [CrossRef] [PubMed]
25. Narayanan, R.M.; Tsang, B.; Bharadwaj, R. Classification and discrimination of birds and small drones using radar micro-Doppler spectrogram images. *Signals* **2023**, *4*, 337–358. [CrossRef]
26. Nemer, I.; Sheltami, T.; Ahmad, I.; Yasar, A.U.H.; Abdeen, M.A. RF-based UAV detection and identification using hierarchical learning approach. *Sensors* **2021**, *21*, 1947. [CrossRef]
27. Zhang, Y. RF-based drone detection using machine learning. In Proceedings of the 2021 2nd International Conference on Computing and Data Science (CDS), Stanford, CA, USA, 28–29 January 2021; pp. 425–428.
28. Medaiyese, O.O.; Syed, A.; Lauf, A.P. Machine learning framework for RF-based drone detection and identification system. In Proceedings of the 2021 2nd International Conference On Smart Cities, Automation & Intelligent Computing Systems (ICON-SONICS), Tangerang, Indonesia, 12–13 October 2021; pp. 58–64.
29. Allahham, M.S.; Khattab, T.; Mohamed, A. Deep learning for RF-based drone detection and identification: A multi-channel 1-D convolutional neural networks approach. In Proceedings of the 2020 IEEE International Conference on Informatics, IoT, and Enabling Technologies (ICIoT), Doha, Qatar, 2–5 February 2020; pp. 112–117.
30. Inani, K.N.; Sangwan, K.; et al. Machine Learning based framework for Drone Detection and Identification using RF signals. In Proceedings of the 2023 4th International Conference on Innovative Trends in Information Technology (ICITIIT), Kottayam, India, 11–12 February 2023; pp. 1–8.

31. Fagiani, F.R.E. Uav Detection and Localization System Using an Interconnected Array of Acoustic Sensors and Machine Learning Algorithms. Master's Thesis, Purdue University, West Lafayette, IN, USA, 2021.
32. Ahmed, C.A.; Batool, F.; Haider, W.; Asad, M.; Hamdani, S.H.R. Acoustic Based Drone Detection Via Machine Learning. In Proceedings of the 2022 International Conference on IT and Industrial Technologies (ICIT), Chiniot, Pakistan, 3–4 October 2022; pp. 1–6.
33. Tejera-Berengue, D.; Zhu-Zhou, F.; Utrilla-Manso, M.; Gil-Pita, R.; Rosa-Zurera, M. Acoustic-Based Detection of UAVs Using Machine Learning: Analysis of Distance and Environmental Effects. In Proceedings of the 2023 IEEE Sensors Applications Symposium (SAS), Ottawa, ON, Canada, 18–20 July 2023; pp. 1–6.
34. Ohlenbusch, M.; Ahrens, A.; Rollwage, C.; Bitzer, J. Robust drone detection for acoustic monitoring applications. In Proceedings of the 2020 28th European Signal Processing Conference (EUSIPCO), Amsterdam, The Netherlands, 18–21 January 2021, pp. 6–10.
35. Al-Emadi, S.; Al-Ali, A.; Al-Ali, A. Audio-based drone detection and identification using deep learning techniques with dataset enhancement through generative adversarial networks. *Sensors* **2021**, *21*, 4953. [CrossRef]
36. Utebayeva, D.; Ilipbayeva, L.; Matson, E.T. Practical study of recurrent neural networks for efficient real-time drone sound detection: A review. *Drones* **2022**, *7*, 26. [CrossRef]
37. Mubarak, A.S.; Vubangsi, M.; Al-Turjman, F.; Ameen, Z.S.; Mahfudh, A.S.; Alturjman, S. Computer vision based drone detection using mask R-CNN. In Proceedings of the 2022 International Conference on Artificial Intelligence in Everything (AIE), Lefkosa, Cyprus, 2–4 August 2022; pp. 540–543.
38. Shang, Y.; Liu, C.; Qiu, D.; Zhao, Z.; Wu, R.; Tang, S. AD-YOLOv5s based UAV detection for low altitude security. *Int. J. Micro Air Veh.* **2023**, *15*, 17568293231190017. [CrossRef]
39. Kabir, M.S.; Ndukwe, I.K.; Awan, E.Z.S. Deep Learning Inspired Vision based Frameworks for Drone Detection. In Proceedings of the 2021 International Conference on Electrical, Communication, and Computer Engineering (ICECCE), Kuala Lumpur, Malaysia, 12–13 June 2021; pp. 1–5.
40. Delleji, T.; Chtourou, Z. An Improved YOLOv5 for Real-time Mini-UAV Detection in No Fly Zones. In Proceedings of the International Conference on Image Processing and Vision Engineering, Online, 22–24 April 2022.
41. Sethu Selvi, S.; Pavithraa, S.; Dharini, R.; Chaitra, E. A Deep Learning Approach to Classify Drones and Birds. In Proceedings of the 2022 IEEE 2nd Mysore Sub Section International Conference (MysuruCon), Mysuru, India, 16–17 October 2022; pp. 1–5.
42. Xu, S.; Wang, X.; Lv, W.; Chang, Q.; Cui, C.; Deng, K.; Wang, G.; Dang, Q.; Wei, S.; Du, Y.; et al. PP-YOLOE: An evolved version of YOLO. *arXiv* **2022**, arXiv:2203.16250.
43. Xu, X.; Jiang, Y.; Chen, W.; Huang, Y.; Zhang, Y.; Sun, X. Damo-yolo: A report on real-time object detection design. *arXiv* **2022**, arXiv:2211.15444.
44. Xie, Y.; Zhang, L.; Yu, X.; Xie, W. YOLO-MS: Multispectral Object Detection via Feature Interaction and Self-Attention Guided Fusion. *IEEE Trans. Cogn. Dev. Syst.* **2023**, *15*, 2132–2143. [CrossRef]
45. Al-Qubaydhi, N.; Alenezi, A.; Alanazi, T.; Senyor, A.; Alanezi, N.; Alotaibi, B.; Alotaibi, M.; Razaque, A.; Abdelhamid, A.A.; Alotaibi, A. Detection of Unauthorized Unmanned Aerial Vehicles Using YOLOv5 and Transfer Learning. *Electronics* **2022**, *11*, 2669. [CrossRef]
46. Pansare, A.; Sabu, N.; Kushwaha, H.; Srivastava, V.; Thakur, N.; Jamgaonkar, K.; Faiz, M.Z. Drone Detection using YOLO and SSD A Comparative Study. In Proceedings of the 2022 International Conference on Signal and Information Processing (IConSIP), Pune, India, 26–27 August 2022; pp. 1–6.
47. Aydin, B.; Singha, S. Drone Detection Using YOLOv5. *Eng* **2023**, *4*, 416–433. [CrossRef]
48. Zhao, J.; Zhang, J.; Li, D.; Wang, D. Vision-based anti-uav detection and tracking. *IEEE Trans. Intell. Transp. Syst.* **2022**, *23*, 25323–25334. [CrossRef]
49. Singha, S.; Aydin, B. Automated drone detection using YOLOv4. *Drones* **2021**, *5*, 95. [CrossRef]
50. Varghese, R.; Sambath, M. YOLOv8: A Novel Object Detection Algorithm with Enhanced Performance and Robustness. In Proceedings of the 2024 International Conference on Advances in Data Engineering and Intelligent Computing Systems (ADICS), Chennai, India, 18–19 April 2024.
51. Wang, C.Y.; Yeh, I.H.; Liao, H.Y.M. YOLOv9: Learning What You Want to Learn Using Programmable Gradient Information. *arXiv* **2024**, arXiv:2402.13616.
52. Han, K.; Wang, Y.; Tian, Q.; Guo, J.; Xu, C.; Xu, C. GhostNet: More Features From Cheap Operations. In Proceedings of the 2020 IEEE/CVF Conference on Computer Vision and Pattern Recognition (CVPR), Seattle, WA, USA, 14–19 June 2020; pp. 1577–1586.
53. Kang, M.; Ting, C.M.; Ting, F.F.; Phan, R.C.W. ASF-YOLO: A novel YOLO model with attentional scale sequence fusion for cell instance segmentation. *Image Vis. Comput.* **2024**, *147*, 105057. [CrossRef]
54. Yu, Z.; Huang, H.; Chen, W.; Su, Y.; Liu, Y.; Wang, X. Yolo-facev2: A scale and occlusion aware face detector. *Pattern Recognit.* **2024**, *155*, 110714. [CrossRef]
55. Li, C.; Li, L.; Jiang, H.; Weng, K.; Geng, Y.; Li, L.; Ke, Z.; Li, Q.; Cheng, M.; Nie, W.; et al. YOLOv6: A Single-Stage Object Detection Framework for Industrial Applications. *arXiv* **2022**, arXiv:2209.02976.
56. Wang, C.; He, W.; Nie, Y.; Guo, J.; Liu, C.; Han, K.; Wang, Y. Gold-YOLO: Efficient Object Detector via Gather-and-Distribute Mechanism. *arXiv* **2023**, arXiv:2309.11331.
57. Wang, A.; Chen, H.; Liu, L.; Chen, K.; Lin, Z.; Han, J.; Ding, G. YOLOv10: Real-Time End-to-End Object Detection. *arXiv* **2024**, arXiv:2405.14458.

58. Jiang, N.; Wang, K.; Peng, X.; Yu, X.; Wang, Q.; Xing, J.; Li, G.; Ye, Q.; Jiao, J.; Han, Z.; et al. Anti-UAV: A large-scale benchmark for vision-based UAV tracking. *IEEE Trans. Multimed.* **2021**, *25*, 486–500. [CrossRef]
59. Svanström, F.; Englund, C.; Alonso-Fernandez, F. Real-time drone detection and tracking with visible, thermal and acoustic sensors. In Proceedings of the 2020 25th International Conference on Pattern Recognition (ICPR), Milan, Italy, 10–15 January 2021; pp. 7265–7272.
60. Coluccia, A.; Fascista, A.; Schumann, A.; Sommer, L.; Dimou, A.; Zarpalas, D.; Méndez, M.; De la Iglesia, D.; González, I.; Mercier, J.P.; et al. Drone vs. bird detection: Deep learning algorithms and results from a grand challenge. *Sensors* **2021**, *21*, 2824. [CrossRef] [PubMed]

remote sensing

MDPI

Article

Multi-Modal Prototypes for Few-Shot Object Detection in Remote Sensing Images

Yanxing Liu [1,2,3,4], Zongxu Pan [1,2,3,4,*], Jianwei Yang [1,2,3,4], Peiling Zhou [1,2,3,4] and Bingchen Zhang [1,2,3,4]

1 Aerospace Information Research Institute, Chinese Academy of Sciences, Beijing 100094, China; liuyanxing21@mails.ucas.ac.cn (Y.L.); yangjianwei20@mails.ucas.ac.cn (J.Y.); zhoupeiling21@mails.ucas.ac.cn (P.Z.); zhangbc@aircas.ac.cn (B.Z.)

2 Key Laboratory of Technology in Geo-Spatial Information Processing and Application System, Chinese Academy of Sciences, Beijing 100190, China

3 School of Electronic, Electrical and Communication Engineering, University of Chinese Academy of Sciences, Beijing 100049, China

4 Key Laboratory of Target Cognition and Application Technology (TCAT), Chinese Academy of Sciences, Beijing 100190, China

* Correspondence: zxpan@mail.ie.ac.cn

Abstract: Few-shot object detection has attracted extensive attention due to the abomination of time-consuming or even impractical large-scale data labeling. Current studies attempted to employ prototype-matching approaches for object detection, constructing class prototypes from textual or visual features. However, single visual prototypes exhibit limited generalization in few-shot scenarios, while single textual prototypes lack the spatial details of remote sensing targets. Therefore, to achieve the best of both worlds, we propose a prototype aggregating module to integrate textual and visual prototypes, leveraging both semantics from textual prototypes and spatial details from visual prototypes. In addition, the transferability of multi-modal few-shot detectors from natural scenarios to remote sensing scenarios remains unexplored, and previous training strategies for FSOD do not adequately consider the characteristics of text encoders. To address the issue, we have conducted extensive ablation studies on different feature extractors of the detector and propose an efficient two-stage training strategy, which takes the characteristics of the text feature extractor into account. Experiments on two common few-shot detection benchmarks demonstrate the effectiveness of our proposed method. In four widely used data splits of DIOR, our method significantly outperforms previous state-of-the-art methods by at most 8.7%.

Keywords: few-shot learning; object detection; remote sensing images

Citation: Liu, Y.; Pan, Z.; Yang, J.; Zhou, P.; Zhang, B. Multi-Modal Prototypes for Few-Shot Object Detection in Remote Sensing Images. *Remote Sens.* **2024**, *16*, 4693. https://doi.org/10.3390/rs16244693

Academic Editors: Qian Du, Yanni Dong and Xiaochen Yang

Received: 4 November 2024
Revised: 10 December 2024
Accepted: 10 December 2024
Published: 16 December 2024

1. Introduction

Benefiting from the rapid development of deep learning, significant progress has been achieved in the realm of remote-sensing object detection. Nonetheless, nearly all deep learning detectors rely heavily on large-scale annotated datasets to achieve satisfactory performance. Unfortunately, collecting sufficient labeled data is time-consuming or even unfeasible, particularly for remote sensing images (RSIs) with large image sizes and complex backgrounds. In this case, existing detectors will suffer from severe overfitting. To mitigate this limitation, few-shot object detection (FSOD) has been proposed and has garnered increasing attention from researchers.

FSOD aims to leverage knowledge learned from data-rich base datasets to enhance the performance of novel classes with limited data. Concretely, most FSOD approaches can be divided into fine-tuning methods and prototype-based methods. Fine-tuning methods [1–4] transfer the knowledge from base classes to novel classes through fine-tuning. Methods based on fine-tuning are more straightforward, yet their detection performance remains unsatisfactory. Prototype-based methods aim to construct a class prototype for each class and detect few-shot classes by matching candidate regions with corresponding class prototypes. Pioneering works [5–8] construct class prototypes from few-shot support images to detect objects in query images, as illustrated in Figure 1a. Inspired by the rapid development of natural language processing, current researchers are exploring the construction of class prototypes from textual class names [9,10] or class descriptions [11], as illustrated in Figure 1b. Owing to the robustness of class prototypes, prototype-based approaches have demonstrated superior detection performance compared to fine-tuning methods in numerous applications.

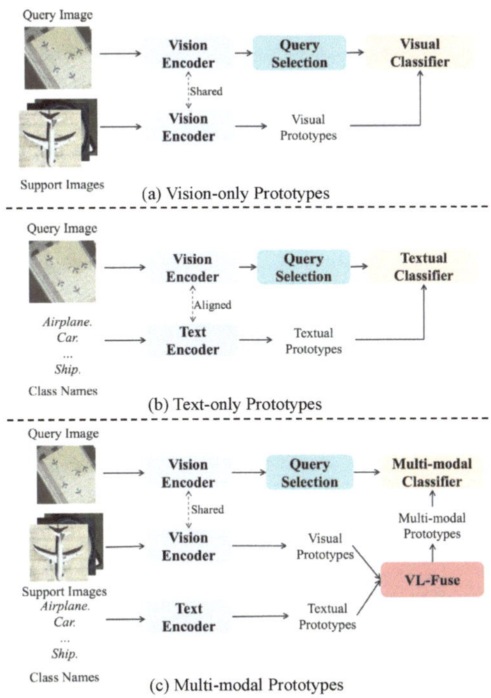

Figure 1. (**a**) Detectors based on vision-only prototypes construct class prototypes from the visual features of few-shot support images. (**b**) Detectors based on text-only prototypes build their class prototypes from class names. (**c**) Our approach constructs multi-modal prototypes to leverage the prior knowledge of textual prototypes and the spatial details of visual prototypes.

However, most prototype-based object detectors [5–11] construct class prototypes from either a single-modal image feature or a class name, as depicted in Figure 1a,b. These single-modal features constrain the feature extraction capabilities of prototypes. Actually, both visual and textual prototypes possess distinct strengths and drawbacks. On the one hand, textual prototypes generated from the text encoder contain rich prior knowledge about target classes and exhibit strong generalization capabilities, as the text encoder has been pre-trained on a large-scale natural language corpus. Nonetheless, since the textual prototypes are domain-agnostic, they cannot provide specific spatial details about target classes in RSIs, thereby limiting their applications in the field of remote-sensing object detection. On the other hand, a picture paints a thousand words. Compared to textual prototypes, visual prototypes generated from RSIs can provide richer details about target classes. Nevertheless, the visual prototypes generated from limited support images lack generalization capabilities in few-shot scenarios, which will significantly degrade detection performance. In summary, single-modal visual and textual prototypes have distinct strengths and limitations.

Therefore, to further boost the performance of FSOD in RSIs, we propose a framework that leverages both visual and textual prototypes and we introduce a prototype aggregation module (PAM) that integrates the generalized prior knowledge of textual prototypes with the spatial details of visual prototypes. To be specific, as illustrated in Figure 1c, few-shot support images and textual class names are processed through their respective feature extractors to derive individual visual and textual prototypes. The highly generalizable textual prototypes and detail-rich visual prototypes are then aggregated by a prototype aggregator to obtain multi-modal prototypes. The produced multi-modal prototype retains both the generalized prior knowledge of textual prototypes and the detailed spatial information of visual prototypes. This multi-modal prototype is then integrated into the encoder and decoder of GroundingDINO [10] to detect targets.

The introduction of pre-trained text encoders makes previous few-shot object detection training strategies no longer suitable for our approach, as these strategies have not taken the text encoder into consideration and the text encoder is inherently different from the vision encoder. Therefore, we have conducted extensive ablation studies on the generalization capability of the pre-trained text encoder for FSOD in RSIs. We surprisingly find that the pre-trained text feature extractor exhibits remarkable robustness in FSOD tasks on RSIs. Even without any fine-tuning of the text encoder, a detector based on the pre-trained text encoder can achieve comparable performance to previous state-of-the-art FSOD methods. We argue that the reason for this is that the text encoder is domain-agnostic. Unlike images that exhibit different characteristics across domains, textual semantic information typically maintains similarity across various domains. Based on ablation studies and previous findings, we propose a novel efficient two-stage training strategy (ETS) for FSOD on RSIs, which is the first to take the characteristics of the pre-trained text feature extractor into account.

By integrating the multi-modal prototypes and the efficient two-stage training strategy with an advanced detector [10], we propose the multi-modal prototypical few-shot object detector (MP-FSDet) and achieve state-of-the-art few-shot object detection performance on DIOR and NWPU VHR-10.v2, which are two widely used datasets for few-shot object detection in RSIs. As shown in Figure 2, our method remarkably outperforms other state-of-the-art detectors, achieving a maximum performance improvement of 8.7% in the remote-sensing FSOD task.

In summary, the contributions of this paper are summarized as follows.

1. We propose a few-shot object detector based on multi-modal prototypes, aiming to enhance few-shot detection performance by combining visual prototypes and textual prototypes.

2. We propose a prototype aggregating method for the construction of multi-modal prototypes, which maintains the spatial details of visual prototypes and semantic prior of textual prototypes.

3. Based on comprehensive analyses and extensive experiments on different components of detectors, we propose an efficient two-stage training strategy (ETS) for FSOD on RSIs, which, to the best of our knowledge, is the first to take the characteristics of the pre-trained text feature extractor into account.

4. By integrating our proposed methods with the state-of-the-art detector [10], we achieve state-of-the-art detection performance on two widely used benchmarks, attaining a maximum performance improvement of 8.7% in the remote-sensing FSOD task.

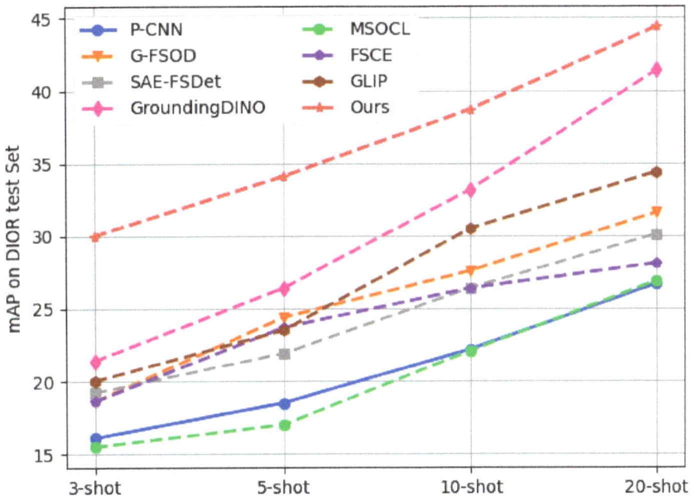

Figure 2. FSOD performance on DIOR test set at different shot numbers. We report the average AP50 on four splits, followed by [8]. As the figure shows, our proposed method is remarkably superior to other state-of-the-art approaches.

2. Related Work

In this chapter, we will introduce relevant works from two aspects. Firstly, we will introduce some generic object detection methods. Secondly, we will discuss various few-shot object detection approaches and analyze the shortcomings of previous research.

2.1. Generic Object Detection

Based on handcraft anchors or reference points, early convolution-based detectors are designed as either two-stage or one-stage models. Two-stage detectors [12,13] typically incorporate a region proposal network (RPN) to propose potential boxes and subsequently classify them in the second stage. One-stage detectors like YOLO v2 [14] and YOLO v3 [15] directly output offsets relative to predefined anchors and classification scores. Currently, a transformer-based end-to-end detector [16] has been proposed, which eliminates hand-designed components like anchors and NMS. To alleviate the computational burden, Deformable DETR [17] predicts 2D anchor offsets and incorporates a deformable attention module that focuses on a specific set around a reference point. By using a contrastive way for denoising training, a mixed query selection method, and a look-forward-twice scheme, DINO [18] further boosts the performance of transformer-based detectors.

Recently, the rapid advancement of natural language processing has enabled researchers to leverage textual information for object detection. Object detection approaches [9,10,19–21] via text queries have exhibited remarkable advancements within natural scenes. They are trained using existing bounding-box annotations and aim at detecting arbitrary classes with the help of generalized text queries. OVR-CNN [19] is a pioneering work that first pre-trains a visual projector on image–caption pairs to learn rich vocabularies and then fine-tunes a detector on the detection dataset. Then, by distilling the knowledge from the CLIP [20] model into an R-CNN-like detector, ViLD [21] further advances open-set object detection. GLIP [9] first formulates object detection as a grounding task and achieves an even stronger performance on fully supervised detection benchmarks without fine-tuning. GroundingDINO [10] utilizes the training strategy of

GLIP as a stronger detector DINO [18] and achieves a state-of-the-art open-set detection performance. The MQ-Det [22] utilizes both textual and visual queries for detection, which is similar to our work. However, it solely utilizes vectors as image queries, thereby failing to preserve the spatial details of the target. Furthermore, MQ-Det [22] is designed for natural scene images, which does not account for the domain gap between remote sensing and natural scene imagery. Since the models are only pre-trained on natural images and only use text embeddings as prototypes, the detection performance of these models is degraded on RSIs.

Along with the rapid development of detectors in natural scenes, object detection methods for aerial images have also achieved tremendous progress [23]. To address the challenge of rotation invariance for feature representation, Mei et al. [24] propose a novel cyclic polar coordinate convolutional layer. Considering the expansive spatial coverage of aerial images and diverse scales of objects, Deng et al. [25] propose a multi-scale object proposal network (MS-OPN), which comprises three proposal branches to predict multi-scale proposals. Due to the unique bird's-eye views of RSIs, objects in aerial images often exhibit arbitrary orientations. To achieve rotation-invariant detection, Cheng et al. [26] propose a Rotation-Invariant CNN (RICNN), which incorporates a novel rotation-invariant layer. Li et al. [27] propose a novel region proposal network that incorporates additional multi-angle anchors to detect objects with arbitrary orientations. Recent studies [28,29] have begun to explore the potential of detecting objects labeled with rotated bounding boxes.

Detectors, as mentioned above, have achieved remarkable performance. However, most of these are prone to overfitting when the labeled data become scarce and this significantly restricts the application of object detection for aerial images.

2.2. Few-Shot Object Detection

With the abomination of time-consuming or even impractical large-scale labeling, FSOD has attracted extensive attention.

Current FSOD methods can be divided into transfer-learning methods [1–3,30] and prototype-based methods [5–7,9,10,22,31]. Transfer-learning methods opt to fine-tune a pre-trained model on limited data for the few-shot task. Flagship works include LSTD [1], TFA [2], FSCE [3], and DeFRCN [30]. Prototype-based methods [5–7,9,10,22,31] aim to build a class prototype for each class and detect novel classes by matching candidate regions to class prototypes. They are trained across tasks for quick adaptation, and they are trained over multiple episodes. Early works [5–7,31,32] only employ few-shot support images as class prototypes. However, the visual prototypes produced by the few-shot images lack generalization. Inspired by the rapid development of natural language processing, the authors of [9,10,22] input class names into a pre-trained text encoder and utilize the resulting text features as class prototypes. As the text encoder is pre-trained on a large-scale textual corpus, textual prototypes generated from it exhibit strong generalization capabilities. Nevertheless, textual prototypes lack specific details about target classes.

Currently, FSOD for aerial images has attracted increasing attention. Li et al. [33] introduce a novel meta-learning approach with multi-scale detection capabilities tailored for the FSOD task in aerial images. Then, Cheng et al. [8] develop a framework named Prototype-CNN, which aims to leverage prototypes for recognizing objects with limited annotations. Similarly, Le et al. [34] also propose a novel prototypical network based on representation learning. Huang et al. [35] address the issue of significant data imbalance between the novel class and base classes by proposing a novel balanced fine-tuning approach and incorporating a shared attention module to exploit the abundant ground information within aerial images. DH-FSDet [36] presents an innovative annotation sampling and preprocessing methodology. To tackle the issue of limited scale diversity in aerial images, MSOCL [37] presents a multiscale object contrastive learning framework. TEMO [11] employs class descriptions as class prototypes. However, the text encoder it utilizes is training from scratch, thereby lacking generalization capabilities. In addition, single-modal prototypes in TEMO [11] cannot provide comprehensive information for target classes. Similarly,

UMFT [38] also proposes a multi-modal transformer to integrate the query features from ViT and the textual features extracted from BERT, aligning multi-modal representations in an end-to-end manner. To overcome the challenges of substantial scale and orientation variations of objects in RSIs, TINet [39] proposes a novel feature pyramid network (FPN) and utilizes prototype features to enhance query features. To rectify the issue of inconsistent label assignments across base training and fine-tuning stages, SAE-FSDet [40] proposes a novel label-consistent classifier named LCC and proposes a novel gradual rpn to enhance the localization performance of novel classes, inspired by the authors of [41,41–43].

Although prototype-based detection approaches have demonstrated remarkable performance in the field of FSOD, single-modality prototypes exhibit distinct strengths and limitations. To further boost the performance of FSOD, we propose a multi-modal prototype that can harness the strengths of both textual and visual prototypes. Moreover, the transferability of detectors based on multi-modal prototypes in the field of remote-sensing FSOD has not been explored. Through a comprehensive analysis, we find that the text encoder is domain-agnostic, while the visual encoder is domain-specific in RSIs, and we propose an efficient two-stage training strategy.

3. Proposed Approach

In this section, we will introduce the proposed MP-FSDet network in detail. Before introducing our approach, we will first present some preliminary knowledge in Section 3.1. Subsequently, we will provide a brief overview of the architecture of our proposed method in Section 3.2. To leverage both the prior knowledge from textual prototypes and the spatial details from visual prototypes, we present a multi-modal prototype construction method introduced in Section 3.3. Furthermore, we have found that the previous training strategy can harm the generalization capability of the text encoder. Thus, we propose an efficient two-stage training strategy in Section 3.4 to address the problem.

3.1. *Preliminary Knowledge*

3.1.1. Problem Setting

FSOD aims to train a detector on a base dataset D_b with abundant annotated data and a novel dataset D_n containing limited data. Classes in D_b are denoted as base classes C_b, and those in D_n are denoted as novel classes C_n. The base classes are fully annotated, containing numerous annotated object instances, while the novel classes have only K-shot annotations. Note that the base classes C_b and the novel classes C_n are non-overlapping, namely $C_b \cap C_n = \emptyset$. For the inference phase in FSOD, the test set contains both base and novel classes ($C_{test} = C_b + C_n$), and the detector is required to detect all objects. As a result, a balanced few-shot dataset D_{few} containing K-shot object annotations for both base and novel classes is constructed for few-shot fine-tuning.

3.1.2. Base Detector

Our model is grounded on GroundingDINO [10], which employs single-modal textual prototypes. It is originally designed for open-set object detection, a task that endeavors to detect arbitrary objects within human inputs, such as category names or referring expressions. However, owing to highly generalized language encoders and large-scale language–image alignment, it demonstrates remarkable performance in the domain of few-shot object detection [22]. The detector first obtains the respective feature maps via the image and text encoders. Subsequently, the textual prototypes generated from the text encoder and image features are fed into the transformer encoder and decoders to obtain the final detection results. The transformer encoder and decoder can be conceptualized as a matcher that matches the feature points on the feature map with the class prototypes of each class, utilizing the class with the highest confidence as the final classification result.

3.2. Multi-Modal Prototypical Few-Shot Object Detector

Textual prototypes that are pre-trained from a large-scale natural language corpus are highly generalizable, while visual prototypes contain detailed spatial information about specific objects. Therefore, we propose a detection framework that aims to combine the generalization properties of textual prototypes with the spatial details of objects in aerial images.

The architecture overview is illustrated in Figure 3. As the figure shows, the textual prototypes $P_t \in \mathbb{R}^{|C| \times D}$ are extracted from the class names by a pre-trained BERT [44] encoder, where $|C|$ is the number of classes. Benefiting from the pre-training of the BERT encoder on large-scale corpora, textual prototypes encompass extensive prior knowledge of targets and exhibit robust generalization capability. Meanwhile, the visual prototypes $P_v \in \mathbb{R}^{|C| \times H_s W_s \times D}$ are extracted from support images by a vision encoder and an RoI pooling layer, where H_s and W_s denote the height and width of support RoI features, respectively. Visual prototypes preserve spatial details of the target within the image. It is noteworthy that, unlike the previous approach [6], which sends the vector after splicing and resizing the support images and foreground masks to the vision encoder, we directly input the original support images into the vision encoder for processing. We argue that consistent pre-processing of support images and query images can benefit the alignment between query features and visual prototypes.

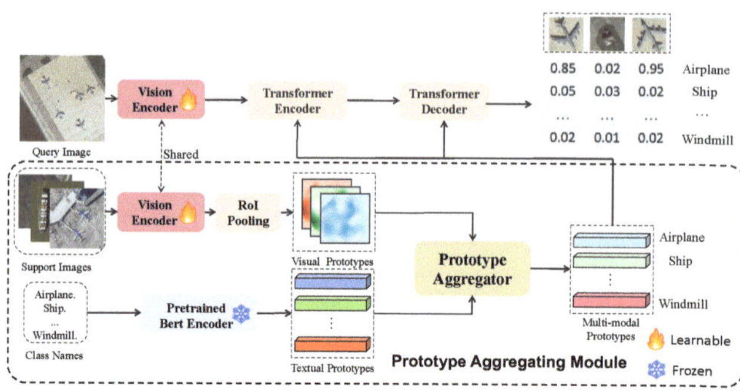

Figure 3. The architecture of our multi-modal prototypical few-shot object detector. It comprises two branches: a query image detection branch (**top**) and a multi-modal prototype construction branch (**bottom**). The upper branch employs the DETR architecture, which is responsible for detecting objects in the query image based on the multi-modal prototypes. The feature enhancer in the branch is a DETR encoder that employs a cross-attention mechanism to enhance the interaction between query features and multi-modal prototypes. The multi-modal decoder in the branch can be regarded as a matcher that matches each query feature with multi-modal prototypes. The bottom branch utilizes the prototype aggregating module (PAM) to construct multi-modal prototypes. The PAM feeds support images and class names into the visual encoder and BERT encoder to obtain visual prototypes and textual prototypes, respectively. The proposed prototype aggregator then aggregates the textual and visual prototypes into multi-modal prototypes. Note that the BERT encoder is frozen in the base training stage to maintain the generalization of textual prototypes.

Subsequently, the textual prototypes P_t and visual prototypes P_v are then fed into the prototype aggregator to obtain multi-modal prototypes $P \in \mathbb{R}^{|C| \times D}$. The prototype aggregator comprises a cross-modal attention mechanism and the details will be introduced in detail in Section 3.3. This module fuses semantic features from the textual prototypes with spatial details from the aerial images to construct multi-modal prototypes. After obtaining multi-modal prototypes, they will be fed into a transformer encoder introduced in GroundingDINO [10], along with the query features $F_q \in \mathbb{R}^{H_q W_q \times D}$. The transformer encoder [10] comprises six mutual cross-attention layers to enhance the interactivity between multi-modal prototypes and query features. At last, the multi-modal prototypes and enhanced query features will be jointly input into the multi-modal decoder [10] to obtain the final detection outputs. The classification layer in the decoder will match the selected object queries with the multi-modal prototypes of each category one by one to obtain the final classification results. The decoder also incorporates several regression layers to refine the positions of object queries.

The training loss of our model is composed as follows:

$$L = L_{cls} + L_{loc} + L_{IoU} + L_{dn} \tag{1}$$

where L_{cls} and L_{loc} represent the classification loss and localization loss of the bounding boxes, respectively. Specifically, we employ the focal loss [45] for the classification loss and the smooth L1 loss [12] for the localization loss. The L_{IoU} is an additional loss employed to enhance the ability of detectors to handle objects with varying scales and shapes. Furthermore, in line with DINO [18], we also incorporate a denoising loss L_{dn} to address the low training efficiency of the DETR [16] model and enhance model convergence, as mentioned in [18,46].

3.3. Prototype Aggregating Module

Inspired by [9], we propose a prototype aggregating module to aggregate visual prototypes and textual prototypes. This module primarily comprises a prototype aggregator, designed to leverage both the spatial detail of visual prototypes and the generalization capability of textual prototypes.

Specifically, we employ a multi-head attention module in the prototype aggregator, where each head computes the cross-modal prototype by aggregating visual features into a text prototype, as shown in Figure 4. The prototype aggregator fuses the visual prototype and textual prototype for each class individually. Without loss of generality, the following description focuses on the construction of a single-class multi-modal prototype.

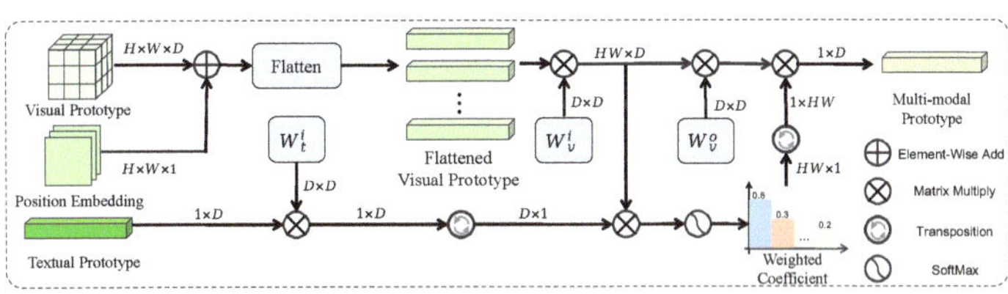

Figure 4. The architecture of the proposed prototype aggregator. The visual prototype, with positional embedding incorporated, is firstly reshaped into $P_v \in \mathbb{R}^{HW \times D}$. The flattened visual prototype and textual prototype are subsequently projected through the input projection weights W_v^i and W_t^i, respectively, to perform feature transformation. W_v^i and W_t^i are the transformation matrices for the input textual features and input visual features of the ith class, respectively. W_v^o represents the output transformation matrix of the visual features and the weighted visual prototypes will ultimately be produced as the multi-modal prototype. Then, the correlation between textual prototypes and visual prototypes is quantified by computing the normalized attention weights. At last, the weighted visual prototype is integrated into the text prototype, resulting in the multi-modal prototype.

The detailed formulas for a multi-modal prototype are presented as follows:

$$O_v = P_v W_v^i, O_t = P_t W_t^i, \tag{2}$$

$$A = \text{SoftMax}(\frac{O_t O_v^T}{\sqrt{D}}), \tag{3}$$

$$P = AP_vW_v^o, \tag{4}$$

where $P_v \in \mathbb{R}^{HW \times D}$, $P_t \in \mathbb{R}^{1 \times D}$ are flattened visual and textual prototypes, respectively, W_v^i and W_t^i are input projection weights, and W_v^o is the output project weights. The visual and textual prototypes are first projected into a high-dimension semantic space by input projection weights, as shown in Equation (2) with O_v and O_t. Then, the correlation matrix A between the semantic textual prototype and the detailed visual prototype is computed by calculating the normalized attention weights, as demonstrated in Equation (3). At last, the multi-modal prototype P is obtained by integrating weighted visual details into the text prototype, as illustrated in Equation (4).

While the proposed PAM appears similar to the feature fusion module in the GLIP [9] model, the two modules serve quite different roles in detectors. The fusion module in GLIP [9] fuses textual features and visual features of query images in the ATSS detection head, while PAM aggregates visual features of support images into textual features before the detection head. Therefore, the proposed PAM and GLIP [9] models are compatible, and the textual prototypes of GLIP [9] can be further enhanced. There are also some differences in the fusion methods between PAM and the fusion module in GLIP [9]. GLIP [9] adopts a bidirectional fusion approach based on early fusion, while our method uses a fusion based on late diffusion.

In addition, note that if each class has multiple samples, directly using the mean of visual samples as a visual prototype can lead to feature degradation due to potential misalignments in the spatial locations of different samples. Therefore, when the number of support images exceeds one, we employ the following formula to align support features from multiple shots. We select the first support feature S_0 of each class as the reference for spatial alignment, so all other support features should be aligned with it. We first compute the similarity between S_i and the S_0 at each spatial location, subsequently normalizing the values to obtain the $\delta(S_0, S_i)$. The detailed formula of $\delta(S_0, S_i)$ is presented as follows:

$$\begin{aligned} \delta(S_0, S_i) &= \sigma((S_0 - E(S_0))(S_0 - E(S_i))^T) \\ i &= 0, 1, ..., K-1, \end{aligned} \tag{5}$$

where $S_i \in \mathbb{R}^{HW \times D}, i = 0, 1, ..., K-1$ represent the feature of the i-th support image, K is the number of support images, $\sigma(.)$ is the SoftMax function, and $E(S_i)$ is the mean of S_i. Subsequently, we spatially align S_i with S_o by multiplying $\delta(S_0, S_i)$ with S_i, and then compute the mean of all aligned support features to obtain the final visual prototype P_v, as shown in Equation (6).

$$P_v = \frac{1}{K} \sum_{i=0}^{K-1} \delta(S_0, S_i) S_i \tag{6}$$

3.4. *Efficient Two-Stage Training Strategy*

Due to the domain gap between aerial images and natural images, detectors trained on natural scene datasets are not good at extracting the features of aerial targets. Therefore, previous few-shot object detectors [8,37] often adopt two-stage training strategies to enhance the feature extraction capabilities of detectors. However, the robustness of pre-trained text encoders such as BERT [44] in few-shot object detection for remote sensing images remains largely unexplored. To fill the blank, we have conducted comprehensive experiments and propose an efficient two-stage training strategy for multi-modal FSOD in RSIs.

In line with prior research, our two-stage training approach also comprises a base training stage followed by a few-shot fine-tuning stage. The base training stage aims to train feature extractors for aerial objects using the sufficiently annotated base dataset D_{base}, while the few-shot fine-tuning stage seeks to transfer the prior knowledge to enhance the performance of few-shot novel classes. The overall framework of ETS is illustrated in Figure 5.

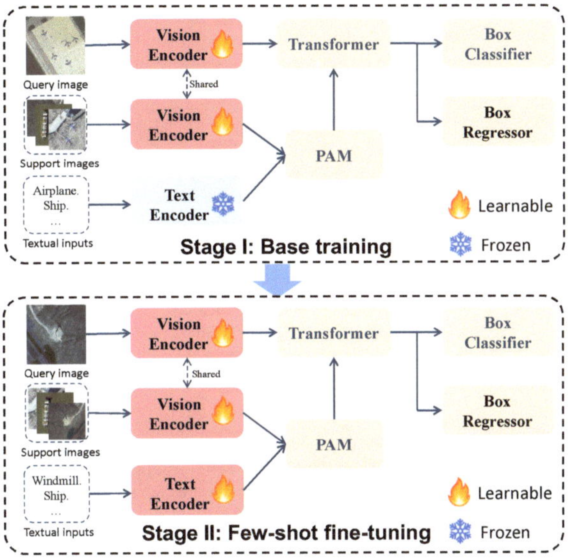

Figure 5. Proposed efficient two-stage training strategy. The modules in red indicate that the modules are learnable at this stage, while the modules in blue indicate that the module is frozen at this stage. Modules in orange are learnable at all stages.

Base training stage. Unlike previous methods [2–4] that train all learnable parameters during the base training stage, our approach trains only the image feature extractor and the DETR [16] encoder and decoder while keeping the text encoder frozen in the stage. The reason for the setting is that we surprisingly find that training the text encoder in the base training stage will harm the detection performance for few-shot novel classes. We argue that this may be because the trained text encoder has a preference for base classes. And we believe that the text encoder is domain-agnostic, implying that the semantic information encoded in the text encoder holds true in both aerial images and natural images. However, in the base training stage, the predominance of base class instances can bias the text encoder, compromising its generalization and diminishing the detection performance for few-shot novel classes. As a result, freezing the text encoder in the base training stage can maintain the generalization capability of the text encoder and can also conveniently reduce training time during large-scale base training.

Few-shot fine-tuning stage. In previous methods [3,4,11], only the image feature extractor is frozen, and all other components are learnable in the few-shot fine-tuning stage. However, our findings suggest that the image feature extractor trained on the base dataset is not optimally suited for the few-shot novel classes. Consequently, to enhance the feature extraction capability of the detector for novel classes, we fine-tune the image feature extractor of the detector with the balanced few-shot set. Since the pre-trained text encoder shows strong robustness, whether the text encoder is trained or not has little impact on the experimental results. Nonetheless, training the text encoder during the fine-tuning stage can still improve performance by about one point, so we unfroze the text encoder during this stage to further improve the detection performance on novel classes.

In summary, during the base training phase, our method maintains the text feature extractor in a frozen state since the textual encoder is domain-agnostic and has been well pre-trained on a large-scale natural language corpus. We are surprised to find that training the text encoder during the base training stage with relatively limited text data actually diminishes the generalization performance of the text encoder. In the fine-tuning stage, we fine-tune all components of the detector to obtain a more robust feature extractor.

4. Experiments and Results

To test the performance of our proposed MP-FSDet, a series of comparative and ablation experiments is conducted. Section 4.1 introduces the details of the experimental setup and evaluation metric. Section 4.2 presents implementation details of the experiments. In Section 4.3, we compare the proposed method with several state-of-the-art detectors, including both generic and few-shot detectors. At last, we carefully analyze the role of each module in the proposed method through ablation studies in Section 4.4.

4.1. Datasets and Evaluation Metrics

DIOR: The DIOR dataset is a large-scale object detection dataset for optical remote sensing images, released by the authors of [23]. In the DIOR dataset, images are sourced from Google Earth, comprising a total of 23,463 images and 192,472 objects distributed across 20 classes. The classes in the DIOR dataset consist of *airplane, airport, baseball field, basketball court, bridge, chimney, dam, expressway service area, expressway toll station, harbor, golf course, ground track field, overpass, ship, stadium, storage tank, tennis court, train station, vehicle,* and *windmill*. The dataset is partitioned into three subsets: a training set, a validation set, and a testing set, comprising 5682, 5863, and 11,738 images, respectively. The images in the DIOR dataset have a spatial resolution ranging from 0.5 to 30 m, with all images being 800×800 pixels in size.

NWPU VHR-10.v2: The NWPU VHR-10.v2 dataset is a high-resolution optical remote sensing dataset introduced by the authors of [47]. It includes 1173 optical images with the size of 400×400. The dataset comprises a total of ten geospatial object classes, which include *airplane, baseball diamond, basketball court, bridge, ground track field, harbor, ship, storage tank, tennis court,* and *vehicle*.

To evaluate the detection performance of our proposed method when there are only a few objects per class, we divide the classes of each dataset into two sets: base class sets and novel class sets. Each class within the base class sets comprises sufficient annotated objects, while each class within the novel class sets contains only K objects. Specifically, we select K = 3, 5, 10, and 20 objects for each class from both the training and validation sets in the K-shot detection task. For the DIOR dataset, we follow the FSOD setup outlined in [8,11] and utilize five different novel/base splits, each comprising 15 base classes and five novel classes as specified in Table 1. For the NWPU VHR-10.v2 dataset, following previous works [40], three categories, *airplane, baseball diamond,* and *tennis court*, are adopted as novel classes, while the others are considered as the base classes.

We evaluate the performance of detectors using mean Average Precision (mAP) introduced in the PASCAL VOC2007 benchmark [48]. The mAP is calculated by averaging 11 precision values as recall ranges from 0 to 1 in increments of 0.1, with an IoU threshold of 0.5.

Table 1. Five different base/novel split settings in the DIOR dataset.

Split	Novel					Base
1	Airplane	Baseball field	Tennis court	Train station	Windmill	Rest
2	Baseball field	Basketball court	Bridge	Chimney	Ship	Rest
3	Airplane	Airport	Expressway toll station	Harbor	Ground track field	Rest
4	Dam	Golf course	Storage tank	Tennis court	Vehicle	Rest
5	Express service area	Overpass	Stadium	Train station	Windmill	Rest

4.2. Implementation Details

Our proposed method is based on GroundingDINO [10], and we adopt the pre-trained detector with the backbone of Swin-T [49] and BERT [44]. In the base training stage, the model is trained by 12 epochs, where the learning rate starts from 1.25×10^{-5} and decreases to 1.25×10^{-6} at 11 epochs. In the fine-tuning stage, the model is fine-tuned by 72 epochs with a learning rate of 1.25×10^{-5}. In both the base training and the fine-tuning stages, we use AdamW as the gradient backward method with a weight decay coefficient of 1×10^{-4}. The training batch size is set to 2 on one NVIDIA GeForce RTX 4090 GPU.

Given the maximum input length of the BERT encoder being 256 tokens, we do not use the class descriptions from the DIOR dataset in TEMO [11] as input text. Instead, we utilize the features of class names encoded by BERT as the textual prototypes. The separators "." between textual prototypes are eliminated during fusion, as we have found that separators lacking semantic information significantly interfere with the performance of prototype aggregation. Regarding visual prototypes, the cropped features of the target objects in support images serve as support features, and the mean features obtained after aligning are utilized as the visual prototypes, as detailed in Section 3.3.

4.3. Experimental Results and Comparisons

4.3.1. Results on the DIOR Dataset

To demonstrate the effectiveness of our proposed method, we compare our approach with several advanced detectors, including detectors based on fine-tuning [3,4,40], visual prototypes [8], and textual prototypes [9–11], as shown in Table 2. We also integrate our MP-FSDet with GLIP-T [9] and GroundingDINO-T [10], naming them MP-GLIP-T and MP-GroundingDINO-T, respectively. Given that GLIP-T [9] and GroundingDINO-T [10] are originally designed for detection in natural scenes, we fine-tune them on the balanced few-shot DIOR dataset to ensure a more equitable comparison.

As illustrated in Table 2, compared with the detectors based on visual prototypes [8] and transfer-learning [3,4,40], detectors based on textual prototypes [9–11] demonstrate superior performance on many data splits. We attribute this to the strong generalization of textual prototypes. The textual prototypes are generated by a massively pre-trained text encoder, thereby inheriting rich prior knowledge about targets from the pre-trained text corpus. The prior knowledge in textual prototypes helps to enhance the generalization of detectors and reduce the risk of overfitting in few-shot scenarios. However, single-modal textual prototypes are highly conceptual and lack specific spatial details of targets, which limits their performance in FSOD. To further enhance the performance of prototype-based few-shot detectors, we combine the semantic-rich textual prototypes with detail-rich visual prototypes through our proposed prototype aggregating module and efficient two-stage training strategy. As the table shows, our method achieves state-of-the-art detection performance on all data splits of the DIOR dataset. The results in the table demonstrate the effectiveness of our proposed MP-FSDet.

In addition to qualitative analysis, we also visualize the detection results of GLIP-T [9], GroundingDINO-T [10], and our proposed method in split1 under three-shot settings, as illustrated in Figure 6. As the figure shows, even when utilizing only three training samples, GLIP-T [9] and GroundingDINO-T [10] demonstrate the capability to detect objects belonging to novel classes. We argue that this ability stems from the fact that both GroundingDINO-T [9] and GLIP-T [10] employ textual prototypes, which possess rich prior knowledge. Yet, since textual prototypes lack spatial details about targets, these two advanced methods still misclassify some novel objects as background or base classes. Concurrently, the absence of spatial details in class prototypes also leads to poor object localization performance. By combining visual prototypes and textual prototypes, our method enhances the spatial information of class prototypes, consequently improving classification and localization accuracy.

Table 2. Few-shot object detection performance on the DIOR test set under 3-, 5-, 10-, and 20-shot settings. The best two performances are highlighted with red and blue. TEMO [11] is solely evaluated on split1, while others [3,4,8,40] are exclusively tested on the remaining four splits; missing data points are denoted by the – symbol.

Shots	Method	Year	Split1	Split2	Split3	Split4	Split5	Mean
3	P-CNN [8]	TGRS2021	-	18.0	14.5	16.5	15.2	16.1
	G-FSOD [4]	ISPRS2023	-	27.6	14.1	16.0	16.7	18.6
	TEMO [11]	TGRS2023	30.9	-	-	-	-	30.9
	SAE-FSDet [40]	TGRS2024	-	28.8	14.0	16.7	17.3	19.2
	FSCE [3]	CVPR2021	-	27.9	13.2	15.6	17.5	18.6
	GLIP-T [9]	CVPR2022	24.4	24.1	23.7	16.5	15.6	20.9
	GroundingDINO-T [10]	ECCV2024	27.8	30.8	24.5	15.2	14.8	22.6
	MP-GLIP-T (Ours)		30.0	25.5	31.0	24.3	27.0	27.6
	MP-GroundingDINO-T (Ours)		42.1	35.7	32.2	28.1	24.0	32.4
5	P-CNN [8]	TGRS2021	-	22.8	14.9	18.8	17.5	18.5
	G-FSOD [4]	ISPRS2023	-	37.5	15.8	23.3	21.0	24.4
	TEMO [11]	TGRS2023	37.2	-	-	-	-	37.2
	SAE-FSDet [40]	TGRS2024	-	32.4	15.6	19.1	20.5	21.9
	FSCE [3]	CVPR2021	-	28.6	14.1	16.2	20.4	19.8
	GLIP-T [9]	CVPR2022	28.2	26.7	28.1	20.3	19.0	20.5
	GroundingDINO-T [10]	ECCV2024	33.1	36.2	29.9	19.4	20.0	27.7
	MP-GLIP-T (Ours)		35.3	29.5	34.5	27.2	29.7	31.2
	MP-GroundingDINO-T (Ours)		46.4	37.3	36.8	30.9	31.2	36.5
10	P-CNN [8]	TGRS2021	-	27.6	18.9	23.3	18.9	22.2
	G-FSOD [4]	ISPRS2023	-	37.5	20.7	26.3	25.8	27.6
	TEMO [11]	TGRS2023	42.8	-	-	-	-	42.8
	SAE-FSDet [40]	TGRS2024	-	37.1	17.4	28.4	22.7	26.4
	FSCE [3]	CVPR2021	-	33.1	15.8	23.8	22.2	23.7
	GLIP-T [9]	CVPR2022	38.5	30.7	33.6	30.6	27.1	32.1
	GroundingDINO-T [10]	ECCV2024	42.4	40.4	33.5	31.2	27.8	35.1
	MP-GLIP-T (Ours)		45.9	32.6	41.0	31.5	34.4	37.1
	MP-GroundingDINO-T (Ours)		53.1	41.0	42.2	36.1	35.5	41.6
20	P-CNN [8]	TGRS2021	-	29.6	22.8	28.8	25.7	26.7
	G-FSOD [4]	ISPRS2023	-	39.8	22.7	32.1	31.8	31.6
	TEMO [11]	TGRS2023	-	-	-	-	-	-
	SAE-FSDet [40]	TGRS2024	-	42.5	21.3	29.9	26.8	30.1
	FSCE [3]	CVPR2021	-	37.6	20.9	28.9	25.0	28.1
	GLIP-T [9]	CVPR2022	42.1	36.3	36.6	33.0	31.5	35.9
	GroundingDINO-T [10]	ECCV2024	48.5	49.3	40.1	39.1	37.1	42.8
	MP-GLIP-T (Ours)		50.0	41.0	40.5	35.9	35.0	40.5
	MP-GroundingDINO-T (Ours)		55.4	51.5	45.5	40.9	39.8	46.6

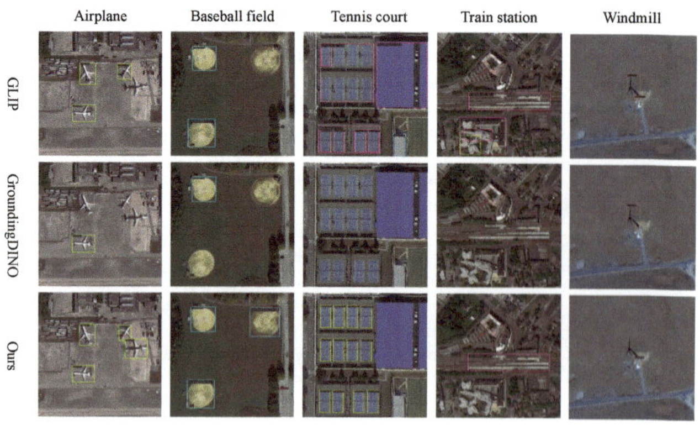

Figure 6. Novel class detection results in SPLIT 1 of the DIOR dataset under three-shot settings. Detection boxes with different colors represent different objects. The first row represents the results of GLIP-T [9], the second row represents the results of GroundingDINO-T [10], and the last row represents the detection results of our proposed method. Note that only the bounding boxes with a confidence greater than 0.3 are visualized.

4.3.2. Results on the NWPU VHR-10 Dataset

Table 3 presents the detection performance of our approach and other comparative methods on the NWPU VHR-10.v2 test set. Similar to the experiments on the DIOR dataset, we still combine our method with GLIP-T [9] and GroundingDINO-T [10]. Note that GLIP-T [9] and GroundingDINO-T [10] have been fine-tuned on the balanced few-shot dataset of NWPU VHR-10.v2.

Table 3. FSOD performance on the NWPU VHR-10.v2 test dataset under 3-, 5-, 10-, and 20-shot settings. The best two performances are highlighted with red and blue.

Method	3-Shot	5-Shot	10-Shot	20-Shot
P-CNN [8]	41.8	49.2	63.3	66.8
G-FSOD [4]	49.1	56.1	71.8	75.4
TEMO [11]	57.6	66.8	75.1	-
SAE-FSDet [40]	58.0	59.4	71.0	85.1
FSCE [3]	41.6	48.8	60.0	79.6
GLIP-T [9]	47.9	57.0	63.2	70.8
GroundingDINO-T [10]	59.1	62.1	84.3	88.1
MP-GLIP-T (Ours)	57.8	65.0	67.4	76.5
MP-GroundingDINO-T (Ours)	77.9	85.3	88.5	88.8

As Table 3 shows, with the combination of textual and visual prototypes and the efficient two-stage training strategy, MP-GroundingDINO-T achieves the mAP of 77.9% with just three samples, surpassing all other detectors by a minimum of 18.8%. This indicates that our approach performs effectively with extremely limited samples. Furthermore, we also observe that our approach converges to saturation performance more rapidly as the number of shots increases. This implies that our method exhibits a lower demand for training data and greater suitability for FSOD compared to other methods.

We visualize the detection results of GLIP-T [9], GroundingDINO-T [10], and our MP-GroundingDINO-T for three novel classes under the three-shot setting, as shown in Figure 7. The results indicate that GLIP-T [9] and GroundingDINO-T [10] suffer from significant misclassification. For instance, GLIP-T [9] misclassifies numerous background regions as foregrounds, while GroundingDINO [10] misclassifies numerous foreground regions as backgrounds. We attribute the poor detection performance to the inherent limitations of single-modal prototypes and the inadequacy of previous training strategies [4,11] to enhance the feature extraction capabilities of the vision encoder while preserving the generalization performance of the text encoder. However, by introducing the multi-modal prototypes and efficient two-stage training strategy, our method can leverage the advantages of detailed visual prototypes and generalized textual prototypes, thereby enhancing the detection performance for few-shot novel classes.

4.4. *Ablation Studies*

4.4.1. Ablation Study for the Components of Our Proposed Method

Table 4 shows the effectiveness of the proposed components in our approach. Given that GroundingDINO-T [10] is originally designed for detection in natural scenes, we fine-tune it on the balanced few-shot DIOR dataset and report the performance in the first row of Table 4. However, even though it utilizes the few-shot dataset to enhance the model's ability to detect the target classes, it still does not perform well on aerial images. Therefore, to introduce more prior knowledge about RSIs and enhance the feature extraction capability of remote-sensing objects, we employ a novel two-stage training strategy named the efficient two-stage training strategy (ETS) for the baseline model. As shown in the second row of Table 4, the incorporation of ETS significantly enhances the detection performance, especially when the number of shots is extremely low. Additionally, to leverage both the semantic information of textual prototypes and the spatial details of visual prototypes, the prototype aggregating module (PAM) is then employed on the detector, which improves the mAP by 9.3% on average. At last, we integrate our proposed ETS and PAM into the baseline detector, which further enhances the performance of FSOD in RSIs. In summary, compared with the baseline detector, our proposed method improves the detection performance by 20.8%, 13.3%, 10.7%, and 6.9% in 3, 5, 10, and 20 shots, respectively, in the first split of the DIOR dataset. In addition, our method also improves the detection performance by 18.8%, 23.2%, 4.2%, and 0.7% in 3, 5, 10, and 20 shots, respectively, in the NWPU VHR-10.v2 dataset. We also find the saturation performance in the 20-shot NWPU VHR-10.v2 dataset, which indicates that with an increasing number of shots, the performance tends to converge. However, our method converges more rapidly, indicating a lower demand for training data.

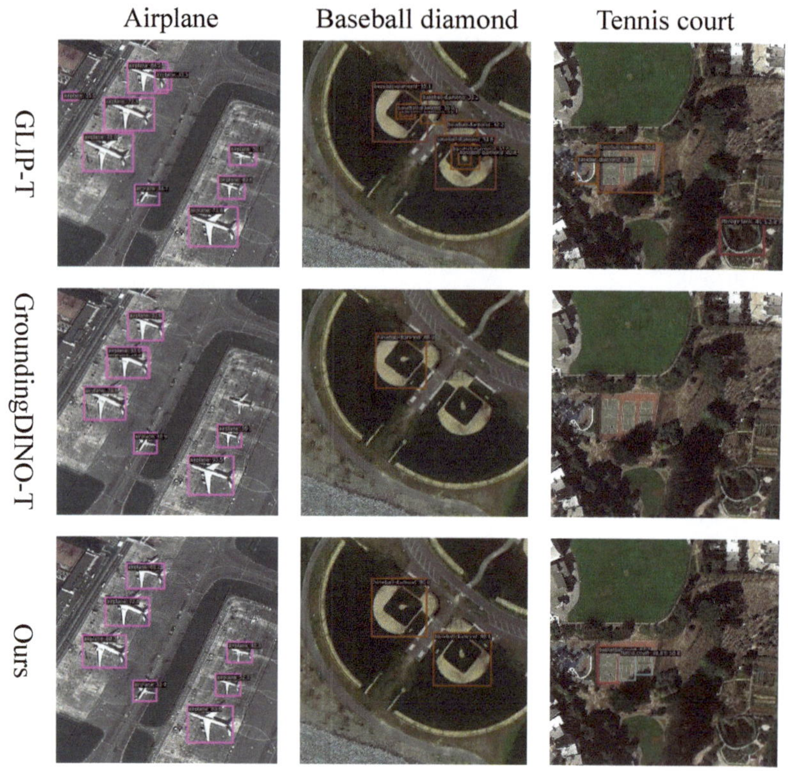

Figure 7. Novel class detection results of the NWPU-VHR10.v2 dataset under three-shot settings. Detection boxes with different colors represent different objects. The first row represents the results of GLIP-T [9], the second row represents the results of GroundingDINO-T [10], and the last row represents the detection results of our proposed method. Note that only the bounding boxes with a confidence greater than 0.3 are visualized.

To demonstrate the effectiveness of each proposed module in our approach, we also conduct extensive ablation studies. All experiments are performed on the DIOR test set in the first base/novel split or the NWPU VHR-10.v2 dataset.

Table 4. Ablation study of each component of our approach on the first split in the DIOR and NWPU VHR-10.v2 datasets.

Method	Components		DIOR				NWPU VHR-10			
	ETS	PAM	3	5	10	20	3	5	10	20
GroundingDINO			27.8	33.1	42.4	48.5	59.1	62.1	84.3	88.1
Ours	✓		38.8	44.7	50.1	53.6	75.2	85.5	86.9	88.4
Ours		✓	40.6	41.8	50.8	53.6	73.4	83.4	87.6	88.5
Ours	✓	✓	**42.1**	**46.4**	**53.1**	**55.4**	**77.9**	**85.3**	**88.5**	**88.8**

4.4.2. Ablation for the Performance of the Prototype Aggregating Module

To further analyze the properties of various prototypes, we conduct a comprehensive study on the first base/novel split of the DIOR dataset, as shown in Table 5. To enhance the feature extraction capability of detectors for aerial images, we have applied the ETS to all detectors in Table 5. As the table illustrates, on the one hand, in the condition of extremely limited shot number, the visual prototypes extracted by the model exhibited poor generalization capabilities, resulting in poor detection performance. Nonetheless, as the number of shots increases, the visual prototypes exhibit enhanced generalization capabilities, leading to a substantial improvement in the detection performance. On the other hand, owing to the robust generalization of textual prototypes, detectors based on textual prototypes can achieve high performance even with extremely limited training data. However, single-modal textual prototypes lack spatial details, which limits their detection performance. At last, by combining visual prototypes and textual prototypes, our proposed approach further enhances the detection performance, achieving a maximum improvement of 3.3%.

Table 5. Abalation study for the performance of different types of prototypes.

Visual Prototype	Textual Prototype	3-Shot	5-Shot	10-Shot	20-Shot
✓		21.3	27.0	42.6	50.6
	✓	38.8	44.7	50.1	53.6
✓	✓	**42.1**	**46.4**	**53.1**	**55.4**

In the above experiments, we find that the visual prototypes perform poorly when the number of training data points is extremely small, possibly due to the difficulty that vision encoders face in adequately extracting semantic features with limited training samples. Due to the rich prior knowledge of the text encoder, detectors based on textual prototypes demonstrate strong detection performance even with extremely limited training data. By combining the spatial information of visual prototypes into textual prototypes, we have achieved better detection performance. We have found that the spatial alignment between textual prototypes and visual prototypes is crucial. When visual prototypes do not retain spatial information or are not aligned with text prototypes, detectors often rely solely on textual prototypes to make predictions.

4.4.3. Ablation for the Performance of the Efficient Two-Stage Training Strategy

In order to analyze the importance of each component in the base training stage or fine-tuning stage, we conduct comprehensive ablation studies on the first base/novel set of the DIOR dataset, as demonstrated in Table 6.

As shown in the first row of Table 6, the detector designed for natural images exhibits poor detection performance in RSIs due to the large domain gap between natural images and aerial images, although it is claimed to be an open-set detector. Subsequently, we fine-tune the baseline detector on the few-shot dataset, and the detector's performance exhibits an improvement, as demonstrated in the second row of Table 6. However, in cases where the number of samples is extremely small, the performance of the detector is still not satisfactory. To further enhance the feature extraction capability of remote-sensing objects, in line with the previous training strategy [4,11], we performed a base training stage on a large-scale base dataset, and the improved detection performance is illustrated in the third row of the table. It is noteworthy that no parameters are frozen during the base training phase. However, we argue that the text encoder is domain-agnostic, while the vision encoder is regarded as domain-specific. Simply training all parameters in the base training stage is inefficient and harms the generalization of the text encoder. Therefore, to further analyze the domain sensitivity of each component in the base training stage, we independently freeze the vision encoder and text encoder during the base training phase, as shown in the fourth and fifth rows of Table 6. Since the primary objective of base training is to enhance the feature extraction capability of the vision encoder, freezing the vision encoder during the base training stage will impair the detector's performance for remote sensing targets, as illustrated in the fourth row of the table. At last, by freezing the text encoder in the base training phase, our method can preserve the generalization of the text encoder and reduce the risk of overfitting to base classes, as demonstrated in the last row of Table 6. The experiment results also corroborate our perspective that the text encoder is domain-agnostic, whereas the vision encoder is domain-specific. Consequently, we believe that leveraging the extensive natural language corpora available on the Internet as prior knowledge can further enhance the capability of FSOD in RSIs.

Table 6. Ablation study for the performance of ETS. The ✗ indicates that the module is frozen during the training process, whereas the ✓ denotes that the module is unfrozen.

Method	Base Training Stage		Fine-Tuning Stage		3-Shot	5-Shot	10-Shot	20-Shot
	Vision Encoder	Text Encoder	Vision Encoder	Text Encoder				
Baseline	✗	✗	✗	✗	11.9	11.9	11.9	11.9
Baseline	✗	✗	✓	✓	21.3	33.1	42.4	48.5
Baseline	✓	✓	✓	✓	40.6	41.8	50.8	53.6
Baseline	✗	✓	✓	✓	35.9	39.2	49.2	51.6
Baseline	✓	✗	✗	✓	39.2	41.5	51.8	53.4
Baseline	✓	✗	✓	✗	41.7	45.7	51.1	54.4
Ours	✓	✗	✓	✓	**42.1**	**46.4**	**53.1**	**55.4**

In addition to conducting experiments during the base training stage, we have also experimented on whether it is necessary to freeze the text encoder and visual encoder during the fine-tuning stage. Although the vision encoder has undergone base training on base datasets, further fine-tuning the vision encoder during adaption stage can still yield some performance improvements during the fine-tuning stage, as shown in the fifth and last rows of Table 6. Since the text encoder already obtains sufficient prior semantic knowledge from a large-scale corpus, additional fine-tuning during this stage has only a negligible impact on performance.

4.4.4. Ablation for Efficiency

To further analyze the computational efficiency of our method, we perform comprehensive experiments on the DIOR dataset, as Table 7 illustrates. All experiments are conducted on a single NVIDIA GeForce RTX 4090 GPU and the training time includes the time of both the base training stage and the fine-tuning stage. It is worth noting that once the model training is complete, the fusion of visual and textual prototypes can be performed offline, thus exerting only a minimal impact on the FLOPs and FPS during inference. Furthermore, as the ETS does not introduce additional parameters and PAM employs a late fusion paradigm that only marginally increases the number of parameters, our method does not differ significantly in the scale of parameters from the baseline. Finally, through our proposed ETS, the model can achieve superior performance with reduced training time.

Table 7. Ablation study for efficiency. All experiments are conducted on one NVIDIA GeForce RTX 4090 GPU.

Method	ETS	PAM	Params	Training Time	FLOPs	FPS
Baseline			175 M	8.0 h	265 G	18.5
Ours	✓		175 M	7.5 h	265 G	18.5
Ours		✓	176 M	11.7 h	268 G	18.0
Ours	✓	✓	176 M	10.9 h	268 G	18.0

4.4.5. Ablation Study for the Construction of Visual Prototypes

We also conduct experiments on the construction method of visual prototypes, as shown in Table 8. We first randomly select a feature from the support features as the visual feature. To reduce randomness, we repeatedly conducted the experiment three times, selecting a different support feature each time and reporting the average results in the first row of Table 8. Subsequently, we present the experimental results obtained by directly employing the mean of all support features and utilizing the features aligned according to Equation (5) as visual prototypes. As the table demonstrates, the performance of randomly sampled visual prototypes is unstable since it is influenced by the quality of the sampled feature. In addition, spatial feature alignment also helps to improve the performance of the few-shot detector.

Table 8. Ablation study for the construction of visual prototypes.

Method	wo/align	w/align	3-Shot	5-Shot	10-Shot	20-Shot
random select	✓		41.8	45.6	52.9	54.5
mean	✓		42.0	45.6	52.6	54.9
Ours		✓	42.1	46.4	53.1	55.4

4.4.6. Failure Cases and Analysis

Although our proposed MP-FSDet has achieved better performance compared with previous methods, there is still room for improvement. We carefully studied the limitations of our method and elaborated on three possible reasons.

(a) The dense distribution of objects: As shown in the first column of Figure 8, the method fails to detect densely distributed airplanes in the airport. Due to the DETR-like methods [10,16,18] employing a fixed number of content queries for matching with feature maps, these approaches perform inadequately when detecting dense objects. To alleviate the failure, it is possible to consider integrating some iterative methods like [50] with our methods to progressively detect dense objects.

(b) Poor feature extraction capability for novel classes: Due to the limited number of objects for the novel classes, the feature extraction capability of the detector is insufficient to detect some hard examples. However, current feature extractors [51,52] exhibit robust feature extraction for remote-sensing objects. Utilizing a more robust feature extractor could potentially enhance the performance of few-shot novel classes.

(c) Large inter-class similarity between classes: As demonstrated by the third column of Figure 8, some classes in the dataset exhibit significant inter-class similarity. High inter-class similarity can result in some objects being misclassified as similar classes, which leads to performance degradation. Some potential works like [53] can be further utilized to address the issue.

Figure 8. Failure cases in the DIOR test set for the first base/novel split under the 10-shot setting. The top row of the figure shows the predictions of our method, while the bottom row shows the ground truths.

5. Discussion

Textual prototypes generated from class names can inherit prior knowledge from a large-scale pre-trained language model. However, in RSIs, many of the distinctions between classes are very subtle. Consequently, textual prototypes that are generated directly from class names may not provide sufficient discriminative features. Therefore, our future work will attempt to use both class descriptions and class names to enhance the detection performance between fine-grained classes.

6. Conclusions

In this article, we propose a novel construction of multi-modal prototypes, termed the prototype aggregating module (PAM), and apply these multi-modal prototypes to a state-of-the-art detector. Multi-modal prototypes simultaneously leverage the advantages of text prototypes and visual prototypes by integrating the generalization properties of text prototypes with the spatial and detailed information of objects in aerial images. Furthermore, we investigate the transferability of pre-trained multi-modal object detectors in remote sensing scenarios. Our comprehensive experimental analyses reveal that the text encoder exhibits domain independence, while the vision coder demonstrates domain dependence. Consequently, we propose an efficient two-stage training strategy (ETS) that can mitigate the risk of overfitting in the text encoder while reducing the overall training time. The experimental results demonstrate that our proposed ETS achieves an average performance improvement of 2.55% compared to previous training strategies. By integrating multi-modal prototypes with the ETS, our proposed approach can further enhance the detection performance by an average of 2.5%.

Author Contributions: Conceptualization, Y.L. and Z.P.; methodology, Y.L., J.Y., and P.Z.; resources, Z.P. and B.Z.; writing—original draft preparation, Y.L. and P.Z.; and supervision, Z.P. and B.Z. All authors have read and agreed to the published version of the manuscript.

Funding: This work was supported by the Youth Innovation Promotion Association, CAS, under number 2022119.

Data Availability Statement: The data presented in this study are openly available in https://github.com/YanxingLiu/RS-Datasets/tree/main/DIOR (accessed on 3 November 2024) at https://doi.org/10.1016/j.isprsjprs.2019.11.023.

Conflicts of Interest: The authors declare no conflicts of interest.

References

1. Chen, H.; Wang, Y.; Wang, G.; Qiao, Y. Lstd: A low-shot transfer detector for object detection. In Proceedings of the AAAI Conference on Artificial Intelligence, New Orleans, LA, USA, 2–7 February 2018; Volume 32.
2. Wang, X.; Huang, T.E.; Darrell, T.; Gonzalez, J.E.; Yu, F. Frustratingly simple few-shot object detection. *arXiv* **2020**, arXiv:2003.06957.
3. Sun, B.; Li, B.; Cai, S.; Yuan, Y.; Zhang, C. Fsce: Few-shot object detection via contrastive proposal encoding. In Proceedings of the IEEE/CVF Conference on Computer Vision and Pattern Recognition, Nashville, TN, USA, 19–25 June 2021; pp. 7352–7362.
4. Zhang, T.; Zhang, X.; Zhu, P.; Jia, X.; Tang, X.; Jiao, L. Generalized few-shot object detection in remote sensing images. *ISPRS J. Photogramm. Remote Sens.* **2023**, *195*, 353–364. [CrossRef]
5. Kang, B.; Liu, Z.; Wang, X.; Yu, F.; Feng, J.; Darrell, T. Few-shot object detection via feature reweighting. In Proceedings of the IEEE/CVF International Conference on Computer Vision, Seoul, Republic of Korea, 27 October–2 November 2019; pp. 8420–8429.
6. Yan, X.; Chen, Z.; Xu, A.; Wang, X.; Liang, X.; Lin, L. Meta r-cnn: Towards general solver for instance-level low-shot learning. In Proceedings of the IEEE/CVF International Conference on Computer Vision, Seoul, Republic of Korea, 27 October–2 November 2019; pp. 9577–9586.
7. Xiao, Y.; Marlet, R. Few-Shot Object Detection and Viewpoint Estimation for Objects in the Wild. In Proceedings of the European Conference on Computer Vision (ECCV), Glasgow, UK, 23–28 August 2020.
8. Cheng, G.; Yan, B.; Shi, P.; Li, K.; Yao, X.; Guo, L.; Han, J. Prototype-CNN for few-shot object detection in remote sensing images. *IEEE Trans. Geosci. Remote Sens.* **2021**, *60*, 1–10. [CrossRef]
9. Li, L.H.; Zhang, P.; Zhang, H.; Yang, J.; Li, C.; Zhong, Y.; Wang, L.; Yuan, L.; Zhang, L.; Hwang, J.N.; et al. Grounded language-image pre-training. In Proceedings of the IEEE/CVF Conference on Computer Vision and Pattern Recognition, New Orleans, LA, USA, 19–20 June 2022; pp. 10965–10975.
10. Liu, S.; Zeng, Z.; Ren, T.; Li, F.; Zhang, H.; Yang, J.; Li, C.; Yang, J.; Su, H.; Zhu, J.; et al. Grounding dino: Marrying dino with grounded pre-training for open-set object detection. *arXiv* **2023**, arXiv:2303.05499.
11. Lu, X.; Sun, X.; Diao, W.; Mao, Y.; Li, J.; Zhang, Y.; Wang, P.; Fu, K. Few-shot object detection in aerial imagery guided by text-modal knowledge. *IEEE Trans. Geosci. Remote Sens.* **2023**, *61*, 1–19. [CrossRef]
12. Ren, S.; He, K.; Girshick, R.; Sun, J. Faster r-cnn: Towards real-time object detection with region proposal networks. *Adv. Neural Inf. Process. Syst.* **2015**, *28*, 1137–1149. [CrossRef]
13. He, K.; Gkioxari, G.; Dollár, P.; Girshick, R. Mask r-cnn. In Proceedings of the IEEE International Conference on Computer Vision, Venice, Italy, 22–29 October 2017; pp. 2961–2969.
14. Redmon, J.; Farhadi, A. YOLO9000: Better, faster, stronger. In Proceedings of the IEEE Conference on Computer Vision and Pattern Recognition, Honolulu, HI, USA, 21–26 July 2017; pp. 7263–7271.
15. Redmon, J.; Farhadi, A. YOLOv3: An Incremental Improvement. *arXiv* **2018**, arXiv:1804.02767. [CrossRef]
16. Carion, N.; Massa, F.; Synnaeve, G.; Usunier, N.; Kirillov, A.; Zagoruyko, S. End-to-end object detection with transformers. In Proceedings of the European Conference on Computer Vision, Glasgow, UK, 23–28 August 2020; Springer: Berlin/Heidelberg, Germany, 2020; pp. 213–229.
17. Zhu, X.; Su, W.; Lu, L.; Li, B.; Wang, X.; Dai, J. Deformable detr: Deformable transformers for end-to-end object detection. *arXiv* **2020**, arXiv:2010.04159.
18. Zhang, H.; Li, F.; Liu, S.; Zhang, L.; Su, H.; Zhu, J.; Ni, L.M.; Shum, H.Y. Dino: Detr with improved denoising anchor boxes for end-to-end object detection. *arXiv* **2022**, arXiv:2203.03605.
19. Zareian, A.; Rosa, K.D.; Hu, D.H.; Chang, S.F. Open-vocabulary object detection using captions. In Proceedings of the IEEE/CVF Conference on Computer Vision and Pattern Recognition, Nashville, TN, USA, 19–25 June 2021; pp. 14393–14402.
20. Radford, A.; Kim, J.W.; Hallacy, C.; Ramesh, A.; Goh, G.; Agarwal, S.; Sastry, G.; Askell, A.; Mishkin, P.; Clark, J.; et al. Learning transferable visual models from natural language supervision. In Proceedings of the International Conference on Machine Learning, PMLR, Virtual, 18–24 July 2021; pp. 8748–8763.
21. Gu, X.; Lin, T.Y.; Kuo, W.; Cui, Y. Open-vocabulary object detection via vision and language knowledge distillation. *arXiv* **2021**, arXiv:2104.13921.
22. Xu, Y.; Zhang, M.; Fu, C.; Chen, P.; Yang, X.; Li, K.; Xu, C. Multi-modal queried object detection in the wild. *Adv. Neural Inf. Process. Syst.* **2024**, *36*.
23. Li, K.; Wan, G.; Cheng, G.; Meng, L.; Han, J. Object detection in optical remote sensing images: A survey and a new benchmark. *ISPRS J. Photogramm. Remote Sens.* **2020**, *159*, 296–307. [CrossRef]
24. Mei, S.; Jiang, R.; Ma, M.; Song, C. Rotation-invariant feature learning via convolutional neural network with cyclic polar coordinates convolutional layer. *IEEE Trans. Geosci. Remote Sens.* **2023**, *61*, 1–13. [CrossRef]

25. Deng, Z.; Sun, H.; Zhou, S.; Zhao, J.; Lei, L.; Zou, H. Multi-scale object detection in remote sensing imagery with convolutional neural networks. *ISPRS J. Photogramm. Remote Sens.* **2018**, *145*, 3–22. [CrossRef]
26. Cheng, G.; Zhou, P.; Han, J. Learning rotation-invariant convolutional neural networks for object detection in VHR optical remote sensing images. *IEEE Trans. Geosci. Remote Sens.* **2016**, *54*, 7405–7415. [CrossRef]
27. Li, Q.; Mou, L.; Liu, Q.; Wang, Y.; Zhu, X.X. HSF-Net: Multiscale deep feature embedding for ship detection in optical remote sensing imagery. *IEEE Trans. Geosci. Remote Sens.* **2018**, *56*, 7147–7161. [CrossRef]
28. Yang, X.; Yang, J.; Yan, J.; Zhang, Y.; Zhang, T.; Guo, Z.; Sun, X.; Fu, K. Scrdet: Towards more robust detection for small, cluttered and rotated objects. In Proceedings of the IEEE/CVF International Conference on Computer Vision, Seoul, Republic of Korea, 27 October–2 November 2019; pp. 8232–8241.
29. Xu, Y.; Fu, M.; Wang, Q.; Wang, Y.; Chen, K.; Xia, G.S.; Bai, X. Gliding vertex on the horizontal bounding box for multi-oriented object detection. *IEEE Trans. Pattern Anal. Mach. Intell.* **2020**, *43*, 1452–1459. [CrossRef]
30. Qiao, L.; Zhao, Y.; Li, Z.; Qiu, X.; Wu, J.; Zhang, C. Defrcn: Decoupled faster r-cnn for few-shot object detection. In Proceedings of the IEEE/CVF International Conference on Computer Vision, Montreal, BC, Canada, 11–17 October 2021; pp. 8681–8690.
31. Fan, Q.; Zhuo, W.; Tang, C.K.; Tai, Y.W. Few-shot object detection with attention-RPN and multi-relation detector. In Proceedings of the IEEE/CVF Conference on Computer Vision and Pattern Recognition, Seattle, WA, USA, 13–19 June 2020; pp. 4013–4022.
32. Lyu, Q.; Wang, W. Compositional prototypical networks for few-shot classification. In Proceedings of the AAAI Conference on Artificial Intelligence, Washington, DC, USA, 7–14 February 2023; Volume 37, pp. 9011–9019.
33. Li, X.; Deng, J.; Fang, Y. Few-shot object detection on remote sensing images. *IEEE Trans. Geosci. Remote Sens.* **2021**, *60*, 1–14. [CrossRef]
34. Le Jeune, P.; Lebbah, M.; Mokraoui, A.; Azzag, H. Experience feedback using representation learning for few-shot object detection on aerial images. In Proceedings of the 2021 20th IEEE International Conference on Machine Learning and Applications (ICMLA), Virtual, 13–16 December 2021; IEEE: New York, NY, USA, 2021; pp. 662–667.
35. Huang, Y.; Peng, J.; Sun, W.; Chen, N.; Du, Q.; Ning, Y.; Su, H. Two-Branch Attention Adversarial Domain Adaptation Network for Hyperspectral Image Classification. *IEEE Trans. Geosci. Remote Sens.* **2022**, *60*, 1–13. [CrossRef]
36. Wolf, S.; Meier, J.; Sommer, L.; Beyerer, J. Double Head Predictor based Few-Shot Object Detection for Aerial Imagery. In Proceedings of the IEEE/CVF International Conference on Computer Vision, Montreal, BC, Canada, 11–17 October 2021; pp. 721–731.
37. Chen, J.; Qin, D.; Hou, D.; Zhang, J.; Deng, M.; Sun, G. Multiscale Object Contrastive Learning-Derived Few-Shot Object Detection in VHR Imagery. *IEEE Trans. Geosci. Remote Sens.* **2022**, *60*, 1–15. [CrossRef]
38. Azeem, A.; Li, Z.; Siddique, A.; Zhang, Y.; Zhou, S. Unified multimodal fusion transformer for few shot object detection for remote sensing images. *Inf. Fusion* **2024**, *111*, 102508. [CrossRef]
39. Liu, N.; Xu, X.; Celik, T.; Gan, Z.; Li, H.C. Transformation-invariant network for few-shot object detection in remote-sensing images. *IEEE Trans. Geosci. Remote Sens.* **2023**, *61*, 1–14. [CrossRef]
40. Liu, Y.; Pan, Z.; Yang, J.; Zhang, B.; Zhou, G.; Hu, Y.; Ye, Q. Few-Shot Object Detection in Remote-Sensing Images via Label-Consistent Classifier and Gradual Regression. *IEEE Trans. Geosci. Remote Sens.* **2024**, *62*, 1–14. [CrossRef]
41. Zheng, Z.; Yang, L.; Wang, Y.; Zhang, M.; He, L.; Huang, G.; Li, F. Dynamic spatial focus for efficient compressed video action recognition. *IEEE Trans. Circuits Syst. Video Technol.* **2023**, *34*, 695–708. [CrossRef]
42. Yang, L.; Zheng, Z.; Wang, J.; Song, S.; Huang, G.; Li, F. Adadet: An adaptive object detection system based on early-exit neural networks. *IEEE Trans. Cogn. Dev. Syst.* **2023**, *16*, 332–345. [CrossRef]
43. Yang, L.; Jiang, H.; Cai, R.; Wang, Y.; Song, S.; Huang, G.; Tian, Q. Condensenet v2: Sparse feature reactivation for deep networks. In Proceedings of the IEEE/CVF Conference on Computer Vision and Pattern Recognition, Nashville, TN, USA, 19–25 June 2021; pp. 3569–3578.
44. Devlin, J.; Chang, M.W.; Lee, K.; Toutanova, K. Bert: Pre-training of deep bidirectional transformers for language understanding. *arXiv* **2018**, arXiv:1810.04805.
45. Lin, T.Y.; Goyal, P.; Girshick, R.; He, K.; Dollár, P. Focal loss for dense object detection. In Proceedings of the IEEE International Conference on Computer Vision, Venice, Italy, 22–29 October 2017; pp. 2980–2988.
46. Li, F.; Zhang, H.; Liu, S.; Guo, J.; Ni, L.M.; Zhang, L. Dn-detr: Accelerate detr training by introducing query denoising. In Proceedings of the IEEE/CVF Conference on Computer Vision and Pattern Recognition, New Orleans, LA, USA, 19–20 June 2022; pp. 13619–13627.
47. Cheng, G.; Han, J.; Zhou, P.; Guo, L. Multi-class geospatial object detection and geographic image classification based on collection of part detectors. *ISPRS J. Photogramm. Remote Sens.* **2014**, *98*, 119–132. [CrossRef]
48. Everingham, M.; Van Gool, L.; Williams, C.K.; Winn, J.; Zisserman, A. The pascal visual object classes (voc) challenge. *Int. J. Comput. Vis.* **2010**, *88*, 303–338. [CrossRef]
49. Liu, Z.; Lin, Y.; Cao, Y.; Hu, H.; Wei, Y.; Zhang, Z.; Lin, S.; Guo, B. Swin transformer: Hierarchical vision transformer using shifted windows. In Proceedings of the IEEE/CVF International Conference on Computer Vision, Montreal, BC, Canada, 11–17 October 2021; pp. 10012–10022.
50. Rukhovich, D.; Sofiiuk, K.; Galeev, D.; Barinova, O.; Konushin, A. Iterdet: Iterative scheme for object detection in crowded environments. In Proceedings of the Structural, Syntactic, and Statistical Pattern Recognition: Joint IAPR International Workshops, S+ SSPR 2020, Padua, Italy, 21–22 January 2021; Proceedings; Springer: Berlin/Heidelberg, Germany, 2021; pp. 344–354.

51. Wang, D.; Zhang, J.; Xu, M.; Liu, L.; Wang, D.; Gao, E.; Han, C.; Guo, H.; Du, B.; Tao, D.; et al. MTP: Advancing remote sensing foundation model via multi-task pretraining. *IEEE J. Sel. Top. Appl. Earth Obs. Remote Sens.* **2024**. [CrossRef]
52. Jiang, W.; Zhang, J.; Wang, D.; Zhang, Q.; Wang, Z.; Du, B. LeMeViT: Efficient vision transformer with learnable meta tokens for remote sensing image interpretation. *arXiv* **2024**, arXiv:2405.09789.
53. Wang, Z.; Yang, B.; Yue, H.; Ma, Z. Fine-Grained Prototypes Distillation for Few-Shot Object Detection. In Proceedings of the AAAI Conference on Artificial Intelligence, Vancouver, BC, Canada, 20–27 February 2024; Volume 38, pp. 5859–5866.

remote sensing

MDPI

Article

A Contrastive-Augmented Memory Network for Anti-UAV Tracking in TIR Videos

Ziming Wang [1,2,3,4], Yuxin Hu [1,2,3,4], Jianwei Yang [1,2,3,4], Guangyao Zhou [1,2,3,4], Fangjian Liu [1,2,3,4] and Yuhan Liu [1,2,3,4,]*

1 Aerospace Information Research Institute, Chinese Academy of Sciences, Beijing 100094, China; wangziming161@mails.ucas.ac.cn (Z.W.); yxhu@mail.ie.ac.cn (Y.H.); yangjianwei20@mails.ucas.ac.cn (J.Y.); zhougy@aircas.ac.cn (G.Z.); liufj@aircas.ac.cn (F.L.)

2 Key Laboratory of Technology in Geo-Spatial Information Processing and Application System, Chinese Academy of Sciences, Beijing 100190, China

3 Key Laboratory of Target Cognition and Application Technology (TCAT), Chinese Academy of Sciences, Beijing 100190, China

4 School of Electronic, Electrical and Communication Engineering, University of Chinese Academy of Sciences, Beijing 100049, China

* Correspondence: liuyuhan@aircas.ac.cn

Abstract: With the development of unmanned aerial vehicle (UAV) technology, the threat of UAV intrusion is no longer negligible. Therefore, drone perception, especially anti-UAV tracking technology, has gathered considerable attention. However, both traditional Siamese and transformer-based trackers struggle in anti-UAV tasks due to the small target size, clutter backgrounds and model degradation. To alleviate these challenges, a novel contrastive-augmented memory network (CAMTracker) is proposed for anti-UAV tracking tasks in thermal infrared (TIR) videos. The proposed CAMTracker conducts tracking through a two-stage scheme, searching for possible candidates in the first stage and matching the candidates with the template for final prediction. In the first stage, an instance-guided region proposal network (IG-RPN) is employed to calculate the correlation features between the templates and the searching images and further generate candidate proposals. In the second stage, a contrastive-augmented matching module (CAM), along with a refined contrastive loss function, is designed to enhance the discrimination ability of the tracker under the instruction of contrastive learning strategy. Moreover, to avoid model degradation, an adaptive dynamic memory module (ADM) is proposed to maintain a dynamic template to cope with the feature variation of the target in long sequences. Comprehensive experiments have been conducted on the Anti-UAV410 dataset, where the proposed CAMTracker achieves the best performance compared to advanced tracking algorithms, with significant advantages on all the evaluation metrics, including at least 2.40%, 4.12%, 5.43% and 5.48% on precision, success rate, success AUC and state accuracy, respectively.

Keywords: anti-unmanned aerial vehicles (UAVs); thermal infrared videos; global instance searching; contrastive-augmented matching; adaptive dynamic memory updating

Citation: Wang, Z.; Hu, Y.; Yang, J.; Zhou, G.; Liu, F.; Liu, Y. A Contrastive-Augmented Memory Network for Anti-UAV Tracking in TIR Videos. *Remote Sens.* **2024**, *16*, 4775. https://doi.org/10.3390/rs16244775

Yanni Dong, Xiaochen Yang and Qian Du

Received: 21 October 2024
Revised: 16 December 2024
Accepted: 19 December 2024
Published: 21 December 2024

1. Introduction

In recent years, advancements in technologies such as intelligent control, microelectronics and sensors have substantially influenced the development of unmanned aerial vehicles (UAVs), particularly in terms of miniaturization, cost efficiency and reduced energy consumption [1–5]. Consequently, due to their portability and flexibility, along with the open-source programming systems, UAVs have received widespread attention and been applied in multiple scenarios [6], including aerial photography [7], geographical mapping, intelligent monitoring [8] and rescue [9,10]. At the same time, however, the advantages of UAVs lead to their increasing abuse, which has brought a significant negative impact on public safety [11]. Therefore, anti-UAV technologies are critically important in both practical applications and research endeavors.

In order to address the public safety issues caused by misused UAVs, multiple anti-UAV methods are introduced, such as radar detection [12], acoustic detection [13] and visual-based methods [14,15]. Among the methods, both radar detection systems and acoustic detection systems are able to perceive targets in a 360-degree surrounding area. But radar systems struggle with detecting small drone targets, especially in complex environments, while acoustic detection systems are sensitive to noise interference in urban environments. Compared to other anti-UAV technologies, visual-based approaches offer higher efficiency, lower power consumption and easier deployment, thereby being extensively adopted to counter the potential threat of illegal drone intrusions [6]. In addition, techniques in thermal infrared (TIR) mode provide stability under different weather conditions and low-light circumstances [16]. As a vital and fundamental step in computer vision, visual object tracking allows the continuous real-time monitoring of particular targets [17]. Combined with the capacity of full-time and full-weather observation afforded by thermal infrared sensors, TIR object tracking offers a promising solution to anti-UAV tasks.

Although TIR object tracking methods are effective for all-time and all-weather anti-UAV tracking tasks, there are still some difficulties due to the TIR videos and the drone targets inside. Unlike RGB videos, frames in TIR videos are usually in grayscale and the resolutions are lower. Grayscale frames in TIR videos provide no color features and insufficient texture features, while low resolution results in inadequate structure and edge features [18,19]. Similarity interference and intensity change are also typical challenges in TIR target tracking tasks [16]. At the meantime, the drone targets in the TIR videos are usually in a smaller scale than the targets in universal tracking tasks, leading to sparse target features. In addition, under some circumstances, drone targets move fast in the video. Therefore, to capture the fast-moving drone targets, sudden camera movements are inevitable [20]. Moreover, in long videos, the background may change greatly, and the appearance of the targets may also shift. In general, the main challenges in TIR anti-UAV tracking include (1) limited target features, due to the quality of TIR videos and the small scale of drone targets; (2) sudden camera movement caused by the fast-moving drone targets; (3) interference from thermal crossover and similar distractors; and (4) target appearance variation and dynamic background change in long-term tracking tasks.

As an important area of computer vision, visual single object tracking is applied in a wide range of social systems and great improvements have been achieved in this field [17]. Generally, current single object trackers can be broadly categorized into two types, including correlation filter trackers and deep learning trackers [21].

Correlation filters were first introduced to the filed of tracking by Bolme et al. in MOSSE [22]. Thanks to their computational efficiency and stability, correlation filter trackers (CF-Trackers) have been improved from various perspectives [21]. CSK [23] introduce fast Fourier transform into the cyclic matrix, enhancing the efficiency of the algorithm and improving the accuracy by converting the image space into a nonlinear space. KCF [24] employs kernel functions to process tracking on high dimensions. To avoid the limitation of single kernel function, MKCF [25] conducts an adaptive fusion of multiple kernel functions to train the correlation filter. STAPLE [26] strengthens feature representation by applying the histogram of oriented gradient (HOG) in feature extraction. To reduce the impact of scale variation, Danelljan et al. propose a discriminative scale space tracker, also known as DSST [27], which trains correlation filters on a scale pyramid representation. SRDCF [28] alleviates boundary effects by employing a spatial weights function to limit the filter coefficients. STRCF [29] introduces the alternative direction method of multipilers (ADMM) to make each sub-problem have a closed solution. To address the boundary problem caused by the usage of cyclic matrix in generating training samples, BACF [30] enlarges the object search area and improves the quality of generated samples. However, the fixed spatial constraints applied to balance the boundary effects ignore the diversity information of the targets. Therefore, ASRCF [31] performs an adaptive spatial regularization and obtain more reliable filter coefficients in the tracking process. Adaptive approaches are also employed in [32], where the authors propose an adaptive spatial-aware CF to increment the weights to the target areas. Although they provide remarkable inference speed, most CF trackers have a lack of flexibility and generalization due to the usage of handcraft features and regularization strategies. Therefore, CF trackers struggle with diverse targets and complex backgrounds [17].

With the breakthroughs [33–35] made by deep learning methods in ImageNet large-scale visual recognition competition (ILSVRC) [36] and visual object tracking (VOT) [37–39], deep learning methods have gained considerable attraction in computer vision fields. The methodology of deep learning and a convolutional neural network (CNN) is also introduced to visual object tracking tasks. A category of deep learning trackers (DL trackers) is integrating features extracted from deep learning networks and traditional CF trackers [40–42]. However, these trackers are not capable of handling various tracking tasks [17]. Therefore, end-to-end trained DL trackers, especially those based on Siamese networks, are mainly investigated, due to better exploitation of the feature representation and feature understanding abilities.

The concept of the Siamese network was first proposed by Tao et al. in SINT [43]. Following that, Bertinetto et al. designed SiamFC [44], employing a fully convolutional structure and a cross-correlation layer to calculate the similarity between the template and searching areas. Inspired by SiamFC, massive Siamese network-based methods are proposed. SiamRPN [45] and SiamRPN++ [46] introduce RPN into Siamese networks, also employing classification and regression prediction in target locating. Apart from the anchor-based approaches, anchor-free methods like SiamCAR [47] and SiamBAN [48] perform classification and regression prediction directly on the response map, leading to significant process on tracking speed. Besides the Siamese networks that search the target in a particular region guided by the previous tracking results, global instance search (GIS) approaches seek targets in the whole image, which alleviates the impact of sudden change. Among them, GlobalTrack [49] introduces the RCNN structure into Siamese networks, conducting a two-stage refinement to enhance the prediction of bounding boxes in global view of the searching image. SiamRCNN [50] proposed a two-stage re-detection and a tracklet mechanism to reduce the disturbance from distractors.

With Transformer [51] showing promising ability in various downstream tasks in computer vision, a number of studies on Transformer-based trackers are proposed [52,52–59]. TransT [53] integrates CNN feature extraction with Transformer feature fusion, exploiting the advantage of the attention mechanism in long-distance dependencies. STARK [52] propose a template updating strategy on the CNN transformer structure to learn spatial and temporal information at the same time. SwinTrack [54] applies swin-transformer as the feature extractor, enhancing object tracking by better feature representation. Unlike the former methods, OSTrack [55] introduces a one-stream tracking framework based on Transformer, instead of the two-stage framework of feature extraction and relation modeling. After that, SeqTrack [56] and ARTrack [57] convert the tracking task into a sequence generation problem, and employ an autoregressive strategy to predict the bounding boxes. AQATrack [59] designs a novel spatial-temporal transformer to learn deep spatial-temporal constraints and perform prediction with the assistance from a spatio-temporal feature fusion module.

Although general visual object tracking has achieved remarkable promotion, different image qualities and target characteristics in TIR frames and RGB frames result in the unsatisfactory performance of the general trackers on TIR tracking tasks. Therefore, trackers for TIR tracking tasks are investigated. MCFTS [60] employs a correlation filter-based ensemble tracker with multilayer convolutional features. Nevertheless, TIR tracking is still confused by scarce benchmarks. To alleviate this problem, Liu et al. propose a TIR pedestrian tracking dataset called PTB-TIR [61] and a large-scale TIR object tracking dataset named LSOTB-TIR [62], which contains 1400 sequences and over 600 K frames with 47 categories of targets. Based on these benchmarks, plenty of trackers for TIR targets are proposed. HSSNet [63], designed by Li et al., utilizes hierarchical convolutional features to obtain a better spatial and semantic feature representation of TIR targets. Liu et al. [64] propose a a multi-level similarity-based Siamese network, namely, MLSSNet, which contains a structure similarity module and a semantic similarity module to compute the similarity for TIR objects in different aspects. Furthermore, MMNet [65] performs multi-level feature matching for TIR targets by adopting a discriminative matching module for inter-class recognition and a fine-grained aware module for intra-class recognition. To reduce the influence of occlusion and distractors, Yuan et al. [66] design a spatial-temporal memory network, collecting high-quality results in previous frames for tracking.

The aforementioned TIR trackers have shown brilliant performance on regular TIR tracking tasks like pedestrian tracking and animal tracking, where the targets are of normal size or the background are quite clean. These trackers often struggle when facing anti-UAV tracking tasks, as the targets are usually smaller and some city backgrounds are clutter [67]. Hence, tracking drones focusing on TIR sensors have received attention from researchers. Jiang et al. [68] propose a large-scale benchmark called Anti-UAV, which contains over 300 video pairs with both visible light and infrared modalities. Huang et al. present Anti-UAV410 [20], consisting of 410 infrared sequences with 438k bounding boxes. SiamSTA [69] designs a two-stage re-detection mechanism inspired by SiamRCNN, reducing the interference of occlusion and camera movement. Shi et al. propose GASiam [70], employing graph attention to support the re-detection of drone targets. UTTracker [14], presented by Yu et al., tackles target appearance variation by a multi-region local tracker. UTTracker also contains a global detection branch to avoid the quick position switch of targets.

Nevertheless, there are still some problems that remain unsolved in TIR anti-UAV tracking. First, usual solutions to camera movement or out-of-view objects are applying a global detection algorithm for assistance, which makes the tracking pipeline more complicated. Second, most trackers tend to put more focus on learning the similarity between ground truth and the templates than the differences between distractors and the templates. This results in weak ability in instance discrimination when dealing with thermal crossover and distractors. Finally, long-term anti-UAV tracking tasks are still challenging, as the target appearances and backgrounds change dynamically in long sequences. Memory modules utilized in current TIR anti-UAV trackers usually rigidly stack previous tracking results together or simply adopt the latest high-quality tracking result as a dynamic template. These approaches cannot combine the common discriminative features in history results and features of recent predictions flexibly.

To address the remaining challenges, in this paper, the anti-UAV tracking task is converted to a global instance search (GIS) problem to perform re-detection targets without other algorithms. The GIS problem is then divided into two subproblems, proposal detection and instance discrimination. Proposal detection gathers regions that are similar to the template, and instance discrimination finds the target among the proposals. Inspired by the ability of contrastive learning in distinguishing inter-class samples [71], a contrastive learning strategy is adopted to better extract the discriminative features of the templates. To maintain high tracking performance in long sequences, a memory module is also necessary. Based on these ideas, a contrastive-augmented memory network (CAMTracker) for TIR anti-UAV tracking tasks is proposed.

Comprehensively speaking, CAMTracker takes the whole image as input to re-detect targets without other algorithms. Generally, CAMTracker contains a Siamese feature extraction backbone and utilizes a two-stage tracking framework, composed of a proposal detection stage and an instance discrimination stage. The proposal detection stage returns possible target regions through an instance-guided region proposal network (IG-RPN). The instance discrimination stage distinguishes the target among the candidates, which is executed mainly by a contrastive-augmented matching module (CAM). In CAM, there is a classification branch, a regression branch and an embedding branch. The branches are supervised by corresponding loss. Specifically, a refined contrastive loss function is proposed to supervise the embedding branch. Furthermore, an adaptive dynamic memory module (ADM) is employed to assist achieving high-quality long-term tracking performance as well. Inspired by the capability of the attention mechanism in capturing key information [51], a attention-like updating strategy is applied in ADM to better extract common features of the initial template and tracking results, while the unique features in recent predictions remain. Experiments on the Anti-UAV410 dataset demonstrate the robustness and effectiveness of CAMTracker when handling long-term UAV tracking in TIR mode with interference of distractors and thermal crossover.

Generally, there are the following contributions in this article:

(1) A contrastive-augmented memory network is proposed for anti-UAV tracking tasks in TIR videos, which employs a GIS tracking structure and defects the interference from camera movement. Comprehensive experiments have shown that the proposed CAMTracker achieves better performance than other mainstream algorithms on benchmarks in the TIR anti-UAV field.
(2) A contrastive-augmented matching module is designed for better instance discrimination. Apart from regular regression and classification prediction, another branch trained by contrastive loss assists the tracker classifying the target.
(3) A refined contrastive loss function is designed to better balance the emphasis on positive and negative samples, thus strengthening the discrimination ability of the tracker.
(4) An adaptive dynamic memory module is proposed to execute memory updating and provide the latest critical features of the targets, thereby improving the tracker's long-term tracking capability.

2. Materials and Methods

The proposed CAMTracker contains a pair of shared-weighted backbones for feature extraction, a two-stage tracking framework consisting of a proposal detection stage and an instance discrimination stage and an adaptive dynamic memory module. Figure 1 provides an overview of the architecture. The proposal detection stage operates coarse detection under the guide of target information and generates possible candidates through an instance-guided region proposal network (IG-RPN). The instance discrimination stage distinguishes the target from the candidates in search frame and performs subtle location, including a contrastive-augmented matching module (CAM) and an adaptive dynamic memory module (ADM).

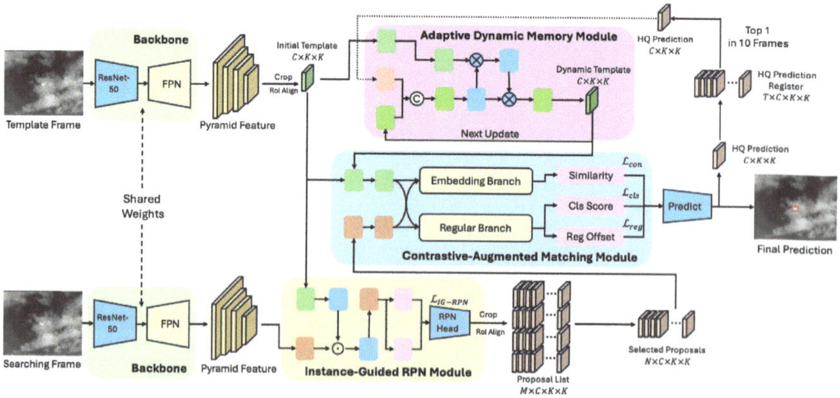

Figure 1. The main architecture of CAMTracker. The whole tracker mainly contains four parts, including a pair of backbones, an instance-guide region proposal network (IG-RPN), a contrastive-augmented matching module (CAM) and an adaptive dynamic memory module (ADM).

2.1. Multi-Scale Siamese Feature Extraction

Considering the insufficient color and texture features in infrared images, together with the small scale of UAV targets, it is challenging to track the targets with only single-scale features. Therefore, a multi-scale Siamese feature extraction module is applied to fuse semantic and spatial characteristics and obtain robust deep features. The Siamese network contains two weight-shared branches, namely, a template branch and a search branch. Each branch is composed of a ResNet and a feature pyramid network (FPN). Given a template frame I_z and a search frame I_x, the frame pair first passes through the backbone network and outputs feature maps from different levels. After that, the FPN takes the multi-level features as input and returns corresponding multi-scale image features $Z = Z_i|i \in OL\} = \phi(\varphi(I_z))$ and $X = \{X_i|i \in OL\} = \phi(\varphi(I_x))$, where φ and ϕ represent the backbone and the FPN, respectively. OL means the output layer set for FPN. Moreover, an RoIAlign algorithm is applied to extract the initial template target feature F_z, which can be calculated as

$$F_z = \delta(Z_i, B_0), \quad (1)$$

where δ denotes the RoIAlign algorithm and B_0 indicates the ground truth box of the target in the initial frame. Z_i means the template frame feature from the level corresponding to the target size. The weight-sharing network acquires semantic consistent features of the template frame and the search frame, thereby enhancing the identification of common objects in the image pair. Moreover, the multi-scale design not only obtains deep semantic features, but preserves the spatial details of the targets, benefiting precise target position locating.

2.2. IG-RPN for Proposal Detection

The proposal detection is mainly executed by an instance-guided region proposal network (IG-RPN), which contains an RPN module and a correlation encoder in front, as Figure 2 depicts, where $\odot$ indicates correlation calculation. The initial target feature from the template branch and the feature of the search frame from the search branch are encoded together by a correlation operation. Then, the encoded correlation feature is inputted into the RPN to generate possible proposals through classification and regression.

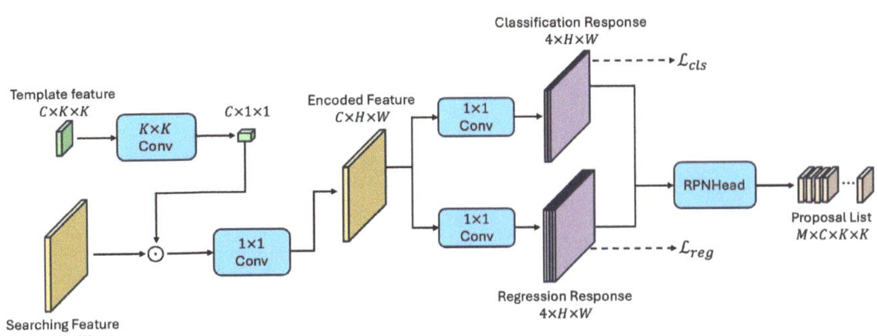

Figure 2. The structure of the IG-RPN, which contains a correlation encoder and an RPN head to generate possible proposals.

Assume that $F_z \in \mathbb{R}^{C\times K\times K}$ is the template feature and $F_x \in \mathbb{R}^{C\times H\times W}$ is the search frame feature from the template branch and the search branch, respectively, where $K \times K$ is the output size of the RoIAlign layer, H, W and C are the height, width and the number of the feature channels of the search frame feature maps. The process of the correlation encoder can be present as follows:

$$\varphi(F_z) = Conv_{K\times K}(F_z), \quad (2)$$

$$Corr(F_z, F_x) = Conv_{1\times 1}(F_x \odot \varphi(F_z)), \quad (3)$$

where $Conv_{K\times K}$ and $Conv_{1\times 1}$ denote convolution operation with the kernel size of $K \times K$ and 1×1, respectively. $\otimes$ indicates the correlation calculation. In the correlation encoder, F_z is first convoluted by a $K \times K$ filter and converted to an integrated target feature with the size of 1×1. After that, correlation between F_x and the integrated target feature $\varphi(F_z)$ is conducted, which outputs a correlation map. The correlation map is then refined by a 1×1 convolution layer and delivered to the RPN. Classification scores and regression results will be presented by the RPN; thereby, the proposal list is formed by the regression bounding boxes with the highest scores.

Conventionally, tracking approaches perform correlation calculation between features of templates and neighborhood search regions to inference the location of targets, where the target sizes in the templates and the search regions are similar. Due to the utilization of global search strategy, there is a considerable gap between the size of the aligned template feature and possible target size in the global features. Therefore, the template feature is converted into a single-pixel feature for the correlation operation to maintain the size balance.

2.3. Contrastive-Augmented Instance Matching

Since there is insufficient color and texture information, it remains challenging to extract discriminative features from thermal infrared frames. Generally, the correlation features calculated from the features of the template and the searching regions are delivered to the heads for classification and regression predictions, both in single-stage Siamese networks and GIS methods. Multiple approaches have been designed to acquire more accurate correlation representation for more precise predictions. However, for single-stage Siamese networks, it is difficult to figure out the crucial information of the infrared drone when there are background noises involved in the searching region. On the other hand, numerous distractors would be introduced to the classification and regression heads in GIS approaches, seriously affecting the stability of the tracker. Therefore, it is unreliable to capture the infrared UAV accounting on the correlation features only.

In order to tackle this challenge, a contrastive-augmented matching module is proposed, with the acronym CAM, which incorporates contrastive learning strategies into the anti-UAV tracker to enhance the ability to distinguish targets from distractors. The structure of CAM is illustrated in Figure 3, mainly consisting of a regular branch and an embedding branch. In Figure 3, Ⓗ means the Hadamard production. Among the branches, the regular branch performs classification and regression in most GIS methods. Meanwhile, the unique embedding branch is responsible for mapping templates and proposals into an embedding space and calculating the similarity scores between the templates and candidates. The similarity scores are used to assist the classification prediction during the inference process.

To be specific, assume there are $Z \in \mathbb{R}^{C\times H\times W}$ and $X_t^i \in \mathbb{R}^{C\times H\times W}$, representing the template feature and the i-th proposal feature in the t-th frame, respectively. C, H and W means the number of channels, height and width of the features. Before entering the two branches, Z and X_t^i pass through two projection layers, respectively. Both projection layers are convolutional layers with 3×3 kernels and padding of 1 pixel. Then, in the regular branch, Hadamard production is applied on the projected template and proposal features to output the modulated feature. The modulated feature passes through a global average pooling layer and a fully connected layer, becoming a $1024 \times 1 \times 1$ vector. The vector is processed by two fully connected layers to obtain the classification score and regression offsets, respectively.

$$Cls_t^i = FC_{cls}(FC(GAP(p_z(Z) \circledH p_x(X_t^i)))), \quad (4)$$
$$Box_t^i = FC_{reg}(FC(GAP(p_z(Z) \circledH p_x(X_t^i)))), \quad (5)$$

where p_z and p_x represent the projection functions that process F_z and F_x, both realized by 3×3 convolution layers with the padding of one pixel. Ⓗ in the equation means the Hadamard production. FC_{cls}, FC_{reg}, FC and GAP denote the fully connected layer for classification, fully connected layer for regression, shared fully connected layer and global average pooling layer, respectively.

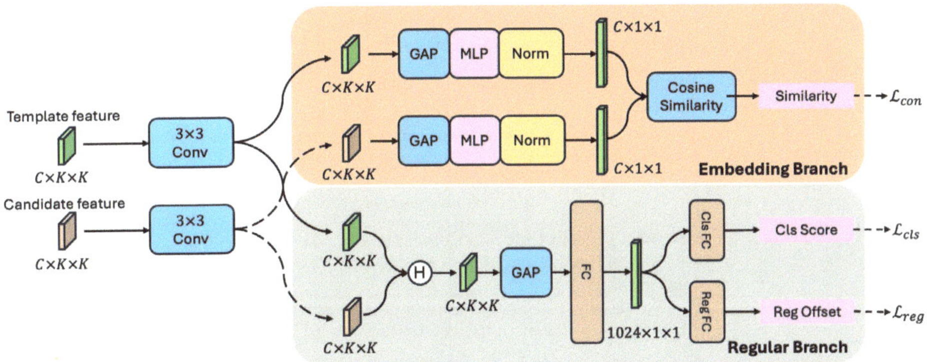

Figure 3. The illustration of CAM. The module is mainly composed of a regular branch and an embedding branch. In the figure, GAP, MLP and Norm means global average pooling, multi-layer perceptron and normalization, respectively.

Meanwhile, the embedding branch calculates the embedding features of the template and candidates, as well as their cosine similarities. As Figure 3 demonstrates, global average pooling is first performed on the projected template and candidate features. Subsequently, the embedding features of particular channels are computed by a multi-layer perceptron, which is composed of two fully connected layers. Finally, the branch calculates cosine similarities between the embedding features of the template and candidates. An embedding feature $f_{embed} \in \mathbb{R}^{C\times1\times1}$ can be obtained through the following function:

$$f_{embed} = FC_2(FC_1(GAP(X))), \quad (6)$$

where X indicates the template feature or candidate features. GAP means global average pooling layer, and FC_1 and FC_2 denote the two full-connection layers in the multi-layer perceptron. Consequently, the cosine similarity between the template embedding feature $\mathbf{z}$ and the i-th candidate embedding feature $\mathbf{x}_i$ in the t-th frame can be derived from

$$s_t^i = \frac{\mathbf{z} \cdot \mathbf{x}_t^i}{\|\mathbf{z}\| \cdot \|\mathbf{x}_t^i\|}, \quad (7)$$

in which $\cdot$ represents the inner production for vectors.

During the process of inference, the module determines whether the candidates are targets depending not only on the classification scores but also the cosine similarities. The similarity scores of the candidates are obtained through a softmax function:

$$Sim_t^i = \frac{exp(s_t^i)}{\sum_{k \in candidates} exp(s_t^k)}, \tag{8}$$

where i means the i-th candidate in the t-th frame. By multiplying the similarity score and the classification score, the final discrimination scores of the candidates in the t-th searching frame can be acquired:

$$Final_t = \{Cls_t^i \times Sim_t^i | i = 1, 2, ..., N\}, \tag{9}$$

where N indicates the number of proposals in the t-th frame. During the tracking process, the highest scored proposal is determined as the predicted result in the frame.

2.4. Offline Training with Refined Contrastive Loss

Aiming to enhance the network's ability to learn stronger semantic similarity and accurate location, CAMTracker is trained offline with a multi-task loss function, which consists of the loss from IG-RPN and the loss from CAM, which can be denoted as

$$\mathcal{L}_{total} = \mathcal{L}_{IG-RPN} + \lambda_{CAM}\mathcal{L}_{CAM}, \tag{10}$$

where $\mathcal{L}_{total}$ means the overall loss for training. $\mathcal{L}_{IG-RPN}$ and $\mathcal{L}_{CAM}$ indicate the loss for the IG-RPN module and the loss for the CAM module. λ_{CAM} means the weight factor to balance $\mathcal{L}_{IG-RPN}$ and $\mathcal{L}_{CAM}$.

For training IG-RPN in the proposal detection stage, a loss function similar to RPN is adpoted, which contains a binary cross-entropy (BCE) function [33] for classification and a smooth-L1 loss [72] for regression:

$$\mathcal{L}_{IG-RPN} = \lambda_c\mathcal{L}_{cls} + \lambda_r\mathcal{L}_{reg}, \tag{11}$$

$$\mathcal{L}_{cls} = \frac{1}{N_c}\sum_i \mathcal{L}_{bce}(y_i, \hat{y}_i), \tag{12}$$

$$\mathcal{L}_{reg} = \frac{1}{N_r}\sum_i \hat{y}_i\mathcal{L}_{smooth-L1}(b_i, \hat{b}_i). \tag{13}$$

In Equations (11)–(13), $\mathcal{L}_{cls}$ and $\mathcal{L}_{reg}$ denote the loss for the classification branch and the regression branch in the IG-RPN module, respectively. λ_c and λ_r are weight factors for $\mathcal{L}_{cls}$ and $\mathcal{L}_{reg}$. In Equations (12) and (13), y_i and b_i indicate the predicted classification score and bounding box offsets, respectively. $\hat{y}_i$ and $\hat{b}_i$ denote the ground truth label and bounding box offsets, respectively. N_c and N_r denote the number of the samples assigned for classification training and regression training, respectively.

In Equation (12), $\mathcal{L}_{bce}$ indicates the BCE loss function, which can be demonstrated as

$$\mathcal{L}_{bce}(y_i, \hat{y}_i) = \hat{y}_i \log y_i + (1 - \hat{y}_i)\log(1 - y_i). \tag{14}$$

In Equation (13), $\mathcal{L}_{smooth-L1}$ indicates the smooth-L1 loss function, which can be shown as

$$\mathcal{L}_{smooth-L1}(b_i, \hat{b}_i) = \sum_{j \in \{x,y,w,h\}} smooth_{L1}(b_i^j - \hat{b}_i^j), \tag{15}$$

where $smooth_{L1}$ is the smooth-L1 function as follows:

$$smooth_{L1}(x) = \begin{cases} \frac{x^2}{2}, & \text{if } |x| < 1, \\ |x| - 1/2, & \text{otherwise.} \end{cases} \tag{16}$$

Similarly, the regular branch in CAM is also trained by a BCE loss and a smooth-L1 loss. Meanwhile, the embedding branch is trained by a refined contrastive loss.

The embedding branch is designed to find an embedding space where the positive proposals and the template have similar embedding features while the embedding of negative proposals are orthogonal with them. In anti-UAV tracking tasks, the negative proposals are not from a convergent class. The diversity of negative proposals is neglected if there is only the BCE loss being applied. Therefore, to perform ideal representing learning for classification, a contrastive-like strategy can be adopted, taking the diversity of negative proposals into consideration by viewing each of them as a class. A softmax-based info-noise contrastive estimation loss (infoNCE Loss) is usually employed for this kind of situation [73]:

$$\mathcal{L}_{info} = -\sum_{\mathbf{x}_+} log \frac{exp(\mathbf{z} \cdot \mathbf{x}_+/\tau)}{exp(\mathbf{z} \cdot \mathbf{x}_+/\tau) + \sum_{\mathbf{x}_-} exp(\mathbf{z} \cdot \mathbf{x}_-/\tau)}, \tag{17}$$

where $\mathbf{z}$ means the query, namely, the normalized template embedding feature in the tracking scenarios. $\mathbf{x}_+$ and $\mathbf{x}_-$ represent the positive candidate sample and negative candidate sample, respectively. $\mathbf{z} \cdot \mathbf{x}_+$ indicate the inner production between $\mathbf{z}$ and $\mathbf{x}_+$, which reflects the similarity between $\mathbf{z}$ and $\mathbf{x}_+$, and τ is the temperature factor to avoid an extreme high inner production. The loss function tends to keep the positive samples close to the template and negative samples far away from the template in the embedding space.

However, infoNCE Loss does not consider the positive and negative samples in a balanced manner. Equation (17) can be written as

$$\begin{aligned} \mathcal{L}_{info} &= -\sum_{\mathbf{x}_+} log \frac{exp(\mathbf{z} \cdot \mathbf{x}_+/\tau)}{exp(\mathbf{z} \cdot \mathbf{x}_+/\tau) + \sum_{\mathbf{x}_-} exp(\mathbf{z} \cdot \mathbf{x}_-/\tau)} \\ &= \sum_{\mathbf{x}_+} log \frac{exp(\mathbf{z} \cdot \mathbf{x}_+/\tau) + \sum_{\mathbf{x}_-} exp(\mathbf{z} \cdot \mathbf{x}_-/\tau)}{exp(\mathbf{z} \cdot \mathbf{x}_+/\tau)} \\ &= \sum_{\mathbf{x}_+} log(1 + \frac{\sum_{\mathbf{x}_-} exp(\mathbf{z} \cdot \mathbf{x}_-/\tau)}{exp(\mathbf{z} \cdot \mathbf{x}_+/\tau)}). \end{aligned} \tag{18}$$

Assume $\mathbf{p}$ is a positive sample and $\mathbf{n}$ is a negative sample; it can be inferred that

$$\begin{aligned} \frac{\partial \mathcal{L}_{info}}{\partial \mathbf{p}} &= \frac{\partial \mathcal{L}_{info}}{\partial exp(\mathbf{z} \cdot \mathbf{p}/\tau)} \frac{\partial exp(\mathbf{z} \cdot \mathbf{p}/\tau)}{\partial \mathbf{p}} \\ &= \frac{\partial log(1 + \frac{\sum_{\mathbf{x}_-} exp(\mathbf{z} \cdot \mathbf{x}_-/\tau)}{exp(\mathbf{z} \cdot \mathbf{x}_+/\tau)})}{\partial \frac{\sum_{\mathbf{x}_-} exp(\mathbf{z} \cdot \mathbf{x}_-/\tau)}{exp(\mathbf{z} \cdot \mathbf{p}/\tau)}} \frac{\partial \frac{\sum_{\mathbf{x}_-} exp(\mathbf{z} \cdot \mathbf{x}_-/\tau)}{exp(\mathbf{z} \cdot \mathbf{p}/\tau)}}{\partial exp(\mathbf{z} \cdot \mathbf{p}/\tau)} \frac{\partial exp(\mathbf{z} \cdot \mathbf{p}/\tau)}{\partial \mathbf{p}} \\ &= -\frac{exp(\mathbf{z} \cdot \mathbf{p}/\tau)}{exp(\mathbf{p} \cdot \mathbf{p}/\tau) + \sum_{\mathbf{x}_-} exp(\mathbf{z} \cdot \mathbf{x}_-/\tau)} \frac{\sum_{\mathbf{x}_-} exp(\mathbf{z} \cdot \mathbf{x}_-/\tau)}{(exp(\mathbf{z} \cdot \mathbf{p}/\tau))^2} \frac{\partial exp(\mathbf{z} \cdot \mathbf{p}/\tau)}{\partial \mathbf{p}} \\ &= -\frac{\sum_{\mathbf{x}_-} exp(\mathbf{z} \cdot \mathbf{x}_-/\tau)}{(exp(\mathbf{z} \cdot \mathbf{p}/\tau))^2 + \sum_{\mathbf{x}_-} exp(\mathbf{z} \cdot \mathbf{x}_-/\tau) exp(\mathbf{z} \cdot \mathbf{p}/\tau)} \frac{\partial exp(\mathbf{z} \cdot \mathbf{p}/\tau)}{\partial \mathbf{p}} \\ &= -\frac{\sum_{\mathbf{x}_-} exp(\mathbf{z} \cdot \mathbf{x}_-/\tau)}{P^2 + \sum_{\mathbf{x}_-} P \cdot exp(\mathbf{z} \cdot \mathbf{x}_-/\tau)} \frac{\partial P}{\partial \mathbf{p}}. \end{aligned} \tag{19}$$

$$\begin{aligned}
\frac{\partial \mathcal{L}_{info}}{\partial \mathbf{n}} &= \frac{\partial \mathcal{L}_{info}}{\partial exp(\mathbf{z}\cdot\mathbf{n}/\tau)}\frac{\partial exp(\mathbf{z}\cdot\mathbf{n}/\tau)}{\partial \mathbf{n}} \\
&= \sum_{\mathbf{x}_+}\frac{\partial log(1+\frac{exp(\mathbf{z}\cdot\mathbf{n}/\tau)}{exp(\mathbf{z}\cdot\mathbf{x}_+/\tau)})}{\partial\frac{exp(\mathbf{z}\cdot\mathbf{n}/\tau)}{exp(\mathbf{z}\cdot\mathbf{x}_+/\tau)}}\frac{\partial\frac{exp(\mathbf{z}\cdot\mathbf{n}/\tau)}{exp(\mathbf{z}\cdot\mathbf{x}_+/\tau)}}{\partial exp(\mathbf{z}\cdot\mathbf{n}/\tau)}\frac{\partial exp(\mathbf{z}\cdot\mathbf{n}/\tau)}{\partial \mathbf{n}} \\
&= \sum_{\mathbf{x}_+}\frac{exp(\mathbf{z}\cdot\mathbf{x}_+/\tau)}{exp(\mathbf{z}\cdot\mathbf{x}_+/\tau)+exp(\mathbf{z}\cdot\mathbf{n}/\tau)}\frac{1}{exp(\mathbf{z}\cdot\mathbf{x}_+/\tau)}\frac{\partial exp(\mathbf{z}\cdot\mathbf{n}/\tau)}{\partial \mathbf{n}} \\
&= \sum_{\mathbf{x}_+}\frac{1}{exp(\mathbf{z}\cdot\mathbf{x}_+/\tau)+exp(\mathbf{z}\cdot\mathbf{n}/\tau)}\frac{\partial exp(\mathbf{z}\cdot\mathbf{n}/\tau)}{\partial \mathbf{n}} \\
&= \sum_{\mathbf{x}_+}\frac{1}{exp(\mathbf{z}\cdot\mathbf{x}_+/\tau)+N}\frac{\partial N}{\partial \mathbf{n}}.
\end{aligned} \tag{20}$$

For simplicity, we use $P = exp(\mathbf{z}\cdot\mathbf{p}/\tau)$ and $N = exp(\mathbf{z}\cdot\mathbf{n}/\tau)$. It can be inferred from Equations (19) and (20) that positive samples have a larger impact on infoNCE loss than negative samples. Therefore, for better balance between positive samples and negative samples, the contrastive loss is refined as

$$\begin{aligned}
\mathcal{L}_{con} &= log(1+\sum_{\mathbf{x}_+}\frac{\sum_{\mathbf{x}_-} exp(\mathbf{z}\cdot\mathbf{x}_-/\tau)}{exp(\mathbf{z}\cdot\mathbf{x}_+/\tau)}) \\
&= log(1+\sum_{\mathbf{x}_+}\sum_{\mathbf{x}_-}\frac{exp(\mathbf{z}\cdot\mathbf{x}_-/\tau)}{exp(\mathbf{z}\cdot\mathbf{x}_+/\tau)}) \\
&= log(1+\sum_{\mathbf{x}_+}\sum_{\mathbf{x}_-}exp((\mathbf{z}\cdot\mathbf{x}_- - \mathbf{z}\cdot\mathbf{x}_+)/\tau))
\end{aligned} \tag{21}$$

and the whole loss for CAM is functioned as

$$\mathcal{L}_{CAM} = \lambda_c\mathcal{L}_{cls} + \lambda_r\mathcal{L}_{reg} + \lambda_{con}\mathcal{L}_{con}, \tag{22}$$

where $\mathcal{L}_{cls}$ and $\mathcal{L}_{reg}$ are similar, as in Equations (12) and (13). λ_c, λ_r and λ_{con} are the corresponding weights of the losses.

2.5. Template Updating with Adaptive Dynamic Memory Module

In long sequences, there may be obvious changes in the appearance of the targets during the tracking procedure. To be aware of the alteration of the target feature, a dynamic template for online tracking is introduced to fully utilize the information of previous tracking results for optimal decision. Traditionally, online tracking algorithms take the latest prediction as the dynamic template, which is easily misled by some hard negative samples. Some construct memory banks to record historical high-quality results for the prediction, which causes considerable consumption on calculation. To address the drawbacks of the conventional methods, inspired by [51], an adaptive updating scheme of maintaining a single dynamic template is designed for the instance discrimination stage. When there comes a new high-quality prediction, the dynamic template is updated with the most recent prediction under the guide of the initial template, which not only follows the latest results, but also takes historical predictions into consideration.

The adaptive dynamic template updating strategy is presented as Figure 4 shows, where Ⓒ indicates concatenation and ⊗ means matrix multiplication. Especially, the initial template $T_{init} \in \mathbb{R}^{C\times K\times K}$ is regarded as the dynamic template at the beginning of online tracking. At the t-th frame of the sequence, the network first distinguishes whether the predicted result $r_t = [x_{min}, y_{min}, x_{max}, y_{max}]$ is a high-quality prediction according to the classification score threshold $thres_c$. If r_t is a high-quality prediction, the target feature $R_t \in \mathbb{R}^{C\times K\times K}$ is obtained through RoIAlign, corresponding to the bounding box predicted in r_t. After that, an integrated feature $F_t \in \mathbb{R}^{2C\times K\times K}$ is obtained by concatenating R_t and

D_t, where $D_t \in \mathbb{R}^{C \times K \times K}$ indicates the dynamic template at the t-th frame. F_t now contains information of historical predictions and the latest high-quality result. Subsequently, F_t is reshaped to the size of $2KK \times C$ and T_{init} is reshaped to the size of $KK \times C$. Then reshaped F_t and reshaped T_{init} are multiplied to get the pixel-to-pixel similarity matrix $w \in \mathbb{R}^{2KK \times KK}$, which is operated by a softmax function on a single dimension for normalization. The process can be denoted as

$$F_t = Concat(R_t, D_t), \tag{23}$$

$$w = Softmax(Res_{2HW \times C}(F_t) \otimes Res_{C \times HW}(T_{init}) / \sqrt{C}), \tag{24}$$

where $Concat$, $Softmax$ and $\otimes$ mean concatenation, softmax function on single dimension and matrix multiplication, respectively. $Res_{M \times N}$ indicates the reshape operation with an output size of $M \times N$. $\sqrt{C}$ is a normalization parameter to control the magnitude in w. In detail, each pixel in w can be formally denoted as

$$w_{ij} = \frac{exp((f_t^{i,\cdot} \odot t_{init}^{\cdot,j})/s)}{\sum_{0 \le k \le 2HW} exp((f_t^{k,\cdot} \odot t_{init}^{\cdot,j})/s)}, \tag{25}$$

where i and j are the coordination of the pixel, and f_t and t_{init} indicate reshaped F_t and T_{init}, respectively. $f_t^{i,\cdot}$ is the i-th row of f_t and $t_{init}^{\cdot,j}$ is the j-th column of t_{init}. $\odot$ means vector dot production.

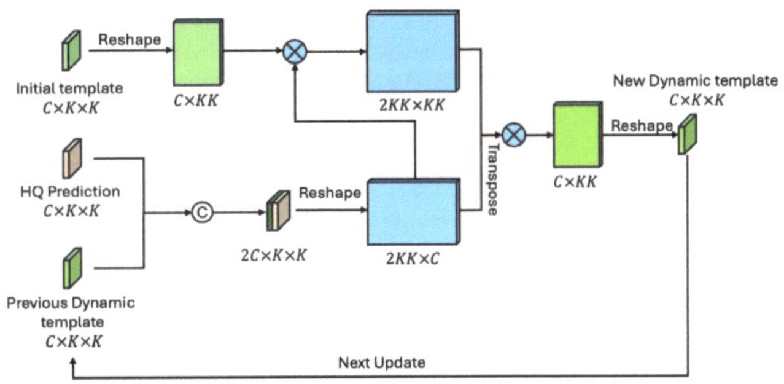

Figure 4. The demonstration of ADM.

Afterwards, w is regarded as a weight map to guide the updating of the dynamic template. Pixels in w demonstrate the similarity between the integrated historical feature and the initial template, hence indicating the common critical information of the historical target features. Therefore, by multiplying f_t and w, target features in historical predictions are captured and refined under the guide of the initial template, which works as the query in the following module:

$$D_{t+1} = Res_{C \times H \times W}(f_t^T \otimes w), \tag{26}$$

where D_{t+1} is the dynamic template in the $(t+1)$-th frame and f_t^T means the transpose of f_t. Through the adaptive updating scheme, almost all high-quality predictions in the sequence are considered for more optimal tracking results. Meanwhile, the information of early predictions are gradually weakened due to the continuous concatenation between the dynamic template and the latest prediction; thus, the information of latest prediction is better emphasized.

Unlike maintaining a memory bank, a single dynamic template costs much less computing consumption. Nevertheless, high-quality predictions in adjacent frames usually share high similarity. Repetitive updating in these frames offers few contributions while occupying computing resources. Therefore, the updating module is activated every 10 frames for more conciseness. The high-quality prediction with the highest classification score during the past 10 frames will be used as the input of ADM.

2.6. Overall Online Tracking Inference with Local Bias

During the online tracking procedure, two CAM modules are performed. One takes the initial template and the proposals as input, while the other takes the dynamic template and the proposals as input. Assume the output scores for the t-th frame are $S^t_{init} \in \mathbb{R}^N$ and $S^t_{dyn} \in \mathbb{R}^N$, where N is the number of proposals. The final classification score is derived as the average score of S^t_{init} and S^t_{dyn}, which is shown as below.

$$Cls_t = \{\frac{1}{2}(S^t_{init}[i] + S^t_{dyn}[i])|i = 1, 2, ..., N\}. \tag{27}$$

At the same time, a simple local bias is adopted during online tracking to better use the temporal information. For each proposal, the distance between the predicted bounding box B_{prop} and the final prediction box B_{last} is considered. Assume the center of B_{prop} and B_{last} is C_{prop} and C_{last}, respectively. A distance score is proposed as below.

$$Dis_{it} = (1 + exp(-\|C_{iprop} - C_{ilast}\|_2/r))/2, i = 1, 2, ..., N, \tag{28}$$

where i means the i-th proposal. r is the width of the square which covers the same area as B_{ilast}. By Equation (28), a score map is constructed to suppress proposals far away from the last prediction. The final score of a proposal is

$$Final_t = \{Cls^i_t \times Dis^i_t|i = 1, 2, ..., N\}. \tag{29}$$

The proposal with highest final score is deemed as the target in the frame.

3. Results

The proposed CAMTracker is evaluated mainly in two datasets: the Anti-UAV-410 dataset [20] for anti-UAV tracking tasks and the LSOTB-TIR dataset [62] for generalized evaluation. Quantitative and qualitative evaluations are conducted on CAMTrackers and other state-of-the-art trackers. An ablation study is performed to verify the effectiveness of the critical modules in CAMTracker.

3.1. Dataset

3.1.1. Anti-UAV410

Anti-UAV410 contains 410 sequences and over 150,000 frames. The numbers of sequences in training, validation and test sets are 200, 90 and 120, respectively. The frame size of all sequences is 640 × 512. Sequences in Anti-UAV410 cover various scenarios, including different light conditions and diverse backgrounds. Meanwhile, the dataset covers multiple difficulties in anti-UAV tracking tasks. Therefore, six attributes of sequences are defined correspondingly, including out-of-view (OV), occlusion (OC), thermal crossover (TC), fast motion (FM), scale variation (SV) and dynamic background clutter (DBC).

Apart from the challenging scenarios, the sequences in the dataset are also labeled in accordance with the target sizes. Four scales are defined based on the length of the diagonal of the bounding boxes, including normal size (NS): [50, inf), medium size (MS): [30, 50), small size (SS): [10, 30) and tiny size (TS): [2, 10).

3.1.2. LSOTB-TIR

As a widely used TIR tracking dataset, there are 1400 sequences and more than 600,000 frames of various image sizes in LSOTB-TIR, where 1280 sequences are for training and 120 sequences are for testing. Four scenarios are contained in LSOTB-TIR according to the camera mounting manners, including surveillance camera, hand-held camera, drone-mounted camera and vehicle-mounted camera. A total of 47 and 22 target classes are included in the training set and test set of LSOTB-TIR, respectively.

3.2. Implementation Details

The proposed approach is implemented on a WSL2 carrying an Ubuntu 22.04.2 system, with an Intel i5-13600KF CPU and an NVIDIA GeForce RTX 4090 GPU. The algorithm is programmed in an environment of Python 3.7, PyTorch 1.13 and CUDA 11.7. ResNet-50 and a feature pyramid network (FPN) are employed as the backbone architecture. The interval of ADM module is set to 10 frames. The score threshold of high-quality tracking results is set to 0.7.

For model training, CAMTracker is optimized by stochastic gradient descent (SGD). The momentum and weight decay are set to 0.9 and 1×10^{-4}. During training, each image pair is generated from the same sequence. The maximum distance between two frames in an image pair is set as 300. The model is trained on Anti-UAV410 for 12 epochs, with an initial learning rate of 0.01. The learning rate descends by 0.1 at the beginning of the eighth and 11th epoch. The weights in Equations (11) and (22) are set as $\lambda_c = 1$, $\lambda_r = 1$ and $\lambda_{con} = 0.01$. The temperature factor in CAM is set to 0.1.

3.3. Evaluation Metrics

Following [74], one-pass evaluation (OPE) is adopted to evaluate the performance of the proposed CAMTracker. The target position in the first frame will be given, then, the algorithm tracks the target in subsequent frames. The evaluation metrics include the precision rate (P), success rate (S_r) and success AUC (S_{auc}) from [74] and the state accuracy (SA) from [68].

3.3.1. Precision Rate

As one of the most popular metrics in the filed of object tracking, the precision rate assesses the tracking accuracy through the center location error (CLE) between the predicted bounding box Box_{pred} and the ground truth bounding box Box_{gt}. The formula for CLE is

$$CLE = \|C_{pred} - C_{gt}\|_2, \tag{30}$$

where C_{pred} and C_{gt} indicate the centers of the predicted bounding box and the ground truth, respectively. If CLE is less than a threshold, it is judged as a successful tracking on the frame, otherwise, it is considered as a failure. Precision is the proportion of successful tracking frames, which can be shown as

$$P = \frac{1}{N}\sum_{t=1}^{N} \epsilon(CLE < thres_p), \tag{31}$$

where N is the number of all sequence frames. D_p means the center distance between the ground truth and the predicted target, while $thres_p$ denotes the threshold. ϵ is an indicator function which returns 1 when the input is true, otherwise, it returns 0. In the experiments, the threshold is conventionally set to 20 pixels.

3.3.2. Success Rate

Although the precision rate gives a summary of the tracking results, the distance between the centers of bounding boxes cannot reflect the tracking performance comprehensively. It is also important that the predicted bounding boxes cover the right regions. The success rate is defined as the rate of the frames where the Intersection over Union (IoU) between Box_{pred} and Box_{gt} is larger than a threshold, which is shown as

$$S_r = \frac{1}{N}\sum_{t=1}^{N} \epsilon(IoU_t > thres_s), \tag{32}$$

where $thres_s$ indicates the threshold, and IoU_t denotes the IoU between the ground truth and the predicted bounding box in the t-th frame. During the experiments, $thres_s$ is conventionally set as 0.5.

3.3.3. Success AUC

Using one particular threshold for the precision rate and success rate may not be representative enough. Therefore, the precision plot and success plot are adopted to show the precision rate and success rate of the tracker performance with different thresholds. The area under curve (AUC) of the success plot is limited; therefore, it is applied as success AUC to evaluate the performance more comprehensively.

3.3.4. State Accuracy

In anti-UAV tasks, it is a common situation that the targets are out of view or occluded. Hence, the presence of the drone targets should be taken into consideration. The state accuracy proposed by [68] introduces the visibility of the targets into the evaluation metrics, and can be calculated by

$$SA = \frac{1}{N}\sum_{t=1}^{N}(IoU_t \times \delta(v_t > 0) + p_t \times (1 - \delta(v_t > 0)), \tag{33}$$

where $\delta()$ denotes an indicator function, outputting 0 when the input is false otherwise 1. v_t represents the visibility flag in the ground truth, while p_t means the predicted visibility results. $p_t = 1$ when the prediction remains empty, else $p_t = 0$. By applying state accuracy, the tracking performance on visible targets is more focused, while the judging ability on the presence of targets can be evaluated.

3.4. *Quantitative Evaluation*

To comprehensively evaluate the performance of the proposed CAMTracker, a comparison with 21 state-of-the-art trackers is performed, among which there are five correlation filter-based trackers, including DSST [27], Staple [26], SRDCF [28], KCF [24] and ECO [75], and 16 deep learning based trackers, including GlobalTrack [49], SiamRCNN [50], AiATrack [76], KeepTrack [77], MixformerV2-B [78], Stark-ST101 [52], SwinTrack-Base, SwinTrack-Tiny [54], DiMP50 [79], PrDiMP50 [80], ToMP50, ToMP101 [81], ATOM [82], SiamBAN [48], SiamCAR [47], TCTrack [83], SeqTrack [56], ROMTrack [84] and AQA-Track [59]. During the experiments, the parameters in the compared trackers are set as the default.

The comprehensive performance of CAMTracker, along with other trackers, is demonstrated in Table 1, where the best, the second best and the third best of each metrics are highlighted with red, green and blue, respectively. Generally speaking, correlation-based trackers struggle in anti-UAV tasks compared to deep learning-based trackers. Among deep learning-based trackers, GIS trackers, including GlobalTrack, SiamRCNN and CAMTracker, outperform the others, apparently. Some trackers that take the whole frame as input, such as AiATrack and KeepTrack, also perform well in the task. In addition, it is worth mentioning that CAMTracker reaches the best quantitative performance on all the metrics, surpassing other trackers by at least 2.40%, 4.12%, 5.43% and 5.48% on precision, success rate, success AUC and state accuracy, respectively. Apart from the statistics, Figure 5 illustrates the precision curve and success curve. As the plots show, CAMTracker keeps leading with different thresholds in both precision and success curves, which, again, proves the effectiveness of the proposed tracker.

Table 1. Overall performance of CAMTracker and other trackers on the test set of Anti-UAV410.

Methods	P (%)	S_r (%)	S_{auc} (%)	SA (%)
DSST [27]	44.40	37.60	29.00	29.09
Staple [26]	38.01	30.17	24.16	24.25
SRDCF [28]	54.04	45.15	34.96	35.17
KCF [24]	41.43	34.89	26.57	26.64
ECO [75]	62.59	55.35	42.61	42.88
ATOM [82]	70.13	65.21	51.02	51.37
SiamBAN [48]	67.34	62.31	47.01	47.32
SiamCAR [47]	64.74	60.45	46.63	46.93
GlobalTrack [49]	86.16	80.75	61.25	61.62
SiamRCNN [50]	82.76	77.70	60.24	60.54
DiMP50 [79]	75.91	71.26	56.27	56.66
PrDiMP50 [80]	75.07	70.33	54.33	54.69
AiATrack [76]	82.35	77.54	59.13	59.56
KeepTrack [77]	80.97	73.09	56.47	56.80
Stark-ST101 [52]	78.62	74.04	56.76	57.15
ToMP50 [81]	73.96	69.63	54.71	55.09
ToMP101 [81]	75.21	70.28	54.72	55.10
TCTrack [83]	60.53	55.21	41.40	41.64
MixformerV2-B [78]	80.66	76.02	59.26	59.65
SwinTrack-Tiny [54]	71.51	67.73	52.79	53.15
SwinTrack-Base [54]	76.49	72.11	55.39	55.74
SeqTrack-B256 [56]	74.00	68.63	52.54	52.87
ROMTrack [84]	67.22	62.22	47.91	48.13
AQATrack [59]	76.57	71.71	55.46	55.81
CAMTracker (ours)	88.56	84.87	66.68	67.10

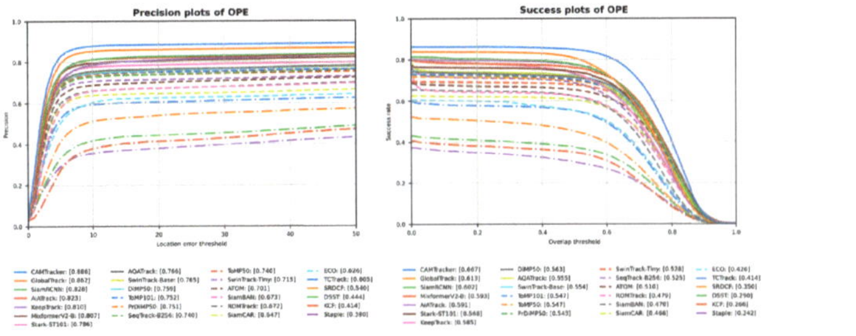

(a) (b)

Figure 5. The overall precision plot (**a**) and success plot (**b**) of CAMTracker and other compared trackers on the test set of Anti-UAV410.

3.5. Attribute-Based Evaluation

To comprehensively depict the effectiveness and robustness of CAMTracker, the performance of the proposed method on sequences with different attributes is analyzed.

The precision and success curves of the trackers under various situations are presented in Figure 6. CAMTracker exceeds the others in all the scenarios. Especially under occlusion (OC) and dynamic background clutter (DBC) scenes, CAMTracker obtains great advantages over the other trackers. In scale variation (SV) and out-of-view (OV) scenarios, CAMTracker slightly leads GlobalTrack, and both of them outperform the others with relatively big gaps.

Comprehensively, the experiment results show the ability and superiority of CAMTracker on tracking drones under diverse challenges in TIR videos.

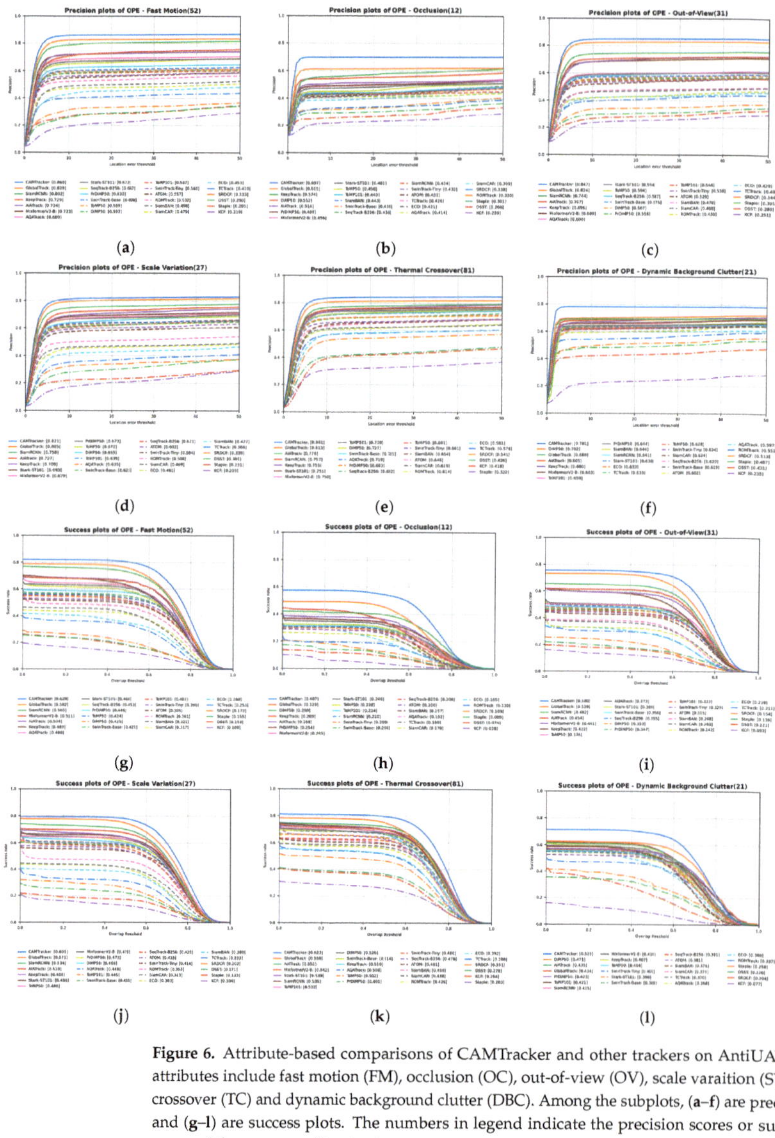

Figure 6. Attribute-based comparisons of CAMTracker and other trackers on AntiUAV-410. The attributes include fast motion (FM), occlusion (OC), out-of-view (OV), scale varaition (SV), thermal crossover (TC) and dynamic background clutter (DBC). Among the subplots, (**a**–**f**) are precision plots, and (**g**–**l**) are success plots. The numbers in legend indicate the precision scores or success AUC scores of the corresponding trackers.

Besides different scenarios, experiments are conducted to assess the performance of the trackers when facing different target sizes, which are mentioned in the *Dataset* part as well. The results are presented in Figures 7 and 8.

According to Figures 7 and 8, it can be inferred that (1) as the target size decreases, the overall performances of the trackers decline as well; (2) when the target is of normal size, CAMTracker only achieves quite a small lead; (3) a big advantage is obtained by CAMTracker in tracking tiny targets, which is 7.20% in precision and 7.70% in success AUC; and 4) the smaller the target size is, the better CAMTracker performs compared with other trackers. The previous experiments have demonstrated the versatility of CAMTracker in confronting heterogeneous challenges, either scenarios or target sizes.

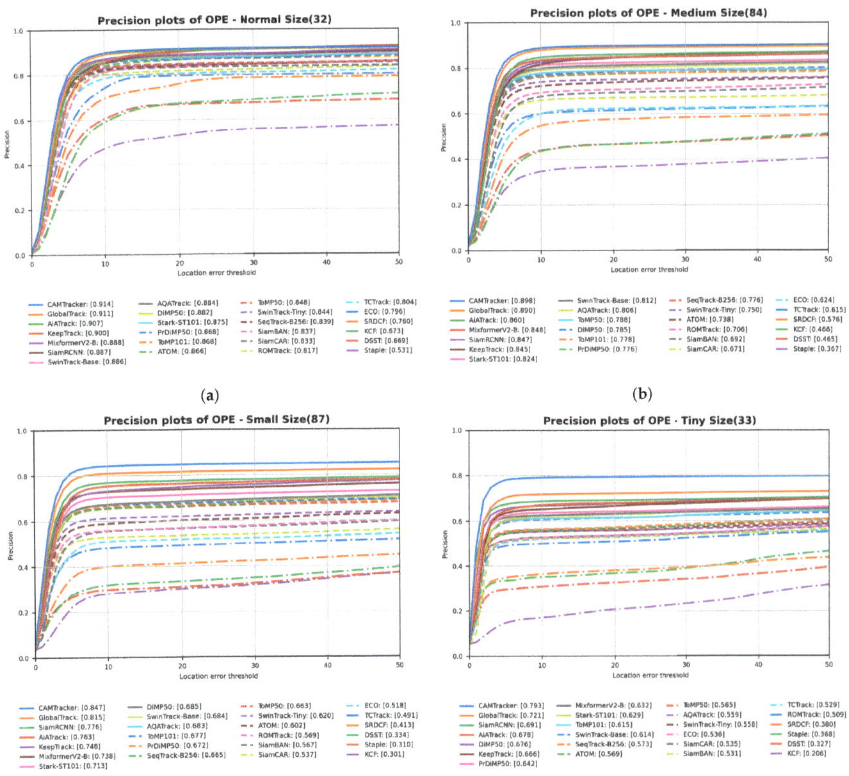

Figure 7. Target size-based comparisons on precision plots of CAMTracker and other trackers on AntiUAV-410. The sizes include normal size (**a**), medium size (**b**), small size (**c**) and tiny size (**d**).

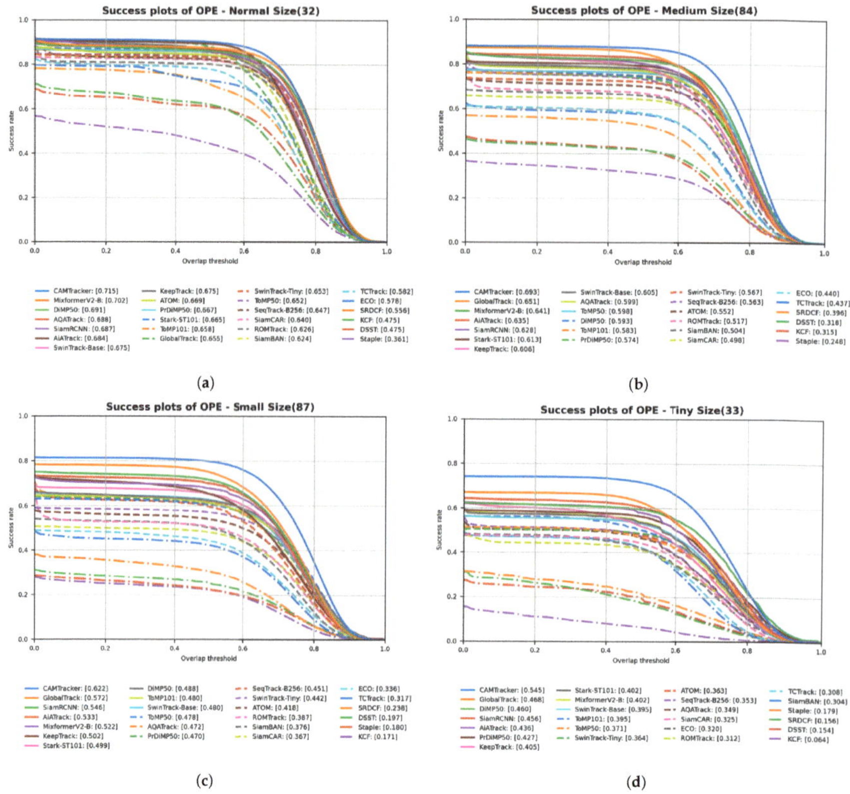

(**a**) (**b**) (**c**) (**d**)

Figure 8. Target size-based comparisons on success plots of CAMTracker and other trackers on AntiUAV-410. The sizes include normal size (**a**), medium size (**b**), small size (**c**) and tiny size (**d**).

3.6. Qualitative Evaluation

Besides the aforementioned quantitative evaluation and attribute evaluation, CAMTracker is also evaluated qualitatively by presenting visual results on particular sequences, aiming to show the performance of CAMTracker in an intuitive way. In this part, six trackers are selected for comparison, including ECO, SiamRCNN, GlobalTrack, Stark-ST101, SwinTrack-Base and MixformerV2-B.

Figure 9 visualizes tracking results on some challenging sequences. The results of the first sequence show the ability of CAMTracker when facing thermal crossover. As the drone remains in front of the building, other trackers gradually miss the target, while the prediction from CAMTracker stays on the ground truth. The second row depicts a typical occlusion scene. When the target becomes obstructed by the tree, all the trackers fail in capturing the drone. CAMTracker, along with GlobalTrack, finds the targets at the first frame the target reappears. The third sequence illustrates the fast motion situation caused by camera movement. Fast motion also leads to a blurred target. It can be found that most trackers cannot follow the target during the fast motion stage, while CAMTracker succeeds in every frame during the camera movement, even the target is fuzzy. The targets in the fourth and fifth row are both in tiny scale. The fourth and fifth sequences show the tracking results under a cluttered background and out-of-view of the target, respectively. Unlike other trackers, CAMTracker achieves much better performance under these two scenarios. The visualized results further prove the capability and superiority of CAMTracker in handling challenging situations in TIR anti-UAV tracking tasks.

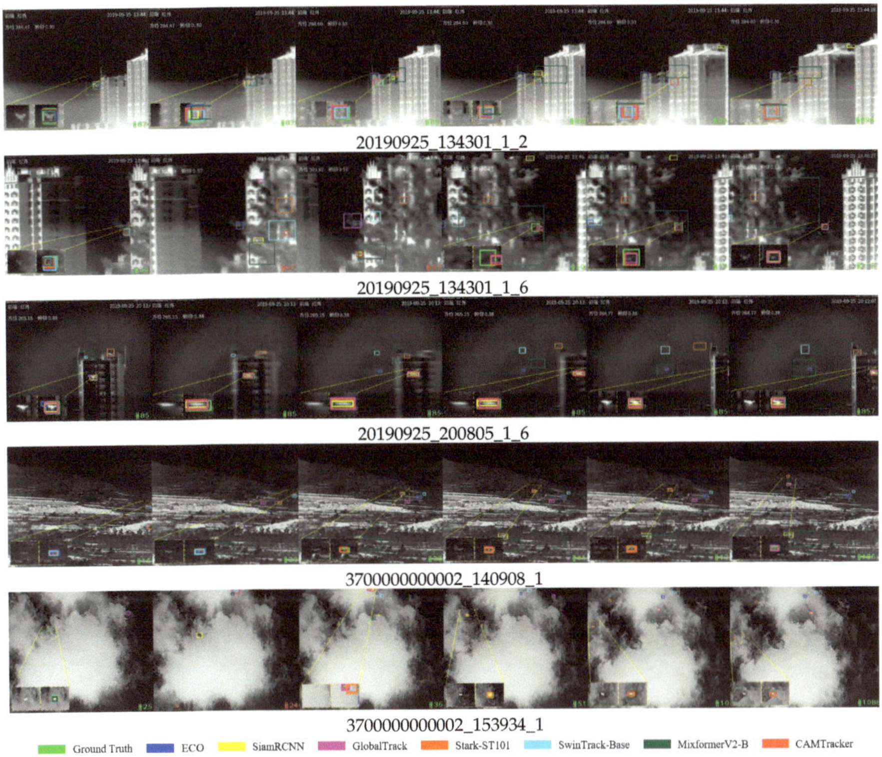

Figure 9. Qualitative evaluation on some challenging sequences.

3.7. Ablation Study

An ablation study on critical modules is conducted on the Anti-UAV410 dataset to verify the effectiveness of proposed components in CAMTracker. The results of the ablation study are presented in Table 2, where the best performances on each metrics are highlighted as bold. GlobalTrack is the baseline to analyze the usefulness of the proposed CAM and ADM. In addition, when CAM is applied, the performances of the tracker with common contrastive loss and the refined one are evaluated, respectively.

Table 2. Overall performance of CAMTracker and other trackers on the test set of Anti-UAV410.

CAM	ADM	Contrastive Loss	*P* (%)	S_r (%)	S_{auc} (%)	*SA* (%)
-	-	-	86.16	80.75	61.25	61.62
-	✓	-	87.63$_{\uparrow 1.47}$	83.80$_{\uparrow 3.05}$	65.33$_{\uparrow 4.08}$	65.73$_{\uparrow 4.11}$
✓	-	common	86.01$_{\downarrow 0.15}$	81.75$_{\uparrow 1.00}$	63.11$_{\uparrow 1.86}$	63.49$_{\uparrow 1.87}$
✓	-	refined	86.46$_{\uparrow 0.30}$	82.38$_{\uparrow 1.63}$	63.93$_{\uparrow 2.68}$	64.32$_{\uparrow 2.70}$
✓	✓	common	87.87$_{\uparrow 1.71}$	84.31$_{\uparrow 3.56}$	66.23$_{\uparrow 4.98}$	66.65$_{\uparrow 5.03}$
✓	✓	refined	**88.56** $_{\uparrow 2.40}$	**84.87**$_{\uparrow 4.12}$	**66.68**$_{\uparrow 5.43}$	**67.10**$_{\uparrow 5.48}$

From Table 2, it can be inferred that both CAM and ADM bring a positive impact on the tracker. The utilization of ADM improves the success AUC by 4.08% and state accuracy by 4.11%, which is a huge advancement. CAM also enhances the performance of the tracker by more than 1.8% on success AUC and state accuracy. Especially, when employing the refined contrastive loss, there is about a 0.9% improvement on success AUC and state accuracy. When combined together, CAM, ADM and the refined contrastive loss contribute 2.40%, 4.12%, 5.43% and 5.48% progress compared to the baseline on precision, success rate, success AUC and state accuracy, respectively.

3.8. Generalization Evaluation

To evaluate the tracking generalization of the proposed CAMTracker in generalized TIR tracking tasks, experiments are conducted on LSOTB-TIR dataset. There are 16 trackers applied in the comparison, including DSST [27], CFNet [85], UDT [86], SRDCF [28], BACF [30], SiamFC [44], SiamMask [87], SiamRPN++ [46], ATOM [82], ECO [75], MDNet [88], ECO-stir [89], ECO-TIR [62], AQATrack [59], SeqTrack [56] and ROMTrack [84]. The precision plot and success plot of the trackers on the test set of LSOTB-TIR are shown in Figure 10. The proposed CAMTracker scores 71.0% on precision rate and 58.7% on success AUC. The results show CAMTracker has an average performance on generalized TIR tracking tasks. ROMTrack, SeqTrack and AQATrack achieve much better performance than other trackers, which shows the ability of transformer-based trackers on tracking generalized objects. The performance of CAMTracker is behind the top tracker by 13.4% in precision rate and 13.2% in success AUC, which shows more studies should be carried out for further improvement.

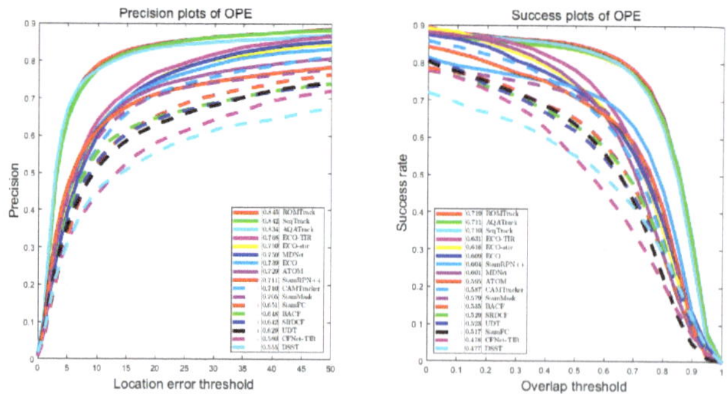

(**a**) (**b**)

Figure 10. The precision plot (**a**) and success plot (**b**) of CAMTracker and other compared trackers on the test set of LSOTB-TIR.

3.9. Evaluation Scenarios with Closed Distractors

During the tracking procedure, it is important to handle the interference from distractors, especially distractors closed to the target. To assess the ability of CAMTracker on handling closed distractors, qualitative experiments on several representative sequences in the LSOTB-TIR dataset are conducted. The tracking results are shown in Figure 11.

In the first sequence, when the target car passes the first closed distractor, CAMTracker is not interfered with. At frame 49, the top half of the image crashes, and the tracker takes a car nearby as the target. But at frame 53, the tracker finds the target again.

In the second sequence, there are two birds. At frame 50, the distractor becomes closed to the target and the tracker is influenced. Then, the tracker finds the target again until another approach comes from the distractor at frame 62. But the tracker is able to re-detect the target.

In the third sequence, the distractor looks almost the same as the target. During the tracking procedure, the tracker is confused several times and cannot successfully predict the target. The qualitative results indicate that the proposed CAMTracker is able to distinguish some of the closed distractors by local bias and the discriminative features. Cars in the first sequences are similar but different, so the tracker can complete the tracking task successfully. In the second sequence, when the tracker is distracted, the distractor becomes blurred due to the quick motion, and the tracker re-detects the target, luckily. In the third sequence, the target and the distractor share the same appearance. The slight change in posture distracts the tracker, leading to tracking failure.

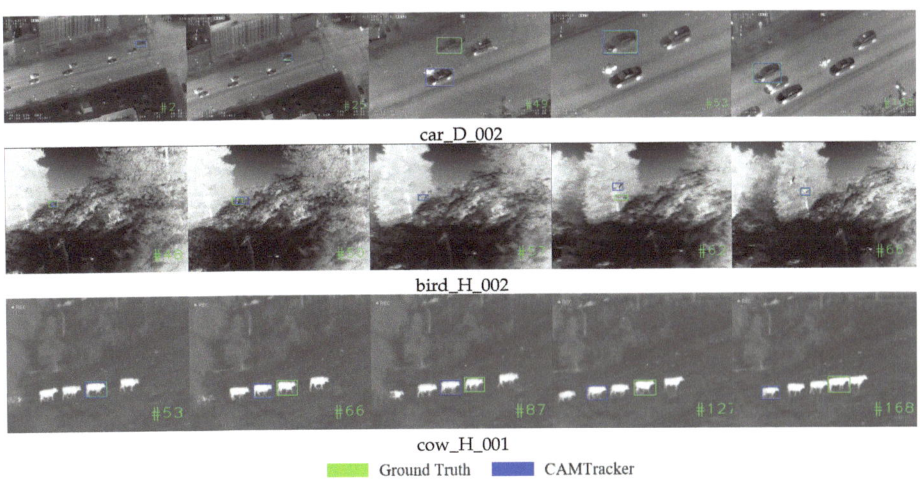

Figure 11. Qualitative evaluation for sequences containing closed distractors.

4. Discussion

As the experiments depict in Section 3, the proposed CAMTracker outperforms the state-of-the-art trackers in TIR anti-UAV tracking tasks, while staying in an averaging position in general TIR tracking tasks. According to Table 1, CAMTracker achieves the best performance on all the four metrics, and leads the second-best tracker for 2.40%, 4.12%, 5.43% and 5.48% on precision, success rate, success AUC and state accuracy, respectively. The high scores in the precision rate show the ability of CAMTracker in locating the target, while the advantage in the success rate and success AUC show the capability of CAMTracker for precisely perceiving the target. The solid lead in the SA score indicates

CAMTracker performs much better than other trackers when facing situations when the target cannot be seen, such as occlusion or out-of-view scenarios.

The attribute-based comparison in Figure 6 shows CAMTracker has a great advantage in scenarios with occlusion, leading by 8.2% in precision rate and 7.8% in success AUC. The performance of CAMTracker on out-of-view sequences leads the second-best tracker by 2.3% in precision rate and 4.1% in success AUC, while surpassing the third-best tracker by 10.3% in precision and 9.8% in success AUC. The results show the advantage of the GIS strategy applied in CAMTracker, which results in more emphasis on re-detection during the tracking procedure and keeping the focus on possible target loss. Compared to other GIS-based trackers, CAMTracker outperforms GlobalTrack in thermal crossover sequences by 2.8% in precision rate and 5.5% in success AUC, which verifies the effectiveness of the CAM module in capturing the discriminative information of drone targets. In dynamic background clutter sequences, CAMTracker has an advantage of 7.9% in precision rate and 5.4% in success AUC. This result proves the effectiveness of the ADM module in long-term tracking tasks.

As shown in Figures 7 and 8, CAMTracker achieves the best performances on targets of all sizes. Specifically, CAMTracker has quite a small advantage in normal-sized targets, only 0.3% in precision rate and 1.3% in success rate. As the target size decreases, the advantage of CAMTracker increases. In tiny target sequences, CAMTracker outperforms the second best tracker by 7.1% in precision rate and 7.7% in success rate. The results show the weakness of trackers for general scenarios when facing small particular targets. The good performance of CAMTracker in tracking tiny targets again proves the ability of the CAM module in extracting the discriminative features of small targets.

Even though the proposed CAMTracker has achieved considerable performance in TIR anti-UAV tracking tasks, there are still some challenges in anti-UAV tracking that remain to be improved. Figure 12 demonstrates two typical challenging sequences that remain to be solved. The first row is from sequence "20190925_124000_1_7" in the Anti-UAV410 dataset. The extreme thermal crossover causes the failure. In addition, the camera does not focus on the target, making the edge of the target more vague. It is even hard for human eyes to recognize the target. The second row is from sequence "3700000000002_133828_2". The main reason of failure is that this sequence is the clutter background, in addition to the tiny size of the target. It is easy for the target to become submerged in the background. Therefore, CAMTracker needs to be improved when facing these kind of scenarios.

Besides the challenging sequences in anti-UAV tasks, the tracking performance of CAMTracker on generalized scenarios is still improvable. According to Figure 10, CAMTracker performs ordinarily compared to other trackers. Unlike in anti-UAV tracking tasks, targets in generalized tracking tasks usually move smoothly and are seldom invisible in the sequences. Therefore, trackers that focus on local regions and spatio-temporal information usually perform better in these scenarios. CAMTracker emphasizes more on a global vision, which introduces more distractors and noise in generalized tracking tasks. In addition, insufficient learning on spatio-temporal constraints in CAMTracker results in a heavy dependence on the target appearance, which restricts the performance in generalized scenarios, especially crowd scenes. Therefore, the balance between local and global regions and the ability to capture spatio-temporal constraints are the main areas in which the proposed CAMTracker needs to improve.

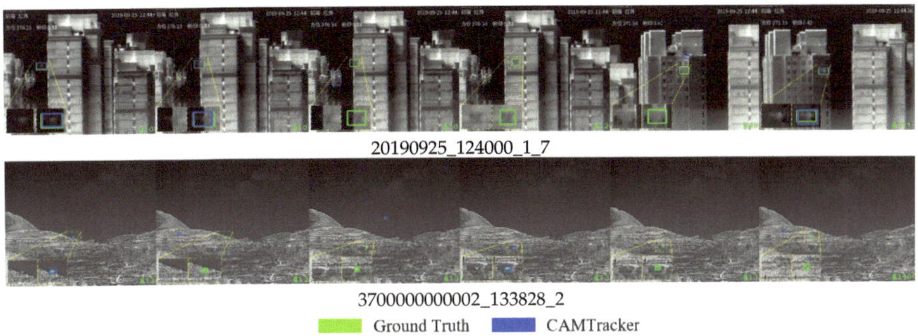

Figure 12. Failure on some challenging sequences.

5. Conclusions

In this article, a contrastive-augmented memory network (CAMTracker) with remarkable performance in long-term anti-UAV tracking tasks in thermal infrared (TIR) videos is proposed. In CAMTracker, a two-stage tracking framework is built to handle camera movement, out-of-view and fast motion situations, where the first stage coarsely searches possible candidate proposals through an instance-guided RPN (IG-RPN). Subsequently, in the second stage, to improve the tracker's discrimination ability when facing distractors and thermal crossover, a contrastive learning strategy and a refined contrastive loss are incorporated into a matching module, namely, a contrastive-augmented matching module. Furthermore, to mitigate model degradation caused by target appearance variation and dynamic backgrounds, a dynamic template is maintained by adaptively combining previous predictions to assist the instance matching in long-term tracking tasks.

Comprehensive experiments have been conducted on the Anti-UAV410 dataset and demonstrated the robustness and effectiveness of the proposed CAMTracker, which achieves 88.56% on precision, 84.87% on success rate, 66.68% on success AUC and 67.10% on state accuracy. The performance surpasses that of most state-of-the-art tracking algorithms by at least 2.40%, 4.12%, 5.43% and 5.48% on precision, success rate, success AUC and state accuracy, respectively, which proves the superiority of the proposed CAMTracker.

Although achieving a promising performance on anti-UAV tracking tasks, CAMTracker still has two main limitations. One is that, as a GIS based tracker, CAMTracker tends to focus on the global vision, therefore introducing more distractors and noise. When facing crowds of similar objects, it is more difficult to distinguish the target. The other limitation is the lack of understanding on spatio-temporal constraints. CAMTracker emphasizes re-detection more to counter challenges in anti-UAV tasks, resulting in the inadequate utilization of spatio-temporal information, which affects the tracking performance in generalized scenarios.

In future studies, approaches to switching trackers between global and local search can be introduced. Moreover, spatio-temporal strategies can be integrated into the tracker to exploit the information in sequences more comprehensively.

Author Contributions: Conceptualization, Z.W. and Y.H.; methodology, Z.W., J.Y. and Y.L.; resources, G.Z. and F.L.; writing—original draft preparation, Z.W.; supervision, Y.H. and Y.L. All authors have read and agreed to the published version of the manuscript.

Funding: This research received no external funding.

Data Availability Statement: Anti-UAV410 dataset is available at https://github.com/HwangBo94/Anti-UAV410 (accessed on 20 December 2024). LSOTB-TIR dataset is available at https://github.com/QiaoLiuHit/LSOTB-TIR (accessed on 20 December 2024).

Conflicts of Interest: The authors declare no conflicts of interest.

References

1. Fan, J.; Yang, X.; Lu, R.; Xie, X.; Li, W. Design and implementation of intelligent inspection and alarm flight system for epidemic prevention. *Drones* **2021**, *5*, 68. [CrossRef]
2. Filkin, T.; Sliusar, N.; Ritzkowski, M.; Huber-Humer, M. Unmanned aerial vehicles for operational monitoring of landfills. *Drones* **2021**, *5*, 125. [CrossRef]
3. McEnroe, P.; Wang, S.; Liyanage, M. A survey on the convergence of edge computing and AI for UAVs: Opportunities and challenges. *IEEE Internet Things J.* **2022**, *9*, 15435–15459. [CrossRef]
4. Wang, Z.; Cao, Z.; Xie, J.; Zhang, W.; He, Z. RF-based Drone Detection Enhancement via a Generalized Denoising and Interference-removal Framework. *IEEE Signal Process. Lett.* **2024**, *31*, 929–933. [CrossRef]
5. Zhou, T.; Xin, B.; Zheng, J.; Zhang, G.; Wang, B. Vehicle Detection Based on YOLOv7 for Drone Aerial Visible and Infrared Images. In Proceedings of the 2024 6th International Conference on Image Processing and Machine Vision, New York, NY, USA, 12–14 January 2024; pp. 30–35.
6. Wang, B.; Li, Q.; Mao, Q.; Wang, J.; Chen, C.P.; Shangguan, A.; Zhang, H. A Survey on Vision-Based Anti Unmanned Aerial Vehicles Methods. *Drones* **2024**, *8*, 518. [CrossRef]
7. Mueller, M.; Smith, N.; Ghanem, B. A benchmark and simulator for uav tracking. In *Proceedings of the Computer Vision–ECCV 2016: 14th European Conference, Amsterdam, The Netherlands, 11–14 October 2016*; Part I 14; Springer: Berlin/Heidelberg, Germany, 2016; pp. 445–461.
8. Huang, B.; Xu, T.; Jiang, S.; Chen, Y.; Bai, Y. Robust visual tracking via constrained multi-kernel correlation filters. *IEEE Trans. Multimed.* **2020**, *22*, 2820–2832. [CrossRef]
9. Jafferis, N.T.; Helbling, E.F.; Karpelson, M.; Wood, R.J. Untethered flight of an insect-sized flapping-wing microscale aerial vehicle. *Nature* **2019**, *570*, 491–495. [CrossRef] [PubMed]
10. Cliff, O.M.; Saunders, D.L.; Fitch, R. Robotic ecology: Tracking small dynamic animals with an autonomous aerial vehicle. *Sci. Robot.* **2018**, *3*, eaat8409. [CrossRef] [PubMed]
11. Svanström, F.; Alonso-Fernandez, F.; Englund, C. Drone detection and tracking in real-time by fusion of different sensing modalities. *Drones* **2022**, *6*, 317. [CrossRef]
12. Li, Y.; Fu, M.; Sun, H.; Deng, Z.; Zhang, Y. Radar-based UAV swarm surveillance based on a two-stage wave path difference estimation method. *IEEE Sens. J.* **2022**, *22*, 4268–4280. [CrossRef]
13. Sun, Y.; Li, J.; Wang, L.; Xv, J.; Liu, Y. Deep Learning-based drone acoustic event detection system for microphone arrays. *Multimed. Tools Appl.* **2024**, *83*, 47865–47887. [CrossRef]
14. Yu, Q.; Ma, Y.; He, J.; Yang, D.; Zhang, T. A unified transformer based tracker for anti-uav tracking. In Proceedings of the IEEE/CVF Conference on Computer Vision and Pattern Recognition, Vancouver, BC, Canada, 18–22 June 2023; pp. 3036–3046.
15. Elsayed, M.; Reda, M.; Mashaly, A.S.; Amein, A.S. LERFNet: An enlarged effective receptive field backbone network for enhancing visual drone detection. *Vis. Comput.* **2024**, 1–14
16. Yuan, D.; Zhang, H.; Shu, X.; Liu, Q.; Chang, X.; He, Z.; Shi, G. Thermal infrared target tracking: A comprehensive review. *IEEE Trans. Instrum. Meas.* **2023**, *73*, 5000419. [CrossRef]
17. Marvasti-Zadeh, S.M.; Cheng, L.; Ghanei-Yakhdan, H.; Kasaei, S. Deep learning for visual tracking: A comprehensive survey. *IEEE Trans. Intell. Transp. Syst.* **2021**, *23*, 3943–3968. [CrossRef]
18. Gao, Z.; Li, D.; Wen, G.; Kuai, Y.; Chen, R. Drone based RGBT tracking with dual-feature aggregation network. *Drones* **2023**, *7*, 585. [CrossRef]
19. Zhang, F.; Peng, H.; Yu, L.; Zhao, Y.; Chen, B. Dual-modality space-time memory network for RGBT tracking. *IEEE Trans. Instrum. Meas.* **2023**, *72*, 1–12. [CrossRef]
20. Huang, B.; Li, J.; Chen, J.; Wang, G.; Zhao, J.; Xu, T. Anti-UAV410: A Thermal Infrared Benchmark and Customized Scheme for Tracking Drones in the Wild. *IEEE Trans. Pattern Anal. Mach. Intell.* **2023**, *46*, 2852–2865. [CrossRef]
21. Kumar, A.; Vohra, R.; Jain, R.; Li, M.; Gan, C.; Jain, D.K. Correlation filter based single object tracking: A review. *Inf. Fusion* **2024**, *112*, 102562. [CrossRef]
22. Bolme, D.S.; Beveridge, J.R.; Draper, B.A.; Lui, Y.M. Visual object tracking using adaptive correlation filters. In Proceedings of the 2010 IEEE Computer Society Conference on Computer Vision and Pattern Recognition, San Francisco, CA, USA, 13–18 June 2010; pp. 2544–2550.
23. Henriques, J.F.; Caseiro, R.; Martins, P.; Batista, J. Exploiting the circulant structure of tracking-by-detection with kernels. In *Proceedings of the Computer Vision–ECCV 2012: 12th European Conference on Computer Vision, Florence, Italy, 7–13 October 2012*; Proceedings, Part IV 12; Springer: Berlin/Heidelberg, Germany, 2012; pp. 702–715.
24. Henriques, J.F.; Caseiro, R.; Martins, P.; Batista, J. High-speed tracking with kernelized correlation filters. *IEEE Trans. Pattern Anal. Mach. Intell.* **2014**, *37*, 583–596. [CrossRef]
25. Tang, M.; Feng, J. Multi-kernel correlation filter for visual tracking. In Proceedings of the IEEE International Conference on Computer Vision, Santiago, Chile, 13–16 December 2015; pp. 3038–3046.

26. Bertinetto, L.; Valmadre, J.; Golodetz, S.; Miksik, O.; Torr, P.H. Staple: Complementary learners for real-time tracking. In Proceedings of the IEEE Conference on Computer Vision and Pattern Recognition, Las Vegas, NV, USA, 26 June–1 July 2016; pp. 1401–1409.
27. Danelljan, M.; Häger, G.; Khan, F.; Felsberg, M. Accurate scale estimation for robust visual tracking. In Proceedings of the British Machine Vision Conference, Nottingham, UK, 1–5 September 2014; Bmva Press: Durham, UK, 2014.
28. Danelljan, M.; Hager, G.; Shahbaz Khan, F.; Felsberg, M. Learning spatially regularized correlation filters for visual tracking. In Proceedings of the IEEE International Conference on Computer Vision, Santiago, Chile, 13–16 December 2015; pp. 4310–4318.
29. Li, F.; Tian, C.; Zuo, W.; Zhang, L.; Yang, M.H. Learning spatial-temporal regularized correlation filters for visual tracking. In Proceedings of the IEEE Conference on Computer Vision and Pattern Recognition, Salt Lake City, UT, USA, 18–22 June 2018; pp. 4904–4913.
30. Kiani Galoogahi, H.; Fagg, A.; Lucey, S. Learning background-aware correlation filters for visual tracking. In Proceedings of the IEEE International Conference on Computer Vision, Venice, Italy, 22–29 October 2017; pp. 1135–1143.
31. Dai, K.; Wang, D.; Lu, H.; Sun, C.; Li, J. Visual tracking via adaptive spatially-regularized correlation filters. In Proceedings of the IEEE/CVF Conference on Computer Vision and Pattern Recognition, Long Beach, CA, USA, 16–20 June 2019; pp. 4670–4679.
32. Tang, F.; Ling, Q. Spatial-aware correlation filters with adaptive weight maps for visual tracking. *Neurocomputing* **2019**, *358*, 369–384. [CrossRef]
33. Krizhevsky, A.; Sutskever, I.; Hinton, G.E. Imagenet classification with deep convolutional neural networks. *Commun. ACM* **2017**, 60, 84–90. [CrossRef]
34. Chatfield, K. Return of the devil in the details: Delving deep into convolutional nets. *arXiv* **2014**, arXiv:1405.3531.
35. Simonyan, K. Very deep convolutional networks for large-scale image recognition. *arXiv* **2014**, arXiv:1409.1556.
36. Russakovsky, O.; Deng, J.; Su, H.; Krause, J.; Satheesh, S.; Ma, S.; Huang, Z.; Karpathy, A.; Khosla, A.; Bernstein, M.; et al. Imagenet large scale visual recognition challenge. *Int. J. Comput. Vis.* **2015**, *115*, 211–252. [CrossRef]
37. Kristan, M.; Pflugfelder, R.; Leonardis, A.; Matas, J.; Porikli, F.; Cehovin, L.; Nebehay, G.; Fernandez, G.; Vojir, T. The vot2013 Challenge: Overview and Additional Results . In Proceedings of the Computer Vision Winter Workshop, Křtiny, Czech Republic, 3–5 February 2014; pp. 59–66.
38. Kristan, M.; Matas, J.; Leonardis, A.; Felsberg, M.; Cehovin, L.; Fernandez, G.; Vojir, T.; Hager, G.; Nebehay, G.; Pflugfelder, R. The visual object tracking vot2015 challenge results. In Proceedings of the IEEE International Conference on Computer Vision Workshops, Washington, DC, USA, 7–13 December 2015; pp. 1–23.
39. Kristan, M.; Leonardis, A.; Matas, J.; Felsberg, M.; Pflugfelder, R.; Cehovin Zajc, L.; Vojir, T.; Bhat, G.; Lukezic, A.; Eldesokey, A.; et al. The sixth visual object tracking vot2018 challenge results. In Proceedings of the European Conference on Computer Vision (ECCV) Workshops, Munich, Germany, 8–14 September 2018.
40. Liu, P.; Liu, C.; Zhao, W.; Tang, X. Multi-level context-adaptive correlation tracking. *Pattern Recognit.* **2019**, *87*, 216–225. [CrossRef]
41. Zhang, J.; Jin, X.; Sun, J.; Wang, J.; Sangaiah, A.K. Spatial and semantic convolutional features for robust visual object tracking. *Multimed. Tools Appl.* **2020**, *79*, 15095–15115. [CrossRef]
42. Zhu, X.F.; Wu, X.J.; Xu, T.; Feng, Z.H.; Kittler, J. Robust visual object tracking via adaptive attribute-aware discriminative correlation filters. *IEEE Trans. Multimed.* **2021**, *24*, 301–312. [CrossRef]
43. Tao, R.; Gavves, E.; Smeulders, A.W. Siamese instance search for tracking. In Proceedings of the IEEE Conference on Computer Vision and Pattern Recognition, Las Vegas, NV, USA, 26 June–1 July 2016; pp. 1420–1429.
44. Bertinetto, L.; Valmadre, J.; Henriques, J.F.; Vedaldi, A.; Torr, P.H. Fully-convolutional siamese networks for object tracking. In *Proceedings of the Computer Vision–ECCV 2016 Workshops: Amsterdam, The Netherlands, 8–10 and 15–16 October 2016*; Proceedings, Part II 14; Springer: Berlin/Heidelberg, Germany, 2016; pp. 850–865.
45. Li, B.; Yan, J.; Wu, W.; Zhu, Z.; Hu, X. High performance visual tracking with siamese region proposal network. In Proceedings of the IEEE Conference on Computer Vision and Pattern Recognition, Salt Lake City, UT, USA, 18–22 June 2018; pp. 8971–8980.
46. Li, B.; Wu, W.; Wang, Q.; Zhang, F.; Xing, J.; Yan, J.S. Evolution of siamese visual tracking with very deep networks. In Proceedings of the IEEE/CVF Conference on Computer Vision and Pattern Recognition, Long Beach, CA, USA, 16–20 June 2019; pp. 15–20.
47. Guo, D.; Wang, J.; Cui, Y.; Wang, Z.; Chen, S. SiamCAR: Siamese fully convolutional classification and regression for visual tracking. In Proceedings of the IEEE/CVF Conference on Computer Vision and Pattern Recognition, Seattle, WA, USA, 13–19 June 2020; pp. 6269–6277.
48. Chen, Z.; Zhong, B.; Li, G.; Zhang, S.; Ji, R. Siamese box adaptive network for visual tracking. In Proceedings of the IEEE/CVF Conference on Computer Vision and Pattern Recognition, Seattle, WA, USA, 13–19 June 2020; pp. 6668–6677.
49. Huang, L.; Zhao, X.; Huang, K. Globaltrack: A simple and strong baseline for long-term tracking. In Proceedings of the AAAI Conference on Artificial Intelligence, New York, NY, USA, 7–12 February 2020; Volume 34, pp. 11037–11044.
50. Voigtlaender, P.; Luiten, J.; Torr, P.H.; Leibe, B. Siam r-cnn: Visual tracking by re-detection. In Proceedings of the IEEE/CVF Conference on Computer Vision and Pattern Recognition, Seattle, WA, USA, 13–19 June 2020; pp. 6578–6588.
51. Vaswani, A. Attention is all you need. In Proceedings of the 31st International Conference on Neural Information Processing Systems, Long Beach, CA, USA, 4–9 December 2017.
52. Yan, B.; Peng, H.; Fu, J.; Wang, D.; Lu, H. Learning spatio-temporal transformer for visual tracking. In Proceedings of the IEEE/CVF International Conference on Computer Vision, Montreal, QC, Canada, 10–17 October 2021; pp. 10448–10457.

53. Chen, X.; Yan, B.; Zhu, J.; Wang, D.; Yang, X.; Lu, H. Transformer tracking. In Proceedings of the IEEE/CVF Conference on Computer Vision and Pattern Recognition, Online, 19–25 June 2021; pp. 8126–8135.
54. Lin, L.; Fan, H.; Zhang, Z.; Xu, Y.; Ling, H. Swintrack: A simple and strong baseline for transformer tracking. *Adv. Neural Inf. Process. Syst.* **2022**, *35*, 16743–16754.
55. Ye, B.; Chang, H.; Ma, B.; Shan, S.; Chen, X. Joint feature learning and relation modeling for tracking: A one-stream framework. In Proceedings of the European Conference on Computer Vision, Tel Aviv, Israel, 23–27 October 2022 Springer: Berlin/Heidelberg, Germany, 2022; pp. 341–357.
56. Chen, X.; Peng, H.; Wang, D.; Lu, H.; Hu, H. Seqtrack: Sequence to sequence learning for visual object tracking. In Proceedings of the IEEE/CVF Conference on Computer Vision and Pattern Recognition, Vancouver, BC, Canada, 18–22 June 2023; pp. 14572–14581.
57. Wei, X.; Bai, Y.; Zheng, Y.; Shi, D.; Gong, Y. Autoregressive visual tracking. In Proceedings of the IEEE/CVF Conference on Computer Vision and Pattern Recognition, Vancouver, BC, Canada, 18–22 June 2023; pp. 9697–9706.
58. Bai, Y.; Zhao, Z.; Gong, Y.; Wei, X. Artrackv2: Prompting autoregressive tracker where to look and how to describe. In Proceedings of the IEEE/CVF Conference on Computer Vision and Pattern Recognition, Seattle, WA, USA, 17–21 June 2024; pp. 19048–19057.
59. Xie, J.; Zhong, B.; Mo, Z.; Zhang, S.; Shi, L.; Song, S.; Ji, R. Autoregressive Queries for Adaptive Tracking with Spatio-Temporal Transformers. In Proceedings of the IEEE/CVF Conference on Computer Vision and Pattern Recognition, Seattle, United States, 17-21 June 2024; pp. 19300–19309.
60. Liu, Q.; Lu, X.; He, Z.; Zhang, C.; Chen, W.S. Deep convolutional neural networks for thermal infrared object tracking. *Knowl.-Based Syst.* **2017**, *134*, 189–198. [CrossRef]
61. Liu, Q.; He, Z.; Li, X.; Zheng, Y. PTB-TIR: A thermal infrared pedestrian tracking benchmark. *IEEE Trans. Multimed.* **2019**, *22*, 666–675. [CrossRef]
62. Liu, Q.; Li, X.; He, Z.; Li, C.; Li, J.; Zhou, Z.; Yuan, D.; Li, J.; Yang, K.; Fan, N.; et al. LSOTB-TIR: A large-scale high-diversity thermal infrared object tracking benchmark. In Proceedings of the 28th ACM International Conference on Multimedia, Seattle, WA, USA, 12–16 October 2020; pp. 3847–3856.
63. Li, X.; Liu, Q.; Fan, N.; He, Z.; Wang, H. Hierarchical Spatial-aware Siamese Network for Thermal Infrared Object Tracking. *Knowl.-Based Syst.* **2019**, *166*, 71–81. [CrossRef]
64. Liu, Q.; Li, X.; He, Z.; Fan, N.; Yuan, D.; Wang, H. Learning Deep Multi-Level Similarity for Thermal Infrared Object Tracking. *IEEE Trans. Multimed.* **2019**, *23*, 2114–2126. [CrossRef]
65. Liu, Q.; Yuan, D.; Fan, N.; Gao, P.; Li, X.; He, Z. Learning Dual-Level Deep Representation for Thermal Infrared Tracking. *IEEE Trans. Multimed.* **2023**, *25*, 1269–1281. [CrossRef]
66. Yuan, D.; Shu, X.; Liu, Q.; He, Z. Aligned Spatial-Temporal Memory Network for Thermal Infrared Target Tracking. *IEEE Trans. Circuits Syst. II Express Briefs* **2023**, *70*, 1224–1228. [CrossRef]
67. Huang, B.; Dou, Z.; Chen, J.; Li, J.; Shen, N.; Wang, Y.; Xu, T. Searching Region-Free and Template-Free Siamese Network for Tracking Drones in TIR Videos. *IEEE Trans. Geosci. Remote Sens.* **2023**, *62*, 5000315. [CrossRef]
68. Jiang, N.; Wang, K.; Peng, X.; Yu, X.; Wang, Q.; Xing, J.; Li, G.; Ye, Q.; Jiao, J.; Han, Z.; et al. Anti-UAV: A large-scale benchmark for vision-based UAV tracking. *IEEE Trans. Multimed.* **2021**, *25*, 486–500. [CrossRef]
69. Huang, B.; Chen, J.; Xu, T.; Wang, Y.; Jiang, S.; Wang, Y.; Wang, L.; Li, J. SiamSTA: Spatio-temporal attention based Siamese tracker for tracking UAVs. In Proceedings of the IEEE/CVF International Conference on Computer Vision, Montreal, QC, Canada, 12–17 October 2021; pp. 1204–1212.
70. Shi, X.; Zhang, Y.; Shi, Z.; Zhang, Y. Gasiam: Graph attention based siamese tracker for infrared anti-uav. In Proceedings of the 2022 3rd International Conference on Computer Vision, Image and Deep Learning & International Conference on Computer Engineering and Applications (CVIDL & ICCEA), Changchun, China, 20–22 May 2022; pp. 986–993.
71. Hu, H.; Wang, X.; Zhang, Y.; Chen, Q.; Guan, Q. A comprehensive survey on contrastive learning. *Neurocomputing* **2024**, *610*, 128645. [CrossRef]
72. Girshick, R. Fast r-cnn. *arXiv* **2015**, arXiv:1504.08083.
73. Oord, A.v.d.; Li, Y.; Vinyals, O. Representation learning with contrastive predictive coding. *arXiv* **2018**, arXiv:1807.03748.
74. Wu, Y.; Lim, J.; Yang, M.H. Online object tracking: A benchmark. In Proceedings of the IEEE Conference on Computer Vision and Pattern Recognition, Portland, OR, USA, 23–28 June 2013; pp. 2411–2418.
75. Danelljan, M.; Bhat, G.; Shahbaz Khan, F.; Felsberg, M. Eco: Efficient convolution operators for tracking. In Proceedings of the IEEE Conference on Computer Vision and Pattern Recognition, Honolulu, HI, USA, 21–26 June 2017; pp. 6638–6646.
76. Gao, S.; Zhou, C.; Ma, C.; Wang, X.; Yuan, J. Aiatrack: Attention in attention for transformer visual tracking. In Proceedings of the European Conference on Computer Vision, Tel Aviv, Israel, 23–27 October 2022; Springer: Berlin/Heidelberg, Germany, 2022; pp. 146–164.
77. Mayer, C.; Danelljan, M.; Paudel, D.P.; Van Gool, L. Learning target candidate association to keep track of what not to track. In Proceedings of the IEEE/CVF International Conference on Computer Vision, Montreal, QC, Canada, 10–17 October 2021; pp. 13444–13454.
78. Cui, Y.; Jiang, C.; Wang, L.; Wu, G. Mixformer: End-to-end tracking with iterative mixed attention. In Proceedings of the IEEE/CVF Conference on Computer Vision and Pattern Recognition, New Orleans, LA, USA, 19–24 June 2022; pp. 13608–13618.

79. Bhat, G.; Danelljan, M.; Gool, L.V.; Timofte, R. Learning discriminative model prediction for tracking. In Proceedings of the IEEE/CVF International Conference on Computer Vision, Seoul, Republic of Korea, 27 October–2 November 2019; pp. 6182–6191.
80. Chen, S.; Qiu, C.; Huang, Y.; Zhang, Z. Robust Probabilistic Discriminative Model Prediction Tracker via Improved Model Update Strategy. 2021. Available online: https://www.semanticscholar.org/paper/Robust-Probabilistic-Discriminative-Model-Tracker-Chen-Qiu/2c1e7fc5edb772ab51583ca5c5a66a87b4060986 (accessed on 20 December 2024). [CrossRef]
81. Mayer, C.; Danelljan, M.; Bhat, G.; Paul, M.; Paudel, D.P.; Yu, F.; Van Gool, L. Transforming model prediction for tracking. In Proceedings of the IEEE/CVF Conference on Computer Vision and Pattern Recognition, New Orleans, LA, USA, 19–24 June 2022; pp. 8731–8740.
82. Danelljan, M.; Bhat, G.; Khan, F.S.; Felsberg, M. Atom: Accurate tracking by overlap maximization. In Proceedings of the IEEE/CVF Conference on Computer Vision and Pattern Recognition, Long Beach, CA, USA, 16–20 June 2019; pp. 4660–4669.
83. Cao, Z.; Huang, Z.; Pan, L.; Zhang, S.; Liu, Z.; Fu, C. TCTrack: Temporal contexts for aerial tracking. In Proceedings of the IEEE/CVF Conference on Computer Vision and Pattern Recognition, New Orleans, LA, USA, 19–24 June 2022; pp. 14798–14808.
84. Cai, Y.; Liu, J.; Tang, J.; Wu, G. Robust object modeling for visual tracking. In Proceedings of the IEEE/CVF International Conference on Computer Vision, Paris, France, 2–6 October 2023; pp. 9589–9600.
85. Valmadre, J.; Bertinetto, L.; Henriques, J.; Vedaldi, A.; Torr, P.H. End-to-end representation learning for correlation filter based tracking. In Proceedings of the IEEE Conference on Computer Vision and Pattern Recognition, Honolulu, HI, USA, 21–26 June 2017; pp. 2805–2813.
86. Wang, N.; Song, Y.; Ma, C.; Zhou, W.; Liu, W.; Li, H. Unsupervised deep tracking. In Proceedings of the IEEE/CVF Conference on Computer Vision and Pattern Recognition, Long Beach, CA, USA, 16–20 June 2019; pp. 1308–1317.
87. Wang, Q.; Zhang, L.; Bertinetto, L.; Hu, W.; Torr, P.H. Fast online object tracking and segmentation: A unifying approach. In Proceedings of the IEEE/CVF Conference on Computer Vision and Pattern Recognition, Long Beach, CA, USA, 16–20 June 2019; pp. 1328–1338.
88. Nam, H.; Han, B. Learning multi-domain convolutional neural networks for visual tracking. In Proceedings of the IEEE Conference on Computer Vision and PATTERN Recognition, Las Vegas, NV, USA, 26 June–1 July 2016; pp. 4293–4302.
89. Zhang, L.; Gonzalez-Garcia, A.; Van De Weijer, J.; Danelljan, M.; Khan, F.S. Synthetic data generation for end-to-end thermal infrared tracking. *IEEE Trans. Image Process.* **2018**, *28*, 1837–1850. [CrossRef]

remote sensing

MDPI

Article

An FPGA-Based Hybrid Overlapping Acceleration Architecture for Small-Target Remote Sensing Detection

Nan Fang [1,2], Liyuan Li [3], Xiaoxuan Zhou [1,2], Wencong Zhang [1,2] and Fansheng Chen [1,3,*]

1 State Key Laboratory of Infrared Physics, Shanghai Institute of Technical Physics, Chinese Academy of Sciences, Shanghai 200083, China; fangnan23@mails.ucas.ac.cn (N.F.); zhouxiaoxuan@mail.sitp.ac.cn (X.Z.); zhangwencong23@mails.ucas.ac.cn (W.Z.)

2 University of Chinese Academy of Sciences, Beijing 100049, China

3 Shanghai Frontier Base of Intelligent Optoelectronics and Perception, Institute of Optoelectronics, Fudan University, Shanghai 200433, China; li_liyuan@fudan.edu.cn

* Correspondence: cfs@mail.sitp.ac.cn

Abstract: Small-object detection in satellite remote sensing images plays a pivotal role in the field of remote sensing. Achieving high-performance real-time detection demands not only efficient algorithms but also low-power, high-performance hardware platforms. However, most mainstream target detection methods currently rely on graphics processing units (GPUs) for acceleration, and the high power consumption of GPUs limits their use in resource-constrained platforms such as small satellites. Moreover, small-object detection faces multiple challenges: the targets occupy only a small number of pixels in the image, the background is often complex with significant noise interference, and existing detection models typically exhibit low accuracy when dealing with small targets. In addition, the large number of parameters in these models makes direct deployment on embedded devices difficult. To address these issues, we propose a hybrid overlapping acceleration architecture based on FPGA, along with a lightweight model derived from YOLOv5s that is specifically designed to enhance the detection of small objects in remote sensing images. This model incorporates a lightweight GhostBottleneckV2 module, significantly reducing both model parameters and computational complexity. Experimental results on the TIFAD thermal infrared small-object dataset show that our approach achieves an average precision (mAP) of 67.8% while consuming an average power of only 2.8 W. The robustness of the proposed model is verified by the HRSID dataset. Combining real-time performance with high energy efficiency, this architecture is particularly well suited for on-board remote sensing image processing systems, where reliable and efficient small-object detection is paramount.

Keywords: FPGA acceleration architecture; remote sensing; lightweight neural networks; yolov5s

Academic Editor: Jon Atli Benediktsson

Received: 22 November 2024
Revised: 23 January 2025
Accepted: 29 January 2025
Published: 31 January 2025

Citation: Fang, N.; Li, L.; Zhou, X.; Zhang, W.; Chen, F. An FPGA-Based Hybrid Overlapping Acceleration Architecture for Small-Target Remote Sensing Detection. *Remote Sens.* **2025**, *17*, 494. https://doi.org/10.3390/rs17030494

1. Introduction

Target detection techniques in remote sensing images have been widely used in urban safety management [1], the monitoring of glacier changes [2], forestry vegetation monitoring [3], disaster prevention and relief [4], and military defense [5]. They have important scientific research values and broad application prospects. Target detection methods can be broadly classified into traditional methods based on manual feature extraction and modern methods based on deep learning. Traditional methods rely on manually designed feature extractors to identify targets, while deep learning methods achieve data-driven feature extraction through convolutional neural networks (CNNs), which can adaptively improve detection performance. With the rapid development of deep learning technology, deep learning-based target detection algorithms have significantly outperformed traditional methods and are widely used in tasks such as image classification and target detection.

In deep learning detection algorithms, the two-stage detection method divides target detection into two steps: first, candidate regions are generated, and then, their positions are refined before classification. Despite the slower detection speed, this method dominates in application scenarios that require high accuracy, such as face recognition and medical image analysis, due to its low detection and recognition error rates and relatively low false alarm rate. In contrast, single-stage detection algorithms simplify the detection process and directly output the probabilities of target categories and position coordinates, significantly improving detection speed, and are therefore, more suitable for tasks such as real-time video surveillance and autonomous driving.

Among single-stage detection algorithms, the YOLO (You Only Look Once) [6] series, as a representative method, directly locates targets and identifies their categories through a single forward propagation, thus achieving efficient real-time detection. The main advantage of YOLO is that it strikes a good balance between speed and accuracy; it is not only suitable for applications demanding high precision but can also run efficiently on resource-constrained embedded platforms. With the evolution of YOLOv4 [7], YOLOv5 [8], and other versions, the model has been significantly improved in small-target detection, parameter optimization, and computational efficiency, making it an ideal choice for real-time target detection.

In addition, with the rapid development of remote sensing and computer technology, the resolution and complexity of remote sensing images have increased dramatically, and the amount of remote sensing data collected by a single satellite data center has reached the terabyte level every day [9]. Traditional image processing methods usually transmit remote sensing images to the ground for processing, but due to transmission delays, the processing speed struggles to keep up with the speed of data acquisition, resulting in a large amount of data not being analyzed and applied in a timely manner [10]. Therefore, combining CNNs with satellite-embedded devices to achieve real-time target detection on satellites has become a promising solution. However, due to the limitations of storage and computational resources of embedded devices, and the large parameters of CNN models, deploying CNNs on embedded devices faces major challenges, which constrains the performance of real-time detection [11].

Usually, satellite-based remote sensing images are captured by satellites or drones and then transmitted to ground GPU platforms for subsequent processing. This approach can take advantage of the strong processing power of ground computing resources to achieve efficient data processing while circumventing the power consumption limitations of satellites. However, downloading data to the ground for processing brings significant transmission delays, which affect the overall system efficiency and make it difficult to meet real-time detection requirements. To address this issue, researchers are increasingly exploring the deployment of CNN models on Application-Specific Integrated Circuits (ASICs) and FPGAs [12]. These specialized hardware possess parallel computing capabilities. With parallel computing capabilities and high performance, these dedicated hardware can significantly accelerate the image processing process and reduce the latency caused by ground transmission, thus supporting efficient real-time detection. In particular, FPGAs, with their low power consumption, high performance, parallel computing capability, and programmability, are an ideal alternative to GPUs in the on-board environment and show great potential for real-time detection in remote sensing applications.

To address these challenges, this paper proposes a network architecture based on the lightweight YOLOv5s [13] model to further optimize small-target detection performance and design a hybrid overlapping architecture to achieve higher computational performance and detection accuracy on FPGA. Our main contributions include the following:

1. We design and propose the lightweight network model GF-YOLO, which is a specialized architecture for detecting small targets in remote sensing images, significantly improving small-target detection.
2. We present a hardware accelerator design based on hybrid overlapping architecture. By reusing computational units, intra-layer fusion, and double buffering mechanisms, we efficiently implement hardware acceleration of network models on programmable logic (PL).
3. We validate our design on the thermal infrared small-target TIFAD. The experimental results demonstrate that our scheme improves the accuracy of small-target detection. In addition, the specialized hardware architecture achieves lower power consumption, providing an important reference for the design of satellite-based image processing systems.

2. Software Optimization

The increased accuracy of deep neural networks has led to an increase in network parameters and computational requirements. Due to the growing demand for low-power and low-latency applications, such as edge computing and terminal computing, lightweight target detection algorithms have been rapidly developed.

Lightweight network design is fundamental for efficient small-target detection. The MobileNet series [14] represents a lightweight network that significantly reduces model parameters and computation volume through deep separable convolution. ShuffleNet [15] further enhances computational efficiency through channel rearrangement operations. In addition, EfficientNet [16] utilizes the Neural Architecture Search (NAS) [17] approach to achieve an optimal balance between model size and performance, making it widely used in lightweight detection tasks.

Optimization strategies for small-target detection aim to enhance the model's ability to detect small-sized targets. The Feature Pyramid Network (FPN) [18] effectively enhances the detection performance of small targets through multi-scale feature fusion. Methods based on attention mechanisms, such as the Squeeze-and-Excitation (SE) module [19] and the Convolutional Block Attention Module (CBAM) [20], improve the model's focus on small target regions by dynamically adjusting feature weights.

The single-stage YOLO algorithm performs well in target detection tasks. Qiu et al. [21] proposed an ASFF-based YOLOv5 method for UAV road detection, integrating the multi-scale feature fusion module of ASFF with the spatial pyramid pooling structure of SPPF to improve target detection performance and address challenges posed by varying feature scales. Li et al. [22] proposed a lightweight M3 YOLOv5 algorithm for multi-target detection and tracking, which enhances the detection of small- and medium-sized targets by replacing the backbone network of YOLOv5 with MobileNetV3 and combining the CA attention mechanism with an optimized neck network. To address the challenges posed by complex environments, uneven target scales, and high computational complexity in small-target detection in UAV infrared images, Wang et al. [23] proposed the lightweight PHSI-RTDETR algorithm, which reduces model parameters and computation while improving detection accuracy by optimizing the feature extraction network and incorporating the AIFI-HiLo attentional mechanism and Slimneck-SSFF feature fusion architecture. Jin et al. [24] proposed the YOLO-TLA model based on YOLOv5, significantly improving small-target detection accuracy while maintaining a lightweight model by adding a small-target detection layer to the neck network, integrating the C3CrossConv module to reduce computational load and parameters, and incorporating a global attention mechanism to enhance target features.

2.1. Basic Module

2.1.1. YOLOv5s

YOLOv5s is a lightweight target detection model in the YOLO family, implemented using pytorch. YOLO, a typical single-stage detector, frames the target detection problem as a regression task by directly predicting the coordinates and category probabilities of the bounding box. It directly outputs the target's category, position coordinates, and confidence score. The YOLOv5 family comprises five models: YOLOv5n, YOLOv5s, YOLOv5m, YOLOv5l, and YOLOv5x. These models share the same network architecture but differ in terms of depth and width [25]. Among these, YOLOv5s strikes a balance between detection accuracy and the number of parameters with a smaller network depth and feature map size, making it more suitable for deployment on embedded platforms than the other models.

The network architecture of YOLOv5s, as shown in Figure 1, is divided into three main components: Backbone, Neck, and Head. The Backbone is responsible for extracting multi-scale features from the input image, incorporating the CBS module, CSP module, and the Spatial Pyramid Pooling Fast Version (SPPF) module. The CBS module consists of a convolutional layer (Conv), a batch normalization layer (BN), and a SiLU activation function. The CSP module integrates the Conv layer with a Bottleneck structure, where the Bottleneck contains a varying number of residual blocks to optimize feature extraction efficiency. The SPPF module enriches feature diversity by performing multi-scale max-pooling on the feature maps and concatenating the outputs at different scales to form a fixed-length feature vector.

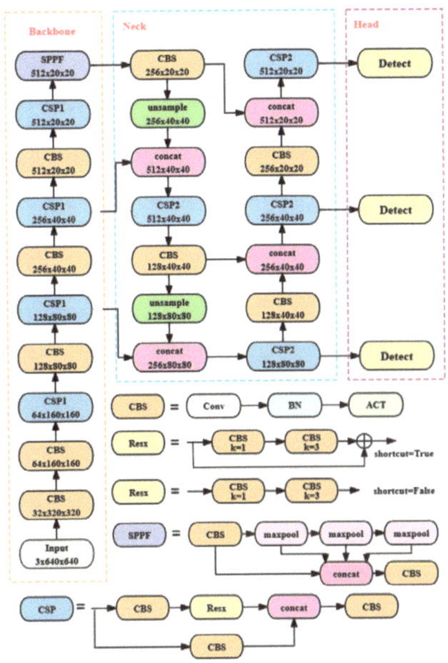

Figure 1. The architecture of YOLOv5s [23].

The Neck component fuses the multi-scale features extracted by the Backbone using a FPN and a Path Aggregation Network (PAN), enhancing feature representation before passing the features to the Head. The Head component serves as the detection unit, responsible for predicting the category and position coordinates of the target. This architecture allows YOLOv5s to significantly reduce the number of parameters and computational complexity while maintaining high detection accuracy, enabling efficient target detection on resource-constrained embedded platforms.

2.1.2. GhostBottleneckV2

The GhostBottleneckV2 module [26], as shown in Figure 2, is a lightweight and efficient convolutional neural network component designed to enhance feature representation while reducing model parameters and computational effort. It integrates the Ghost module, an improved bottleneck structure, and a Decoupled Fully Connected (DFC) attention mechanism.

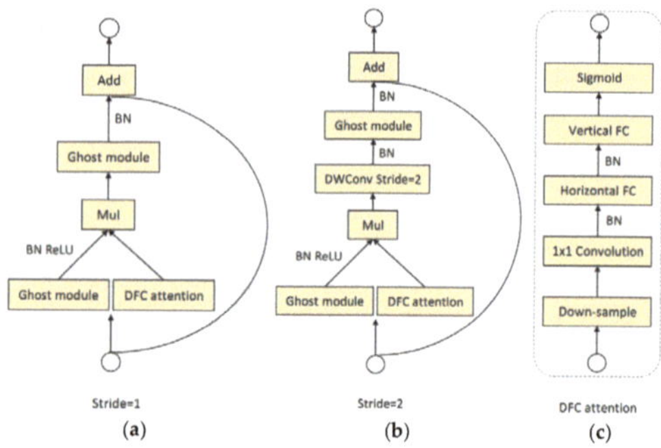

Figure 2. Bottleneck structure diagram of GhostNetV2 [26]: (**a**) bottleneck with a step length of 1; (**b**) Bottleneck with a step length of 2; (**c**) DFC attention. The Ghost module and DFC attention operate as two parallel branches, each extracting information from a different perspective.

The Ghost module aims to approximate the output of a traditional convolution by generating redundant features, thereby reducing computational complexity. It first generates a portion of the base features using standard convolution and then produces additional features through a series of inexpensive linear operations to approximate the full convolution output. This method significantly reduces the number of parameters and the computational load while preserving model performance.

In the GhostBottleneckV2 module, an improved bottleneck structure further optimizes information flow. The traditional bottleneck structure typically includes a descending convolutional layer, a deep convolutional layer, and an ascending convolutional layer, but this structure may result in information loss and high computational complexity. In contrast, the GhostBottleneckV2 module replaces both the descending and ascending convolutional layers with the Ghost module, further reducing computational complexity while retaining more feature information. Additionally, a residual connection is introduced to promote stable gradient flow and enhance model training.

The DFC attention mechanism is employed in the GhostBottleneckV2 module to further enhance feature representation. The DFC mechanism separately weights the channel and spatial dimensions of the feature map by decoupling channel and spatial attention. In comparison to the traditional SE attention mechanism, the DFC mechanism enables a finer adjustment of feature map weights, improving the model's focus on key features while reducing computational complexity. This is particularly important for small-target detection, as small-target features can be overwhelmed by complex background information.

The GhostBottleneckV2 module significantly reduces the number of parameters and computational load while maintaining efficient feature extraction capabilities. This makes the model well suited for deployment on resource-constrained embedded devices, achieving superior performance in small-target detection tasks. By incorporating the GhostBottleneckV2 module at key locations in the network, the overall efficiency and accuracy of the model are effectively improved.

2.2. GF-YOLO

In this study, we propose a novel network model, GF-YOLO, for small-target detection in remote sensing images. This model is based on the YOLOv5s architecture, with key innovations including the replacement of the CSP module in the Backbone with the GhostBottleneckV2 module and the substitution of the CSP module in the Feature Fusion Network (Neck) with the GhostSE module. Lowering the parameter count and computational complexity, these modifications make the model more suitable for deployment on platforms with restricted resources.

2.2.1. Backbone

In this study, we perform a comprehensive optimization of the Backbone network in the GF-YOLO model to reduce computational complexity and enhance small-target detection performance. The specific structure is shown in Figure 3. The traditional YOLOv5s model extensively employs the CSP module in the Backbone, which offers strong feature extraction capabilities. However, the large number of convolution operations increases both model parameters and computational requirements, which hinders deployment on resource-constrained hardware platforms.

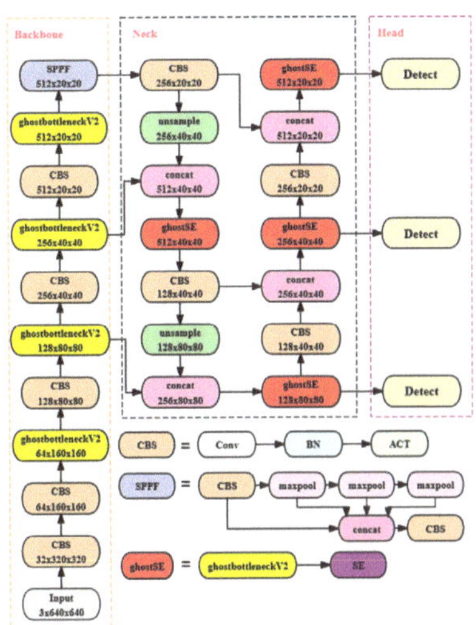

Figure 3. GF-YOLO structure.

To solve this problem, we substitute the CSP module in the Backbone with the lightweight GhostBottleneckV2 module. This module approximates the output of traditional convolution by generating redundant features, thereby reducing the number of convolution operations. Specifically, the GhostBottleneckV2 module generates part of the basic feature map using standard convolution and then produces additional features through a series of inexpensive linear operations, approximating the full convolution output. With this approach, there is a notable reduction in parameters and computational overhead, yet feature expressiveness remains intact.

This modification offers several advantages:

1. Reduced Computational Complexity: Utilizing the GhostBottleneckV2 module lessens the reliance on standard convolution, markedly lowering the model's FLOPs (Floating Point Operations Per Second) and accelerating inference speed.
2. Reduced Model Size: The decrease in the number of parameters allows the model to occupy less storage space, facilitating deployment on a resource-constrained platform.
3. Retained Feature Extraction Capability: Despite its lightweight design, the GhostBottleneckV2 module effectively extracts image features, ensuring the model's detection performance.

The experimental data indicate that replacing the Backbone's CSP module with the GhostBottleneckV2 module significantly lowers the GF-YOLO model's number of parameters and computational demands, all while maintaining detection precision and improving the detection of small targets. These results substantiate the success of our Backbone optimization.

2.2.2. Neck

In the Feature Fusion Network (Neck), the traditional YOLOv5s model also uses the CSP module to integrate multi-scale features. However, this direct application of the CSP module results in high computational complexity and insufficient enhancement of small-target features. Therefore, we propose the introduction of the GhostSE module in the Neck to replace the original CSP module.

The GhostSE module is an enhanced component that integrates the SE attention mechanism with the GhostBottleneckV2 module. The SE mechanism adaptively adjusts the weights of each channel by modeling the relationships between channels, thereby highlighting important features and suppressing less significant ones. In small-target detection, the target often occupies fewer pixels, and its features are easily overwhelmed by the rich background information. The introduction of the SE mechanism enhances the focus on small-target features, thereby improving detection sensitivity.

The specific process is that the GhostSE module uses the GhostBottleneckV2 module instead of the CSP module with more convolutional operations to extract the features in the image, thus reducing the computational complexity. Then, the SE attention mechanism is used to recalibrate the extracted features to enhance the representation of key features. Such a design strikes a balance between a lightweight model and fine feature extraction. The GhostSE module replaces the CSP module on the neck. The introduction of the SE mechanism enhances the response of the model to small-target features, makes its representation in the feature map more prominent, and improves the detection accuracy. Moreover, the GhostSE module inherits the lightweight characteristics of the Ghost module, which provides stronger feature expression ability while reducing the number of parameters and computational complexity.

3. Hardware Optimization

Compared with CPUs and GPUs, FPGAs are very efficient in performing arithmetic operations, which can greatly improve energy efficiency and power consumption, and have advantages in accelerating CNN on lightweight hardware platforms. Moreover, FPGAs provide a reconfigurable custom hardware architecture that accelerates a variety of computing tasks by building dedicated pipelines, parallel processing units, data reuse mechanisms, and custom bit widths. Therefore, in recent years, the use of FPGA to accelerate CNN research has received considerable attention.

Ding et al. [27] introduced REQ-YOLO, a resource-aware system weight quantization framework designed to meet the real-time and energy-efficiency requirements of deep neural networks in autonomous systems. By employing a block–cyclic matrix approach and a heterogeneous weight quantization technique, combined with an efficient FPGA hardware implementation, REQ-YOLO effectively compresses the YOLO model while maintaining a minimal loss of accuracy, enabling efficient target detection in resource-constrained environments. Yu et al. [28] proposed an FPGA accelerator architecture for the YOLOv5 model, which reduces off-chip data transfer times by employing two parallel approaches and a pipelined architecture with ping-pong buffers. They further optimize pipeline performance by rescheduling Multiply–Accumulate (MAC) operations. Zhang et al. [29] developed an FPGA implementation of YOLOv2 optimized for resource-constrained IoT edge devices. Their approach improves data transmission and computational efficiency through scalable cross-layer data flow strategies, filter-level data reuse, and multi-level sliding buffer-optimized convolutional computation. Additionally, Zhang et al. [30] designed an FPGA acceleration scheme combining software and hardware to optimize the YOLOv4 tiny-target detection model. This scheme integrates fixed-point static quantization, a batch normalization layer, and a large-scale pipelined design. Zheng et al. [31] proposed a YOLO-based target detection algorithm deployed on an FPGA platform. By implementing model quantization and optimizing hardware operations, they fully utilized the parallel computing capabilities of FPGAs to achieve detection accuracy comparable to conventional algorithms, while consuming significantly less power than CPUs and GPUs and delivering fast inference speeds suitable for environmental detection in mobile devices.

3.1. Core Architecture

3.1.1. Layer Fusion

To further enhance the inference performance of the GF-YOLO model on FPGAs, we introduce Layer Fusion as an optimization technique [32]. Layer Fusion aims to reduce unnecessary computational workload and data transfer by merging adjacent network layers, thereby minimizing computational complexity and improving the execution efficiency of the model on embedded platforms.

First, to reduce data transmission and storage overhead, we merge the Batch Normalization (BN) layer with the adjacent convolutional layer. During inference, the parameters (mean and variance) of the BN layer are fixed, allowing the computation of the BN layer to be incorporated into the convolutional layer by applying equivalent transformations to the convolutional layer's weights and biases. Specifically, for the output of the convolutional layer y = $Wx + b$, the weights and biases after fusion can be expressed as

$$W' = W \cdot \frac{\gamma}{\sqrt{\sigma^2 + \varepsilon}}, \mathrm{b}' = (\mathrm{b} - \mu) \cdot \frac{\gamma}{\sqrt{\sigma^2 + \varepsilon}} + \beta$$

where γ and β are the scaling and translation parameters of the BN layer, and μ and σ^2 represent the mean and variance, respectively. ε is a small constant added to avoid division by zero. This fusion eliminates the need for a separate BN layer, thereby reducing the storage and computational overhead associated with intermediate results.

Secondly, we substitute the resource-intensive Scaled Exponential Linear Unit (SiLU) activation function with the more efficient Rectified Linear Unit (ReLU). The SiLU activation function, also referred to as Swish, requires exponential computations, making it computationally demanding. On resource-constrained platforms like FPGAs, exponential operations require more hardware resources and computation time. In contrast, the ReLU function, defined as $f(x) = \max(0,x)$, is simple to compute and demands fewer hardware resources, while providing comparable performance to SiLU in most cases.

These optimizations greatly simplify the computational flow of the GF-YOLO model, enhancing its computational efficiency. When deployed on FPGAs, the Layer Fusion technique significantly reduces the amount of access to external memory and decreases the bandwidth requirements for data transmission. Additionally, replacing Si-LU with ReLU lowers the complexity of the computation unit, further conserving hardware resources.

3.1.2. Double Buffer

To further enhance the computational efficiency and resource utilization of the system, we introduce the double buffering mechanism into the accelerator design [33]. The double buffering mechanism reduces the overall processing latency by parallelizing data transfer and computation, allowing these processes to overlap. In the concrete implementation, we configure two independent buffers for each acceleration module. While the data in one buffer is being processed, the other buffer is simultaneously preloaded with the next batch of data. As a result, after processing the current data, the accelerator can immediately switch to the new, preloaded data, thereby eliminating the time overhead associated with waiting for data to be loaded.

This mechanism has been effectively applied in operations such as convolutional layers, upsampling, and maximum pooling. Since these operations have high data throughput and bandwidth requirements, the introduction of double buffering significantly enhances the efficiency of data transfer and reduces the idle time of computational units. With this design, we achieve a high degree of overlap between computation and data transfer, resulting in substantial performance improvements. Additionally, the double buffering mechanism alleviates the pressure on storage bandwidth, further optimizing the utilization of hardware resources.

3.2. *Hybrid Overlay Architecture*

Neural network accelerators are generally designed using two main architectures: single-engine Processing Element (PE) architectures and streaming architectures. In a single PE architecture, different layers across the network share the same PE. This architecture is energy-efficient and consumes minimal hardware resources by reusing the same hardware modules to perform computations across all layers. It is particularly suited for resource-constrained environments, as it minimizes the demand for hardware resources. However, since all layers must use the same PE unit sequentially, network computations must be queued, resulting in relatively high overall computation latency. This may not meet the demands of applications that require high real-time performance. In contrast, streaming architectures allocate dedicated PEs for each layer of the network, enabling data to be transferred and processed in a pipelined fashion between layers. The advantage of this architecture is its ability to support low-latency, end-to-end inference, which significantly boosts computational speed, making it ideal for real-time processing applications. However, streaming architecture assigns independent hardware resources to each layer, leading to increased energy consumption, hardware costs, and power demands. Additionally, the network architecture becomes more rigid and less flexible, which may render it unsuitable for resource-constrained edge computing platforms.

To strike a balance between latency, energy efficiency, and hardware resource utilization, this paper proposes a hybrid overlapping architecture. In this design, we implement various types of operators on the Programmable Logic (PL) of FPGAs to accelerate the inference process of the GF-YOLO model. GF-YOLO is a lightweight target detection model optimized for edge devices. However, its convolutional layers are computationally intensive and frequently invoked, consuming a significant portion of the model's computation time. To address this, we design independent acceleration modules for convolutional operations (Conv), each equipped with a dedicated PE array capable of processing a large number of convolution operations in parallel. Designed to support multiple kernel sizes, input channels, and output channels, these convolutional accelerators make full use of the parallel computing capabilities of FPGAs to improve computational efficiency.

At the same time, the upsample and maxpool operations, although low in computational complexity, play an important role in the model. By exploiting the hardware parallelism of FPGAs, we design dedicated accelerators for these operations. Through pixel-level parallel processing, the upsampling and maxpooling accelerators can handle multiple pixel computations simultaneously, significantly improving the execution efficiency of these operations and further reducing overall computational latency.

In the Processing System (PS), we implement a software application in C++ that dynamically controls the invocation of each hardware accelerator based on the model's structure. Through accelerator drivers, the software application coordinates the workflow between convolution, upsampling, and max-pooling accelerators, ensuring efficient data transfer and processing between the modules. This integration of software and hardware enables efficient inference across the entire model.

The block diagram of the data flow is shown in Figure 4:

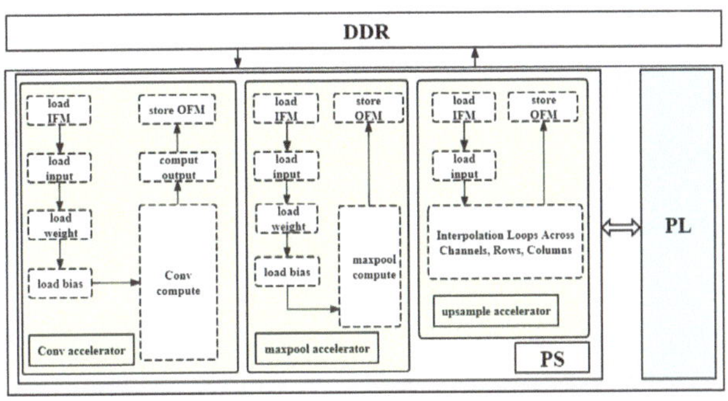

Figure 4. Data flow diagram.

In summary, our proposed hybrid overlap acceleration architecture takes full advantage of FPGA's parallel processing capability and flexible resource allocation. The redundant use of hardware resources is reduced by reusing computing units. Layer fusion optimizes data flow and reduces data transmission delay. The double buffering mechanism further improves data throughput and ensures continuous data processing. These optimization strategies allow GF-YOLO to achieve efficient hardware acceleration on FPGA, significantly reduce computational complexity, and meet the real-time inspection needs of resource-constrained platforms. After the experiment, our hybrid overlapping architecture successfully facilitates the efficient inference of the GF-YOLO model on resource-constrained edge devices. This architecture effectively balances energy efficiency, hardware resource utilization, and real-time performance, providing a viable solution for deploying complex deep learning models on edge computing platforms.

4. Discussion

4.1. Dataset

4.1.1. TIFAD

The Thermal Infrared Flying Aircraft Dataset (TIFAD) [34–36] is the first three-band space-based aircraft dataset constructed using on-orbit images from the SDGSAT-1 Thermal Infrared Sensor (TIS). The dataset contains more than 100 confirmed vehicle samples, filling a significant gap in the field of on-orbit vehicle detection using high spatial resolution (30 m) thermal infrared imagery. The SDGSAT-1 TIS has a mapping bandwidth of 300 km and covers three detection spectral ranges: B1 (8–10.5 μm), B2 (10.3–11.3 μm), and B3 (11.5–12.5 μm). These bands are synthesized into a trispectral pseudo-color image for better visualization. The TIS consists of multiple long linear detectors (M1, M2, M3, and M4) that efficiently capture the thermal radiation characteristics of vehicles in orbit.

Since vehicles occupy only 1×1 to 2×2 pixels in thermal infrared images, traditional shape-based detection methods are often ineffective. To address this limitation, an improved wake detection algorithm based on the asymptotic probabilistic Hough transform is applied to the TIFAD. This algorithm filters aircraft samples from over 325 GB of SDGSAT-1 TIS remotely sensed imagery, confirming their presence based on wake information. This approach not only improves the accuracy of vehicle detection but also enhances processing efficiency, ensuring effective data screening and sample labeling.

The establishment of the TIFAD provides valuable data support for the all-weather monitoring of global aircraft targets. It also highlights the broad application potential of thermal infrared remote sensing (TIRS) technology in airborne motion target detection. The dataset lays a solid foundation for the further development of TIS-based aircraft detection algorithms, which will improve airspace safety management and enhance civil flight protection capabilities. Additionally, the data quality and diversity of TIFAD offer an important resource for academic research and industrial applications in related fields, driving innovation in thermal infrared target detection technology. By providing high spatial resolution and multi-spectral information, TIFAD supports researchers and engineers in developing more accurate and efficient aircraft detection methods and accelerates the application of thermal infrared remote sensing in key areas such as military surveillance, air traffic management, and environmental monitoring.

4.1.2. HRSID Dataset

The HRSID dataset [37] is a high-resolution SAR image dataset, released in 2020, designed for ship detection, instance segmentation, and semantic segmentation tasks. The dataset, which crops 1 m to 5 m images into SAR images of 800×800 pixels, contains 5604 cropped SAR images and 16,951 ship instances in different resolutions, polarizations, and sea states. Each image in the dataset contains an average of three ships. The dataset includes 54.5% of small ships, 43.5% of medium ships and 2% of large ships.

In the experiment, we divided the dataset into a training set and a test set, with a ratio of 8:2. Specifically, the training set contains 4483 images for the model to learn and adjust its parameters. The remaining 1121 images constituted a test set, which was used to verify the accuracy and efficiency of the model.

4.2. Ablation Experiments

To assess the performance of the proposed GF-YOLO model, we carried out detailed ablation experiments on the TIFAD. The results presented in Table 1 indicate that integrating the GhostBottleneckV2 module into the model's backbone achieves a 17.8% reduction in parameters and a 31.8% decrease in computational load (GFLOPs) compared to the baseline YOLOv5s model. More importantly, the model's mean average precision (mAP) is improved by 5.9%, indicating enhanced detection performance while reducing model complexity. Subsequently, we further optimized the model by incorporating the GhostSE module into the neck part of the network. This modification leads to an additional reduction of 26.2% in the number of parameters and a corresponding decrease of 40.5% in computational effort. These optimizations reduce the model's storage needs and computational overhead, making it more apt for deployment in environments with limited resources. They also improve the model's ability to detect targets, especially small ones in complex scenarios. In conclusion, the ablation results demonstrate that GF-YOLO effectively balances enhancing model efficiency with maintaining detection performance, highlighting its applicability for practical uses.

Table 1. Ablation experiments.

Model	mAP	Param (M)	GFLOPs
YOLOv4-tiny	53.2%	5.91	3.43
YOLOv5l	62.3%	46.63	57.31
YOLOv5s	61.9%	7.07	8.19
YOLOv5s + GhostBottleneckV2	65.2%	5.81	5.85
GF-YOLO	**67.8%**	**5.22**	**4.87**

Through a comprehensive series of ablation experiments, we validate the effectiveness of the GF-YOLO framework, which is optimized based on the YOLOv5s architecture. The experimental results show that GF-YOLO reduces both the number of parameters and the computational complexity of the model, while maintaining or even improving detection performance. These optimizations not only enhance the operational efficiency of the model, making it more suitable for deployment in resource-constrained environments but also reduce inference time, thereby improving the response time in real-world applications. Moreover, despite the reduction in model complexity, GF-YOLO demonstrates superior performance in small-target detection tasks. This is particularly important in application scenarios that require fine-grained recognition. The improved small-target detection capability suggests that GF-YOLO effectively enhances the model's ability to capture fine-grained features while maintaining high detection accuracy.

4.3. Comparisons with Other Methods on HRSID

To make our model more convincing, we performed a comparative analysis on the HRSID dataset with several existing methods, including SSD512 [38], YOLOv8s [39], YOLOX-Tiny [40], Qin's method [41], AMANet [42], and YOLOv8-FDF [43]. Table 2 shows the results of our model compared with other small-target detection methods. As for the mAP, compared with several baseline YOLO methods (such as YOLOv8s and YOLOX-Tiny), our model has an improvement of 2.2% and 5.48%. Our model performed best in terms of GFLOP and number of parameters, and our accuracy improved by 1.8% and 14.4% compared to lightweight models AMANet and YOLOv8-FDF. Compared with several existing small-target detection models, our model has advantages in map, parameters, and GFLOPs, which proves that our model is a lightweight model for small-target detection, which is suitable for resource-limited spaceborne hardware platforms.

Visualized results of GF-YOLO detection on the HRSID dataset are shown in Figure 5, including target bounding boxes, ship labels, and confidence scores for model predictions. Through the observation of the detection results, we can see that our model has good detection accuracy for multi-object scenes, and there is no missing detection. In complex scenes such as ports, GF-YOLO maintains good detection performance and can accurately extract features from images to identify and detect ship targets in images. In summary, our model has good detection accuracy in SAR images, and for complex scenes with little difference, our model can extract features well, which proves the robustness of our model.

Table 2. Experimental results of different methods on HRSID.

Model	mAP	Param (M)	GFLOPs
SSD512	85.46	24.4	214.0
YOLOv8s	90.1	11.1	14.2
YOLOX-Tiny	87.72	5.8	11.9
Qin'method	75.3	53.5	-
AMANet	91.4	17.3	24.7
YOLOv8-FDF	78.8	6.2	12.3
GF-YOLO	**93.2**	**5.22**	**4.78**

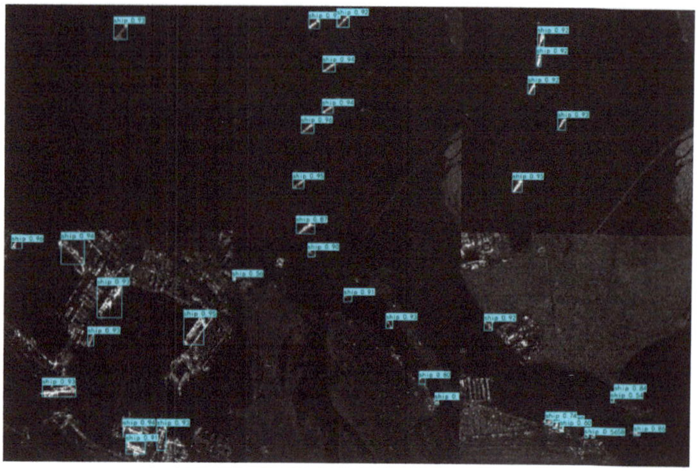

Figure 5. GF-YOLO detection plot on the HRSID.

4.4. Hardware Accelerator Performance Evaluation

Our proposed hardware accelerator is implemented using synthesizable High-Level Synthesis (HLS), where the Processing Element (PE) array size is user-defined. We evaluated the performance of the accelerator using Vivado 2019.2, Vivado HLS 2019.2, Xilinx Vitis IDE 2019.2, and the Xilinx Zynq-7020 hardware platforms.

As shown in Table 3, we set the clock frequency to 100 MHz in our tests, and the resource utilization of the accelerator is detailed below.

Table 3. Resource utilization rate.

Resource	LUT	LUTRAM	FF	BRAM	DSP
Utilization	46,970	13,296	40,935	107	205
Available	53,200	17,400	106,400	140	220
Utilization (%)	**88.2**	**76.4**	**38.4**	**76.4**	**93.1**

Table 3 details the usage and utilization of each type of resource. Specifically, the accelerator used 205 DSP units, representing 93.1% of the available DSP resources. These DSP units are mainly utilized to perform convolutional operations, reflecting the accelerator's ability to efficiently handle computationally intensive tasks. However, the utilization of LUTs reaches 88.2%, which is close to the upper resource limit, indicating strain on the logic implementation. In addition, the utilization rate of LUTRAM is 76.4%, showing high resource consumption that may limit the potential for further functionality expansion or design optimization. In contrast, the utilization rate of Flip-Flops (FF) is only 38.4%, much lower than the available resources, suggesting ample capacity in register usage. This could be leveraged to improve performance by increasing pipeline stages or introducing more parallel processing units. The utilization rate of BRAM is 76.4%, which, while high, still allows some margin for storing additional intermediate data or optimizing data access strategies. Overall, the high utilization rates of DSP and LUT resources indicate a significant advantage in achieving high-performance convolutional operations but also suggest the need for adjustments in logic resource allocation and optimization strategies to improve resource utilization efficiency and performance.

To fully evaluate the energy efficiency of the accelerator when executing the GF-YOLO model, we conducted experiments on different hardware platforms and summarized the results in Table 4.

Table 4. Comparison of different hardware platforms.

Platform	Throughputs (GOPs)	Power Consumption (W)	GOPs/W
Intel i9-13900	48.1	56.2	0.85
RTX4060	104.5	90.5	1.15
Ours	**10.2**	**2.8**	**3.64**

The experimental results show that the accelerator significantly outperforms traditional CPU and GPU platforms in terms of the GOPs/W metric. Specifically, the Intel i9-13900 processor achieves a throughput of 48.1 GOPs with a power consumption of 56.2 Watts, resulting in a GOPs/W of 0.85. The NVIDIA RTX 4060 graphics card achieves a throughput of 104.5 GOPs with a power consumption of 90.5 Watts, corresponding to a GOPs/W of 1.15. In comparison, our designed accelerator achieves a throughput of 10.2 GOPs with a power consumption of only 2.8 Watts, resulting in a GOPs/W of up to 3.64, which is almost four times that of the CPU platform and about three times that of the GPU platform.

These results indicate that the accelerator offers significant advantages in terms of energy efficiency, especially for energy-sensitive application scenarios where the need for extremely high computational throughput is less critical. Although the absolute throughput of the accelerator is lower than that of high-performance CPU and GPU platforms, its superior energy efficiency makes it valuable for applications in mobile devices, edge computing, and systems that need to run for extended periods without relying on high power supplies. Additionally, the high GOPs/W value reflects the accelerator's ability to achieve efficient computational performance with lower energy consumption when performing specific deep learning tasks, which is important for advancing green computing and sustainable development goals. It can be concluded that the accelerator performs well in the energy-efficiency evaluation of our model, and although there is room for improvement in throughput, its potential to reduce energy consumption provides strong support for future hardware design and application deployment.

Figure 6 shows the detection results of our model on the remote sensing dataset TIFAD. It can be observed that GF-YOLO maintains good detection performance in complex scenes and in the presence of streak noise, accurately identifying small targets in thermal infrared images. This validates the effectiveness of the model in thermal infrared small-target detection tasks and demonstrates its robustness to challenges such as low contrast and noise interference. Additionally, GF-YOLO is able to accurately distinguish targets from noise in complex backgrounds, demonstrating its advantages in extracting key features and suppressing background interference. By implementing these optimizations, the model's storage requirements and computational costs are decreased, facilitating its deployment in settings with limited resources. Additionally, these changes enhance the model's ability to detect targets, particularly small ones within complex environments. In conclusion, the ablation experiments show that GF-YOLO effectively harmonizes model efficiency improvements with detection performance, highlighting its potential for practical use.

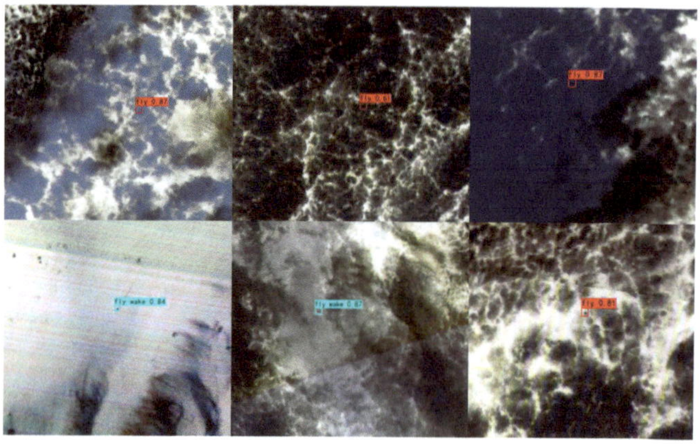

Figure 6. GF-YOLO detection plot on the TIFAD.

5. Conclusions

In this paper, we present and implement GF-YOLO, a highly efficient model for detecting small targets in remote sensing images and assess its comprehensive performance alongside a specialized hardware accelerator. Through detailed ablation experiments conducted on the TIFAD, which is based on data from the SDGSAT-1 thermal infrared sensor, we demonstrate the GF-YOLO model's effectiveness in refining the model structure and enhancing detection performance. The optimized model achieves a 26.2% reduction in parameter count and a 40.5% decrease in computational complexity, all while maintaining high detection accuracy. These enhancements not only substantially lower the model's storage requirements and computational overheads, making it more suitable for deployment in environments with limited resources, but also improve its ability to detect small targets in complex scenarios, thereby demonstrating improved detection accuracy and robustness.

In order to improve the performance of the hardware accelerator, we designed a hybrid overlay architecture. The evaluation results show that the accelerator achieves a GOPs/W metric of 3.64, which is higher than traditional CPU (0.85 GOPs/W) and GPU (1.15 GOPs/W) platforms, demonstrating a considerable energy efficiency advantage. Although the accelerator's absolute throughput (10.2 GOPs) is lower than that of high-performance CPU and GPU platforms, its significant energy efficiency advantage makes it promising for applications in mobile devices, edge computing, and systems that require long-term operation with stringent energy consumption constraints.

This study shows that through model optimization and hardware acceleration, the overall performance and energy efficiency of the detection system are significantly improved. The GF-YOLO model achieves a balance between high accuracy and low complexity through efficient structural optimization, while the dedicated hardware accelerator improves the operational efficiency of the system by optimizing resource utilization and increasing energy efficiency. These results not only have great significance for the development of aircraft detection technology based on thermal infrared remote sensing but also lay a solid foundation for the deployment of deep learning models in practical applications.

In future research, we can further optimize the resource allocation of the accelerator, improve the performance of the accelerator under high throughput requirements, and expand the adaptability of our model to more complex scenarios and different datasets, thereby broadening its application scope. In addition, the adoption of more advanced model compression and acceleration technologies can also further improve the overall performance and energy efficiency of the system, providing a more reliable technical basis for the intelligent object detection system on board.

Author Contributions: Conceptualization, N.F. and L.L.; methodology, N.F.; software, N.F.; validation, L.L., X.Z. and W.Z.; formal analysis, N.F.; investigation, N.F.; resources, F.C.; data curation, W.Z.; writing—original draft preparation, N.F.; writing—review and editing, L.L.; visualization, N.F.; supervision, L.L.; project administration, F.C. All authors have read and agreed to the published version of the manuscript.

Funding: This work is supported by the Strategic Priority Research Program of the Chinese Academy of Sciences [Grant no. XDA19010102], and the National Natural Science Foundation of China [Grant no. 62405062].

Data Availability Statement: Data is unavailable due to privacy or ethical restrictions.

Acknowledgments: We thank the International Research Center of Big Data for Sustainable Development Goals for providing us with data.

Conflicts of Interest: The authors declare no conflict of interest.

References

1. Hong, T.; Liang, H.; Yang, Q.; Fang, L.; Kadoch, M.; Cheriet, M. A real-time tracking algorithm for multi-target UAV based on deep learning. *Remote Sens.* **2022**, *15*, 2. [CrossRef]
2. Cook, K.L.; Andermann, C.; Gimbert, F.; Adhikari, B.R.; Hovius, N. Glacial lake outburst floods as drivers of fluvial erosion in the Himalaya. *Science* **2018**, *362*, 53–57. [CrossRef] [PubMed]
3. Wei, W.; Polap, D.; Li, X.; Woźniak, M.; Liu, J. Study on remote sensing image vegetation classification method based on decision tree classifier. In Proceedings of the 2018 IEEE Symposium Series on Computational Intelligence (SSCI), Bangalore, India, 18–21 November 2018; IEEE: Piscataway, NJ, USA, 2018; pp. 2292–2297.
4. Voudouri, K.A.; Ntona, M.M.; Kazakis, N. Snowfall Variation in Eastern Mediterranean Catchments. *Remote Sens.* **2023**, *15*, 1596. [CrossRef]
5. Zhu, M.; Xu, Y.; Ma, S.; Li, S.; Ma, H.; Han, Y. Effective airplane detection in remote sensing images based on multilayer feature fusion and improved nonmaximal suppression algorithm. *Remote Sens.* **2019**, *11*, 1062. [CrossRef]
6. Redmon, J.; Divvala, S.; Girshick, R.; Farhadi, A. You only look once: Unified, real-time object detection. In Proceedings of the 2016 IEEE Conference on Computer Vision and Pattern Recognition, Las Vegas, NV, USA, 27–30 June 2016; pp. 779–788. [CrossRef]
7. Bochkovskiy, A.; Wang, C.-Y.; Liao, H.-Y.M. Yolov4: Optimal speed and accuracy of object detection. *arXiv* **2020**, arXiv:2004.10934.
8. Jocher, G. YOLOv5 by Ultralytics 2020. Available online: https://github.com/ultralytics/yolov5 (accessed on 22 November 2022).
9. Ma, Y.; Wu, H.; Wang, L.; Huang, B.; Ranjan, R. Remote sensing big data computing: Challenges and opportunities. *Future Gener. Comput. Syst.* **2015**, *51*, 47–60. [CrossRef]
10. Wertz, P.; Hespeler, B.; Kiessling, M.; Hagmanns, F.-J. Next generation high data rate downlink subsystems based on a flexible APSK modulator applying SCCC encoding. In Proceedings of the IEEE International Workshop on Tracking, Telemetry and Command Systems for Space Applications, Noordwijk, The Netherlands, 13–16 September 2016; pp. 1–7.
11. Furano, G.; Meoni, G.; Dunne, A.; Moloney, D.; Ferlet-Cavrois, V.; Tavoularis, A.; Byrne, J.; Buckley, L.; Psarakis, M.; Voss, K.-O.; et al. Towards the Use of Artificial Intelligence on the Edge in Space Systems: Challenges and Opportunities. *IEEE Aerosp. Electron. Syst. Mag.* **2020**, *35*, 44–56. [CrossRef]
12. Amara, A.; Amiel, F.; Ea, T. FPGA vs. ASIC for low power applications. *Microelectron. J.* **2006**, *37*, 669–677.

13. Zhang, Y.; Guo, Z.; Wu, J.; Tian, Y.; Tang, H.; Guo, X. Real-time vehicle detection based on improved yolo v5. *Sustainability* **2022**, *14*, 12274. [CrossRef]
14. Howard, A.G.; Zhu, M.; Chen, B.; Kalenichenko, D.; Wang, W.; Weyand, T.; Andreetto, M.; Adam, H. MobileNets: Efficient Convolutional Neural Networks for Mobile Vision Applications. *arXiv* **2017**, arXiv:1704.04861.
15. Zhang, X.; Zhou, X.; Lin, M.; Sun, J. Shufflenet: An extremely efficient convolutional neural network for mobile devices. In Proceedings of the IEEE Conference on Computer Vision and Pattern Recognition, Salt Lake City, UT, USA, 18–23 June 2018.
16. Tan, M.; Le, Q. Efficientnet: Rethinking model scaling for convolutional neural networks. In Proceedings of the International Conference on Machine Learning, Long Beach, CA, USA, 9–15 June 2019.
17. Ghiasi, G.; Lin, T.-Y.; Le, Q.V. NAS-FPN: Learning Scalable Feature Pyramid Architecture for Object Detection. In Proceedings of the 2019 IEEE/CVF Conference on Computer Vision and Pattern Recognition (CVPR), Long Beach, CA, USA, 15–20 June 2019; pp. 7029–7038.
18. Lin, T.-Y.; Dollár, P.; Girshick, R.; He, K.; Hariharan, B.; Belongie, S. Feature pyramid networks for object detection. In Proceedings of the IEEE Conference on Computer Vision and Pattern Recognition, Honolulu, HI, USA, 21–26 July 2017.
19. Hu, J.; Shen, L.; Sun, G. Squeeze-and-excitation networks. In Proceedings of the IEEE Conference on Computer Vision and Pattern Recognition, Salt Lake City, UT, USA, 18–23 June 2018.
20. Woo, S.; Park, J.; Lee, J.-Y.; Kweon, I.S. CBAM: Convolutional Block Attention Module. In Proceedings of the European Conference on Computer Vision (ECCV), Munich, Germany, 8–14 September 2018.
21. Qiu, M.; Huang, L.; Tang, B.H. ASFF-YOLOv5: Multielement Detection Method for Road Traffic in UAV Images Based on Multiscale Feature Fusion. *Remote Sens.* **2022**, *14*, 3498. [CrossRef]
22. Li, X.; Liu, Z.; Hu, C.; Xu, C. Light-Weight Multi-Target Detection and Tracking Algorithm Based on M3-YOLOv5. In Proceedings of the 2023 42nd Chinese Control Conference (CCC), Tianjin, China, 24–26 June 2023; pp. 8159–8164.
23. Wang, S.; Jiang, H.; Li, Z.; Yang, J.; Ma, X.; Chen, J.; Tang, X. PHSI-RTDETR: A Lightweight Infrared Small Target Detection Algorithm Based on UAV Aerial Photography. *Drones* **2024**, *8*, 240. [CrossRef]
24. Ji, C.-L.; Yu, T.; Gao, P.; Wang, F.; Yuan, R.-Y. Yolo-tla: An Efficient and Lightweight Small Object Detection Model based on YOLOv5. *J. Real-Time Image Proc.* **2024**, *21*, 141. [CrossRef]
25. Jocher, G.; Stoken, A.; Chaurasia, A.; Borovec, J.; Kwon, Y.; Michael, K.; Liu, C.; Fang, J.; Skalski, P.; Hogan, A.; et al. *ultralytics/yolov5: v6.0-YOLOv5n 'Nano' Models, Roboflow Integration, TensorFlow Export, OpenCV DNN Support*; Zenodo: Geneva, Switzerland, 2021.
26. Han, K.; Wang, Y.; Tian, Q.; Guo, J.; Xu, C.; Xu, C. GhostNet: More Features From Cheap Operations. In Proceedings of the 2020 IEEE/CVF Conference on Computer Vision and Pattern Recognition (CVPR), Seattle, WA, USA, 13–19 June 2020; pp. 1577–1586.
27. Ding, C.; Wang, S.; Liu, N.; Xu, K.; Wang, Y.; Liang, Y. Reqyolo: A resource-aware, efficient quantisation framework for object detection on fpgas. In Proceedings of the 2019 ACM/SIGDA International Symposium on Field-Programmable Gate Arrays, Seaside, CA, USA, 24–26 February 2019; pp. 33–42.
28. Yu, H.; Li, S. A higher performance accelerator for resource-limited FPGA to deploy deeper object detection networks. In Proceedings of the 2022 IEEE 16th International Conference on Anti-Counterfeiting, Security, and Identification (ASID), Xiamen, China, 2–4 December 2022.
29. Zhang, Z.; Mahmud, M.A.P.; Kouzani, A.Z. Resource-constrained FPGA implementation of YOLOv2. *Neural Comput. Appl.* **2022**, *34*, 16989–17006. [CrossRef]
30. Zhang, F.; Li, Y.; Ye, Z. Apply yolov4-tiny on an fpga-based accelerator of convolutional neural network for object detection. *J. Phys. Conf. Ser.* **2022**, *2303*, 012032. [CrossRef]
31. Zheng, X.; He, T. Reduced-Parameter YOLO-like Object Detector Oriented to Resource-Constrained Platform. *Sensors* **2023**, *23*, 3510. [CrossRef]
32. Neill, J.O.; Steeg, G.V.; Galstyan, A. Compressing deep neural networks via layer fusion. *arXiv* **2020**, arXiv:2007.14917.
33. Tan, X.; Shen, X.-W.; Ye, X.-C.; Wang, D.; Fan, D.-R.; Zhang, L.; Li, W.-M.; Zhang, Z.-M.; Tang, Z.-M. A non-stop double buffering mechanism for dataflow architecture. *J. Comput. Sci. Technol.* **2018**, *33*, 145–157. [CrossRef]
34. Hu, Z.; Li, X.; Li, L.; Su, X.; Yang, L.; Zhang, Y.; Hu, X.; Lin, C.; Tang, Y.; Hao, J.; et al. Wide-swath and high-resolution whisk-broom imaging and on-orbit performance of SDGSAT-1 thermal infrared spectrometer. *Remote Sens. Environ.* **2024**, *300*, 113887. [CrossRef]
35. Li, L.; Zhou, X.; Hu, Z.; Gao, L.; Li, X.; Ni, X.; Chen, F. On-orbit monitoring flying aircraft day and night based on SDGSAT-1 thermal infrared dataset. *Remote Sens. Environ.* **2023**, *298*, 113840. [CrossRef]
36. Zhou, X.; Li, L.; Yu, J.; Gao, L.; Zhang, R.; Hu, Z.; Chen, F. Multimodal aircraft flight altitude inversion from SDGSAT-1 thermal infrared data. *Remote Sens. Environ.* **2024**, *308*, 114178. [CrossRef]
37. Wei, S.; Zeng, X.; Qu, Q.; Wang, M.; Su, H.; Shi, J. HRSID: A high-resolution SAR images dataset for ship detection and instance segmentation. *IEEE Access* **2020**, *8*, 120234–120254. [CrossRef]

38. Liu, W.; Anguelov, D.; Erhan, D.; Szegedy, C.; Reed, S.; Fu, C.-Y.; Berg, A.C. SSD: Single shot MultiBox detector. In Proceedings of the European Conference Computer Vision, Amsterdam, The Netherlands, 11–14 October 2016; pp. 21–37.
39. Jocher, G.; Chaurasia, A.; Qiu, J. YOLO by Ultralytics, February 2023. Available online: https://github.com/ultralytics/ultralytics (accessed on 22 November 2022).
40. Wang, H.; Han, D.; Cui, M.; Chen, C. NAS-YOLOX: A SAR ship detection using neural architecture search and multi-scale attention. *Connect. Sci.* **2023**, *35*, 1–32. [CrossRef]
41. Qin, W.; Chi, H.; Zhu, X.; Qin, Q.; Liang, Z. Enhanced DETR Structure for Ship Detection in SAR Images. In Proceedings of the 2024 43rd Chinese Control Conference (CCC), Kunming, China, 28–31 July 2024; pp. 7929–7934. [CrossRef]
42. Ma, X.; Cheng, J.; Li, A.; Zhang, Y.; Lin, Z. AMANet: Advancing SAR Ship Detection With Adaptive Multi-Hierarchical Atten-tion Network. *IEEE Access* **2024**, *12*, 105952–105967. [CrossRef]
43. Jiang, W.; Han, D.; Han, B.; Wu, Z. YOLOv8-FDF: A Small Target Detection Algorithm in Complex Scenes. *IEEE Access* **2024**, *12*, 119223–119237. [CrossRef]

remote sensing

MDPI

Article

Advancing Fine-Grained Few-Shot Object Detection on Remote Sensing Images with Decoupled Self-Distillation and Progressive Prototype Calibration

Hao Guo [1,2,3,4], Yanxing Liu [1,2,3,4], Zongxu Pan [1,2,3,4,*] and Yuxin Hu [1,2,3,4]

1 Aerospace Information Research Institute, Chinese Academy of Sciences, Beijing 100094, China; guohao221@mails.ucas.ac.cn (H.G.); liuyanxing21@mails.ucas.ac.cn (Y.L.); yxhu@mail.ie.ac.cn (Y.H.)
2 Key Laboratory of Technology in Geo-Spatial Information Processing and Application System, Chinese Academy of Sciences, Beijing 100190, China
3 School of Electronic, Electrical and Communication Engineering, University of Chinese Academy of Sciences, Beijing 100049, China
4 Key Laboratory of Target Cognition and Application Technology, Beijing 100190, China
* Correspondence: zxpan@mail.ie.ac.cn

Abstract: In data-scarcity scenarios, few-shot object detection (FSOD) methods exhibit a notable advantage in alleviating the over-fitting problem. Currently, research on FSOD in the field of remote sensing is advancing rapidly and FSOD methods based on the fine-tuning paradigm have initially displayed their excellent performance. However, existing fine-tuning methods often encounter classification confusion issues. This is potentially because of the shortage of explicit modeling for transferable common knowledge and the biased class distribution, especially for fine-grained targets with higher inter-class similarity and intra-class variance. In view of this, we first propose a decoupled self-distillation (DSD) method to construct class prototypes in two decoupled feature spaces and measure inter-class correlations as soft labels or aggregation weights. To ensure a robust set of class prototypes during the self-distillation process, we devise a feature filtering module (FFM) to preselect high-quality class representative features. Furthermore, we introduce a progressive prototype calibration module (PPCM) with two steps, compensating the base prototypes with the prior base distribution and then calibrating the novel prototypes with adjacent calibrated base prototypes. Experiments on MAR20 and customized SHIP20 datasets have demonstrated the superior performance of our method compared to other existing advanced FSOD methods, simultaneously confirming the effectiveness of all proposed components.

Keywords: remote sensing images; few-shot object detection; fine-grained object detection; self-distillation; prototype calibration

Academic Editors: Qian Du, Yanni Dong and Xiaochen Yang

Received: 17 December 2024
Revised: 25 January 2025
Accepted: 28 January 2025
Published: 31 January 2025

Citation: Guo, H.; Liu, Y.; Pan, Z.; Hu, Y. Advancing Fine-Grained Few-Shot Object Detection on Remote Sensing Images with Decoupled Self-Distillation and Progressive Prototype Calibration. *Remote Sens.* **2025**, *17*, 495. https://doi.org/10.3390/rs17030495

1. Introduction

As one of the important tasks in remote sensing image interpretation and recognition, object detection in remote sensing images (RSIs) aims to obtain the semantic category and location information of specific remote sensing targets. With the refinement of practical application scenarios and the progress of high-resolution remote sensing imagery acquisition technology, fine-grained object detection methods for RSIs have gradually become a research hotspot. On the one hand, this problem is challenging, since RSIs are quite different from general natural images in terms of the complicated background, multi-scale objects, randomly oriented objects and fewer discriminative feature details. On the other hand, from a top-down perspective, compared with general objects, these fine-grained remote sensing objects tend to exhibit higher inter-class similarity and intra-class variability. As shown in Figure 1, F-15, F-16, and F-22 all belong to Aircraft, they have most of the similar external characteristics, such as the symmetric fuselage and wings, only differing in the fuselage aspect ratio, the shape of tails, the engine position and other details. Meanwhile, for the same type of aircraft, samples often have diverse coatings and wing states, etc. The same applies to Ship.

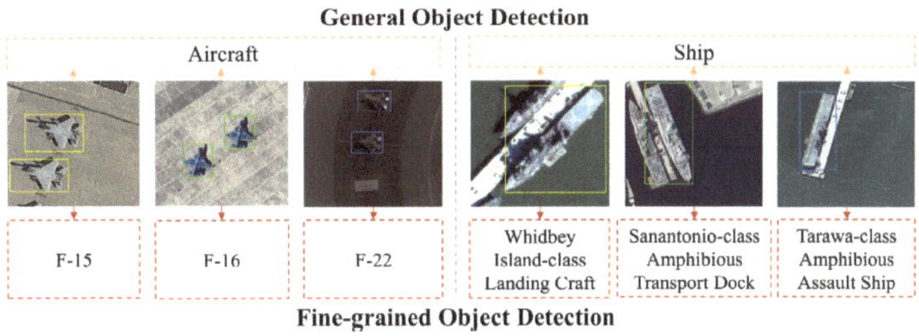

Figure 1. Comparison between general object detection and fine-grained object detection on RSIs.

Moreover, it has to be admitted that the accumulation of fine-grained remote sensing samples is difficult, especially for these newly developed and high-value targets. This is due to the influence of the inherent scarcity of fine-grained target sources, the satellite revisit period, and meteorological illumination conditions. However, most traditional object detection methods based on deep learning need to be optimized by training on massive samples. Correspondingly, these traditional methods are prone to serious overfitting under the condition of limited samples. To solve this contradiction, few-shot object detection (FSOD) methods are introduced.

In few-shot scenes, the problem of the model's poor ability to distinguish different categories is intensified. As shown in Figure 2, TIDE toolbox [1] uses the improvement in the mean Average Precision (mAP) after fixing various errors to measure the importance of each error type. We use it to identify detection errors in our baseline [2] and find that the main error type is Classification Error of localizing correctly but misclassifying novel classes, while Missed Error ranks second. We attribute the superficial reason to the fact that the common knowledge is not sufficiently transferred from the learned base classes to unseen novel classes and sample scarcity leads to obvious biased distributions.

Although [3] carries out the implicit knowledge transfer by freezing the backbone, the Region Proposal Network (RPN) and the Region of Interest (RoI) feature extractor during the fine-tuning stage, it mostly relies on the shared parameters and lacks the explicit extraction and transfer of the shared knowledge. Visually similar categories usually show higher semantic correlations [4], so relations between them can be regarded as a guide to establish connections and differences between the unseen and seen classes. SRR-FSD [5] also emphasizes the constant semantic relation and aligns the visual representations with corresponding class embeddings; however, it needs to introduce additional label prior relations. Moreover, the knowledge distillation mechanism has become an efficient strategy. It can tackle the shortage of few-shot models' poor perception ability for the common knowledge. By combining with the knowledge distillation, those FSOD methods train the teacher network to assist the student network in learning more robust representations. Some studies introduce the self-distillation mechanism for FSOD, training the student model as its own teacher for self-feedback and self-optimization [6–8].

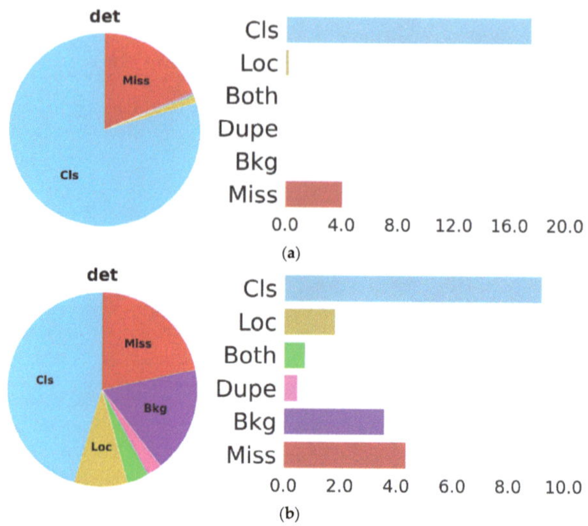

Figure 2. Analysis of our baseline. (**a**) Error analysis on the MAR20 dataset under the split1 10-shot setting. (**b**) Error analysis on SHIP20 dataset under the split1 10-shot setting. Det indicates the detection results. Classification Error (Cls) indicates localized correctly but classified incorrectly. Localization Error (Loc) indicates classified correctly but localized incorrectly. Both Cls and Loc Error (Both) indicate classified incorrectly and localized incorrectly. Duplicate Error (Dupe) indicates another higher-scoring detection already matched. Background Error (Bkg) indicates detected background as foreground. Missed Error (Miss) indicates all undetected ground truths (false negatives) not already covered by Cls or Loc.

FSOD methods with self-distillation have made some progress, but some problems still exist: (a) The features required for the classification and regression branches have different property preferences. However, the current un-decoupled single-branch self-distillation methods ignore this disparity and show no extra concern for the regression branch, as shown in Figure 3. (b) The memory bank dynamically calculates class prototypes during the entry and exit of RoI features. However, without evaluating whether these features are qualified to represent category attributes with high quality, the interference of personalized outlier features cannot be avoided. (c) Class prototypes are the core for representing the feature distribution of few-shot samples. Nevertheless, the misalignment between the learned class prototypes and the true feature distributions may widen during fine-tuning and further cause the blurred category edge problem.

To address the above problems, we propose to implement a decoupled self-distillation method based on the memory bank, yielding similarity probabilities between RoI features and class prototypes as soft labels for the classifier and aggregation weights for the regressor. To be specific, with the help of the affine layer and the gradient decoupled layer, classification-related and localization-related features are separated, and they serve as candidates to participate in the prototype construction of the above two types. Instead of directly throwing all feature vectors into the memory bank in a queue manner, we initially select a part of high-confidence feature vectors and enable the feature filtering module to play the role of preliminary screening. Additionally, in our study, we find that the base class distribution preserved in the pre-training phase is crucial for the anchoring of base prototypes in the fine-tuning stage. To maximize its effectiveness, we devise a two-step progressive prototype calibration module to leverage the base–novel correlations and gradually calibrate the distribution deviation.

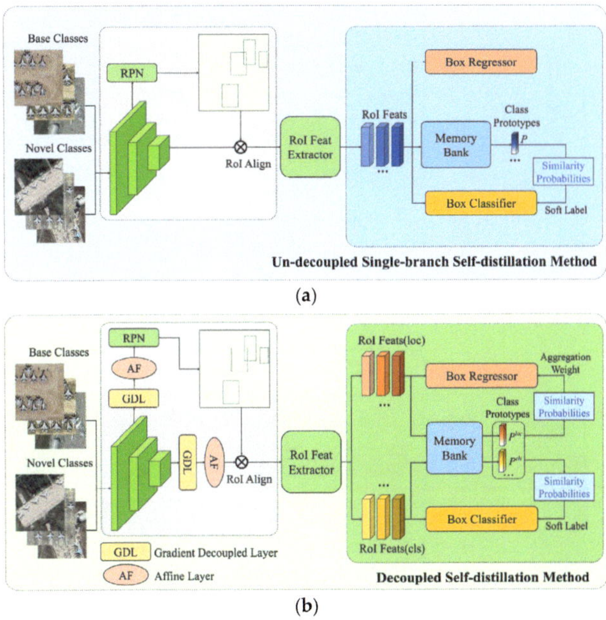

Figure 3. Comparison of different self-distillation methods. (**a**) The un-decoupled single-branch self-distillation method does not adopt the gradient decoupled layer and the affine layer. It directly inputs the shared RoI features into the classifier, the regressor and the memory bank to execute the self-distillation operation only for the classification branch. (**b**) The decoupled self-distillation method (Ours) first decouples the backbone from both the RCNN and the Region Proposal Network (RPN). Then, it, respectively, inputs the decoupled classification-related and localization-related RoI features into two branches. At the same time, both sets of features are input into the memory bank to extract two types of class prototypes by category. Then, the similarity probabilities between the input RoI features and these class prototypes can be calculated and act as soft labels or aggregation weights for self-distillation.

The main contributions of this article are concisely summarized as follows:

- We propose a decoupled self-distillation method based on a memory bank. This method automatically learns the distinctive characteristics of the base and novel classes by dynamically calculating the classification-related and localization-related class prototypes. This set of class prototypes is in return used to measure the similarity between the input feature and each category. The calculated similarity probabilities act as teachers to instruct the classifier and regressor by introducing two distillation losses. Thus, both of them are optimized to adapt to the inter-class heterogeneity and promote intra-class consistency. In addition, this method facilitates the integration of other modules.

- We devise a feature filtering module to screen feature samples with low confidence of class probabilities, which tends to improve the overall quality of all class prototypes. This module sets a higher admission standard for the memory bank based on the predicted class probabilities from the classifier.
- Moreover, a two-step progressive prototype calibration module based on distance metrics is also integrated. This module not only compensates for the distribution shift of base prototypes with prior base distribution, but also reduces the distribution misalignment for novel prototypes with calibrated base prototypes.
- We conduct extensive experiments on two remote sensing fine-grained datasets, MAR20 and SHIP20. The former is a public aircraft recognition dataset, and the latter is a ship detection dataset we reorganized with no overlap between subcategories and parent categories. The experimental results provide compelling evidence for the superiority of our method. Further, all proposed modules are applied only in the fine-tuning phase.

2. Related Works

2.1. General Object Detection

The traditional general object detection methods based on deep learning can be divided into the single-stage and the two-stage methods. Single-stage detectors usually directly produce final predictions from the feature map without extra operations, and they often have a faster inference speed [9–11]. Two-stage detectors use the region proposal network to generate object proposals, which are further put into the classifier and the regressor. The representative algorithms include Faster R-CNN [12] and its derivatives [8,10].

Afterwards, the anchor-free object detection methods are gradually emerging, such as CornerNet [13], CenterNet [14], and FCOS [15]. Furthermore, the transformers are introduced to replace the traditional convolutional neural network (CNN) along with the anchor mechanism for the end-to-end detection. As a result, DETR [16] and Deformable DETR [17] have attracted more attention.

2.2. Few-Shot Object Detection

At present, most FSOD methods follow the two-stage training paradigm. This paradigm utilizes the knowledge learned from a data-rich base-class dataset (the first stage) to improve the detection performance on a data-limited novel-class dataset (the second stage). Normally, FSOD can be broadly grouped into two branches: meta-learning based and fine-tuning based methods. The meta-learning methods can adapt to novel classes by iteratively training on a series of few-shot episode tasks. Specifically, Yan et al. [18] propose the Predictor-head Re-modeling Network (PRN) that uses novel-class samples to learn class-attentive vectors. Those vectors are then used as the meta-knowledge to implement the channel-level aggregation with query features. Fan et al. [19] propose an attention-based RPN aiming at guiding the network to generate more proposals related to query objects. Zhang et al. [20] propose an image-level meta-learning method. It extracts class-specific features by the encoder and uses the class-agnostic decoder to predict. However, such methods greatly increase the computational cost and exhibit poor interpretability [21].

Fine-tuning methods usually have simpler training procedures, a faster convergence rate and stronger generalization ability. They also show a slight advantage in two benchmark datasets [22], PASCAL VOC and MS COCO. Wang et al. [3] first propose a two-stage fine-tuning approach by freezing the backbone, RPN and RoI feature extractor. Sun et al. [23] introduce a contrastive branch to maximize the intra-class agreement and the inter-class disagreement, which learns contrastive-aware proposal embeddings. Wu et al. [24] propose an object pyramid as an auxiliary branch. It enables the model to capture multi-scale objects. Qiao et al. [2] propose a decoupled Faster R-CNN (DeFRCN) that contains a gradient decouple layer and an offline prototype calibration block.

2.3. Few-Shot Object Detection in Remote Sensing

Few-shot object detection methods in the remote sensing field have experienced a rapid growth over the past few years, resulting in a series of highly effective algorithms. For example, Wang et al. [25] introduce a context information refinement module to enhance inherent features and leverage transfer learning. Xiao et al. [26] design a lightweight feature attention highlight module. It contains cascaded coarse and fine highlights to provide more information. Cheng et al. [27] propose to generate region proposals of higher quality by applying a prototype-guided RPN network. Liu et al. [28] propose a transformation-invariant network (TINet) to solve the spatial misalignment problem caused by orientation variations in objects in the query and support images. Zhang et al. [29] design a metric-based discriminative loss to learn a more discriminative classifier. They also propose a representation compensation module to alleviate the catastrophic forgetting problem by decoupling the representation learning of the prior knowledge and the novel knowledge. Based on YOLOv5, Zhang et al. [30] propose a segmentation assistance (SA) module to guide the attention of the network toward positive objects. In addition, they design a triplet head to mitigate the issue of forgetting the base-class knowledge.

3. Methodology

3.1. Problem Definition

We follow the standard FSOD definition introduced in [3]. Given two class sets: base classes C_b with abundant annotated samples and novel classes C_n with k-shot annotated samples per category (usually $k \leq 30$). Normally, the above two sets do not intersect ($C_b \cap C_n = \varnothing$) and cover all categories of targets ($C_b \cup C_n = C_{total}$). Our training procedure complies with the two-stage training paradigm. During the base-training stage, our model is pre-trained on sufficient instances from C_b. During the fine-tuning stage, our model is trained jointly on k-shot instances per category from both C_b and C_n.

3.2. Overview of the Framework

We construct the proposed method based on the foundational framework of DeFRCN. As illustrated in Figure 4, our method first decouples classification-related and localization-related features by an affine layer and scales the backpropagation gradients by a gradient decoupled layer. Then, it performs decoupled self-distillation (DSD) based on a memory bank for the two subtask branches, respectively. During this procedure, our method integrates a feature filtering module (FFM) and a progressive prototype calibration module (PPCM) to build a set of robust class prototypes. This set of prototypes is utilized for continuously self-optimizing the classifier and the regressor.

In the base training stage, we assume that the retained base distribution of each category follows the Gaussian distribution. The mean and variance of this distribution can be calculated from the abundant base-class samples. In the fine-tuning stage, we adopt the DSD method to deal with these two types of RoI feature vectors and mine more instructive information from few samples. We tactfully devise a feature filtering module (FFM) to controls the quality of the features allowed into the memory bank. Moreover, we propose a progressive prototype calibration module (PPCM) to calibrate the base prototypes with the base distribution and then calibrate the novel prototypes with the calibrated base prototypes. Consequently, these class prototypes are robust enough to represent the class distribution centers and can be utilized as references to model the inter-class relations. This is crucial to generate a positive or negative feedback for the detector. The architecture is detailed in the following sections.

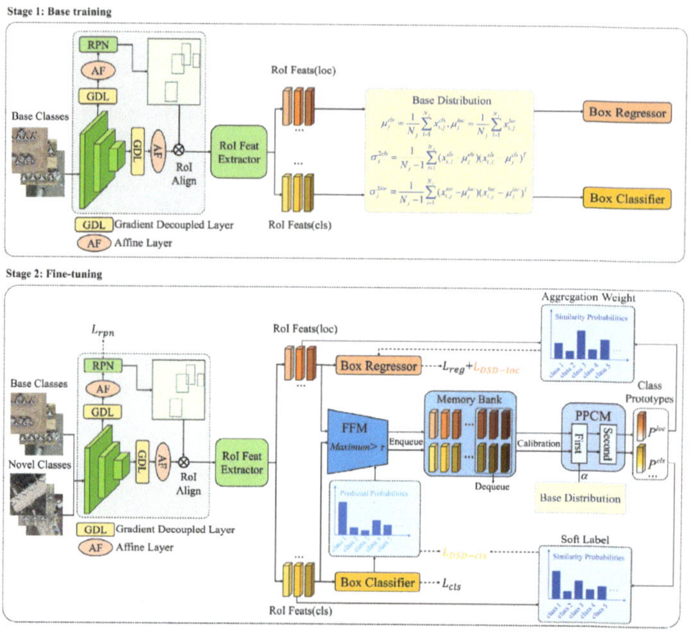

Figure 4. Overview of the proposed DSPPC. In the base training stage, the base distributions of all base classes have been preserved. In the fine-tuning stage, our proposed DSD method enables the classification and regression branches to distill representative class prototypes with the memory bank. These prototypes are then used to yield similarity probabilities as soft labels or aggregation weights for self-instruction, which is facilitated by two distillation losses, $L_{DSD\text{-}cls}$ and $L_{DSD\text{-}loc}$. FFM ensures that low-quality RoI features do not participate in the dynamic calculation of class prototypes. PPCM achieves the two-step prototype calibration from the base to the novel classes.

3.3. The Decoupled Self-Distillation Method

As stated in [2], the classification branch requires translation invariant features whereas the regression branch requires translation covariant features. These two types of features need to be treated in a different way. Therefore, our DSD method contains two branches owing to the gradient decoupled layer (GDL) and the affine layer (AL) after the backbone. The GDL modulates the backward gradients via imposing a scale factor on the gradients from the RPN and the RCNN. The AL carries out a linear transformation for the classification-related and the localization-related RoI features. Our method has not changed the original transmission branches of the above two types of RoI features, but it additionally delivered these RoI features to the memory bank for the subsequent prototype extraction. Class prototypes can be further divided into base prototypes and novel prototypes. The feature distribution of few samples can be represented by this set of relatively stable class prototypes $P = \left\{[P^{cls}_{C_b}, P^{cls}_{C_n}], [P^{loc}_{C_b}, P^{loc}_{C_n}]\right\} \in \mathbb{R}^{C_{total} \times d}$, where d represents the feature dimension. The final form of the prototype representation is specifically illustrated in Section 3.5.

Then, we stand on an underlying assumption that categories of high similarity in the classification-related feature space tend to share the feature regions of common concern; the same holds for the localization-related feature space [7]. Therefore, in the fine-tuning stage, we use cosine similarities to quantize the inter-class similarity between the input foreground features and the class prototypes in the above two feature spaces. These similarities $Sim_{i,j}^{\cdot} \in \mathbb{R}^{N \times C_{total}}$ are initialized to -1 and then calculated as follows:

$$Sim_{i,j}^{cls} = \begin{cases} cos(f_i^{cls}, P_j^{cls}), i \in \{1,2,\ldots,N\}, j \in C_n \\ cos(f_i^{cls}, P_j^{cls}), i = \{1,2,\ldots,N\}, j = c_{gt} \\ -1, others \end{cases} \quad (1)$$

$$Sim_{i,j}^{loc} = \begin{cases} cos(f_i^{loc}, P_j^{loc}), i \in \{1,2,\ldots,N\}, j \in C_n \\ cos(f_i^{loc}, P_j^{loc}), i = \{1,2,\ldots,N\}, j = c_{gt} \\ -1, others \end{cases} \quad (2)$$

$$cos(x_1, x_2) = \frac{x_1 \cdot x_2{}^T}{\|x_1\|_2 \cdot \|x_2\|_2} \quad (3)$$

where N represents the total number of samples stored in the memory bank, $||\cdot||_2$ denotes the $L2$ norm of the feature vectors, i denotes the i-th foreground sample, f_i^{cls} denotes the classification-related RoI feature of the i-th foreground sample, f_i^{loc} denotes the localization-related RoI feature of the i-th foreground sample, j denotes the class index of the prototype, c_{gt} denotes the ground-truth class index of the i-th foreground sample. There are three situations for $Sim_{i,j}^{cls}$. To enhance the model's perception of the base–novel correlations and reduce the competition from the non-ground-truth base classes and unrelated categories, we define the similarity between f_i^{cls} or f_i^{loc} and these irrelevant base prototypes to be a small constant value. But we retain the cosine similarity values associated with the novel classes C_n and the ground-truth class c_{gt}. In order to intensively map the cosine similarity into the probability distribution, we apply a softmax function with a temperature scaling parameter t for normalization:

$$Prob_{i,j}^{cls} = softmax(\frac{Sim_{i,j}^{cls}}{t}), i \in \{1,2,\ldots,N\}, j \in C_{total} \quad (4)$$

$$Prob_{i,j}^{loc} = softmax(\frac{Sim_{i,j}^{loc}}{t}), i \in \{1,2,\ldots,N\}, j \in C_{total} \quad (5)$$

Most importantly, our method adopts two different ways to make the above similarity probabilities influence the prediction results. For the classification branch, we grab the predicted class probabilities $Pred_{i,j} \in \mathbb{R}^{N \times C_{total}}$ from the classifier and evaluate the Kullback–Leibler (KL) divergence between $Pred_{i,j}$ and $Prob_{i,j}^{cls}$, where $Prob_{i,j}^{cls}$ plays the role of a teacher:

$$L_{DSD\text{-}cls} = \sum_{\substack{i \in \{1,2,\ldots,N\} \\ j \in C_{total}}} Prob_{i,j}^{cls} log \frac{Prob_{i,j}^{cls}}{Pred_{i,j}} \quad (6)$$

For the localization branch, our method manages to increase the aggregation weights for these categories closely similar to c_{gt}, and tilt attention towards the novel classes. The motivation is that based on the decoupled localization-related feature spaces, similar categories tend to share the localization-related commonalities, such as the similar boundary regions. For a localization-related RoI feature of a certain base class, our DSD method leverages the predictions from the regressors of the ground-truth base class along with the highly similar novel classes. Essentially, this process is potentially equivalent to training the regressors of these novel classes with this RoI feature. Consequently, the regressors of the novel classes are slightly optimized by these additional training samples from similar base classes that focus on the common feature regions. Hence, the DSD method does not only calculate the regression loss on c_{gt} of the foreground samples, but also weights the regression loss by category based on $Prob_{i,\cdot}^{loc}$:

$$L_{DSD\text{-}reg} = \frac{1}{N}\sum_{j=1}^{C_{total}} Prob_{i,j}^{loc} \sum_{i\in\{x,y,w,h\}} Smooth_{L1}(p_{i,j} - g_i) \quad (7)$$

where $p_{i,j}$ denotes the predicted bounding box coordinates and g_i denotes the bounding box ground truth.

The total loss function is represented as follows:

$$L_{total} = L_{rpn} + L_{cls} + L_{reg} + \gamma_1 L_{DSD\text{-}cls} + \gamma_2 L_{DSD\text{-}reg} \quad (8)$$

where L_{rpn} is the loss of RPN, L_{cls} is the cross-entropy loss for the bounding box classification, and L_{reg} is the Smooth L1 loss for the bounding box regression, and γ_1 and γ_2 denote the factors for the self-distillation losses.

3.4. The Feature Filtering Module

To obtain a set of representative class prototypes, the usual practice is to set the length of the memory bank to a fixed value, and the memory bank allocates separate storage spaces of the same length for each category. Each category has two queues for classification-related and localization-related RoI features, which can be expressed as Q_j^{cls} and Q_j^{loc}. Then, f_i^{cls} and f_i^{loc} are paired to enqueue and dequeue the memory bank $M = \{Q_j^{cls}, Q_j^{loc} | j \in C_{total}\}$ according to the first-in first-out order.

Based on the above process, we insist on the view that RoI feature vectors showing distinct characteristics of categories should make a more significant contribution to the synthesis of class prototypes. To be specific, when a RoI feature vector has a confidence score below a certain threshold for each category, it is not reliable enough to support the prototype construction of any category. Therefore, we design a front-loaded feature filtering module for the memory bank, simple but effective.

The feature filtering module (FFM) also takes advantage of the predicted class probabilities $Pred_{i,j}$. The standard for enqueuing is regulated by a confidence threshold τ after the warm-up iterations. For the i-th foreground sample of class c_{gt}, the filtering process can be expressed as follows:

$$Q_{c_{gt}}^{cls} = \{f_i^{cls} | \max_{j\in C_{total}} (Pred_{i,j}) > \tau, i = \{1, 2, \ldots, N\}\} \quad (9)$$

$$Q_{c_{gt}}^{loc} = \{f_i^{loc} | f_i^{cls} \; in \; Q^{cls}, i = \{1, 2, \ldots, N\}\} \quad (10)$$

There is a correspondence relationship among f_i^{cls}, c_{gt} and $Pred_{i,j}$. If the maximum value of the predicted class probabilities is below the confidence threshold, it suggests that the current RoI feature is considerably distant from the feature representation centers of all categories and does not exhibit the representative characteristics of any category. Thus, it is not considered a high-quality candidate for building the two types of class prototypes. Eventually, we align Q_j^{loc} with Q_j^{cls}.

The update procedures of the memory bank can be summarized in the following two stages: (a) During the early warm-up iterations, when the number of the stored RoI features of a certain category has not reached the maximum length of the memory bank, the new RoI features of this category are enqueued into Q_j^{loc} and Q_j^{loc} in pairs. Once the number reaches the maximum length, those earlier enqueued RoI features of this category are dequeued in pairs to make room for the new ones. The number of the stored RoI features thereafter remains at the maximum length. It should be noted that considering the potential instability of the detector in the early training iterations, the calculation of class prototypes, the feature filtering operation and the self-distillation operation are not performed during this stage. (b) After the warm-up iterations, the confidence threshold τ is set for the current batch of the input RoI features. Meanwhile, the stored RoI features of each category are prepared to calculate the base and novel prototypes and perform the subsequent self-distillation operations.

3.5. The Progressive Prototype Calibration Module

Inspired by [31], we integrate a progressive prototype calibration module (PPCM) to refine both the base and the novel prototypes learned from the few-shot samples in the fine-tuning stage, as shown in Figure 5.

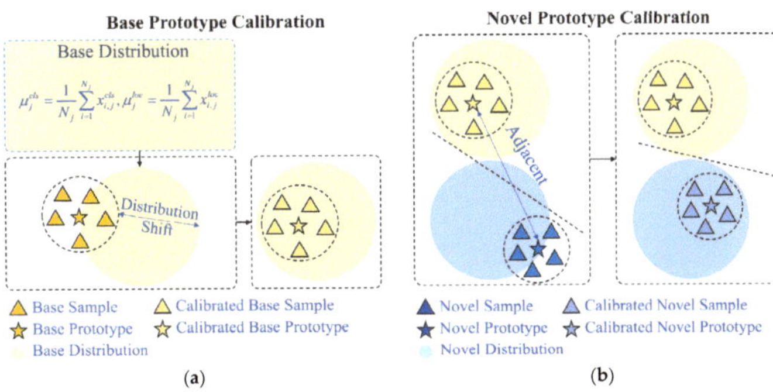

Figure 5. Illustration of our proposed PPCM. It contains two steps: base prototype calibration and novel prototype calibration. (**a**) In the first step, our module partially utilizes the base distributions preserved in the base training stage to compensate for the base prototypes in the fine-tuning stage. (**b**) In the second step, our module makes use of the top-k adjacent base prototypes to calibrate novel prototypes, both of which are dynamically updated.

Since the base training procedure is based on sufficient samples for each base class, the model is capable of fully interacting with the diverse base-class samples and learning the common characteristics between categories. In other words, the learned empirical base distributions are approximately consistent with the expected base distributions, following the Gaussian distribution $\{\Omega_j^{cls}, \Omega_j^{loc} | \Omega_j^{cls} \sim N(\mu_j^{cls}, \sigma_j^{cls}), \Omega_j^{loc} \sim N(\mu_j^{loc}, \sigma_j^{loc}), j \in C_b\}$:

$$\mu_j^{cls} = \frac{1}{N_j}\sum_{i=1}^{N_j} x_{i,j}^{cls}, \mu_j^{loc} = \frac{1}{N_j}\sum_{i=1}^{N_j} x_{i,j}^{loc}, j \in C_b \tag{11}$$

$$\sigma_j^{2cls} = \frac{1}{N_j - 1}\sum_{i=1}^{N_j} (x_{i,j}^{cls} - \mu_j^{cls})(x_{i,j}^{cls} - \mu_j^{cls})^T, \sigma_j^{2loc} = \frac{1}{N_j - 1}\sum_{i=1}^{N_j} (x_{i,j}^{loc} - \mu_j^{loc})(x_{i,j}^{loc} - \mu_j^{loc})^T, j \in C_b \tag{12}$$

where $x_{i,j}^{cls}$ and $x_{i,j}^{loc}$ denote the i-th positive RoI features of base class j in the base training stage, N_j denotes the total number of positive RoI features of base class j, and μ_j^{cls}, μ_j^{loc}, σ_j^{2cls}, and σ_j^{2loc} separately denote the means and variances of base class j.

When it comes to the fine-tuning stage, we primarily average the candidate RoI samples stored in the memory bank by category to obtain the intermediate form of both base and novel prototypes:

$$P_j^{cls} = Mean(Q_j^{cls}), P_j^{loc} = Mean(Q_j^{loc}), j \in C_{total} \tag{13}$$

Then, we attempt to use the retained base distributions to calibrate the base prototypes. We compensate for the distribution shift of base classes in the first step to ultimately produce a set of base prototypes:

$$P_j^{cls\prime} = \alpha\mu_j^{cls} + (1-\alpha)P_j^{cls}, j \in C_b \tag{14}$$

$$P_j^{loc\prime} = \alpha\mu_j^{loc} + (1-\alpha)P_j^{loc}, j \in C_b \tag{15}$$

where hyperparameter α controls the calibration strength for the base prototypes.

In the second step, we refine the biased novel prototype with calibrated base prototypes, which is verified in [32] that the feature representations of similar categories usually share mean and variance. It is notable that we do not follow the previous practice of averaging all the k nearest calibrated base prototypes and applying them to novel classes. Instead, we measure the base–novel similarity relations based on the Euclidean distance, and then we aggregate the k nearest calibrated base prototypes by Exponential Decay Weighting (EDW). Given a biased novel prototype of class n, it is calibrated as follows:

$$P_n^{cls\prime} = \frac{P_n^{cls} + \sum_{j \in S_k^{cls}} e^{-\lambda d_j^{cls}} P_j^{cls\prime}}{1 + \sum_{j \in S_k^{cls}} e^{-\lambda d_j^{cls}} P_j^{cls\prime}} \tag{16}$$

$$P_n^{loc\prime} = \frac{P_n^{loc} + \sum_{j \in S_k^{loc}} e^{-\lambda d_j^{loc}} P_j^{loc\prime}}{1 + \sum_{j \in S_k^{loc}} e^{-\lambda d_j^{loc}} P_j^{loc\prime}} \tag{17}$$

where $S_k^{cls} = \{j | d_j^{cls} \in topk(-d_j^{cls}), j \in C_b\}$ and $S_k^{loc} = \{j | d_j^{loc} \in topk(-d_j^{loc}), j \in C_b\}$ denote the sets of its k nearest base classes, $d_j^{cls} = ||P_n^{cls} - P_j^{cls\prime}||^2$ and $d_j^{loc} = ||P_n^{loc} - P_j^{loc\prime}||^2$ denote the Euclidean distances, and λ controls the exponential decay rate.

Subsequently, our model calculates the similarity probabilities based on the above calibrated prototypes, together with the distillation losses in Section 3.3.

4. Experiments and Results

To evaluate the effectiveness of the proposed method, we use the same evaluation metrics as the mainstream FSOD works, and perform a series of comparisons on two fine-grained remote sensing datasets, MAR20 [33] and customized SHIP20.

4.1. Datasets and Metrics

(a) MAR20. The MAR20 dataset is a public aircraft recognition dataset; it comprises 3842 high-resolution remote sensing images, 22,341 instances and 20 different types of fine-grained aircrafts from Google Earth. For comparative experiments, we divide these aircrafts into 3 different base–novel splits, each of which consists of 15 base classes and 5 novel classes, as shown in Table 1. For fine-tuning, we use the 3-, 5-, 10-shot settings on the train set.

(b) SHIP20. Considering that the existing ship object detection datasets usually use multi-level labels, there is an overlapping problem of parent and subclass labels, which is not conducive to the fine-grained few-shot task. Therefore, we reorganize a custom ship dataset, SHIP20, based on [34–36]. It contains 9877 images, 29,337 instances and 20 types of ships. The image size is 800 × 800, and the data sources are Google Earth, GF-2 satellite, etc. The specific examples are shown in Figure 6. We also divide the ships into three different splits, as shown in Table 2. For the fine-tuning stage, we use the 3-, 5-, 10-shot settings on the train set.

(c) Evaluation Metrics. Our main evaluation metric is the mean Average Precision of the novel-class set with an IoU (Intersection over Union) threshold. The calculation of the Average Precision for a certain novel class is expressed as follows:

$$AP = \frac{1}{11} \sum_{R \in \{0,0.1,0.2,\ldots,1\}} P(R) \quad (18)$$

where R indicates the recall value and $P(R)$ indicates the value of the Precision-Recall curve when recall is determined. The nAP50 and nAP75 denote the average of AP with an IoU of 0.5, 0.75 for all novel classes. In addition, to assess the model complexity and efficiency, we also assess the metrics of Params, FLOPs (Floating-point Operations), the training time per iteration, the inference time per image, and the memory usage of our model.

Table 1. Three different base–novel splits for the MAR20 dataset.

Split	Novel Classes					Base Classes
1	FA-18	TU-95	KC-10	SU-34	SU-24	others
2	E-8	TU-22	F-15	KC-135	F-22	others
3	SU-35	C-130	C-17	C-5	F-16	others

Table 2. Three different base–novel splits for the SHIP20 dataset.

Split	Novel Classes					Base Classes
1	Passenger Ship	Engineering Ship	WhidbeyIsland.	Perry.	Sanantonio.	others
2	Barge	Oil Tanker	Nimitz.	Tarawa.	Arleigh Burke.	others
3	Liquid Cargo Ship	Dry Cargo Ship	Ticonderoga.	Austen.	Tail-OX.	others

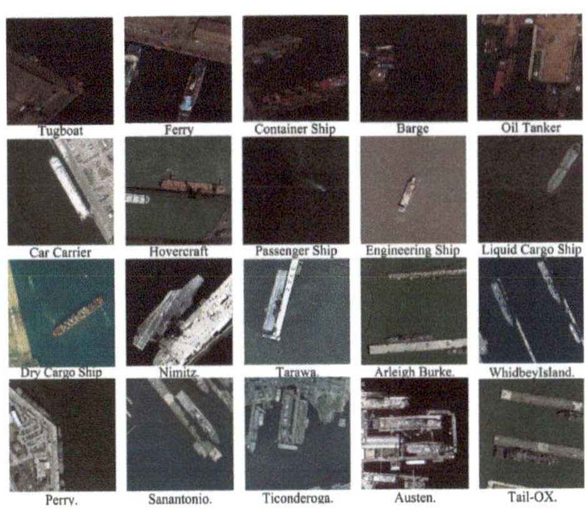

Figure 6. Examples of 20 categories of ship targets in the SHIP20 dataset.

4.2. Implementation Details

As mentioned above, we choose DeFRCN [2] as our baseline to evaluate our methods, the training strategy and training settings of our model are consistent with the baseline. Specifically, our experiments are conducted by using the PyTorch 1.6.0 framework on an NVIDIA 2080Ti GPU (11 GB), with a batch size of 4. For base training, the maximum iteration is set to 15,000 with an initial learning rate of 0.01, decreased by 0.1 at 10,000 and 13,300 iterations. For fine-tuning, the initial learning rate is set to 0.01, decreased by 0.1 at 500 iterations before the end of training. Additionally, the data augmentation strategy is not used in training or inference. To make a fair comparison, we use ResNet101 as the backbone for all nine comparison methods, including four meta-learning methods and five fine-tuning methods. For MAR20, we utilize 2511 images from the test set for evaluation and the rest for training, and for SHIP20, 3951 images for evaluation and the rest for training.

Regarding the hyperparameters in DSD method, we set the similarity value for irrelevant base classes to -1, and the queue length of memory bank to 2048, the temperature scaling parameter t to 0.1, and γ_1, γ_2 to 0.1, 0.7 by default. According to the ablation results, for the hyperparameters in FFM, we set the confidence threshold τ to 0.3, 0.3, 0.4 under 3-, 5-, 10-shot settings, respectively, for two datasets across all splits. For the hyperparameters in PPCM, we select the parameter α to be 0.1, λ to be 0.4 for two datasets across all splits. While the parameter k requires a case-by-case analysis, for the 3-shot and 10-shot settings, k should be, respectively, set to 1 and 2 for both datasets, for the 5-shot setting, k should be set to 1 for MAR20 and 2 for SHIP20. Lastly, the lengths of the warm-up iterations are also separately determined to 600 for MAR20 and 800 for SHIP20. Table 3 lists the specific parameter configurations on two datasets under different shot settings.

Table 3. The hyperparameter configurations on two datasets.

Hyperparameter	All Splits of MAR20			All Splits of SHIP20		
	3-Shot	**5-Shot**	**10-Shot**	**3-Shot**	**5-Shot**	**10-Shot**
τ	0.3	0.3	0.4	0.3	0.3	0.4
α	0.9	0.9	0.9	0.9	0.9	0.9
λ	0.4	0.4	0.4	0.4	0.4	0.4
k	1	1	2	1	2	2
The warm-up length	600	600	600	800	800	800

4.3. Results on MAR20

Table 4 presents the comparative results on the MAR20 dataset. It can be seen that the proposed method DSPPC shows a notable performance leap on three splits of MAR20. For the 3-, 5-, 10-shot settings on three splits, our method achieves average improvements of 10.7%, 11.3%, 8.5% over the baseline and leads the second-best method ECEA by 5.8%, 6.9%, 6.5% on average, reaching state-of-the-art performance on MAR20 test set. When comparing by row, our method produces a greater enhancement on the 3-shot and 5-shot settings on MAR20. Also, due to the different detection difficulties of novel classes from various splits, the improvements may be uneven for different splits [37]. For instance, our method exhibits a greater gain on the split 1 of MAR20 from the comparison between columns. In more detail, it can be seen from Figure 7 that under the 3-, 5-, 10-shot settings, although our model does not achieve the highest AP on every novel class, it actually keeps the AP of each novel class at a high level (over 44%). This result demonstrates that our method more uniformly improves the detection performance for each novel class.

Table 4. Experimental results on the MAR20 test set under 3-, 5-, 10-shot settings with IoU of 0.5. Results in bold denote the best performance and the underlined ones represent the second best. The superscript † indicates that the results are reproduced by MMFewShot [38].

Shot	Type	Method	Year	Split1	Split2	Split3	Mean
3	Meta-learning	Meta R-CNN † [18]	2019	24.4	35.8	25.9	28.7
		Attention RPN † [19]	2020	15.5	32.7	33.1	27.1
		VFA [39]	2023	30.2	35.3	34.9	33.5
		FPD [40]	2024	27.5	39.1	38.8	35.1
	Fine-tuning	TFA w/cos † [3]	2020	16.4	20.7	18.8	18.6
		MPSR † [24]	2020	18.7	19.0	15.3	17.7
		FSCE † [23]	2021	32.5	37.8	42.3	37.5
		DeFRCN [2]	2021	38.3	41.9	48.1	42.8
		ECEA [41]	2023	<u>45.1</u>	<u>43.6</u>	**54.4**	<u>47.7</u>
		DSPPC (Ours)	-	**58.8**	**48.7**	<u>52.9</u>	**53.5**
5	Meta-learning	Meta R-CNN † [18]	2019	31.9	41.6	29.4	34.3
		Attention RPN † [19]	2020	28.2	34.7	36.7	33.2
		VFA [39]	2023	31.8	44.4	44.9	40.4
		FPD [40]	2024	32.1	42.0	47.8	40.6
	Fine-tuning	TFA w/cos † [3]	2020	17.9	22.9	24.4	21.7
		MPSR † [24]	2020	27.6	21.4	27.6	25.5
		FSCE † [23]	2021	35.5	39.7	48.3	41.2
		DeFRCN [2]	2021	42.2	49.6	50.3	47.4
		ECEA [41]	2023	<u>48.4</u>	<u>50.7</u>	<u>56.2</u>	<u>51.8</u>
		DSPPC (Ours)	-	**62.1**	**56.8**	**57.1**	**58.7**
10	Meta-learning	Meta R-CNN † [18]	2019	38.4	47.6	31.7	39.2
		Attention RPN † [19]	2020	34.1	38.8	37.3	36.7
		VFA [39]	2023	42.3	50.9	46.9	46.7
		FPD [40]	2024	42.9	50.8	52.3	48.7
	Fine-tuning	TFA w/cos † [3]	2020	21.5	26.7	29.6	25.9
		MPSR † [24]	2020	32.8	28.9	32.6	31.4
		FSCE † [23]	2021	40.6	44.9	47.5	44.3
		DeFRCN [2]	2021	51.5	52.0	52.8	52.1
		ECEA [41]	2023	<u>53.1</u>	<u>52.2</u>	<u>57.0</u>	<u>54.1</u>
		DSPPC (Ours)	-	**64.1**	**58.8**	**58.9**	**60.6**

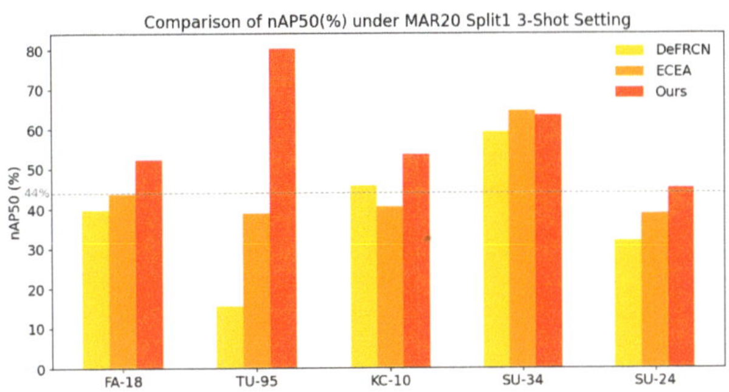

Figure 7. Comparison of nAP50 with the baseline and the second-best method ECEA for each novel class under MAR20 split1 3-shot setting.

Visualization results of MAR20 are presented in Figure 8, where we compare the prediction results of our model with two other fine-tuning models, the baseline DeFRCN and the second-best ECEA. We select four pairs of easily confused aircrafts and refer to the first row, ground truth (GT), to evaluate the performances of three models. Under the same confidence threshold, our model correctly predicts all targets of E-3 versus E-8 and F-15 versus F-16 shown in the first and fourth columns. It also can partially distinguish between SU-34 and SU-35, as well as KC-10 and KC-135, to a greater degree than the other two models, as shown in the second and third columns. The visualization results demonstrate that our model yields commendable detection results on both base classes and novel classes of the MAR20 dataset.

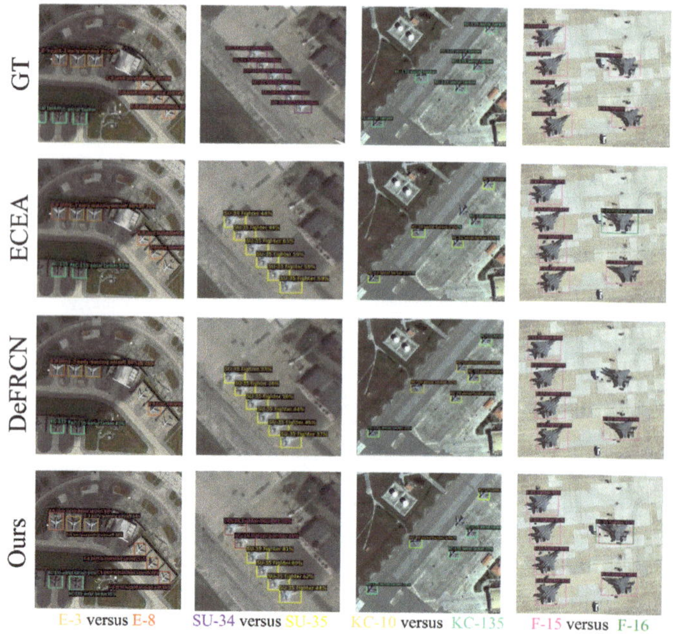

Figure 8. Visualization of prediction results of DeFRCN, ECEA and our model on the MAR20 dataset under the split1 10-shot setting.

In order to intuitively show the effect of the DSPPC method for correcting classification confusion errors, we further present the confusion matrix and the sub-confusion matrix on novel classes of the baseline and the proposed method on MAR20 split1 in Figure 9. The confusion matrix is generated based on the model's prediction results of samples across 20 classes in the MAR20 test set. To more rigorously evaluate the model's discriminative ability among novel classes, we have only retained the prediction results of the classifier and the regressor exclusively on these five novel classes. Next, we conduct inference on samples from just these five novel classes in the test set, and calculate the sub-confusion matrix based on these prediction results.

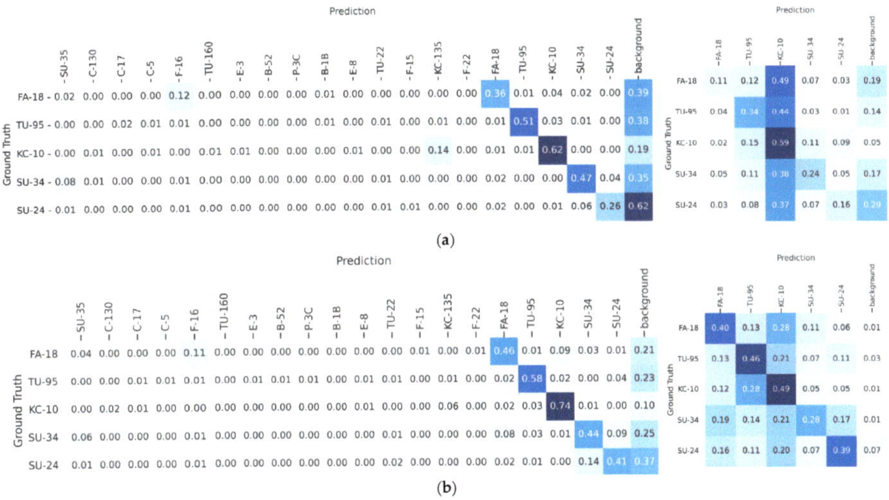

Figure 9. Comparison of the confusion matrix and the sub-confusion matrix on novel classes on MAR20 split1 of the baseline (top) and the proposed method (bottom) under 10-shot setting. The color gradient shifts from light blue to dark blue, mapping the values from small to large. (**a**) Confusion matrix (left) and sub-confusion matrix (right) on novel classes of the baseline. (**b**) Confusion matrix (left) and sub-confusion matrix (right) on novel classes of the proposed method.

According to the comparison of the confusion matrix, our method effectively reduces the error cases of misclassifying novel classes as background or base classes, which indicates that our method strengthens the feature representation for novel classes. Further, within the novel-class set, the comparison of the sub-confusion matrix on the right also indicates that rather than showing a relatively higher classification accuracy for any single category (e.g., KC-10), our method produces relatively balanced correct results across all five categories. This suggests that the proposed method also performs effectively in distinguishing these novel fine-grained aircrafts.

4.4. Results on SHIP20

Table 5 shows the experimental results on the SHIP20 test set. Similarly, our proposed method achieves the best performance under all shot settings, and also consistently maintains the best performance across all three splits. It brings average improvements of 4.4%, 4.1%, 3.7% under 3-, 5-, 10-shot settings separately compared with the baseline, which is also the second-best method. These results also showcase the benefits of our proposed method in tackling scenarios with low-shot samples. Moreover, it can be noticed that improvements on the SHIP20 dataset is generally smaller than those on the MAR20 dataset, this difference might be attributed to the fact that the detection errors on SHIP20 are more multi-type, and the proposed method has not adopted additional optimization strategies for tiny objects or multi-scale objects, such as Passenger Ship.

Table 5. Experimental results on the SHIP20 test set under 3-, 5-, 10-shot settings with IoU of 0.5. Results in bold denote the best performance and the underlined ones represent the second best. The superscript † indicates that the results are reproduced by MMFewShot [38].

Shot	Type	Method	Year	Split1	Split2	Split3	Mean
3	Meta-learning	Meta R-CNN † [18]	2019	23.7	20.5	26.8	23.7
		Attention RPN † [19]	2020	7.1	14.0	11.2	10.8
		VFA [39]	2023	23.7	21.1	27.3	24.0
		FPD [40]	2024	25.3	26.5	30.4	27.4
	Fine-tuning	TFA w/cos † [3]	2020	13.4	15.1	9.9	12.8
		MPSR † [24]	2020	16.9	21.1	14.5	17.5
		FSCE † [23]	2021	24.7	23.1	21.9	23.2
		DeFRCN [2]	2021	26.6	<u>27.6</u>	<u>31.7</u>	<u>28.6</u>
		ECEA [41]	2023	<u>27.7</u>	25.4	28.5	27.2
		DSPPC (Ours)	-	**31.2**	**30.9**	**37.0**	**33.0**
5	Meta-learning	Meta R-CNN † [18]	2019	31.5	27.1	27.7	28.8
		Attention RPN † [19]	2020	11.4	16.6	12.1	13.4
		VFA [39]	2023	32.0	30.6	27.4	30.0
		FPD [40]	2024	33.8	32.1	35.3	33.7
	Fine-tuning	TFA w/cos † [3]	2020	26.0	19.8	14.7	20.2
		MPSR † [24]	2020	17.0	29.1	15.0	20.4
		FSCE † [23]	2021	31.7	27.6	27.5	28.9
		DeFRCN [2]	2021	35.3	<u>32.3</u>	37.2	<u>34.9</u>
		ECEA [41]	2023	<u>36.1</u>	29.8	<u>37.6</u>	34.5
		DSPPC (Ours)	-	**39.5**	**36.9**	**40.6**	**39.0**
10	Meta-learning	Meta R-CNN † [18]	2019	42.0	32.2	32.0	35.4
		Attention RPN † [19]	2020	17.9	17.1	15.7	16.9
		VFA [39]	2023	38.2	36.2	36.2	36.9
		FPD [40]	2024	<u>42.2</u>	<u>37.6</u>	38.2	39.3
	Fine-tuning	TFA w/cos † [3]	2020	32.7	23.3	27.3	27.8
		MPSR † [24]	2020	29.4	37.8	30.1	32.4
		FSCE † [23]	2021	40.6	34.1	38.2	37.6
		DeFRCN [2]	2021	41.0	35.5	<u>41.8</u>	<u>39.4</u>
		ECEA [41]	2023	41.3	32.9	39.2	37.8
		DSPPC (Ours)	-	**45.9**	<u>38.8</u>	**44.6**	**43.1**

We use TIDE toolbox to analyze the prediction results of our model, as shown in Figure 10. It is obvious that the mAP growths after fixing Classification, Missed, Localization and Both Errors have decreased. This indicates that our method has worked in correcting these error cases, especially for Missed Errors. However, Dupe and Background Errors have slightly elevated.

Next, we provide the visualization results of SHIP20 in Figure 11, also comparing with the baseline and ECEA. In particular, we present the prediction results on two novel-class targets in the same image, as shown in the first column. The baseline model just misidentifies WhidbeyIsland Class Landing Craft as Sanantonio Class Amphibious Transport Dock, while our model and ECEA are able to perceive the slight differences between these two categories. In the second column, taking GT as a reference, our model produces the optimal bounding box position. For the third and fourth columns, although with some false positive examples, the prediction results from our model are more complete and more accurate than those of others.

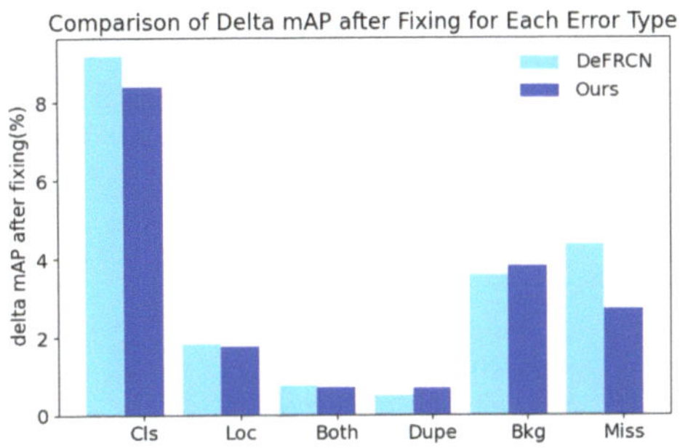

Figure 10. Error analysis comparison between the baseline and the proposed method. The prediction results on the test set of SHIP20 come from the model trained under the split1 10-shot setting.

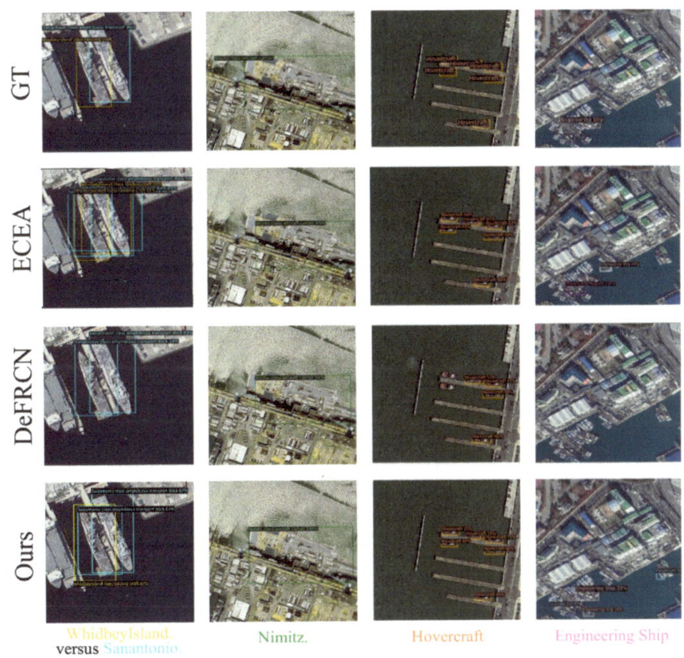

Figure 11. Visualization of prediction results of DeFRCN, ECEA and our model on SHIP20 dataset under the split1 10-shot setting.

As shown in Figure 12, we also provide the comparison of the confusion matrix and the sub-confusion matrix on novel classes between the baseline and the proposed method on SHIP20 split1. It is worth noting that according to the comparison of the confusion matrix on the left, our model increases the discrimination ability for the novel-class ships to a certain extent, because more positive samples are correctly pulled back from background or base classes. Similarly, the sub-confusion matrix on SHIP20 split1 reveals that our method exhibits a higher average classification accuracy on five novel classes, indicating greater reliability in handling novel fine-grained ships over the baseline. Nonetheless, we also find that the classification confusion problem among novel classes for SHIP20 still exists, and further approaches to amplify their differentiation need to be explored.

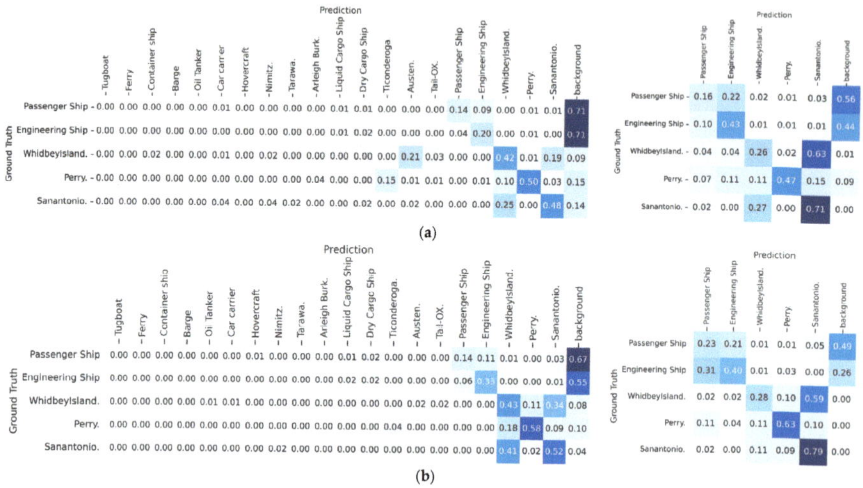

Figure 12. Comparison of the confusion matrix and the sub-confusion matrix on novel classes on SHIP20 split1 of the baseline (top) and the proposed method (bottom) under 10-shot setting. The color gradient shifts from light blue to dark blue, mapping the values from small to large. (**a**) Confusion matrix (left) and sub-confusion matrix (right) on novel classes of the baseline. (**b**) Confusion matrix (left) and sub-confusion matrix (right) on novel classes of the proposed method.

4.5. Model Complexity and Efficiency

Lastly, we have explored the comparative analysis of the model complexity and efficiency between our proposed method and other methods. As clearly presented in Table 6, our model has a larger number of parameters (approximately 16% more than the baseline) and more computation operations mainly caused by the elaborate RoI head. As for the training time per iteration, our model also takes almost 0.07 s more than the baseline to train in each iteration. However, for both the inference time per image and the GPU memory usage metrics, our model is able to exhibit certain advantages compared to ECEA, while maintaining a higher performance on the metric of nAP. Lastly, it should be emphasized that, in our method, since the memory bank stores class prototypes by category, the GPU memory usage increases linearly with the number of categories in the fine-tuning stage.

Table 6. Comparative analysis of model complexity and efficiency. Results in bold denote the best performance and the underlined ones represent the second best.

Model	Params (M)	FLOPs (G)	Training Time	Inference Time	GPU Memory Usage (G)
DeFRCN	**52.10**	**351.24**	**0.46**	**0.128**	**6.5**
ECEA	<u>55.40</u>	<u>359.45</u>	<u>0.50</u>	0.139	7.8
Ours	60.50	359.63	0.53	<u>0.133</u>	<u>7.5</u>

The unit of time measurement is s.

4.6. Ablation Studies

We conduct comprehensive ablation experiments on MAR20 and SHIP20 using DeFRCN as our baseline to delve deeper into the contribution of each proposed component and the impact of different hyperparameters

4.6.1. Ablation Study of Three Proposed Components

The ablation performances on MAR20 and SHIP20 are displayed in Tables 7 and 8. The FFM module and the PPCM module can be integrated into the DSD method either individually or in combination. As stated above, the DSD method aims to explicitly decouple the classification and regression branches and distill instructive inter-class and intra-class information for self-optimization, which is the foundation of the FFM module and the PPCM module. Meanwhile, it relies on these representative class prototypes, which requires the assistance of the FFM module and the PPCM module.

Table 7. Ablation experiments of the three proposed components on MAR20 split1. DSD denotes the decoupled self-distillation method, FFM denotes the feature filtering module, and PPCM denotes the progressive prototype calibration module. The metric is nAP50 (%).

DSD	FFM	PPCM	3-Shot	5-Shot	10-Shot
			38.3	42.2	51.5
✓			52.8 (+14.5)	53.7 (+11.5)	59.6 (+8.1)
✓	✓		55.9 (+17.6)	58.2 (+16.0)	61.0 (+9.5)
✓		✓	53.5 (+15.2)	61.6 (+19.4)	62.5 (+11.0)
✓	✓	✓	58.8 (+20.5)	62.1 (+19.9)	64.1 (+12.6)

✓ indicates the use of this component.

Table 8. Ablation experiments of the three proposed components on SHIP20 split1. DSD denotes the decoupled self-distillation method, FFM denotes the feature filtering module, and PPCM denotes the progressive prototype calibration module. The metric is nAP50 (%).

DSD	FFM	PPCM	3-Shot	5-Shot	10-Shot
			26.6	35.3	41.0
✓			28.8 (+2.2)	37.6 (+2.3)	43.5 (+2.5)
✓	✓		29.5 (+2.9)	38.5 (+3.2)	44.2 (+3.2)
✓		✓	30.7 (+4.1)	39.1 (+3.8)	44.5 (+3.5)
✓	✓	✓	31.2 (+4.6)	39.5 (+4.2)	45.9 (+4.9)

✓ indicates the use of this component.

Compared to the baseline of the first row in Table 7, the DSD method can independently provide a steady gain of 8.1~14.5% on the MAR20 dataset. When the FFM module is solely combined with the DSD method, the improvements are raised to 9.5~17.6% in the third row. When the PPCM module is solely combined with the DSD method, the improvements are raised to 11.0~19.4% in the fourth row. When both are combined, the improvements are further raised to 12.6~20.5%, which indicates that the proposed three components are able to effectively complement each other and exert a promoting effect.

Compared to the baseline of the first row in Table 8, the DSD method can provide an improvement of 2.2~2.5% on the SHIP20 dataset. When the FFM module is solely combined with the DSD method, the improvements are promoted to 2.9~3.2%. When the PPCM module is solely combined with the DSD method, the improvements can be raised to 3.5~4.1%. When both are combined, the improvements are further promoted to 4.2~4.6%, which further proves the effectiveness of the three components.

Through a comprehensive comparison of Tables 7 and 8, we draw a preliminary conclusion that the three components are capable of advancing the baseline and achieve superior results on two fine-grained remote sensing datasets, separately and jointly. This improvement is even more pronounced on aircraft objects and on the low-shot settings.

4.6.2. Ablation Study of the Decoupled Self-Distillation Method

Self-distillation methods have different forms as illustrated in Figure 3. They are mainly divided into the decoupled method versus the un-decoupled method and the single-branch method versus the two-branch method. Given that the un-decoupled two-branch self-distillation method is not advocated because of its ineffectiveness, we have conducted ablation experiments on the other three forms. The proposed DSD method follows the form of the decoupled two-branch methods.

As shown in Table 9, we compare our proposed method with the other forms. In particular, the results of the second and third rows exhibit a significant drop compared to the baseline. These results indicate the decoupling operations are quite essential for FSOD. For a deeper analysis, the reason is related to two aspects: (a) According to [2], the gradient from RPN plays a negative role in optimizing the backbone, while the gradient from RCNN has a positive effect. Since the above two methods do not modulate the backpropagation gradients from the RPN and RCNN to varying degrees, the overfitting risk of the few-shot detector is significantly increased. (b) The proven consensus is that the classification-related features focus more on capturing local salient regions, while the localization-related features pay more attention to the global boundaries. When they are placed in the same feature space and delivered to the classifier and the regressor, it causes a decline in the few-shot detector's performance. Moreover, comparing the second row with the third row, it should be noted that the adverse impact becomes significantly more pronounced when the gradient decoupled layer is not used. Evidently, the gradient decoupling operation exerts a more substantial effect on the few-shot overfitting problem.

Table 9. Ablation experiments of different self-distillation methods on MAR20 split1. Baseline indicates using decoupled RoI features without self-distillation; Un-decoupled + Single-branch (w/o AL and GDL) indicates using un-decoupled RoI features without GDL and AL for the single-branch self-distillation; Un-decoupled + Single-branch (w/o GDL) indicates using un-decoupled RoI features with AL but without GDL for the single-branch self-distillation; Decoupled + Single-branch indicates using decoupled RoI features for the single-branch self-distillation; Decoupled + Two-branch indicates using decoupled RoI features for the two-branch self-distillation. Results in bold denote the best performance.

Method	3-Shot		5-Shot		10-Shot	
	nAP50	nAP75	nAP50	nAP75	nAP50	nAP75
Baseline	38.3	28.2	42.2	27.4	51.5	36.5
Un-decoupled + Single-branch (w/o AL and GDL)	13.9	10.9	21.6	17.3	23.2	19.0
Un-decoupled + Single-branch (w/o GDL)	17.4	14.8	22.1	17.8	25.3	20.0
Decoupled + Single-branch	50.9	42.2	52.6	36.9	58.9	47.0
Decoupled + Two-branch (Ours)	**52.8**	**43.8**	**53.7**	**43.0**	**59.6**	**48.0**

Moreover, the comparison between the first and fourth rows shows the effectiveness of the classification-related self-distillation branch. In the fourth row, the decoupled single-branch self-distillation method can solely bring a prominent enhancement on both nAP50 and nAP75 metrics. The difference between the methods in the fourth and fifth rows is whether the weighted self-distillation loss is calculated for the regression branch. Before this loss is introduced, the loss of the regression branch is calculated only on the ground-truth class. The comparison between the fourth and fifth rows shows that the similarity-driven weighted self-distillation loss for the regression branch can further obtain another increase of 0.7~1.9% on nAP50 and 1.0~6.1% on nAP75. The overall results demonstrate the superiority of the proposed DSD method, it is capable of distilling more instructive information for two-branch tasks with limited samples.

Within the DSD method, we also perform an ablation study on the similarity values for irrelevant base classes, as shown in Table 10. Three approaches of taking the values are considered for comparison. According to the experimental results in the first and second rows, we can assume that the original similarity-driven probabilities are qualified to produce a positive impact on the detector. Furthermore, by scaling down the similarity values for irrelevant base classes, we encourage the detector to concentrate less on the commonalities between base classes and more on exaggerating the disparity between high and low similarity values. The results in the third and fourth rows show that the DSD (scaled sim) and DSD (-1) approaches yield greater average improvements on MAR20 and SHIP20, and we adopt the DSD (-1) approach ultimately.

Table 10. Ablation experiments of the similarity value for irrelevant base classes on MAR20 split1 and SHIP20 split1. DSD (sim) denotes that the original cosine similarity values are used; DSD (scaled sim) denotes that the original cosine similarity values are scaled down by a factor of 10 for irrelevant base classes; DSD (-1) denotes that the original cosine similarity values are replaced by a constant of -1. The metric is nAP50 (%). Results in bold denote the best performance and the underlined ones represent the second best.

Value	MAR20				SHIP20			
	3-Shot	5-Shot	10-Shot	Mean	3-Shot	5-Shot	10-Shot	Mean
Baseline	38.3	42.2	51.5	44.0	26.6	35.3	41.0	34.3
DSD (sim)	47.3	48.2	55.9	50.5	27.5	36.1	43.2	35.6
DSD (scaled sim)	52.6	53.8	58.7	<u>55.0</u>	27.9	37.2	43.2	<u>36.1</u>
DSD (-1)	52.8	53.7	59.6	**55.4**	28.8	37.6	43.5	**36.6**

4.6.3. Ablation Study of the Feature Filtering Module

We have conducted the ablation studies, respectively, for 3-, 5-, 10-shot settings on MAR20 split1 and SHIP20 split1 to determine their values of the hyperparameter τ. As shown in Table 11, the bold results represent the best results. On this basis, it can be observed that the τ value of the low-shot settings should be set smaller than that of the high-shot settings. We set the τ value to 0.3, 0.3, 0.4 for the 3-, 5-, 10-shot settings on the two datasets, respectively. Based on the DSD method, this module can bring additional increases of 1.4~4.5% for MAR20 and 0.7~0.9% for SHIP20 under 3-, 5-, 10-shot settings.

Table 11. Ablation experiments of the hyperparameters τ in the feature filtering module on the two datasets. The metric is nAP50 (%). Results in bold denote the best performance.

τ	MAR20			SHIP20		
	3-Shot	5-Shot	10-Shot	3-Shot	5-Shot	10-Shot
0.2	54.7	55.0	60.9	29.1	38.4	43.6
0.3	**55.9**	**58.2**	59.1	**29.5**	**38.5**	44.0
0.4	52.0	57.5	**61.0**	28.0	38.1	**44.2**
0.5	51.7	56.2	60.7	27.0	37.9	43.3

Furthermore, we have sampled every 40 iterations during the training process of MAR20 split1 and SHIP20 split1 to record the proportion of filtered RoI features under 3-shot setting, as a representative case. As shown in Figure 13, it can be observed, combined with Table 11, that when τ is set to a smaller value ($\tau \leq 0.3$), a small proportion of low-quality RoI samples are filtered out, enhancing the overall quality of the candidate RoI samples in the memory bank. However, when τ is set to a larger value ($\tau \geq 0.4$), a significantly higher proportion of RoI feature samples are discarded, which can be particularly detrimental to the training of the detector under extremely few-shot conditions.

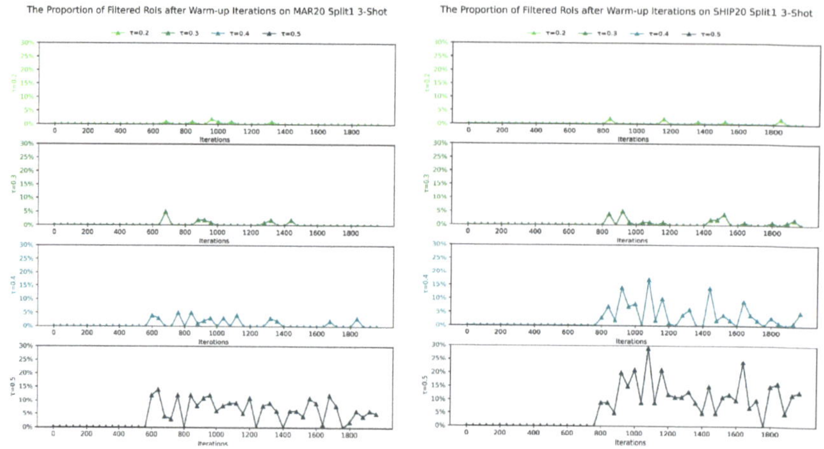

(a) (b)

Figure 13. Visualization of the proportion of the filtered RoI features after the warm-up iterations on MAR20 split1 and SHIP20 split1 under 3-shot setting. (**a**) MAR20. (**b**) SHIP20.

With respect to the class distribution within the memory bank, as mentioned previously, the memory bank accumulates an adequate number of RoI feature samples for each category during the early iterations. Taking the 10-shot setting as an example, the numbers of the stored RoI feature samples for different categories are shown in Figure 14 by sampling every 40 iterations. Once this number for a certain category reaches its maximum length, it triggers the independent dequeuing for this category. The final manifestation is that each category holds the same number of samples. This process is equivalent to establishing a longer sliding window across training iterations, and it allows the use of the samples accumulated over multiple iterations for the class prototype representation. Therefore, the class distribution of the memory bank is relatively stable during training.

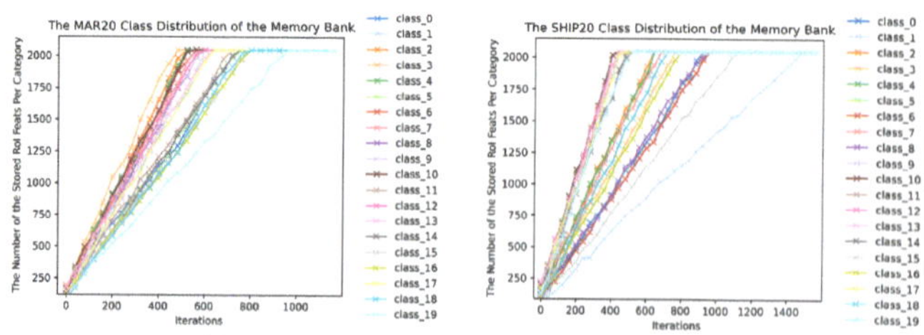

(a) (b)

Figure 14. Visualization of the class distribution in the memory bank at the early training iterations. (**a**) MAR20. (**b**) SHIP20.

4.6.4. Ablation Study of the Progressive Prototype Calibration Module

Inside the PPCM, there are three hyperparameters α, k and λ, where α represents the calibration strength for shifted base prototypes in the first-step calibration, k represents the selected number of the nearest calibrated base prototypes used for the second-step calibration and λ is to adjust the exponential decay rate of distance-driven weights in the second-step calibration.

To explore the impact of different values of the parameter α, we incrementally take its value to strengthen the anchoring effect of base distributions towards base prototypes and observe the model's performance. Meanwhile, we keep the second-step calibration fixed. The ablation results are shown in Table 12, where $\alpha = 0.0$ in the first row indicates that the first-step calibration is not enabled. Based on this foundation, our experiments reveal that within a moderate range of the calibration strength, the first-step calibration exerts a beneficial effect, for example, the best performances are achieved with $\alpha = 0.1$ for two datasets. Moreover, one intriguing observation is that certain immoderate calibrations on base class prototypes can overemphasize the anchoring impact of the base distributions, for instance, it yields worse performances than the first row when $\alpha = 0.3$ for MAR20 or $\alpha = 0.2$ for SHIP20, possibly due to insufficient attention to the novel feature distributions developed in the fine-tuning stage.

Table 12. Ablation experiments of the hyperparameter α in the progressive prototype calibration module on MAR20 split1 and SHIP20 split1. The metric is nAP50 (%). Results in bold denote the best performance.

α	MAR20			SHIP20		
	3-Shot	5-Shot	10-Shot	3-Shot	5-Shot	10-Shot
0.0	57.6	61.2	62.6	30.9	39.1	44.6
0.1	**58.8**	**62.1**	**64.1**	**31.2**	**39.5**	**45.9**
0.2	58.5	62.0	64.0	30.6	37.7	42.7
0.3	56.7	58.5	61.4	-	-	-

It is important to note that, with a certain strength of the first-step calibration, the distances between base prototypes and their corresponding base distributions are adjusted to a specific numerical range. When base prototypes are under-calibrated and over-calibrated, this anchored distance range may become inappropriate for the specific dataset. However, by conducting a sweep experiment to determine the optimal value of parameter α, an appropriate anchored distance range can be defined, ensuring the effectiveness of the first-step calibration.

Due to the mutual influence between the two parameters λ and k, we conduct joint ablation experiments for them on MAR20 split1 and SHIP20 split1, maintaining the calibration strength of the first step unchanged. On the basis of the best results shown in Tables 13 and 14, the parameter λ is uniformly set to 0.4 under all settings for two datasets. As for the parameter k, we set $k = 2$ under 3-shot setting and $k = 1$ under 10-shot setting for the two datasets. However, there is a slight difference under 5-shot setting, we severally adopt $k = 1$ for MAR20 and $k = 2$ for SHIP20 in the end.

To further confirm the promoting effect of the first step (base prototype calibration) on the second step (novel prototype calibration), we also have conducted the following experiments. We maintain the base prototypes completely calculated from the memory bank without the calibration of the base distribution, and then compare the detector's performance on MAR20 split1, shown in Table 15. The results in the second and the third rows prove that the first-step calibration is able to exclusively bring an enhancement of 1.0~1.7%, and the second-step calibration can result in an increase of 1.6~2.7% under 3-,

5-, 10-shot settings. Nevertheless, compared with the fourth row, neither calibrating only the base prototypes nor calibrating novel prototypes with uncalibrated base prototypes is enough to bring the maximum improvements. But after accepting the compensation for the base distribution in the first step then calibrating in the second step, remarkable gains can be achieved, which illustrates the complementarity of the two-step calibration in enhancing the few-shot detector's performance.

Table 13. Ablation experiments of the hyperparameters λ and k in the progressive prototype calibration module on MAR20 split1. The metric is nAP50 (%). Results in bold denote the best performance.

Hyperparameter	3-Shot			5-Shot			10-Shot		
	λ = 0.2	λ = 0.4	λ = 0.6	λ = 0.2	λ = 0.4	λ = 0.6	λ = 0.2	λ = 0.4	λ = 0.6
$k = 1$	55.1	56.5	51.7	61.5	**62.1**	61.7	62.7	**64.1**	62.8
$k = 2$	54.0	**58.8**	57.5	60.4	61.1	61.0	62.4	63.6	62.2
$k = 3$	51.8	53.0	58.0	61.6	61.9	61.4	62.4	63.2	62.3
$k = 4$	55.8	57.6	52.0	61.4	61.7	60.8	62.2	63.1	61.5

Table 14. Ablation experiments of the hyperparameters λ and k in the progressive prototype calibration module on SHIP20 split1. The metric is nAP50 (%). Results in bold denote the best performance.

Hyperparameter	3-Shot			5-Shot			10-Shot		
	λ = 0.2	λ = 0.4	λ = 0.6	λ = 0.2	λ = 0.4	λ = 0.6	λ = 0.2	λ = 0.4	λ = 0.6
$k = 1$	29.7	30.9	30.1	38.4	39.2	36.6	44.0	**45.9**	44.4
$k = 2$	30.1	**31.2**	30.6	38.6	**39.5**	39.0	44.3	45.1	44.1
$k = 3$	29.2	30.2	30.0	37.9	39.2	39.0	43.6	44.3	43.7
$k = 4$	30.4	30.6	30.5	37.7	38.0	37.2	44.3	43.8	43.8

Table 15. The effect of the first step and the second step of the PPCM module on MAR20 split1. The metric is nAP50 (%).

PPCM		3-Shot	5-Shot	10-Shot
First	Second			
		55.9	58.2	61.0
✓		57.6 (+1.7)	59.7 (+1.5)	62.0 (+1.0)
	✓	57.6 (+1.7)	61.2 (+2.7)	62.6 (+1.6)
✓	✓	58.8 (+2.9)	62.1 (+3.9)	64.1 (+3.1)

✓ indicates the execution of this calibration step.

Furthermore, taking the second row as a reference, we revisit the results presented in Table 13 to analyze the effect of the second-step calibration with various parameter configurations It can be seen that the majority of the calibration results under 5, 10-shot settings surpass the reference result; however, the calibration results under 3-shot setting are more dependent on the selection of the optimal hyperparameters.

Figure 15 also provides a more intuitive comparison of the distance between novel prototypes and their empirical distribution centers with and without PPCM. The novel distribution centers are obtained by introducing all novel category samples. For the classification-related feature space, our PPCM module significantly reduces this distance gap. For the localization-related feature space, the upper limit of this distance is much smaller than that in the classification feature space, and our proposed module enables this distance to decrease or increase in a smaller range for the five novel classes.

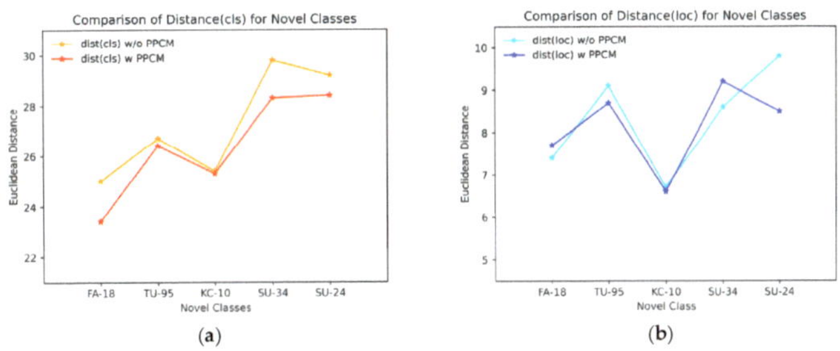

Figure 15. Comparison of the distance between novel prototypes and their empirical distribution centers with and without PPCM on MAR20 split1 under 3-shot setting: (**a**) in the classification-related feature space; (**b**) in the localization-related feature space.

4.6.5. Ablation Study of the Warm-Up Iterations

To study the effect of the warm-up iterations under low-shot and high-shot settings, we have conducted the ablation experiments based on the aforementioned method, as shown in Figure 16. On this basis, we determine to use 600 and 800 warm-up iterations for MAR20 split1 and SHIP20 split1, respectively, under different shot settings. Despite equivalent results for 800 and 1000 warm-up iterations in Figure 16d, we consider that extending the warm-up stage without yielding additional gains is unwarranted. Further analysis suggests that a lower number of warm-up iterations may prevent the model from adequately learning the feature distributions and accumulating candidate samples. As the number of warm-up iterations grows much larger, the intervention of the FFM module and the PPCM module is delayed. Meanwhile, low-quality candidate features are over-incorporated into the memory bank, and the misalignment of feature distributions potentially becomes more pronounced.

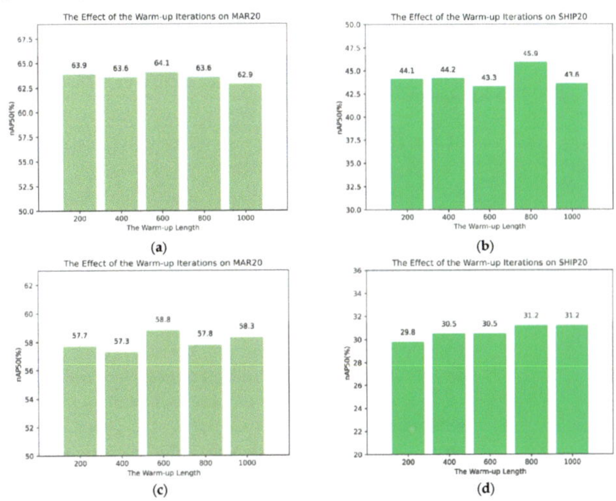

Figure 16. Visualization of the effect of the warm-up iterations. (**a**) MAR20 10-shot. (**b**) SHIP20 10-shot. (**c**) MAR20 3-shot. (**d**) SHIP20 3-shot.

5. Discussion

This paper proposes a novel few-shot object detection method that introduces a decoupled self-distillation method integrated with two well-designed modules. Our proposed method has managed to mitigate the misclassification problem according to the figures of the confusion matrix in Figures 9 and 12 and has demonstrated its superiority according to the comparison results with other methods in Tables 4 and 5. Through further analyses, our method is able to correct multiple error types according to Figure 10 to varying degrees. However, a certain degree of classification confusion still remains within novel classes and the inadaptability to large differences in object scales is synchronously revealed on SHIP20 when comparing the results on the two datasets. These above issues should be further studied. As for the model complexity and efficiency, which we have compromised to ensure a higher level of prediction accuracy, it remains an aspect for potential optimization.

For verification, we conduct a series of ablation experiments on MAR20 and SHIP20 datasets, Tables 7 and 8 give evidence of the effectiveness of each component and the impressive improvements by uniting them. Utilizing the DSD method can produce improvements of 8.1~14.5% and 2.2~2.5% under 3-, 5-, 10-shot settings on MAR20 and SHIP20, respectively. Further, compared with other self-distillation methods in different forms, our DSD method still takes the lead as shown in Table 9. This progress should be attributed to its powerful ability to mine inter-class correlations for knowledge sharing in two decoupled feature spaces. After integrating the FFM module into the DSD method, the minimum improvements can rise up to 9.5% and 2.9% on MAR20 and SHIP20, because more distinctive features are selected for the prototype construction. Eventually, the PPCM module refines these class prototypes by two steps to lift the minimum improvements to 12.6% and 4.2%. The two-step calibration can ensure a greater enhancement by means of the anchoring effect of prior base distributions on base prototypes and the calibration effect of correlated base prototypes on novel prototypes. Despite these improvements, the still high proportion of Classification Error also catches our attention. We also subjectively associate this problem with the novel confusion. Therefore, we attempt to further explore a more advanced correlational modeling network as well as the prototype differentiation approaches. In addition, another aspect worthy of further exploration is to enhance the PPCM module's adaptability to ensure consistency in the hyperparameter selection, especially in extremely few-shot conditions.

6. Conclusions

This study has focused on explicitly leveraging the transferable common knowledge between base classes and novel classes and calibrating the biased distribution in data-scarcity conditions. We propose a fine-grained few-shot object detection method for RSIs which mainly includes three key components: the decoupled self-distillation (DSD) method, the feature filtering module (FFM) and the progressive prototype calibration module (PPCM). We first introduce the DSD method for both the classification-related and the localization-related RoI features to distill class prototypes with transferable commonalities, which are further utilized to explicitly model inter-class correlations and yield similarity probabilities as soft labels or aggregation weights. Simultaneously, the proposed FFM module is integrated for the purpose of selecting high-quality feature vectors as candidates for the memory bank and then constructing robust class prototypes. Moreover, a two-step PPCM module is placed after the memory bank in order to compensate for the distribution shift of base classes and the distribution misalignment for novel classes in two decoupled feature spaces. Extensive experiments and comparisons on two fine-grained remote sensing datasets, MAR20 and SHIP20, show that the proposed method outperforms current advanced FSOD methods by achieving average improvements of 10.7%, 11.3%,

8.5% over the baseline on MAR20 splits and 4.4%, 4.1%, 3.7% on SHIP20 splits under 3-, 5-, 10-shot settings. Also, all three components work effectively and cooperate with each other.

However, combining the comprehensive evaluation and the in-depth analyses of ablation studies, our method still has potential for further enhancement. For instance, our future work will extend our method from several aspects: (1) improving the adaptability to large scale differences; (2) introducing prototype differentiation approaches; (3) enhancing the consistency in the hyperparameter selection of the PPCM module.

Author Contributions: Conceptualization, H.G.; methodology, H.G.; software, H.G. and Y.L.; validation, H.G.; formal analysis, H.G. and Y.L.; investigation, H.G.; resources, H.G., Y.L. and Z.P.; data curation, H.G. and Z.P.; writing—original draft preparation, H.G.; writing—review and editing, H.G., Y.L. and Z.P.; visualization, H.G.; supervision, Y.H.; project administration, Z.P. and Y.H.; funding acquisition, Z.P. All authors have read and agreed to the published version of the manuscript.

Funding: This work was supported by the Youth Innovation Promotion Association, Chinese Academy of Sciences (CAS) under grant number 2022119.

Data Availability Statement: The public dataset MAR20 is openly available at MAR20, reference number [33]. The customized dataset SHIP20 is derived from following public domain resources available at FAIR1M, HRSC2016, and xView, reference number [34–36].

Conflicts of Interest: The authors declare no conflicts of interest.

References

1. Bolya, D.; Foley, S.; Hays, J.; Hoffman, J. Tide: A general toolbox for identifying object detection errors. In Proceedings of the Computer Vision–ECCV 2020: 16th European Conference, Glasgow, UK, 23–28 August 2020; pp. 558–573.
2. Qiao, L.; Zhao, Y.; Li, Z.; Qiu, X.; Wu, J.; Zhang, C. Defrcn: Decoupled faster r-cnn for few-shot object detection. In Proceedings of the IEEE/CVF International Conference on Computer Vision, Montreal, QC, Canada, 10–17 October 2021; pp. 8681–8690.
3. Wang, X.; Huang, T.E.; Darrell, T.; Gonzalez, J.E.; Yu, F. Frustratingly simple few-shot object detection. In Proceedings of the 37th International Conference on Machine Learning, Virtual, 13–18 July 2020; pp. 9919–9928.
4. Khandelwal, S.; Goyal, R.; Sigal, L. Unit: Unified knowledge transfer for any-shot object detection and segmentation. In Proceedings of the IEEE/CVF Conference on Computer Vision and Pattern Recognition, Nashville, TN, USA, 20–25 June 2021; pp. 5951–5961.
5. Zhu, C.; Chen, F.; Ahmed, U.; Shen, Z.; Savvides, M. Semantic relation reasoning for shot-stable few-shot object detection. In Proceedings of the IEEE/CVF Conference on Computer Vision and Pattern Recognition, Nashville, TN, USA, 20–25 June 2021; pp. 8782–8791.
6. Luan, Y.; Zhao, H.; Yang, Z.; Dai, Y. MSD: Multi-self-distillation learning via multi-classifiers within deep neural networks. *arXiv* **2019**, arXiv:1911.09418. [CrossRef]
7. Wu, S.; Pei, W.; Mei, D.; Chen, F.; Tian, J.; Lu, G. Multi-faceted distillation of base-novel commonality for few-shot object detection. In Proceedings of the European Conference on Computer Vision, Tel Aviv, Israel, 23–27 October 2022; pp. 578–594.
8. Xiong, W. CD-FSOD: A benchmark for cross-domain few-shot object detection. In Proceedings of the ICASSP 2023-2023 IEEE International Conference on Acoustics, Speech and Signal Processing (ICASSP), Rhodes Island, Greece, 4–10 June 2023; pp. 1–5.
9. Ross, T.-Y.; Dollár, G. Focal loss for dense object detection. In Proceedings of the IEEE Conference on Computer Vision and Pattern Recognition, Honolulu, HI, USA, 21–26 July 2017; pp. 2980–2988.
10. Liu, W.; Anguelov, D.; Erhan, D.; Szegedy, C.; Reed, S.; Fu, C.-Y.; Berg, A.C. Ssd: Single shot multibox detector. In Proceedings of the Computer Vision–ECCV 2016: 14th European Conference, Amsterdam, The Netherlands, 11–14 October 2016; pp. 21–37.
11. Farhadi, A.; Redmon, J. Yolov3: An incremental improvement. In Proceedings of the Computer Vision and Pattern Recognition, Salt Lake City, UT, USA, 18–22 June 2018; pp. 1–6.
12. Ren, S.; He, K.; Girshick, R.; Sun, J. Faster R-CNN: Towards real-time object detection with region proposal networks. *IEEE Trans. Pattern Anal. Mach. Intell.* **2016**, *39*, 1137–1149. [CrossRef] [PubMed]
13. Law, H.; Deng, J. Cornernet: Detecting objects as paired keypoints. In Proceedings of the European conference on computer vision (ECCV), Munich, Germany, 8–14 September 2018; pp. 734–750.
14. Duan, K.; Bai, S.; Xie, L.; Qi, H.; Huang, Q.; Tian, Q. Centernet: Keypoint triplets for object detection. In Proceedings of the IEEE/CVF International Conference on Computer Vision, Seoul, Republic of Korea, 27 October–2 November 2019; pp. 6569–6578.
15. Tian, Z.; Shen, C.; Chen, H.; He, T. FCOS: Fully convolutional one-stage object detection. *arXiv* **2019**, arXiv:1904.01355. [CrossRef]

16. Carion, N.; Massa, F.; Synnaeve, G.; Usunier, N.; Kirillov, A.; Zagoruyko, S. End-to-end object detection with transformers. In Proceedings of the European Conference on Computer Vision, Glasgow, UK, 23–28 August 2020; pp. 213–229.
17. Zhu, X.; Su, W.; Lu, L.; Li, B.; Wang, X.; Dai, J. Deformable detr: Deformable transformers for end-to-end object detection. *arXiv* **2020**, arXiv:2010.04159. [CrossRef]
18. Yan, X.; Chen, Z.; Xu, A.; Wang, X.; Liang, X.; Lin, L. Meta r-cnn: Towards general solver for instance-level low-shot learning. In Proceedings of the IEEE/CVF International Conference on Computer Vision, Seoul, Republic of Korea, 27 October–2 November 2019; pp. 9577–9586.
19. Fan, Q.; Zhuo, W.; Tang, C.-K.; Tai, Y.-W. Few-shot object detection with attention-RPN and multi-relation detector. In Proceedings of the IEEE/CVF Conference on Computer Vision and Pattern Recognition, Seattle, WA, USA, 13–19 June 2020; pp. 4013–4022.
20. Zhang, G.; Luo, Z.; Cui, K.; Lu, S.; Xing, E.P. Meta-detr: Image-level few-shot detection with inter-class correlation exploitation. *IEEE Trans. Pattern Anal. Mach. Intell.* **2022**, *45*, 12832–12843. [CrossRef]
21. Demirel, B.; Baran, O.B.; Cinbis, R.G. Meta-tuning loss functions and data augmentation for few-shot object detection. In Proceedings of the IEEE/CVF Conference on Computer Vision and Pattern Recognition, Vancouver, BC, Canada, 17–24 June 2023; pp. 7339–7349.
22. Xin, Z.; Chen, S.; Wu, T.; Shao, Y.; Ding, W.; You, X. Few-shot object detection: Research advances and challenges. *Information Fusion* **2024**, *107*, 102307. [CrossRef]
23. Sun, B.; Li, B.; Cai, S.; Yuan, Y.; Zhang, C. Fsce: Few-shot object detection via contrastive proposal encoding. In Proceedings of the IEEE/CVF Conference on Computer Vision and Pattern Recognition, Nashville, TN, USA, 20–25 June 2021; pp. 7352–7362.
24. Wu, J.; Liu, S.; Huang, D.; Wang, Y. Multi-scale positive sample refinement for few-shot object detection. In Proceedings of the Computer Vision–ECCV 2020: 16th European Conference, Glasgow, UK, 23–28 August 2020; pp. 456–472.
25. Wang, Y.; Xu, C.; Liu, C.; Li, Z. Context information refinement for few-shot object detection in remote sensing images. *Remote Sens.* **2022**, *14*, 3255. [CrossRef]
26. Xiao, Z.; Zhong, P.; Quan, Y.; Yin, X.; Xue, W. Few-shot object detection with feature attention highlight module in remote sensing images. In Proceedings of the 2020 International Conference on Image, Video Processing and Artificial Intelligence, Shanghai, China, 21–23 August 2020; pp. 217–223.
27. Cheng, G.; Yan, B.; Shi, P.; Li, K.; Yao, X.; Guo, L.; Han, J. Prototype-CNN for few-shot object detection in remote sensing images. *IEEE Trans. Geosci. Remote Sens.* **2021**, *60*, 5604610. [CrossRef]
28. Liu, N.; Xu, X.; Celik, T.; Gan, Z.; Li, H.-C. Transformation-invariant network for few-shot object detection in remote-sensing images. *IEEE Trans. Geosci. Remote Sens.* **2023**, *61*, 5625314. [CrossRef]
29. Zhang, T.; Zhang, X.; Zhu, P.; Jia, X.; Tang, X.; Jiao, L. Generalized few-shot object detection in remote sensing images. *ISPRS J. Photogramm. Remote Sens.* **2023**, *195*, 353–364. [CrossRef]
30. Zhang, J.; Hong, Z.; Chen, X.; Li, Y. Few-Shot Object Detection for Remote Sensing Imagery Using Segmentation Assistance and Triplet Head. *Remote Sens.* **2024**, *16*, 3630. [CrossRef]
31. Wang, Q.-W.; Zhou, D.-W.; Zhang, Y.-K.; Zhan, D.-C.; Ye, H.-J. Few-shot class-incremental learning via training-free prototype calibration. In Proceedings of the 37th International Conference on Neural Information Processing Systems, New Orleans, LA, USA, 10–16 December 2023; pp. 15060–15076.
32. Yang, S.; Liu, L.; Xu, M. Free lunch for few-shot learning: Distribution calibration. *arXiv* **2021**, arXiv:2101.06395. [CrossRef]
33. Wenqi, Y.; Gong, C.; Meijun, W.; Yanqing, Y.; Xingxing, X.; Xiwen, Y.; Junwei, H. MAR20: A benchmark for military aircraft recognition in remote sensing images. *Natl. Remote Sens. Bull.* **2024**, *27*, 2688–2696. [CrossRef]
34. Sun, X.; Wang, P.; Yan, Z.; Xu, F.; Wang, R.; Diao, W.; Chen, J.; Li, J.; Feng, Y.; Xu, T. FAIR1M: A benchmark dataset for fine-grained object recognition in high-resolution remote sensing imagery. *ISPRS J. Photogramm. Remote Sens.* **2022**, *184*, 116–130. [CrossRef]
35. Liu, Z.; Yuan, L.; Weng, L.; Yang, Y. A high resolution optical satellite image dataset for ship recognition and some new baselines. In Proceedings of the International Conference on Pattern Recognition Applications and Methods, Porto, Portugal, 24–26 February 2017; pp. 324–331.
36. Lam, D.; Kuzma, R.; McGee, K.; Dooley, S.; Laielli, M.; Klaric, M.; Bulatov, Y.; McCord, B. xView: Objects in context in overhead imagery. *arXiv* **2018**, arXiv:1802.07856. [CrossRef]
37. Zhang, Y.; Zhang, B.; Wang, B. Few-shot object detection with self-adaptive global similarity and two-way foreground stimulator in remote sensing images. *IEEE J. Sel. Top. Appl. Earth Obs. Remote Sens.* **2022**, *15*, 7263–7276. [CrossRef]
38. MMFewShot Contributors. OpenMMLab Few Shot Learning Toolbox and Benchmark. 2021. Available online: https://github.com/open-mmlab/mmfewshot (accessed on 30 January 2025).
39. Han, J.; Ren, Y.; Ding, J.; Yan, K.; Xia, G.-S. Few-shot object detection via variational feature aggregation. In Proceedings of the AAAI Conference on Artificial Intelligence, Washington, DC, USA, 7–14 February 2023; pp. 755–763.

40. Wang, Z.; Yang, B.; Yue, H.; Ma, Z. Fine-Grained Prototypes Distillation for Few-Shot Object Detection. In Proceedings of the AAAI Conference on Artificial Intelligence, Vancouver, BC, Canada, 26–27 February 2024; pp. 5859–5866.
41. Xin, Z.; Wu, T.; Chen, S.; Zou, Y.; Shao, L.; You, X. ECEA: Extensible Co-Existing Attention for Few-Shot Object Detection. *arXiv* **2023**, arXiv:2309.08196. [CrossRef] [PubMed]

remote sensing

MDPI

Article

Automatic Detection of War-Destroyed Buildings from High-Resolution Remote Sensing Images

Yu Wang [1,2], Yue Li [1,2] and Shufeng Zhang [1,2,*]

1 Science and Technology on Integrated Logistic Support Laboratory, National University of Defense Technology, Changsha 410073, China
2 College of Intelligence Science and Technology, National University of Defense Technology, Changsha 410073, China
* Correspondence: sfzhang@nudt.edu.cn

Abstract: Modern high-intensity armed conflicts often lead to extensive damage to urban infrastructure. The use of high-resolution remote sensing images can clearly detect damage to individual buildings which is of great significance for monitoring war crimes and damage assessments that destroy civilian infrastructure indiscriminately. In this paper, we propose SOCA-YOLO (Sampling Optimization and Coordinate Attention–YOLO), an automatic detection method for destroyed buildings in high-resolution remote sensing images based on deep learning techniques. First, based on YOLOv8, Haar wavelet transform and convolutional blocks are used to downsample shallow feature maps to make full use of spatial details in high-resolution remote sensing images. Second, the coordinate attention mechanism is integrated with C2f so that the network can use the spatial information to enhance the feature representation earlier. Finally, in the feature fusion stage, a lightweight dynamic upsampling strategy is used to improve the difference in the spatial boundaries of feature maps. In addition, this paper obtained high-resolution remote sensing images of urban battlefields through Google Earth, constructed a dataset for the detection of objects on buildings, and conducted training and verification. The experimental results show that the proposed method can effectively improve the detection accuracy of destroyed buildings, and the method is used to map destroyed buildings in cities such as Mariupol and Volnovaja where violent armed conflicts have occurred. The results show that deep learning-based object detection technology has the advantage of fast and accurate detection of destroyed buildings caused by armed conflict, which can provide preliminary reference information for monitoring war crimes and assessing war losses.

Keywords: armed conflict; destroyed buildings; high resolution remote sensing images; object detection; convolutional neural network

Academic Editor: Jon Atli Benediktsson

Received: 28 November 2024
Revised: 21 January 2025
Accepted: 29 January 2025
Published: 31 January 2025

Citation: Wang, Y.; Li, Y.; Zhang, S. Automatic Detection of War-Destroyed Buildings from High-Resolution Remote Sensing Images. *Remote Sens.* **2025**, *17*, 509. https://doi.org/10.3390/rs17030509

1. Introduction

In contemporary armed conflicts, the extensive utilization of thermal weapons not only poses a grave threat to live targets but also presents a significant peril to the natural landscape and human environment [1]. Particularly, the devastation of urban infrastructure incurs substantial costs in terms of both material resources and humanitarian consequences such as triggering a "refugee wave" [2]. Consequently, the indiscriminate destruction of civilian infrastructure including residential homes, commercial buildings, and cultural landscapes is deemed both inhumane and constitutes a war crime.

Since the beginning of the 21st century, the international community has been in turmoil due to various conflicts such as the Iraq War, Syrian War, and civil war in Myanmar, which have inflicted significant damage on mankind's living environment and profoundly impacted global peace and development [3–5]. The ongoing Russian–Ukrainian conflict and the new round of Israeli–Palestinian conflict have also resulted in extensive destruction of non-military facilities like urban infrastructure and civilian buildings. Concerningly, the blockade of war zones and deteriorating security conditions pose challenges for United Nations peacekeeping efforts [6]. Restricted access for relevant personnel hampers field mapping activities while blocked information impedes the international community's supervision and mediation guidance within these areas. Therefore, there is an urgent need to conduct rapid and accurate mapping of destroyed buildings in war zones to facilitate timely intervention and protection of humanitarian rights [7].

The development of earth observation technology provides people with an observation perspective that is not restricted by the region so that the destroyed buildings can be observed more objectively and comprehensively in armed conflicts. However, the resolution of early remote sensing images is not high, and the application of destroyed building detection mainly focuses on the disaster assessment after the occurrence of natural disasters such as earthquakes and mudslides [8]. Currently, Interferometric Synthetic Aperture Radar (InSAR) technology and regional spectral change measurement have been used to perform rough regional damage perception [9]. Janalipour et al. [10] proposed an automatic building damage detection framework based on the LiDAR data after a disaster, which combines texture features with average digital projectors and can greatly improve the detection accuracy under the conditions of effective texture feature extraction. Huang et al. [11] combined synthetic aperture radar images with different simultaneous phases for coherence calculation, combined with open-source building vectors for classification extraction of destroyed buildings, and obtained regional detection results consistent with events. Based on optical remote sensing images, Ghandourj et al. [12] proposed a method to estimate building damage by using shadow features and gray co-occurrence matrix features and conducted building damage assessments in areas affected by the Syrian war near Damascus. In addition, some scholars have studied the collaborative detection of collapsed buildings by pre-disaster and post-disaster data from the perspective of multi-source data [13,14]. However, due to the diversity of data and the limitation of low resolution, there was no unified method for the detection of destroyed buildings during this period. Most of the methods were highly complex, and the classification features depended on artificial design, so the damage degree of buildings could only be roughly estimated.

Nowadays, countries all over the world have developed and launched high-resolution military, civilian, and commercial remote sensing satellites, such as Worldview, SPOT, and GF-2. The resolution of remote sensing images has been improved to the submeter level, enabling high-resolution and fine-grained imaging of individual buildings. Data resources are no longer the bottleneck restricting the acquisition of information in war zones. Therefore, how to quickly and accurately detect destroyed building individuals from wide-area remote sensing images has become the focus of attention [15].

Fortunately, advances in computer vision and artificial intelligence technology have made it possible to quickly detect destroyed buildings from a massive and large range of high-resolution remote sensing images. Many high-performance methods have been used or are being used in this field for research and application, the most typical cases are convolutional neural networks (CNNs) (e.g., Faster-RCNN [16], SSD [17], YOLOs [18–26]) and the detection with Transformers (e.g., ViT-FRCNN [27], RT-DET [28]). Ji et al. [29] took the lead in using a VGG [30] network to detect collapsed buildings after earthquakes in remote sensing images, and after fine-tuning, the detection accuracy of the network was effectively improved. Aiming at the classification of collapsed buildings, Wu et al. [31] used the improved U-Net [32] network to segment collapsed buildings at the pixel level, and the classification accuracy reached 0.792. Shi et al. [33] propose an improved YOLOv4 [20] algorithm for detecting collapsed buildings in aerial images after earthquakes. However, the actual post-disaster scene is complex and diverse, the collapsed buildings and the background are easily confused, there are still some difficulties in extracting robust features, and the detection accuracy is still not high. To solve these problems, Bai et al. [34] proposed a pyramid-pool modular semi-twin network for detecting destroyed buildings and improved the detection accuracy by adding residual blocks with expansion convolution and extrusion excitation blocks into the network. Ma et al. [35] took ShuffleNet v2 [36] as the backbone network of YOLOv3 [19] and introduced a generalized intersection over Union (GIoU) [37] loss to improve the detection accuracy of the model under a complex background. Overall, the object detection method based on deep learning has been proven to be feasible in detecting destroyed buildings in satellite images. It is worth noting that the YOLO series algorithms have been widely adopted in recent years due to their advantages of both accuracy and efficiency. In addition, attention mechanisms [38–40], feature pyramids [41–43], sampling methods [44,45], and other methods are often used to improve the detection performance of networks in specific scenarios.

However, most of the current research focuses on the detection of destroyed buildings caused by natural disasters such as earthquakes and mudslides, and there are no studies on the detection of destroyed buildings in the context of armed conflict. Figure 1 shows the difference in collapse between buildings destroyed by earthquake and war. Earthquake usually leads to overall structural damage to buildings. In particular, a seismic wave will cause the building to lose the support of the facade and collapse as a whole, as shown in Figure 1a. In armed conflict, the damage caused by air strikes, shelling, and rockets to buildings is often localized and point-like. Large buildings often have partially or completely collapsed roofs and relatively intact facades, as shown in Figure 1b, while small buildings show irregular ruins, as shown in Figure 1c.

(a) (b) (c)

Figure 1. Buildings destroyed by earthquakes or war. The red boxes represent the buildings destroyed by the earthquake, and the green boxes represent the war-destroyed buildings. (**a**) Buildings destroyed by the earthquakes (source: [35]); (**b**) large buildings destroyed by war (source: Google Earth); (**c**) small buildings destroyed by war (source: Google Earth).

Since the detection of destroyed buildings in high-resolution remote sensing images is highly dependent on the robustness of spatial detail feature extraction, we have made targeted improvements to CNNs and proposed an effective detection method. The main contributions of this work are as follows: (1) obtaining high-resolution Google images of the Russian–Ukrainian conflict area, and making the first dataset for the detection of destroyed buildings by visual interpretation and expert annotation; (2) in view of the task's dependence on spatial details, the Haar wavelet downsampling module (HWD), lightweight dynamic upsampling module (LDU), and coordinate attention mechanism (CA) were used to improve the convolutional neural network, effectively improving the detection accuracy of destroyed buildings; and (3) a comprehensive assessment and mapping of building damage in typical cities during the Russian–Ukrainian conflict is presented, and the advantages and limitations of our method for monitoring destroyed buildings in high-resolution optical remote sensing images are discussed.

2. Dataset

2.1. Remote Sensing Data Acquisition

In order to verify the effectiveness of the proposed method, we chose Mariupol, a city severely damaged by the Russian–Ukrainian conflict, and used its Google Earth images as the source of the production dataset. From February 2022 to May 2022, both sides fought with a large number of air missiles and ground artillery in the course of the armed conflict. In addition to military installations, residential and commercial buildings in the city were severely damaged. According to a report by UN High Commissioner for Human Rights Michelle Bachelet [46], the 82-day armed conflict resulted in the destruction of 90% of apartment buildings and 60% of private homes in Mariupol, with many neighborhoods razed to the ground.

Therefore, in this study, we obtained images with a resolution of about 0.3 m from Google Earth provided by Maxar, which covers the main urban area of Mariupol and contains a large number of destroyed buildings, as shown in Figure 2.

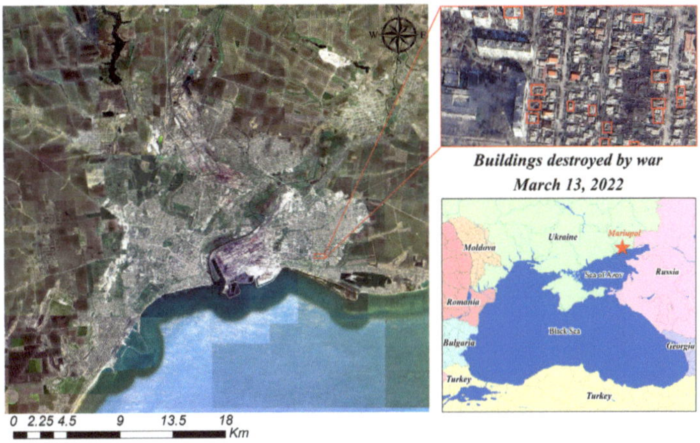

Figure 2. Collection area of the dataset, the red boxes in the upper right sub-image indicate the labels of the destroyed building. (Source: Google Earth.)

2.2. Dataset Production

Since the size of the original image is too large to be directly input into the CNN, the image needs to be pre-sliced. Meanwhile, in order to ensure the integrity of each individual building in the sample image, the image was cut into 640 × 640-pixel image slices according to the overlap degree of 25%, and the destroyed building was marked by Labelimg Tools [47]. Figure 3 shows the pre-slicing process in the production of the dataset and the labeling information of the destroyed buildings.

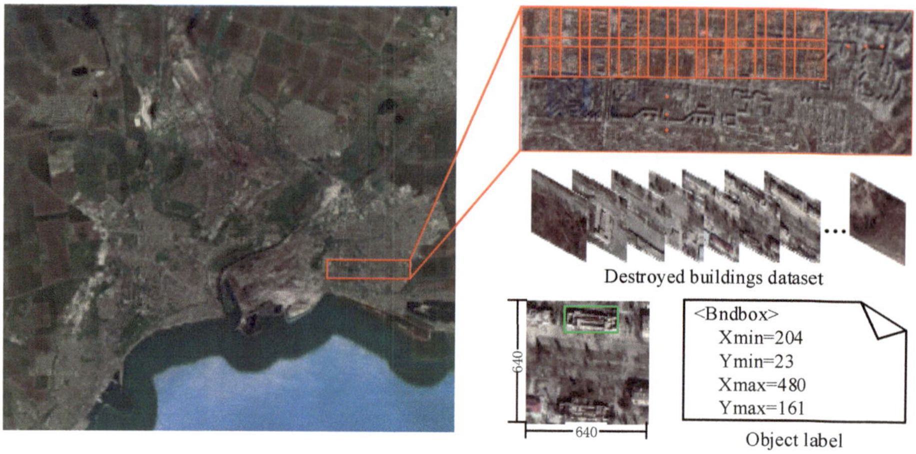

Figure 3. The production process of the destroyed buildings dataset; The red boxes in the upper right sub-image indicate the sliding windows, the green box in the lower right sub-image indicates the label of the destroyed building. (Source: Google Earth.)

In general, we cut the collected high-resolution remote sensing images of Mariupol into 14,023 slice samples, screened out 4251 sample images containing destroyed buildings, and marked 11,604 destroyed building objects. In addition, we collected the same resolution remote sensing images of the Vornovakha region as test data to verify that the model can stably detect destroyed buildings in different regions.

3. Methodology

In order to effectively take advantage of the rich detail information of high-resolution remote sensing images, we improved the sampling method and feature expression of the CNN, including the extraction of shallow spatial features and the key screening and utilization of deep semantic features of the CNN. The network structure of SOCA-YOLO is shown in Figure 4.

In the process of feature extraction, geometric features such as the shape and texture of the object are crucial for detecting destroyed buildings in high-resolution remote sensing images. In order to avoid excessive loss of shallow spatial details, we introduced the Haar wavelet 2D decomposition to replace the early stage of the backbone network for downsampling [44]. In addition, we combined the coordinate attention [40] with the C2f module in the backbone network to enhance feature representation with spatial detail information earlier, promoting cross-scale feature fusion.

In the stage of feature fusion, it is usually necessary to upsample the feature map and connect it with the shallow feature map to enrich the multi-scale object information and improve the detection performance. In this section, we introduce a lightweight dynamic upsampling module [45] into the network. By combining point sampling and offset, a feature map with more spatial position perception can be obtained and the details of semantic features in the reconstructed feature map can be better retained.

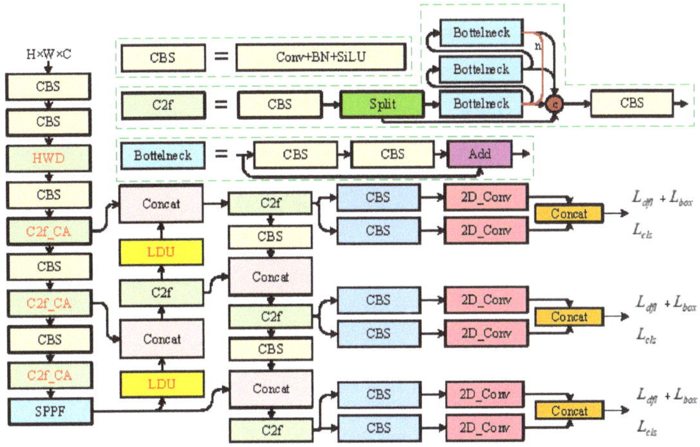

Figure 4. The network structure of SOCA-YOLO.

3.1. Haar Wavelet Downsample

The structure of the Haar wavelet downsampling (HWD) module is shown in Figure 5, which is composed of the Haar wavelet transform (HWT) and the convolution block (CBS).

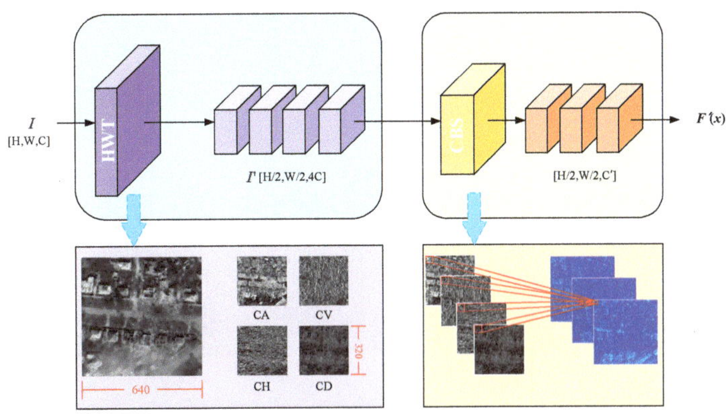

Figure 5. Structure of Haar wavelet downsampling module.

Assume that the input image I has C channels of size $H \times W$. Row i can be represented as $I_i = [I_{i,0}, I_{i,1}, \cdots I_{i,W-1}]$. According to the sequence of channels, each row of the image is transformed by the one-dimensional Haar wavelet:

$$a_{i,j} = \frac{I_{i,2j} + I_{i,2j+1}}{\sqrt{2}}, \quad j = 0, 1, \cdots, \tfrac{W}{2} - 1 \tag{1}$$

$$d_{i,j} = \frac{I_{i,2j} - I_{i,2j+1}}{\sqrt{2}}, \quad j = 0, 1, \cdots, \tfrac{W}{2} - 1 \tag{2}$$

where $a_{i,j}$ and $d_{i,j}$ are the low-frequency and high-frequency coefficients of pixels (i,j), respectively.

Furthermore, $a_{i,j}$ and $d_{i,j}$ are made into low-frequency matrix A and high-frequency matrix D, and then each column of them is transformed by a one-dimensional Haar wavelet:

$$CA_{k,l} = \frac{A_{2k,l} + A_{2k+1,l}}{\sqrt{2}}, \quad k = 0,1,\cdots,\tfrac{H}{2}-1, \quad l = 0,1,\cdots,\tfrac{W}{2}-1 \tag{3}$$

$$CH_{k,l} = \frac{A_{2k,l} - A_{2k+1,l}}{\sqrt{2}}, \quad k = 0,1,\cdots,\tfrac{H}{2}-1, \quad l = 0,1,\cdots,\tfrac{W}{2}-1 \tag{4}$$

$$CH_{k,l} = \frac{D_{2k,l} + D_{2k+1,l}}{\sqrt{2}}, \quad k = 0,1,\cdots,\tfrac{H}{2}-1, \quad l = 0,1,\cdots,\tfrac{W}{2}-1 \tag{5}$$

$$CD_{k,l} = \frac{D_{2k,l} - D_{2k+1,l}}{\sqrt{2}}, \quad k = 0,1,\cdots,\tfrac{H}{2}-1, \quad l = 0,1,\cdots,\tfrac{W}{2}-1 \tag{6}$$

where k and l represent row and column indexes, respectively. CA, CH, CV, and CD represent the approximate component, horizontal detail component, vertical detail component, and diagonal detail component, respectively. These components form a temporary feature set $I'[H/2, W/2, 4C]$. Since this can be seen as a lossless coding process, the four components obtained after decomposition can contain more spatial features. In particular, the high-frequency detail features that damage the building are preserved. With X' as the input of the CBS block, after convolution, batch normalization, and nonlinear activation of the SiLU function, the output feature set $F'(x)$ of HWD is obtained. After HWD, the image can be downsampled, while the spatial details are fully preserved, and the application of the CBS block also enhances the training stability of the model.

Figure 6 shows the visualization results of four downsampling methods, including Average pooling, Maximum pooling, Strided convolution, and HWD. It is evident that the image with Haar wavelet downsampling has clearer texture and shape features. In particular, the spatial details of the destroyed buildings in the red boxes are preserved to a greater extent.

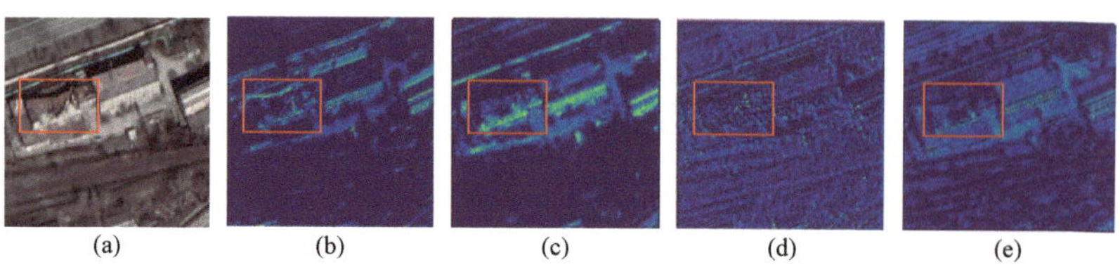

Figure 6. Visualization of different downsampling methods. The red box marks a damaged building in the input image. (**a**) Input image, (**b**) result of average pooling, (**c**) result of maximum pooling, (**d**) result of step convolution, (**e**) result of HWD.

3.2. Coordinate Attention

To further enhance the ability of the CNN to extract and locate the features of destroyed buildings, we constructed a C2f_CA module integrating coordinate attention, as shown in Figure 7. The CA module was input into Boottleneck through two stages of coordinate information generation and coordinate attention calculation to enhance spatial features and analyze dependency of inter-channel and inter-position dependency. The model's space awareness and feature representation ability are improved by stacking multiple CA_Bottleneck.

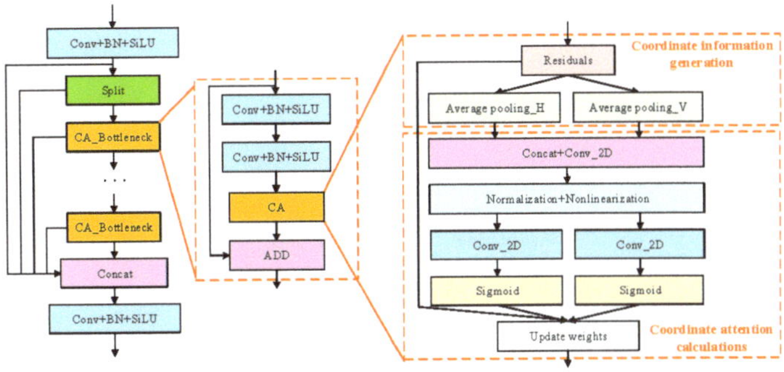

Figure 7. Structure of C2f_CA module.

In the generation stage of coordinate information, the CNN carries out one-dimensional spatial pooling in the horizontal and vertical directions of the input feature map, respectively, so that the attention module can capture accurate spatial position perception in different directions. In other words, while effectively capturing long-range dependencies in one spatial direction, the network can also obtain precise location information in the other direction, as shown in Equations (7) and (8):

$$z_c^h(h) = \frac{1}{W}\sum_{0\le i<W} x_c(h,i) \tag{7}$$

$$z_c^w(w) = \frac{1}{H}\sum_{0\le i<H} x_c(j,w) \tag{8}$$

where $z_c^h(h)$ and $z_c^w(w)$, respectively, represent the output by the c channel with height H in the vertical direction and the c channel with width W in the horizontal direction; $x_c(h,i)$ is the i pixel value in row h of channel c for the input feature map; and $x_c(j,w)$ is the j pixel value in column w of channel c for the feature map.

In the coordinate attention calculation stage, the channel dimension of the feature graph output in the previous stage is combined to make it have two independent spatial direction feature perceptions at the same time. The intermediate feature graph is further generated through 2D convolution, normalization, and nonlinearization:

$$f = \sigma\left(F_1\left(\left[z^h, z^w\right]\right)\right) \tag{9}$$

where f is the middle feature map; σ is the h_swish activation function; F_1 is a convolution operation; $[\cdot,\cdot]$ is the splicing channels; and z^h and z^w are the feature maps generated in the vertical and horizontal directions, respectively, in the previous stage.

The intermediate feature map is decomposed again into vertical and horizontal vectors f^h and f^w, and the number of channels is adjusted by convolution and activated by a Sigmoid function to generate the attention weights g^h and g^w in the corresponding directions. Finally, the attention feature map is obtained by weighting the two directions with the input feature map:

$$y(i,j) = x(i,j) \times g^h(i) \times g^w(j) \tag{10}$$

where $y(i,j)$ is the output attention feature map, $x(i,j)$ is the input feature map, and $g^h(i)$ and $g^w(j)$ are the vertical and horizontal attention weights, respectively.

With embedding coordinate attention in C2f, different weights can be obtained to different spatial positions of input feature maps, enhancing or suppressing feature information. More importantly, the spatial context of local features can be better understood by the network, improving the utilization efficiency and recognition accuracy of spatial structure information.

3.3. *Lightweight Dynamic Upsampling*

The lightweight dynamic upsampling module consists of two parts: the Dynamic sampling point generator and Dynamic upsampling calculator, whose structures are shown in Figures 8a and 8b, respectively.

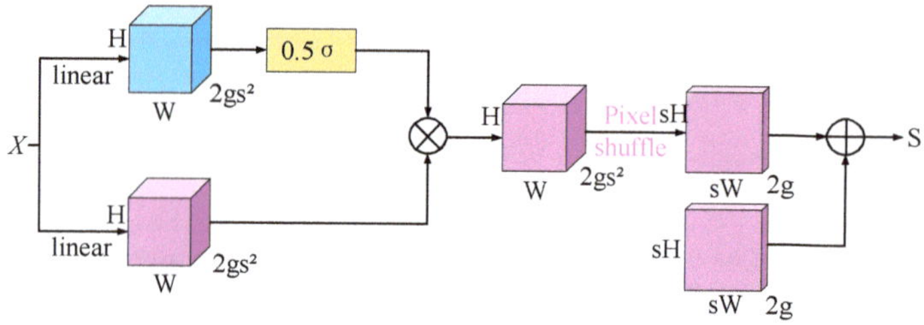

(**a**) Dynamic sampling point generator

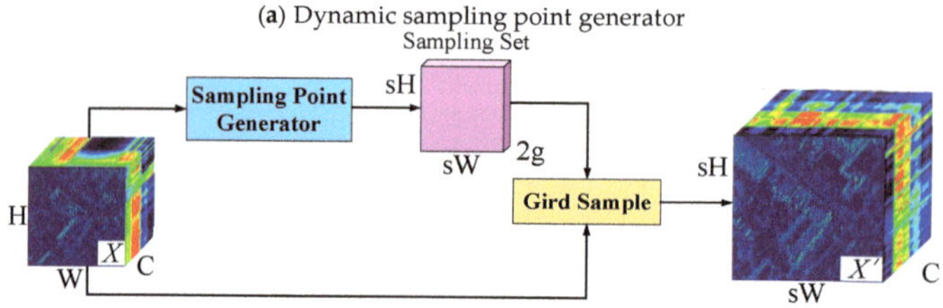

(**b**) Dynamic upsampling calculator

Figure 8. The structure of lightweight dynamic upsampling module.

The input feature map is denoted by X, whose size is $C \times H_1 \times W_1$, and the sample set S has size $2 \times H_2 \times W_2$, where C is the number of channels, H is the height of the image, W is the width of the image, and 2 in the sample set S is the x and y coordinates. Given the premise that the upsampling rate is s, the upsampling process can be expressed as

$$X' = grid\_sample(X, S) \tag{11}$$

where X' is the upsampling output.

Based on this, a linear offset layer O with the number of input channels C and the number of output channels $2s^2$ is introduced into the sampling module on Dysample. The coordinate S of each point in the feature graph X' can be obtained by

$$S = O + G \tag{12}$$

$$O = linear(X) \tag{13}$$

where G is the corresponding sampling point and O is its corresponding offset.

When O is non-0, the domain information of the sampling point can be taken into account. However, too large offset results in mixing other semantic cluster information when sampling the edge points of the semantic cluster, leading to the boundary being indivisible. Therefore, multiplying the scaling factor γ in Equation (13) is used to limit the range of offset. After scaling and adjusting the offset, the coordinates are reordered by normalization and pixel recombination, and the final position coordinates are obtained by adding the positions of the corresponding sampling points. Finally, in order to make the offset process more flexible, the scaling factor is generated point by point by way of linear projection to adapt to different feature distributions:

$$O = 0.5 \cdot sigmoid(linear_1(X) \times linear_2(X)) \tag{14}$$

In this case, the offset range is [0, 0.5]. Compared with the traditional upsampling method, the lightweight dynamic upsampling module can retain the geometric feature information of the feature graph, and be used against boundary discontinuity and detail ambiguity caused by simple upsampling.

3.4. Slicing-Aided Hyper Inference

To detect destroyed buildings rapidly in a wide range of remote sensing images, the slice-assisted super inference strategy [48] is adopted in the process of network inference. The method cuts a wide range of remote sensing images into manageable sub-images, detects the destroyed building in sub-images one by one, and splices the detection results. It can greatly optimize memory usage and simplify the process of reasoning to visualize the geospatial location distribution of destroyed buildings.

4. Experimental Setting

4.1. Experimental Environment and Dataset Configuration

The experimental environment of this paper is as follows: Windows 10 operating system and Pytorch deep learning framework. The computer is configured as 12th Gen Intel(R) Core (TM) i7-12700H, 2.30 GHZ; 16 GB RAM, NVIDIA GeForce RTX 3050 graphics card, and 4G video memory.

In this work, we use 80% of the samples as the training set and the remaining 20% as the validation set. The data distribution is shown in Table 1.

Table 1. Distribution of experimental datasets.

	The Number of Sample Images	The Number of Destroyed Buildings	Average Number of Destroyed Buildings per Image
Train	3400	9348	2.75
Valid	851	2256	2.65

There are various types of scenes and objects in the dataset, including a high-density low-rise building area, low-density high-rise building area, industrial park, non-residential area, etc. At the same time, the size and shape of destroyed buildings differ greatly, and the characteristics of damage are not inconsistent (such as wall facade collapse and roof collapse), as shown in Figure 9, where the green boxes represent the ground truth of the destroyed building. Therefore, detecting destroyed buildings in complex and diverse scenarios is a challenging task.

Scene: Dense building region **Objects:** Small house

Scene: Sparse building region **Objects:** Large apartment

Scene: Industrial region **Objects:** Factory

Scene: Commercial region **Objects:** Large irregular building

Figure 9. The various scenes and various objects contained in the dataset. The green boxes represent the ground truth of the destroyed building.

4.2. Evaluation Indicators

The experiment verifies the effectiveness of the algorithm according to the evaluation indexes commonly used in object detection, including recall (R), precision rate (P), Average Precision (AP), $F1$ score, model size, parameter number, GFLOPs, and inference time. As shown in Formulas (15)–(18). Among them, the recall rate reflects the ability to detect the object from the image, and the higher the recall rate, the less missed detection. Precision reflects the ability to detect the correct object, and the higher the precision, the less false detection. AP is the definite integral of the $P-R$ curve, the $F1$ score is the harmonic average of recall and precision; they are all comprehensive indexes to evaluate the performance of model detection. model size, parameter number, GFLOPs, and inference time measure the efficiency of the model in real-world applications.

$$Recall = \frac{TP}{TP + FN} \tag{15}$$

$$Precision = \frac{TP}{TP + FP} \tag{16}$$

$$AP = \int_0^1 P(R)dR \tag{17}$$

$$F1 = \frac{2 \cdot Recall \cdot Precision}{Recall + Precision} \tag{18}$$

where TP is the correctly predicted positive samples; FN is the incorrectly predicted negative samples; FP is the incorrectly predicted positive samples, and $P(R)$ is the curve of accuracy and recall rate. The larger the average precision, the better the detection performance of the model.

5. Result

5.1. Comparison of Different Improvement Strategies

Considering that HWD modules and the coordinate attention mechanism have various embeddedness positions and embeddedness ways, we designed several sets of comparative experiments to discuss the validity and rationality of the embeddedness positions and methods in the above methods.

(1) The embedding position of HWD

Previous experiments have proved that HWD can fully retain the spatial detail features of destroyed buildings when downsampling shallow feature maps. Therefore, in this section, we discuss whether consecutively embedding different numbers of HWD modules in a backbone network contributes to network performance. To be specific, we still use

YOLOv8 as the baseline. HWD replaces the first C2f layer (Group A experiment, ours), the first two C2f layers (Group B experiment), and all C2f layers (Group C experiment) in the backbone network according to the sequence. The comparison experiment results are shown in Table 2.

Table 2. Experimental results of embedding HWD in different positions.

	Recall	Precision	AP50	Model Size/M	Number of Parameters	GFLOPs	Inference Time/ms
Baseline	0.661	0.667	0.705	5.96	3,005,843	8.2	5.2
Group A	0.698	0.660	0.712	5.95	3,002,771	7.8	5.6
Group B	0.687	0.652	0.708	5.88	2,970,131	7.3	6.3
Group C	0.660	0.634	0.682	4.73	2,380,563	6.3	7.0

It can be seen that the HWD module can significantly reduce the model size, model parameters, and required GFLOPs, which has the great advantage of being lightweight. When the HWD was used to replace the first C2f layer, the AP50 of the model increased by 0.7%, but with the continuous embedding of the HWD, the AP50 decreased. The reason is that the HWD can improve the expression of shallow spatial features such as edges and textures in the initial stage of the network and provide more effective feature input for subsequent layers. However, the description of deep feature maps is more dependent on semantic features, and the HWD is weak in extracting semantic features. Therefore, the continuous use of the HWD to replace the C2f layer will destroy the compatibility and consistency of the network structure, and thus reduce the detection accuracy.

(2) The embedding method of CA

To discuss the effectiveness of different embedding methods of the CA, we designed two sets of comparative experiments based on the Group A experiment. The first method is to take the CA as a single layer of the network (Group D experiment) and embed it into the last layer of the network's detection head, enhancing the detection head's focus on important features. The second method is to fuse the CA with a C2f layer in the backbone to constitute a C2f_CA module (Group E experiment, ours). The comparative experimental results are shown in Table 3.

Table 3. Experimental results of embedding CA in different ways.

	Recall	Precision	AP50	Model Size/M	Number of Parameters	GFLOPs	Inference Time/ms
Baseline	0.661	0.667	0.705	5.96	3,005,843	8.2	5.2
Group A	0.698	0.660	0.712	5.95	3,002,771	7.8	5.6
Group D	0.681	0.662	0.718	5.97	3,014,491	7.8	6.0
Group E	0.692	0.669	0.723	5.99	3,012,747	7.8	6.1

It can be seen that the AP50 of Group E is 0.5% higher than that of Group D. This is because C2f_CA can make full use of spatial information to enhance feature representation in the early stage of feature extraction and then make more effective use of global context information to generate richer spatial concern graphs on feature maps. In the follow-up inspection process, it is more helpful to accurately locate the destroyed building.

5.2. Ablation Experiment

Here, we design a group of ablation experiments to examine the effect of each component, as shown in Table 4. Experimental parameters were set as epochs: 100; initial learning rate: 0.01; IoU threshold: 0.5; batch size: 16; input image size: 640 × 640; the optimizer adopts SGD; no pre-training weights are used.

Table 4. The results of the ablation experiment.

HWD	C2f_CA	LDU	Recall	Precision	AP50	Model Size/M	Number of Parameters	GFLOPs	Inference Time/ms
×	×	×	0.661	0.667	0.705	5.96	3,005,843	8.2	5.2
√	×	×	0.691	0.660	0.712	5.95	3,002,771	7.8	5.6
×	√	×	0.657	0.671	0.708	5.41	2,695,075	6.9	5.6
×	×	√	0.688	0.649	0.710	5.39	2,696,715	6.8	5.3
√	√	×	0.692	0.669	0.723	5.99	3,012,747	7.8	6.1
√	√	√	0.670	0.669	0.730	6.01	3,025,099	7.9	6.0

(AP50 represents the detection accuracy of the destroyed building under the condition that the IoU is 0.5).

It can be seen that adding the HWD, C2f_CA, and LDU separately improves the detection accuracy of the network, but the reasons are different. By retaining more spatial details at the beginning of the network calculation, the model with the HWD improved recall by 3.0%, indicating that it learned to more fully express the characteristics of the destroyed building. The model with C2f_CA can focus on the accurate expression of features, ensure the network learns more robust features, and improve the precision of the network. The introduction of the LDU can help the model to distinguish feature edges more effectively in the later stage, mainly in the improvement of recall.

In addition, in the ablation experiment, we observed that after adding the HWD, CA, and LDU to the baseline successively, the AP50 is successively increased by 0.7%, 1.8%, and 2.5%. It proves that the HWD module provides more spatial details for the subsequent feature extraction, and on this basis, C2f_CA can focus more on the expression of important features, and LDU can further distinguish the boundary distinction during the sampling process on the feature map. The information exchange of each module plays an active role in improving the detection accuracy of destroyed buildings.

At the same time, it is observed that the introduction of each module does not bring too much parameter number and calculation overhead, the model size is basically maintained at about 6.0 M, the number of model parameters is only increased by 0.59% compared with that of the baseline, and GFLOPs decreases from 8.2 to 7.9. It shows that our method can significantly improve the detection accuracy of destroyed buildings, as well as limit the increase in parameter number and algorithm complexity.

5.3. Visual Inspection

Figure 10 shows the detection results of each model in the ablation experiment under different scenarios. For large individual buildings, the improved model can more fully assess the size of the box, envelop destroyed buildings more completely, and improve confidence. Furthermore, it can also improve the recall rate in dense and small destroyed building scenes.

To illustrate the influence of each module on the attention area and degree of the network, we visualize the attention of the network detection layer for different sizes of objects, as shown in Figure 11.

For large destroyed buildings, we visualize the attention of the large-scale object detection layer. It is found that the attention of the baseline has inaccurate positioning and excessive noise response. With the HWD, the noise response near the object is significantly reduced. With the C2f_CA and LDU, the attention is further focused on the destroyed area, indicating that the introduction of each module can make the network pay more attention to the important features of the destroyed building area. In the detection of small destroyed buildings, heatmaps have the same form. Particularly, C2f_CA and LDU modules enable the network to reconstruct large images with more details on the basis of enhanced attention and improve the detection accuracy of small destroyed buildings.

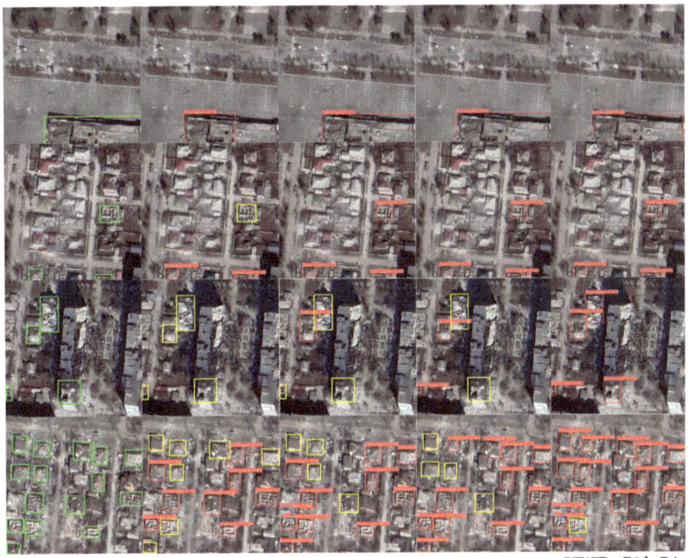

Figure 10. Visualization results of ablation experiments. The green boxes represent the ground truth of the destroyed building, the red boxes represent the correctly detected destroyed buildings, the yellow boxes represent the undetected destroyed buildings, and the blue circles represent the false alarm objects.

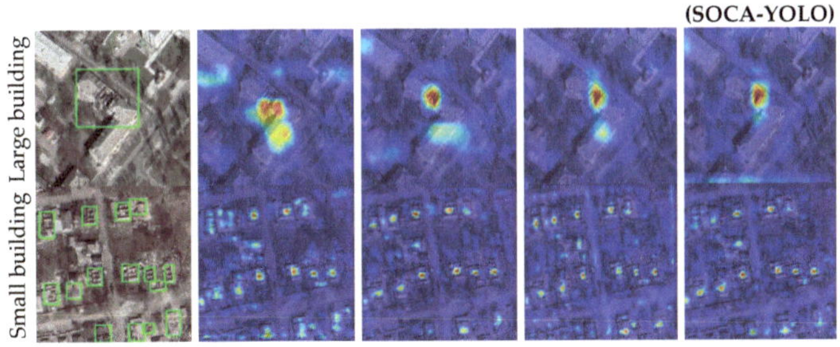

Figure 11. Attention visualization of CNN detection heads. The green boxes represent the ground truth of the destroyed building.

5.4. Comparison with Other Methods

To evaluate the comprehensive performance of the proposed method in terms of detection accuracy, model size, and detection efficiency, we compared the current mainstream same model size YOLO series algorithms, including YOLOv3-tiny, YOLOv5-n, YOLOv6-n, et al. As shown in Table 5. During the experiment, the parameter settings of each algorithm are consistent with those in Section 5.2. From the comparison results, it can be seen that the AP50 of SOCA-YOLO reaches 0.73, which is 0.3% higher than that of the second-best model, YOLOv3-tiny, but the model volume is much smaller than it. Overall, SOCA-YOLO achieves a balance of model size, number of parameters, GFLOPs, and inference time while ensuring high-precision detection, and its comprehensive performance is the best.

Table 5. The detection results of various methods.

	Recall	Precision	AP50	Model Size/M	Number of Parameters	GFLOPs	Inference Time/ms
YOLOv3-tiny [19]	**0.695**	0.661	0.727	23.2	12,128,178	18.9	6.3
YOLOv5-n [21]	0.684	0.652	0.701	5.03	2,503,139	7.1	5.1
YOLOv6-n [22]	0.666	0.661	0.700	8.29	4,233,843	11.8	5.2
YOLOv8-n [24]	0.661	0.667	0.705	5.96	3,005,843	8.2	**5.0**
YOLOv9-t [25]	0.669	0.644	0.688	**4.41**	**1,970,979**	7.6	6.7
YOLOv10-n [26]	0.673	0.658	0.707	5.47	2,694,806	8.2	6.5
YOLOv11-n [24]	0.662	0.664	0.704	5.22	2,582,347	**6.3**	5.5
SOCA-YOLO	0.670	**0.669**	**0.730**	6.01	3,025,099	7.9	6.0

Bold represents the optimal value, underlining represents the sub-optimal value, and wavy lines represents the third optimal value.

5.5. Generalization Ability Test

To test the generalization ability of the model on high-resolution remote sensing images outside the dataset, we selected high-resolution remote sensing images from the Vornovakha region of Ukraine as test data, which were also sourced from Google Earth. The image size was 34,318 × 27,634 and was taken on 2 September 2022, containing a large number of buildings destroyed by the war. In the process of the generalization ability test, a slice-assisted super reasoning strategy was adopted to conduct sliding window detection on the entire remote sensing image and Mosaic it into a large-scale image. The detection results are shown in Figure 12, where the red boxes represent the correctly detected destroyed buildings, the yellow boxes represent the undetected destroyed buildings, and the blue boxes represent the false alarm objects. According to the results of the generalization ability test, as shown in Table 6, the overall recall rate is 0.708, indicating that there is a certain number of missed tests. In addition, the missing objects were mostly small destroyed buildings (recall rate was 0.655), indicating that the detection ability of the model for small destroyed buildings needs to be improved. On the other hand, the overall precision is 0.986 with fewer false alarms, indicating that the model has good robustness for feature extraction of destroyed buildings. The comprehensive index F1 score is 0.824, which verifies that the proposed method has a certain generalization ability for the detection of destroyed buildings under different scenarios.

Table 6. The results of generalization ability test.

Objects	Recall	Precision	F1 Score
Large destroyed buildings	1.000	0.941	0.967
Small destroyed buildings	0.655	0.950	0.775
ALL	0.708	0.986	0.824

According to the visualization results of geospatial information, the destroyed buildings on 2 September 2022 were mainly distributed on the east side of the railway line. The main positions of the warring parties in the city were located on the east side of the railway line, and the destroyed buildings in the north and the south were especially concentrated. The above results show that the method proposed in this paper can provide an important reference for rapid disaster assessment and humanitarian relief.

Figure 12. Detection results of destroyed buildings in Vornovakha (2 September 2022). The red boxes represent the correctly detected destroyed buildings, the yellow boxes represent the undetected destroyed buildings, and the blue boxes represent the false alarm objects.

6. Conclusions

In this study, we designed SOCA-YOLO, an object detection algorithm combining Haar wavelet downsampling, lightweight dynamic upsampling, and a coordinate attention mechanism to detect war-destroyed building objects in high-resolution remote sensing images. In the stage of feature extraction, Haar wavelet is used to decompose high-resolution remote sensing images in two dimensions to realize downsampling and retain the details of remote sensing images to a greater extent. In addition, coordinate attention is combined with C2f to make full use of spatial feature information, and a coordinate attention module is introduced in the early stage of feature extraction to achieve feature optimization and focus to a greater extent. Finally, a lightweight dynamic upsampling module is used in the feature fusion phase to further enrich and accurately detail information on the feature map.

To evaluate the effectiveness of the method, we collected high-resolution optical remote sensing images of the Mariupol region, Ukraine, and produced the world's first datasets for the detection of buildings destroyed by war. The proposed method was verified by ablation experiments, and the validity and reasons for the embedding positions and methods of HWD and CA were discussed through multiple sets of comparative experiments. Compared with the classical deep learning object detection algorithm, the results show that the proposed method has better comprehensive performance in detection accuracy, model size, and inference speed.

It is worth noting that SOCA-YOLO mainly focuses on detecting individual destroyed buildings by making full use of spatial details and enhancing spatial perception. However, it still has some limitations: it is limited by the size and region of the training samples, the architectural styles of different countries and regions have obvious differences, and the generalization performance of the model needs to be further improved. In addition, the model has limited recognition of difficult objects, some destroyed buildings that look like ruins are still missed, and some structurally complex buildings are susceptible to false detection. In addition, the lightweighting and reasoning speed of the model need to be improved. In our following work, we will consider classifying destroyed buildings according to the degree of damage. By combining a feature recombination strategy and difficult object mining, fine-grained identification of buildings with different degrees of damage has become another exciting issue.

Author Contributions: Conceptualization, Y.L. and Y.W.; methodology, Y.L. and Y.W.; software, Y.W. and S.Z.; validation, Y.L. and S.Z.; formal analysis, Y.W.; investigation, S.Z.; resources, S.Z. and Y.W.; data curation, Y.W.; writing—original draft preparation, Y.W.; writing—review and editing, Y.W., S.Z. and Y.L.; visualization, Y.W.; supervision, S.Z.; project administration, Y.L. and Y.W. All authors have read and agreed to the published version of the manuscript.

Funding: This research received no external funding.

Data Availability Statement: The data presented in this study are available on request from the corresponding author due to (specify the reason for the restriction).

Acknowledgments: We would like to thank Google Earth and MAXAR for providing high-resolution optical remote sensing image data. We would also like to thank ultralytics and its users for their contributions to the YOLO series of object detection algorithms. Finally, thanks to the reviewers and editors who put forward valuable suggestions for improving the quality of this paper.

Conflicts of Interest: The authors declare no conflicts of interest.

References

1. Serhii, A.S.; Vyshnevskyi, V.I.; Olena, P.B. The Use of Remote Sensing Data for Investigation of Environmental Consequences of Russia-Ukraine War. *J. Landsc. Ecol.* **2022**, *15*, 36–53. [CrossRef]
2. ICRC. *Urban Services During Protracted Armed Conflflict: A Call for a Better Approach to Assisting Affected People*; International Committee of the Red Cross: Geneva, Switzerland, 2015.
3. Alacevich, M. Planning Peace: The European Roots of The Post-War Globel Development Cchallenge. *Past Present* **2018**, *239*. [CrossRef]
4. Kashtelan, S.O. About analysis of global trends of modern armed struggle. *Collect. Sci. Work. Mil. Inst. Kyiv Natl. Taras Shevchenko Univ.* **2023**, *79*, 7–12. [CrossRef]
5. A history of the Israeli-Palestinian conflict. *Choice Rev. Online* **2010**, *47*, 47–3990. [CrossRef]
6. Sarjoon, A.; Yusoff, M.A. The United Nations peacekeeping operations and challenges. *Acad. J. Interdiscip. Stud.* **2019**, *8*, 202–211. [CrossRef]
7. Mello, P.A. The Oxford handbook of United Nations peacekeeping operations. *Political Stud. Rev.* **2017**, *15*, 106–107. [CrossRef]
8. Dong, L.; Shan, J. A comprehensive review of earthquake-induced building damage detection with remote sensing techniques. *ISPRS J. Photogramm. Remote Sens.* **2013**, *84*, 85–99. [CrossRef]
9. Boloorani, A.D.; Darvishi, M.; Weng, Q.; Liu, X. Post-War Urban Damage Mapping Using InSAR: The Case of Mosul City in Iraq. *ISPRS Int. J. Geo-Inf.* **2021**, *10*, 140. [CrossRef]
10. Janalipour, M.; Mohammadzadeh, A. A novel and automatic framework for producing building damage map using post-event LiDAR data. *Int. J. Disaster Risk Reduct.* **2019**, *39*, 101238. [CrossRef]
11. Huang, Q.; Jin, G.; Xiong, X.; Ye, H.; Xie, Y. Monitoring Urban Change in Conflict from the Perspective of Optical and SAR Satellites: The Case of Mariupol, a City in the Conflict Between RUS and UKR. *Remote Sens.* **2023**, *15*, 3096. [CrossRef]
12. Ghandour, A.J.; Jezzini, A.A. Post-War Building Damage Detection. *Proceedings* **2018**, *2*, 359. [CrossRef]
13. Zitzlsberger, G.; Podhoranyi, M. Monitoring of Urban Changes With Multimodal Sentinel 1 and 2 Data in Mariupol, Ukraine, in 2022/23. *IEEE J. Sel. Top. Appl. Earth Obs. Remote Sens.* **2024**, *17*, 5245–5265. [CrossRef]
14. Aimaiti, Y.; Sanon, C.; Koch, M.; Baise, L.G.; Moaveni, B. War Related Building Damage Assessment in Kyiv, Ukraine, Using Sentinel-1 Radar and Sentinel-2 Optical Images. *Remote Sens.* **2022**, *14*, 6239. [CrossRef]
15. Gupta, R.; Goodman, B.; Patel, N.; Hosfelt, R.; Sajeev, S.; Heim, E.; Doshi, J.; Lucas, K.; Choset, H.; Gaston, M. Creating xBD: A Dataset for Assessing Building Damage from Satellite Imagery. In Proceedings of the IEEE/CVF Conference on Computer Vision and Pattern Recognition (CVPR) Workshops, Long Beach, CA, USA, 16–20 June 2019; Available online: https://arxiv.org/pdf/1911.09296 (accessed on 11 October 2024).
16. Ren, S.; He, K.; Girshick, R.; Sun, J. Faster R-CNN: Towards Real-Time Object Detection with Region Proposal Networks. *IEEE Trans. Pattern Anal. Mach. Intell.* **2017**, *39*, 1137–1149. [CrossRef]
17. Liu, W.; Anguelov, D.; Erhan, D.; Szegedy, C.; Reed, S.; Fu, C.Y.; Berg, A.C. SSD: Single shot MultiBox detector. In Proceedings of the European Conference on Computer Vision, Amsterdam, The Netherlands, 11–14 October 2016; pp. 21–37. [CrossRef]
18. Redmon, J.; Farhadi, A. YOLO9000: Better faster stronger. In Proceedings of the IEEE Conference On Computer Vision and Pattern Recognition, Honolulu, HI, USA, 21–26 July 2017; pp. 7263–7271.
19. Farhadi, A.; Redmon, J. YOLOv3: An incremental improvement. *arXiv* **2018**, arXiv:1804.02767.
20. Bochkovskiy, A.; Wang, C.Y.; Liao, H.Y.M. YOLOv4: Optimal speed and accuracy of object detection. *arXiv* **2020**, arXiv:2004.10934.
21. Yolov5. Available online: https://github.com/ultralytics/yolov5 (accessed on 24 October 2024).
22. Li, C.; Li, L.; Jiang, H.; Weng, K.; Geng, Y.; Li, L.; Ke, Z.; Li, Q.; Cheng, M.; Nie, W.; et al. YOLOv6: A single-stage object detection framework for industrial applications. *arXiv* **2022**, arXiv:2209.02976.
23. Wang, C.Y.; Bochkovskiy, A.; Liao, H.Y.M. YOLOv7: Trainable bagof- freebies sets new state-of-the-art for real-time object detectors. *arXiv* **2022**, arXiv:2207.02696.
24. Ultralytics. Available online: https://github.com/ultralytics/ultralytics (accessed on 24 October 2024).
25. Wang, C.Y.; Yeh, I.H.; Mark Liao, H.Y. YOLOv9: Learning What You Want to Learn Using Programmable Gradient Information. In *Computer Vision—ECCV 2024*; Leonardis, A., Ricci, E., Roth, S., Russakovsky, O., Sattler, T., Varol, G., Eds.; Lecture Notes in Computer Science; Springer: Cham, Switzerland, 2025; Volume 15089. [CrossRef]
26. Wang, A.; Chen, H.; Liu, L.; Chen, K.; Lin, Z.; Han, J.; Ding, G. YOLOv10: Real-Time End-to-End Object Detection. *arXiv* **2024**, arXiv:2405.14458.
27. Beal, J.; Kim, E.; Tzeng, E.; Park, D.H.; Zhai, A.; Kislyuk, D. Toward transformer-based object detection. *arXiv* **2020**, arXiv:2012.09958.
28. Zhao, Y.; Lv, W.; Xu, S.; Wei, J.; Wang, G.; Dang, Q.; Liu, Y.; Chen, J. DETRs Beat YOLOs on Real-time Object Detection. In Proceedings of the 2024 IEEE/CVF Conference on Computer Vision and Pattern Recognition (CVPR), Seattle, WA, USA, 17–21 June 2024; pp. 16965–16974. [CrossRef]

29. Ji, M.; Liu, L.; Zhang, R.; Buchroithner, M.F. Discrimination of Earthquake-Induced Building Destruction from Space Using a Pretrained CNN Model. *Appl. Sci.* **2020**, *10*, 602. [CrossRef]
30. Simonyan, K.; Zisserman, A. Very deep convolutional networks for large-scale image recognition. *arXiv* **2014**, arXiv:1409.1556.
31. Wu, C.; Zhang, F.; Xia, J.; Xu, Y.; Li, G.; Xie, J.; Du, Z.; Liu, R. Building Damage Detection Using U-Net with Attention Mechanism from Pre- and Post-Disaster Remote Sensing Datasets. *Remote Sens.* **2021**, *13*, 905. [CrossRef]
32. Ronneberger, O.; Fischer, P.; Brox, T. U-Net: Convolutional Networks for Biomedical Image Segmentation. In *Medical Image Computing and Computer-Assisted Intervention–MICCAI 2015: 18th International Conference, Munich, Germany, 5–9 October 2015*; Springer International Publishing: Berlin/Heidelberg, Germany, 2015. [CrossRef]
33. Shi, L.; Zhang, F.; Xia, J.; Xie, J.; Zhang, Z.; Du, Z.; Liu, R. Identifying Damaged Buildings in Aerial Images Using the Object Detection Method. *Remote Sens.* **2021**, *13*, 4213. [CrossRef]
34. Bai, Y.; Hu, J.; Su, J.; Liu, X.; Liu, H.; He, X.; Meng, S.; Mas, E.; Koshimura, S. Pyramid Pooling Module-Based Semi-Siamese Network: A Benchmark Model for Assessing Building Damage from xBD Satellite Imagery Datasets. *Remote Sens.* **2020**, *12*, 4055. [CrossRef]
35. Ma, H.; Liu, Y.; Ren, Y.; Yu, J. Detection of Collapsed Buildings in Post-Earthquake Remote Sensing Images Based on the Improved YOLOv3. *Remote Sens.* **2020**, *12*, 44. [CrossRef]
36. Ma, N.; Zhang, X.; Zheng, H. ShuffleNet v2: Practical guidelines for efficient CNN architecture design. In Proceedings of the European Conference on Computer Vision, Munich, Germany, 8–14 September 2018; pp. 116–131.
37. Rezatofighi, H.; Tsoi, N.; Gwak, J.Y. Generalized intersection over union: A metric and a loss for bounding box regression. In Proceedings of the IEEE Conference on Computer Vision and Pattern Recognition, Los Angeles, CA, USA, 16–19 June 2019; pp. 658–666.
38. Lu, W.; Wei, L.; Nguyen, M. Bitemporal Attention Transformer for Building Change Detection and Building Damage Assessment. *IEEE J. Sel. Top. Appl. Earth Obs. Remote Sens.* **2024**, *17*, 4917. [CrossRef]
39. Liu, C.; Sepasgozar, S.M.; Zhang, Q.; Ge, L. A novel attention-based deep learning method for post-disaster building damage classification. *Expert Syst. Appl.* **2022**, *202*, 117268. [CrossRef]
40. Hou, Q.; Zhou, D.; Feng, J. Coordinate Attention for Efficient Mobile Network Design. In Proceedings of the 2021 IEEE/CVF Conference on Computer Vision and Pattern Recognition (CVPR), Virtual, 19–25 June 2021; pp. 13708–13717. [CrossRef]
41. Lin, T.Y.; Dollár, P.; Girshick, R.; He, K.; Hariharan, B.; Belongie, S. Feature Pyramid Networks for Object Detection. In Proceedings of the 2017 IEEE Conference on Computer Vision and Pattern Recognition (CVPR), Honolulu, HI, USA, 21–26 July 2017; pp. 936–944. [CrossRef]
42. Tan, M.; Pang, R.; Le, Q.V. EfficientDet: Scalable and Efficient Object Detection. In Proceedings of the IEEE/CVF Conference on Computer Vision and Pattern Recognition, Seattle, WA, USA, 13–19 June 2020; pp. 10778–10787. [CrossRef]
43. Zhang, H.; Zhang, Y.; Wang, D.; Ma, G. Damaged Building Object Detection From Bi-temporal Remote Sensing Imagery: A Cross-task Integration Network and Five Datasets. *IEEE Trans. Geosci. Remote Sens.* **2024**, *60*, 5648827. [CrossRef]
44. Xu, G.; Liao, W.; Zhang, X.; Li, C.; He, X.; Wu, X. Haar wavelet downsampling: A simple but effective downsampling module for semantic segmentation. *Pattern Recognit.* **2023**, *143*, 109819. [CrossRef]
45. Liu, W.; Lu, H.; Fu, H.; Cao, Z. Learning to Upsample by Learning to Sample. In Proceedings of the 2023 IEEE/CVF International Conference on Computer Vision (ICCV), Paris, France, 2–6 October 2023; pp. 6004–6014. [CrossRef]
46. High Commissioner Updates the Human Rights Council on Mariupol, Ukraine. Available online: https://www.ohchr.org/en/statements/2022/06/high-commissioner-updates-human-rights-council-mariupol-ukraine (accessed on 11 October 2024).
47. Labelimg. Available online: https://github.com/HumanSignal/labelImg (accessed on 7 October 2024).
48. Akyon, F.C.; Altinuc, S.O.; Temizel, A. Slicing Aided Hyper Inference and Fine-Tuning for Small Object Detection. In Proceedings of the 2022 IEEE International Conference on Image Processing (ICIP), Bordeaux, France, 16–19 October 2022; pp. 966–970. [CrossRef]